S0-AKJ-210

TECHNIQUES AND INSTRUMENTATION IN ANALYTICAL CHEMISTRY – VOLUME 7

ELECTROANALYSIS

THEORY AND APPLICATIONS IN AQUEOUS AND NON-AQUEOUS MEDIA AND IN AUTOMATED CHEMICAL CONTROL

TECHNIQUES AND INSTRUMENTATION IN ANALYTICAL CHEMISTRY

Volume 1 **Evaluation and Optimization of Laboratory Methods and Analytical Procedures. A Survey of Statistical and Mathematical Techniques**
by D.L. Massart, A. Dijkstra and L. Kaufman

Volume 2 **Handbook of Laboratory Distillation**
by E. Krell

Volume 3 **Pyrolysis Mass Spectrometry of Recent and Fossil Biomaterials. Compendium and Atlas**
by H.L.C. Meuzelaar, J. Haverkamp and F.D. Hileman

Volume 4 **Evaluation of Analytical Methods in Biological Systems**
Part A. Analysis of Biogenic Amines
edited by G.B. Baker and R.T. Coutts
Part B. Hazardous Metals in Human Toxicology
edited by A. Vercruysse

Volume 5 **Atomic Absorption Spectrometry**
edited by J.E. Cantle

Volume 6 **Analysis of Neuropeptides by Liquid Chromatography and Mass Spectrometry**
by D.M. Desiderio

Volume 7 **Electroanalysis. Theory and Applications in Aqueous and Non-Aqueous Media and in Automated Chemical Control**
by E.A.M.F. Dahmen

TECHNIQUES AND INSTRUMENTATION IN ANALYTICAL CHEMISTRY – VOLUME 7

ELECTROANALYSIS

THEORY AND APPLICATIONS IN AQUEOUS AND NON-AQUEOUS MEDIA AND IN AUTOMATED CHEMICAL CONTROL

E.A.M.F. Dahmen

Emeritus Professor, Twente University of Technology, Enschede, The Netherlands

ELSEVIER

Amsterdam – Oxford – New York – Tokyo 1986

ELSEVIER SCIENCE PUBLISHERS B.V.
Sara Burgerhartstraat 25
P.O. Box 211, 1000 AE Amsterdam, The Netherlands

Distributors for the United States and Canada:

ELSEVIER SCIENCE PUBLISHING COMPANY INC.
52, Vanderbilt Avenue
New York, NY 10017, U.S.A.

Library of Congress Cataloging-in-Publication Data

Dahmen, E. A. M. F. (Edouard Albert Marie Fernand),
1914-
Electroanalysis : theory and applications in aqueous and non-aqueous media and in automated chemical control.

(Techniques and instrumentation in analytical chemistry ; v. 7)
Includes bibliographies and indexes.
1. Electrochemical analysis. I. Title. II. Series.
QD115.D25 1986 543'.0871 86-8992
ISBN 0-444-42534-9 (U.S.)

ISBN 0-444-42534-9 (Vol. 7)
ISBN 0-444-41744-3 (Series)

Printed in The Netherlands

CONTENTS

PREFACE

The tremendous success of the International Congress on Polarography, held at Prague from August 25th to 29th 1980 in honour of Prof. Jaroslav Heyrovský, illustrates once more the great importance of Electroanalysis, not only from a theoretical point of view but also as a source of experimental techniques of increasing applicational potential.

Since the appearance of the well known books by Kolthoff, Lingane and Delahay in the early 1950s, thousands of papers on electroanalysis appeared; however, at present there are few up-to-date comprehensive textbooks on Electroanalysis. It is the aim of this book to fill that gap to a certain extent, by presenting on the one hand a systematic treatment of electroanalysis and its commonly used techniques on a more explanatory basis, and on the other by illustrating the practical applications of these techniques in chemical control in industry and of health and environment. As such control today requires the increasing introduction of automation and computerization, electroanalysis with its direct input and/or output of electrical signals often has advantages over other techniques, especially because recent progress in electronics and computerization have greatly stimulated new developments in the electroanalysis techniques themselves.

I have tried to make the book self-contained, apart from the assumption that the reader has an adequate knowledge of basic physical chemistry and will consult the selected bibliography when necessary. Moreover, having kept practice constantly in mind, I hope I have succeeded in building a welcome bridge between electroanalytical chemistry as a science and electrochemical analysis as an art.

I am most indebted to Prof. Dr. F. Verbeek of the State University of Ghent, Belgium, for his kindness in criticizing many topics in the manuscript, and to that University in general for providing library facilities.

Also, I would like to pay tribute to my former co-workers, especially those in the field of titration in non-aqueous media: Dr. H. B. van der Heijde, Dr. N. van Meurs and Dr. M. Bos, of whom the last mentioned also deserves my additional appreciation not only for his work on electroanalysis in solvents of relatively low dielectric constants, but also for his development of automated and computerized methods of electroanalysis. Some other investigators, such as Dr. H. Donche (State University of Ghent, Belgium) and Dr. B. H. van der Schoot (Twente University of Technology, The Netherlands), most obligingly provided me with details of their recent work even before it had been published. My sincere thanks are also due to Mr. A. A. Deetman for his literature research on automation of electroanalysis and to his employers AKZO Zout Chemie, Hengelo, The Netherlands, for their kind permission to carry out this task. Further, I am indebted to many firms and their representatives in The

Netherlands for kindly providing up-to-date documentation and photographs of apparatus as well as for their generous permission to use this information and reproduce pictures for illustrative material; among these firms I would particularly like to mention Metrohm, Mettler, Orion, PARC, Philips, Polymetron, Radiometer, Tacussel and Yokagawa Electrofact. Finally, I am very grateful to Mrs. A. G. M. Smit-Keyzers for her primary check of the English text, and not least to my dear wife Mrs. M. Th. Dahmen-Smit for her support and endurance, which were a great help in completing this work.

LIST OF SYMBOLS

In the practical system of international measuring units according to Giorgi, the MKSA system (SI = système international), there are seven basic units: m (meter), kg (kilogram), s (second), A (ampere), K (Kelvin), mol (mol), cd (candela), and many derived units such as °C (centigrade), Hz (hertz), N (newton), Pa (pascal), J (joule), W (watt), C (coulomb), V (volt), F (farad), Ω (ohm), S (siemens) and Wb (weber); multiples and parts are given according to the following list of preceding letters:

T	10^{12}	deci	d	10^{-1}
G	10^{9}	centi	c	10^{-2}
M	10^{6}	milli	m	10^{-3}
k	10^{3}	micro	μ	10^{-6}
h	10^{2}	nano	n	10^{-9}
da	10	pico	p	10^{-12}

All measuring units are indicated by roman symbols. Except for greek letters, physical and physico-chemical properties are, although in the figures represented by roman symbols, in the text and the equations consistently indicated by italic symbols in accordance with the following list.

A	area; atomic weight
a	activity of component or ion, e.g., a_i, a_{i^+} or a_{i^-} ion activity
$a_{\pm}$	mean ion activity
C	capacitance; bulk concentration
c	local concentration, e.g., at electrode surface; velocity of light
D	diffusion constant
d	diameter; distance; density
E	energy, e.g., E_A = activation energy; electrode potential
E^0	standard electrode potential
E_j	junction potential
$E_{\frac{1}{2}}$	half-wave potential
F	faraday (96,487 C)
f	frequency (Hz); activity coefficient, e.g., f_i, f_{i^+} or f_{i^-} ion activity coefficient
$f_{\pm}$	mean ion activity coefficient
G	free enthalpy
g	gas; gravitational acceleration
h	height, e.g., h_{eff} of a dme
I	current intensity, mostly replaced by i because I is often used for di/dE and I_p the concerning peak value; ionic product, e.g., I_{HgCl_2}; ionic strength

I	information quantity (bits)
I_t	information generating velocity (Bd)
i	current, e.g., i_l = limiting, i_{res} = residual, i_d = diffusion-limited, i_k = kinetically controlled part of i_l
i_C	charging current
i_F	faraday current
j	current density
j^0	exchange current density
K	reaction constant, e.g., K_a, K_b, $K_{a'}$, K_h, etc.
K_s	ionic product of solvent
K_w	ionic product of water
k	Boltzmann constant ($1.381 \cdot 10^{-23}$ J K^{-1} $molecule^{-1}$); reaction rate constant, e.g., k_f and k_b = forward and backward reaction rate constant, respectively; general constant
$k_{s,h}$	specific heterogeneous (or intrinsic) rate constant
L	conductance; inductance; film thickness of MTFE
l	liquid; length
M	molarity; molecular weight
m	molality; mercury flow-rate of dme; measurableness, e.g., m_{tot}, m_d, m_a, m_n
N	normality; Avogadro number ($6.022 \cdot 10^{23}$ molecules mol^{-1})
n	number of electrons in a redox system
P	power
p	(gas) pressure; potential gradient
Q	total electric charge
q	single electric charge
R	gas constant (8.31431 J K^{-1} mol^{-1})
r	radius; quality of regulation
r_p	regulatability
S or S_0	solubility product, e.g., S_{AgCl} or $S_{0\,AgCl}$
s	solid
T	absolute temperature (K); time constants with respect to stages in analysis, e.g., T_d, T_a, T_p
t	time; temperature (°C)
t_0^+	transport or transference number of cation
t_0^-	transport or transference number of anion
U	voltage (see also V); solution flow velocity (Levich)
u_i, u_{i+}, u_{i-}	ion mobility
V	volume; (sometimes) voltage
v	velocity
W	weight
Z	impedance
z_i, z_{i+}, z_{i-}	ion charge number
α	dissociation degree; (charge) transference coefficient (at electrode)
δ	diffusion layer

ε	dielectric constant (in vacuum $\varepsilon = \varepsilon_0$ with proportionality factor $f = 1/\varepsilon_0 = 1$ in cgs and $= 9 \cdot 10^9$ in MKSA system)
η	equiv. cm^{-3}; overpotential, total η_t also specified as η_D, η_A, η_R, η_c for diffusion, activation, reaction, crystallization, respectively
θ	conductivity cell constant
κ	conductivity (electric specific conductance)
Λ	equivalent (ionic) conductivity
λ	titration parameter
μ	dipole moment
μ_i	chemical potential of component i
ν	kinematic viscosity
ν, ν^+, ν^-	number of ions in an electrolyte
π	circle circumference to diameter ratio; type of electron or bond
ϱ	resistivity (electric specific resistance); density of solution (ϱ_0 of solvent)
σ	surface tension
σ_x	standard deviation; σ_x^2 variance
τ	time constant; specific time indication or period, e.g., pulse time in pulse and square-wave polarography, transition time in chronopotentiometry, in stripping voltammetry deposition, rest and stripping steps
Φ	galvani potential
ϕ	phase angle shift vs. E_{ac}
χ	time relation factor with respect to controlled current density chronopotentiometry
ω	angular frequency

GENERAL INTRODUCTION

Electroanalysis as a representative of the wet-chemical methods has many attractive advantages, such as

selectivity and sensitivity, notwithstanding its inexpensive equipment;
ample choice of possibilities; and
direct accessibility, especially to
 electronic and hence automatic control even at distance;
 automatic data treatment; and
 simple insertion, if desirable, into a process-regulation loop.

There may be circumstances in which an electroanalytical method, as a consequence of the additional chemicals required, has disadvantages in comparison with instrumental techniques of analysis; however, the above-mentioned advantages often make electroanalysis the preferred approach for chemical control in industrial and environmental studies. Hence, in order to achieve a full understanding of what electroanalysis can do in these fields:

first, it will be treated more systematically in Part A;

second, some attention will be paid in Part B to electroanalysis in non-aqueous media in view of its growing importance; and

finally, the subject will be rounded off in Part C by some insight into and some examples of applications to automated chemical control.

A. ELECTROANALYSIS (SYSTEMATIC TREATMENT)

Chapter 1

Introduction

Electroanalysis consists of chemical–analytical techniques in which an essential or at least an indispensable role is played by electrochemistry.

In general, one is dealing with electrochemistry if electrical potentials are internally building up or externally being applied across two separate poles in contact with a liquid or exceptionally with a solid. The latter situation is mainly of interest in industry, e.g., for the electrolytic production of sodium metal from molten salt and for the electrolytic recovery of aluminium metal from bauxite in fused cryolite. More frequently, industrial electrolysis is applied to a liquid as such (e.g., when producing hydrogen and oxygen gas from water) or to a solute in a liquid (e.g., when recovering sodium hydroxide, hydrogen gas and chlorine gas from a salt solution in a mercury cell and more recently in a (cation-exchange) membrane cell, or when reducing aromatic nitro compounds to amines, etc.). Other important electrolytic processes are electrorefining (e.g., of copper), continuous removal of impurities from water and galvanotechnical applications such as electroplating, anodizing aluminium, galvanoplasty (e.g., in the mass production of gramophone records), etc.

1.1. DEFINITIONS AND SELECTED BIBLIOGRAPHY

In electroanalysis, the techniques are pre-eminently based on processes that take place when two separate poles, the so-called electrodes, are in contact with a liquid electrolyte, which usually is a solution of the substance to be analysed, the analyte. By means of electrometry, i.e., by measuring the electrochemical phenomena occurring or intentionally generated, one obtains signals from which chemical–analytical data can be derived through calibration. Often electrometry (e.g., potentiometry) is applied in order to follow a reaction that goes to completion (e.g., a titration), which essentially represents a stoichiometric method, so that the electrometry merely acts as an end-point indicator of the reaction (which means a potentiometric titration). The electrochemical phenomena in electroanalysis, whether they take place in the solution or at the electrodes, are often complicated and their explanation requires a systematic treatment of electroanalysis.

BIBLIOGRAPHY

P. Delahay, New Instrumental Methods in Electrochemistry, Interscience, New York, 1954.

J. J. Lingane, Electroanalytical Chemistry, Interscience, New York, 2nd ed., 1958.

A. I. Vogel, A Textbook of Quantitative Inorganic Analysis Including Elementary Instrumental Analysis, Longmans, London, 3rd ed., 1961.
G. Charlot, J. Badoz-Lambling and B. Trémillon, Electrochemical Reactions, Elsevier, Amsterdam, 1962.
A. J. Bard (Editor), Electroanalytical Chemistry, Marcel Dekker, New York, from 1966 (11 Vols.).
J. O'M. Bockris and A. K. N. Reddy, Modern Electrochemistry, Plenum Press, New York, 1970 (2 Vols.).
J. O'M. Bockris and D. M. Dražić, Electrochemical Science, Taylor and Francis, London, 1972.
W. J. Moore, Physical Chemistry, Longmans, London, 5th ed., 1972 (paperback).
H. H. Willard, L. L. Merritt and J. A. Dean, Instrumental Methods of Analysis, Van Nostrand, New York, 5th ed., 1974 (paperback).
Z. Galus, Fundamentals of Electrochemical Analysis, Ellis Horwood, Chichester, 1976.
A. J. Bard and L. R. Faulkner, Electrochemical Methods, Fundamentals and Applications, Wiley, New York, 1980.
A. M. Bond, Modern Polarographic Methods in Analytical Chemistry, Marcel Dekker, Basle, 1980.

1.2. SYSTEMATICS AND STANDARD ABBREVIATIONS

1.2.1. Systematics

When a solid material has been placed in an electrolytic solution a certain electrical potential may be built up at the contact surface; however, this single potential cannot be measured in the absolute sense, nor can an electrical current be forced through the electrode without the aid of a second electrode. Therefore, electrometry in an electrolyte always requires two electrodes, the poles or terminals of the electroanalytical cell, and can be carried out by means of either non-faradaic or faradaic methods.

Non-faradaic methods of electroanalysis, with a zero net electrical current, are represented by:

(a) Conductometric analysis, i.e., determination of the analyte through measurement of the electrical conductance of the solution between inert electrodes, preferably of the same type.

(b) Normal potentiometric analysis, i.e., determination of the analyte through measurement of the potential difference (or emf, the electromotive force) across an indicator electrode and a reference electrode.

Faradaic methods of electroanalysis, with a non-zero net electrical current, are represented by:

(a) Voltammetric analysis, i.e., determination of the kind of analyte and its concentration through measurement of the voltam(pero)metric curve; the latter can be obtained either by imposing a varying voltage across the two electrodes of an electrochemical cell and measuring the resulting change in current (amperometry) or by forcing a varying current through and measuring the resulting change of potential (potentiometry). Both poles may be indicator electrodes, usually of the same type, which then represents biamperometry or bipotentiometry, respectively; if there is one indicator electrode one simply speaks of amperometry or potentiometry; the latter term, however, requires a descriptor in order to distinguish its faradaic character, e.g., controlled-current potentiometry. If the kind of analyte and its (two-dimensional) voltammetric curve are previously known, one-dimensional amperometry or potentiometry will suffice to determine the analyte concentration, provided that a suitable constant potential or current, respectively, can be selected.

(b) Electrogravimetric analysis, i.e., determination of the analyte through measurement of the weight of a solid deposited electrolytically on an electrode.

(c) Coulometric analysis, i.e., determination of the analyte through measurement of the amounts of electricity (number of coulombs) required for complete chemical reaction of the analyte.

There are a few other analytical methods in which electrochemistry plays an essential role, such as (paper) electrophoresis, isotachophoresis, electrography and electrochromatography (according to Fujinaga); as they belong to analytical separation techniques, they are beyond the scope of this book.

As far as the above-mentioned methods are concerned, the following remarks can be made:

1. The so-called indicator electrodes must be considered as microelectrodes, which means that the active surface area is very small compared with the volume of the analyte solution; as a consequence, the electrode processes cannot perceptibly alter the analyte concentration during analysis in either non-faradaic potentiometry or faradaic voltammetry.

2. The (working) electrodes in the other techniques should preferably possess large active surface areas in order to permit an appreciable current throughput, often also with the purpose of avoiding long analysis times; thus, in non-faradaic conductometry, and also in the faradaic techniques of electrogravimetry and coulometry (both of which require exhaustive bulk electrolysis), the electrical current can be high, as the large electrode surfaces prevent unacceptably high current densities.

Finally, it should be realized that modern electroanalysis is most fruitfully stimulated by the harmonious combination of ionics, electrodics and electronics (terms first proposed by Bockris), representing knowledge and experience of the phenomena and processes that occur in ionic solutions, at electrodic surfaces and in electronic circuits, respectively.

1.2.2. Standard abbreviations

In addition to SI units ("Système International"), the still occasionally used cgs system and the list of symbols at the beginning of the book, the following standard abbreviations are used:

ASV	anodic stripping voltammetry
CV	cyclic voltammetry
DME (dme)	dropping mercury electrode
DPP	differential pulse polarography
emf	electromotive force
HMDE (hmde)	hanging mercury drop electrode
IDE	ideal depolarized electrode
IPE	ideal polarized electrode
ISE	ion-selective electrode
LSV	linear sweep voltammetry
MFE	mercury film electrode
MTFE	mercury thin-film electrode
NPP	normal pulse polarography
RDE	rotating disk electrode
RRDE	rotating ring-disk electrode
SCE	saturated calomel electrode
SHE	standard hydrogen electrode
SMDE	static mercury drop electrode
VDME (vdme)	vibrating dropping mercury electrode.

1.3. ELECTROCHEMICAL CELLS

As an introduction to the working of electrochemical cells, we shall consider first the various types of electrodes and then cell features such as potentials at the electrode solution interface and solution electrolysis, if any.

1.3.1. Electrodes

An electrode is an electronic conductor in contact with an ionic conductor, the electrolyte. The electrode reaction is an electrochemical process in which charge transfer at the interface between the electrode and the electrolyte takes place, and two types of reaction can occur, viz.:

(a) The electrode merely acts as an electron donor or acceptor and so is an inert or passive electrode (redox electrode);

(b) The electrode acts as an ion donor or acceptor and so is a participating or active electrode.

On the basis of this and in connection with the electrode performance in practical applications, a more detailed classification of electrodes can be given (see page 7).

1.3.2. Cell features

An electrode in contact with an electrolyte represents a half-cell; two half-cells when combined form an electrochemical cell with two terminals, the one with the higher potential being the positive pole (+ pole) and the one with the

Electrodes

Indicator electrodes (polarizable)

Normal electrodes
(a) Redox (inert) electrodes (normal type and gas electrodes)
(b^1) Electrodes of the 1st kind (metal/metal ion electrode)
(b^2) Electrodes of the 2nd kind (metal/insoluble salt or metal/metal oxide electrodes)
(b^3) Electrodes of the 3rd kind [e.g., mercury electrode with fixed amount of mercury(II) chelate]

Membrane electrodes
(a) solid membrane electrodes
(b) liquid membrane electrodes

Reference electrodes (non-polarizable)
(a) constructed as a reference type (e.g., calomel electrodes)
(b) simply used as a reference type [e.g., glass electrode in a pH buffer or Ag–$AgCl_{(s)}$ at a constant chloride concentration]

Remarks

1. Well functioning indicator electrodes, i.e., those which show ion-selective Nernstian behaviour in non-faradaic methods, represent the ISEs.
2. Most electrodes are stationary, such as a mercury pool electrode (SMDE) or the HMDE; others are non-stationary, such as the DME, MFE, MTFE, RDE and RRDE.
3. In the application of tubular electrodes and electrodes in flow cells, there may often be hydrodynamic complications, especially in voltammetry.

lower potential being the negative pole (– pole). Under non-faradaic conditions the cell is called a galvanic (or voltaic) cell; it provides a definite voltage across the terminals, the emf or more recently specified as the (electro)-chemical potential. Under faradaic conditions the cell is called an electrolytic cell, which either by internal* electrolysis delivers electrical energy via chemical conversion and still can be considered as a galvanic cell, or by a kind of reversed electrolysis consumes electrical energy forced through by imposing across the terminals a voltage higher than the decomposition voltage of the cell.

*Internal electrolysis is the term applied by Sand[1,2] to an electrogravimetric analysis proceeding spontaneously without the application of an external voltage, i.e., by the short-circuited galvanic cell.

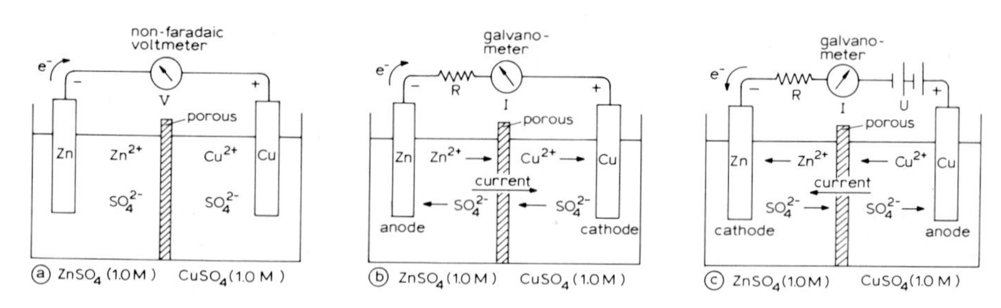

Fig. 1.1. Daniell cell, (a) at zero current, (b) delivering current-energy, (c) consuming current-energy.

The three afore-mentioned possibilities are illustrated in Fig. 1.1 for the well known Daniell cell (the arrows of current flow are conventional, i.e., the flow direction of the positive charge which is opposite to the actual flow of electrons as far as metallic conductance and charge transfer at the electrodes are concerned).

The following explanation can be provided. With Cu^{2+} ions there is a tendency for them to be reduced to Cu metal and precipitated on the electrode, which is reflected by a positive standard reduction potential (+0.34 V). For Zn metal there is a tendency for it to be oxidized to Zn^{2+} ions and dissolved in the electrolyte, which is reflected by a negative standard reduction potential (−0.76 V). In fact, with Zn one could speak of a positive oxidation potential for the electrolyte versus the electrode, as was often done formerly; however, some time ago it was agreed internationally that hence forward the potentials must be given for the electrode versus the electrolyte; therefore, today lists of electrode potentials in handbooks etc. always refer to the standard reduction potentials (see Appendix); moreover, these now have a direct relationship with the conventional current flow directions.

In case (a), the galvanic cell under non-faradaic conditions, one obtains an emf of $0.34 - (-0.76) = 1.10$ V across the Cu electrode (+ pole) and the Zn electrode (− pole). In case (b), the galvanic cell with internal electrolysis, the electrical current flows in the same direction as in case (a) and the electrical energy thus delivered results from the chemical conversion represented by the following half-reactions and total reaction, repsectively:

$$\begin{array}{ll} \text{half-reaction 1:} & Cu^{2+} + 2e^- \rightarrow Cu \text{ (cathodic reduction)} \\ \text{half-reaction 2:} & Zn - 2e^- \rightarrow Zn^{2+} \text{ (anodic oxidation)} \\ \hline \text{total reaction:} & Cu^{2+} + Zn \rightarrow Cu + Zn^{2+} \end{array}$$

Each half-reaction possesses its own redox system and from these two systems the cell reaction or redox reaction is built up.

The quantitativeness of the electrochemical conversion is governed by Faraday's laws:

1. The amounts of substances deposited or liberated at the electrodes of a cell are directly proportional to the quantity of electricity passing through.

2. The amounts of specific substances deposited or liberated by the same quantity of electricity are directly proportional to their chemical equivalents.

Now, 1 mole of electrons carries a charge of 96,487 C; therefore, in electrolysis the amount of chemical conversion may be determined from the number of faradays that have passed through.

In case (c), a voltage opposite to and higher than the emf* of the galvanic cell is imposed; as a consequence, the current flow and hence also the electrochemical reactions are reversed, which means that half-reaction 1 becomes an anodic oxidation and half-reaction 2 is a cathodic reduction, so that Zn is deposited instead of Cu.

Further, it can be seen from Fig. 1.1 that under all conditions prevailing Cu is the positive and Zn the negative pole; however, in case (b) Cu is the cathode (reduction) and Zn the anode (oxidation). Considering the flow direction within the electrolyte, one usually finds that the anode is upstream and the cathode downstream. It is also clear that by the electrochemical conversions the original galvanic cell is depleted in case (b), but can be restored by the external electrical energy source in case (c).

Commercially available galvanic cells include primary (non-regenerable) cells (e.g., the mostly cylindrical Zn–MnO_2 dry cells and the often flat, compact and mostly round mercury cell), secondary (regenerable) cells (e.g., the Pb–PbO_2, the Ni–Cd and other storage batteries which can be topped up) and fuel cells (delivering electrical energy mainly from a redox process between H_2, CO or hydrocarbons and air oxygen).

Another galvanic cell of highly practical and theoretical importance is the so-called standard cell (see Section 2.2.2), use of which has to be made as a calibration standard in non-faradaic potentiometry. For this purpose, the saturated Weston cell is the most accepted as its emf is reproducible, precisely known, only slightly temperature dependent in the region around 25° C (1.01832 V) and insensitive to unexpected current flows, if any.

REFERENCES

1 H. J. S. Sand, Analyst (London), 55 (1930) 309.

2 H. H. Willard, L. L. Meritt and J. A. Dean, Instrumental Methode of Analysis, Van Nostrand, New York, 5th ed., 1974, p. 690.

*In some cell types, especially those in which electrolysis generates gas at an electrode, the phenomenon of overvoltage may occur, which means that the voltage to be imposed must be higher than the emf plus an overvoltage; the term overpotential must be strictly used for the single electrode.

Chapter 2

Non-faradaic methods of electrochemical analysis

The most important non-faradaic methods are conductometric analysis and (normal) potentiometric analysis; in the former we have to deal essentially with the ionics and in the latter mainly with the electrodics. Strictly, one should assign a separate position to high-frequency analysis, where not so much the ionic conductance but rather the dielectric and/or diamagnetic properties of the solution are playing a role. Nevertheless, we shall still consider this techniques as a special form of conductometry, because the capacitive and inductive properties of the solution show up versus high-frequency as a kind of AC resistance (impedance) and, therefore, as far as its reciprocal is concerned, as a kind of AC conductance.

2.1. CONDUCTOMETRIC ANALYSIS

This method is primarily based on measurement of the electrical conductance of a solution from which, by previous calibration, the analyte concentration can be derived. The technique can be used if desired to follow a chemical reaction, e.g., for kinetic analysis or a reaction going to completion (e.g., a titration), as in the latter instance, which is a conductometric titration, the stoichiometry of the reaction forms the basis of the analysis and the conductometry, as a mere sensor, does not need calibration but is only required to be sufficiently selective.

2.1.1. Conductometry

2.1.1.1. *Electrolytic conductance*

If a voltage is applied across two platinum electrodes (usually platinum sheets coated with platinum black) placed in an electrolytic solution, an electric current will be transferred to an extent that is in accordance with the amounts and the mobilities of the free positive and negative ions present in the solution. Under the precautions required (see Section 2.1.1.2), the experiment obeys Ohm's law, which here can be described by

$$R = \varrho l / A \qquad (2.1)$$

where R is the resistance in Ω. If l (length of liquid column between Pt sheets) is expressed in cm and A (cross-sectional area of that column) in cm^2, then ϱ represents the resistivity (specific resistance) of the solution expressed in Ω cm.

In view of theoretical considerations one prefers the formula for the reciprocal of the resistance, i.e. the conductance

$$L = \frac{1}{R} = \kappa \cdot \frac{A}{l} \tag{2.2}$$

where $1/R$ is expressed in Ω^{-1} (or mho = siemens), so that $\kappa = 1/\varrho$ represents the conductivity (specific conductance) of the solution expressed in $\Omega^{-1}\,cm^{-1}$ (or mho cm^{-1}).

In fact, the conductivity κ can be thought of as the conductance of 1 cm^3 of the electrolyte solution. Now, let us suppose that 1 cm^3 would contain 1 g-equiv. of electrolyte and let us call its conductivity the equivalent conductivity, Λ, then the relation

$$\Lambda = \frac{1000}{C} \cdot \kappa \tag{2.3}$$

holds, where κ represents the conductivity of a solution of concentration C (in g-equiv. l^{-1}), and Λ is expressed in $\Omega^{-1}\,cm^2\,equiv.^{-1}$. As part of the current is transferred by the cation and the remainder by the anion, we can simply write for a salt solution of monovalent cation and anions

$$\Lambda = \Lambda^+ + \Lambda^- \tag{2.4}$$

where Λ^+ and Λ^- represent the equivalent ionic conductivities of the cation and anion, respectively.

Arrhenius postulated in 1887 that an appreciable fraction of electrolyte in water dissociates to free ions, which are responsible for the electrical conductance of its aqueous solution. Later Kohlrausch plotted the equivalent conductivities of an electrolyte at a constant temperature against the square root of its concentration; he found a slow linear increase of Λ with increasing dilution for so-called strong electrolytes (salts), but a tangential increase for weak electrolytes (weak acids and bases). Hence the equivalent conductivity of an electrolyte reaches a limiting value at infinite dilution, defined as

$$\Lambda_0 = \Lambda_0^+ + \Lambda_0^- \tag{2.5}$$

where Λ_0^+ and Λ_0^- represent the equivalent ionic conductivities at infinite dilution. The idea that at a certain temperature in aqueous solution constant values might be assigned to Λ_0^+ and Λ_0^- led Kohlrausch to his law of the independent migration of ions. Arrhenius attributed the above results to the shift of the chemical equilibrium between the undissociated electrolyte and the dissociated ions, i.e., an increase in the degree of dissociation with increasing dilution up to complete dissociation for infinite dilution; thus he calculated the degree of dissociation, α, from conductivity measurements according to the equation

$$\alpha = \Lambda/\Lambda_0 \tag{2.6}$$

so that

$$\Lambda = \alpha\Lambda_0 = \alpha(\Lambda_0^+ + \Lambda_0^-) \tag{2.7}$$

This concept of Arrhenius received striking confirmation in Ostwald's dilution law, e.g., for the dissociation constant of a weak acid HA:

$$K = \frac{[H^+][A^-]}{[HA]} = \frac{\alpha^2 C}{(1 - \alpha)} \quad \text{or} \quad K = \frac{\alpha^2}{(1 - \alpha)V} \tag{2.8}$$

for 1 equiv. of HA in V l of solution (or $C = 1/V$). This is the classical representation of the law, as later it was recognized that the activity coefficients of the dissociation participants and the interionic interferences (on Λ_0) cannot be neglected; as a consequence, the K values calculated for different dilutions by means of the classical equation show minor deviations which the values from the correct equation do not.

Eqn. 2.5 suggests that the respective contributions of the cation and anion to the transfer of an electric current are represented by

$$t_0^+ = \frac{\Lambda_0^+}{\Lambda_0} \tag{2.9}$$

and

$$t_0^- = \frac{\Lambda_0^-}{\Lambda_0} \tag{2.10}$$

where t_0^+ and t_0^- are the transport or transference numbers of the cation and anion, respectively, for inifinite dilution. Their values were determined for the first time by Hittorf (1854–1859) in an electrolytic cell divided into three compartments and constructed in such a way that the net concentration changes in the neighbourhood of the electrodes, caused by passage of a current, can be measured. The principle of Hittorf's method is illustrated in Fig. 2.1 for $AgNO_3$ solution.

If for (c) compared with (b) 1 faraday (96,487 C) has passed, then 1 equiv. of Ag has been dissolved at the anode and 1 equiv. of Ag has been precipitated at the cathode, so that the concentration changes will be as follows:

within the catholyte:		within the anolyte:	
removed	1 equiv. Ag^+	supplied	1 equiv. Ag^+
supplied	t_0^+ equiv. Ag^+	removed	t_0^+ equiv. Ag^+
net removal	$(1 - t_0^+) = t_0^-$ equiv. Ag^+	net supply	$(1 - t_0^+) = t_0^-$ equiv. Ag^+
removed	t_0^- equiv. NO_3^-	supplied	t_0^- equiv. NO_3^-
net concentration decrease	t_0^- equiv. $AgNO_3$	net concentration increase	t_0^- equiv. $AgNO_3$

Hence the decrease of $AgNO_3$ concentration within the catholyte is exactly equal to its increase within the anolyte, which for this symmetrical type of cell is to be expected; therefore, only one of the two needs to be measured in order to determine the transference numbers. From the transference numbers and the limiting equivalent conductivity Λ_0, one obtains the equivalent ionic conductivities $\Lambda_0^+ = t_0^+ \Lambda_0$ and $\Lambda_0^- = t_0^- \Lambda_0$.

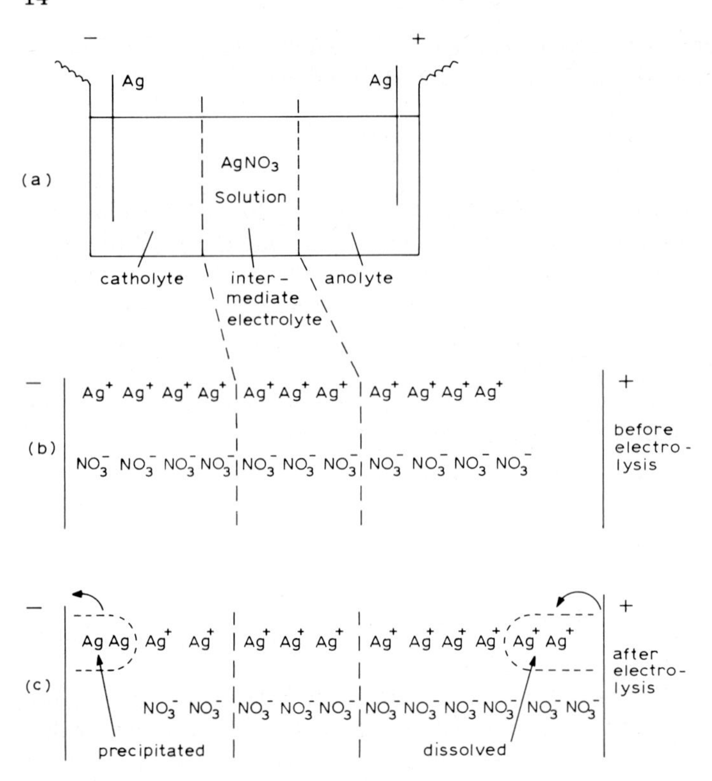

Fig. 2.1. Cell according to Hittorf.

Another method for obtaining these values is to determine the mobilities of ions, from which the ratio Λ^+/Λ^- can be calculated. This is based on the measurement of the absolute velocities of the cations and anions under the influence of a potential gradient, as originally suggested by Lodge (1886) and applied later by Masson and many others. For instance, Masson (1899)[1] carried out experiments with 10% KCl solution in gelatine gel, the principle of which is illustrated in Fig. 2.2.

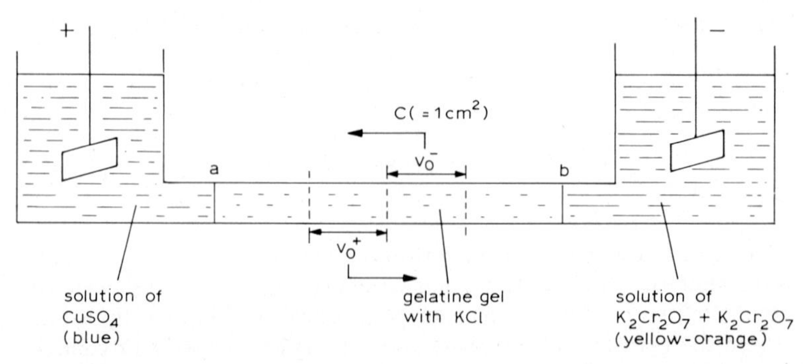

Fig. 2.2. Ion velocity measurement (Masson).

We start from the following conditions: KCl dissociation degree α, KCl concentration η (equiv. cm^{-3}), potential gradient p (V cm^{-1}), resistivity ϱ (Ω cm), cross-section of glass tube $c = 1\,\text{cm}^2$, and velocity of cation v_0^+ (cm s^{-1}) and of anion v_0^- (cm s^{-1}). If one considers the cross-section c, then in 1 sec all K^+ cations in part v_0^+ of the tube will pass to the right and all Cl^- anions in part v_0^- to the left; so these ions will transfer in 1 s a charge of $\alpha\eta(v_0^+ + v_0^-)F$ coulumbs to the right, which means a current intensity (ampères) of

$$I = \alpha\eta(v_0^+ + v_0^-)F \tag{2.11}$$

By applying Ohm's law, one finds for the passage of this current

$$I = p/\varrho = p\eta\Lambda = p\eta\alpha(\Lambda_0^+ + \Lambda_0^-) \tag{2.12}$$

and from eqns. 2.11 and 2.12 one obtains

$$(v_0^+ + v_0^-)F = p(\Lambda^+ + \Lambda^-) \tag{2.13}$$

and hence

$$v_0^+ = \frac{p\Lambda_0^+}{F} \quad \text{and} \quad v_0^- = \frac{p\Lambda_0^-}{F} \tag{2.14}$$

or

$$\frac{\Lambda_0^+}{\Lambda_0^-} = \frac{v_0^+}{v_0^-} \tag{2.15}$$

The values of v_0^+ for K^+ and v_0^- for Cl^- are equal to the shift per second of the colour boundaries a and b, respectively, because the Cu^{2+} ions are forced to move with the same velocity as the K^+ ions, and the CrO_4^2 or $Cr_2O_7^{2-}$ ions with the same velocity as the Cl^- ions; if not, gaps of positive and negative charges, respectively, would occur. One could object that the results in gelatine gel are not representative of a normal aqueous solution; therefore, measurements in gels of different gelatine contents have been made, so that the values in pure water could be obtained by extrapolation.

Further, in order to arrive at the absolute velocity, the so-called mobility* of the ions, v_0^+ and v_0^-, must be divided by the potential gradient, p; so

$$u_{i^+} = v_0^+/p \quad \text{and} \quad u_{i^-} = v_0^-/p \tag{2.16}$$

In fact, u_i is the limiting velocity of the ion in an electric field of unit strength, with the dimensions cm s^{-1} per V cm^{-1} or cm^2 V^{-1} s^{-1}. By inserting eqn. 2.14 into eqn. 2.16 we find

*Sometimes in the literature the equivalent ionic conductivity at infinite dilution is erroneously termed ion mobility; however, eqn. 2.17 clearly shows the interesting linear relationship between both properties with the faraday as a factor.

$$u_{i+} = \Lambda_0^+/F \quad \text{and} \quad u_{i-} = \Lambda_0^-/F \tag{2.17}$$

with the dimensions (see also eqn. 2.3) $\Omega^{-1}cm^2 equiv.^{-1}$ $(C\ equiv.^{-1})^{-1}$ or $cm^2 C^{-1}\Omega^{-1}$; applying Ohm's law ($V = C s^{-1}\Omega$) we obtain $cm^2 V^{-1} s^{-1}$ or $cm^2 C^{-1}\Omega^{-1}$ again.

In order to provide more insight into transference numbers, ionic conductivities and ion mobilities, some data collected by MacInnes[2] are given in Table 2.1 and 2.2; the data for Λ_0 were taken from the Handbook of Chemistry and Physics, 61st ed.; all measurements were made at 25° C in aqueous solutions.

TABLE 2.1

CATION TRANSFERENCE NUMBERS AND EQUIVALENT CONDUCTIVITIES ($\Omega^{-1}cm^2$ $equiv.^{-1}$) IN AQUEOUS SOLUTIONS AT 25° C

$\Lambda(NaOH)$ = 238.0 at 0.01 N, 244.7 at 0.001 N and 247.8 at infinite dilution.

Electrolyte	Infinite solution	Concentration ($equiv.\,l^{-1}$)						
		0.01		0.05		0.1		0.2
		t^+	Λ	t^+	Λ	t^+	Λ	t^+
HCl	426.16	0.8251	412.00	0.8292	399.09	0.8314	391.32	0.8337
NaCl	126.45	0.3918	118.51	0.3876	111.06	0.3854	106.74	0.3821
KCl	149.86	0.4902	141.27	0.4899	133.37	0.4898	128.96	0.4894
NH_4Cl	149.7	0.4907	141.28	0.4905	133.29	0.4907	128.75	0.4911
KNO_3	144.96	0.5084	132.82	0.5093	126.31	0.5103	120.40	0.5120
Na_2SO_4	129.9	0.3848	112.44	0.3829	97.75	0.3828	89.98	0.3828
K_2SO_4	–	0.4829	–	0.4870	–	0.4890	–	0.4910

TABLE 2.2

EQUIVALENT IONIC CONDUCTIVITIES AND ION MOBILITIES AT INFINITE DILUTION IN AQUEOUS SOLUTIONS AT 25° C

Ion	Λ_0 ($\Omega^{-1}cm^2 equiv.^{-1}$)	u ($cm\,s^{-1}$ per $V\,cm^{-1}$)
H_3O^+	349.82	$3.625 \cdot 10^{-3}$
Li^+	38.69	$4.010 \cdot 10^{-4}$
Na^+	50.11	$5.193 \cdot 10^{-4}$
K^+	73.52	$7.619 \cdot 10^{-4}$
NH_4^+	73.4	$7.61 \cdot 10^{-4}$
$\frac{1}{2}Ca^{2+}$	59.50	$6.166 \cdot 10^{-4}$
OH^-	198	$2.05 \cdot 10^{-3}$
Cl^-	76.34	$7.912 \cdot 10^{-4}$
NO_3^-	71.44	$7.404 \cdot 10^{-4}$
$\frac{1}{2}SO_4^{2-}$	79.88	$8.27 \cdot 10^{-4}$
AcO^-	40.9	$4.24 \cdot 10^{-4}$

Apart from the direct check on the mutual relationships which the tables afford, they also illustrate the following interesting conclusions:

(1) the exceptionally high mobilities of H_3O^+ and, to a lesser extent, of OH^-, caused by a series of proton transfers between neighbouring water molecules, confirm the special properties of acids and bases, respectively;

(2) the remarkable increase in mobility in the sequence of Li^+, Na^+, K^+, etc., is reversibly proportional to the atomic numbers of the alkali metals; Stokes' law indicates that there is a decrease of size in that order for the entire ionic moiety as it is moved by the electric driving force; this is concerned with the decreasing degree of hydration of the alkali metal ions, as already shown in 1906 by Buchböck[3] and later by others (they added non-electrolytes such as sugars or alcohols and determined their concentration changes in the anolyte and catholyte, from which the difference in the amounts of water supplied by the anions and carried off by the cations was calculated);

(3) the mobilities of Cl^- and NO_3^- are nearly equal to the mobility of K^+, which explains the use of KCl or KNO_3 bridges in order to avoid diffusion potentials (cf., Section 2.2.1).

So far, the data mentioned were measured at 25° C as is usual in electrochemical practice. However, it should not be forgotten that the ion mobilities increase considerably with temperature (see the Smithsonian table of equivalent conductivities as different temperatures in the Handbook of Chemistry and Physics, 61st ed.), although with the same trends for the various ions; therefore, the change in transference numbers remains small and shows a tendency to approach a value of 0.5 at higher temperatures.

2.1.1.2. *Practice of conductometry*

In general, a conductance cell in one arm of a Wheatstone bridge circuit is used. It may be of a different type, as shown schematically in Fig. 2.3 (A, B, C and D).

The electrodes consist of platinum discs, slightly platinized* and mounted in glass tubes which are placed in the glass vessels. The types of Wheatstone bridge commercially available for conductometry and known as conductivity bridges can be used in either the resistance mode or the conductance mode; the choice between these modes depends on the character of the solution under investigation and on the performance of the conductance cell.

As stated previously, the resistance in the cell is $R = \varrho l/A$ (eqn. 2.1) and the conductance is its reciprocal, $L = 1/R = \kappa A/l$ (eqn. 2.2), so that

$$l/A = \kappa R = \theta \tag{2.18}$$

the so-called cell constant expressed in cm^{-1}, and

$$L = \kappa/\theta \tag{2.19}$$

*For platinizing procedure, see ref. 4.

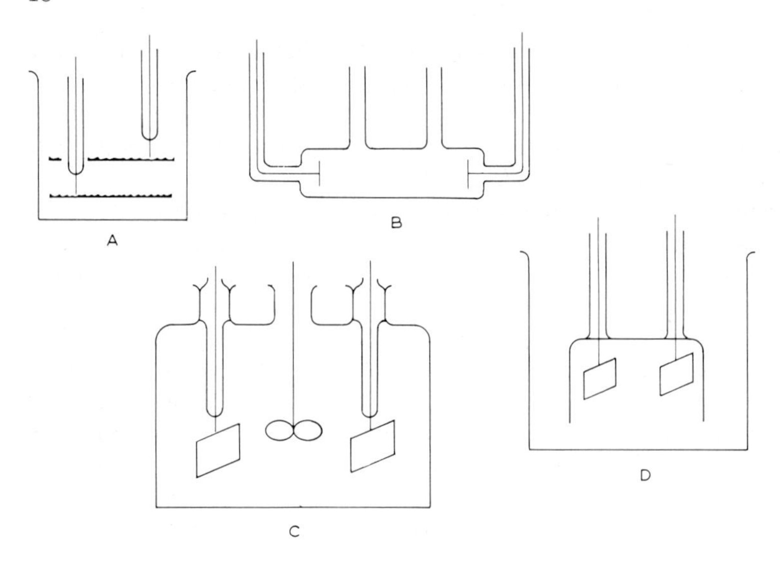

Fig. 2.3. Conductometry cells.

In fact, one can construct conductance cells with accurately known values of l and A in which the conductivity of standard electrolytes can be calibrated; however, in analytical practice cells with less restricted shape requirements are applied as their cell constants can be adequately established by measurements on a standard electrolyte (generally KCl) of known conductivity (see Fig. 2.3). Suppliers of commercial cells usually give the cell constants.

As far as conductometry is concerned, there remain a few complications caused by processes at the electrodes, e.g., electrolysis above the decomposition voltage of the electrolyte with some liberation of decomposition products at the electrode, or apparent capacitance and resistance effects as a consequence of polarization of the electrode and exchange of electrons at its surface. In order to reduce these complications the following measures are taken:

(1) the cell constant must be comparatively high (1000–50,000 Ω);

(2) the electrode surfaces should be large (note: the apparently controversial requirements 1 and 2 may reconciled by using a longitudinal cell of small diameter but with widened ends admitting large electrode surfaces);

(3) platinization of the electrodes increases their effective surface area, but one should guard against catalytic reduction;

(4) in addition to the precautions 1–3, which greatly diminish the polarization of the electrodes and the risk of electrolysis, which in a few cases even allows a direct current method, the use of an alternating current, as originally advised by Kohlrausch, remains the most reliable precaution in conductometry; by means of periodic reversion of the current the polarization, although not completely cancelled, will be equal for both electrodes of the same

surface area and thus becomes compensated; however, as a consequence of the electric double-layer at the electrode/electrolyte interfaces some capacitive effects on the measurement can remain.

Hence in face we have to deal with the impedance:

$$Z = \sqrt{R^2 + \frac{1}{\omega^2 C^2}} \tag{2.20}$$

as R represents the electrolytic resistance under investigation, its determination can be obtained most explicitly by taking

$$\omega = 2\pi f \tag{2.21}$$

and hence the ac frequency f (Hz) comparatively high. The measurement becomes even more reliable if the remaining capacitive effect is also compensated; as a capacitor within an ac circuit produces a positive phase shift, the cell capacitance effect can be balanced by means of an adjustable capacitor parallel to the resistance of the bridge arm adjacent to the cell; thus, by means of a phase detector together with a null meter, a more correct zero adjustment of the Wheatstone bridge is obtained. For the highest precision of balance indication the use of oscilloscopes is recommended. Of course, the range of the adjustable capacitor depends on the approximate value of R of the conductance and its appropriate frequency together with the cell capacitance, and so can be calculated; e.g., to compensate a series capacitance of 100 μF at 1000 Hz with a resistance of 1000 Ω one needs only 300 pF.

High-frequency (H.F.) conductometry offers another possibility for the effective elimination of the complicating effects at the electrodes in conductometry. In fact the H.F. method is an electrodeless technique as there is no direct electric contact with the liquid of the conductance cell, which is made part of, or is coupled to, a circuit oscillating at several MHz; for this purpose the measuring cell is placed between metal plates (e.g., a rectangular glass cell between two plates attached to the walls of the cell, being a capacity cell) or within a coil (e.g., a cylindrical glass cell with a coil wound around, being a inductive cell).

Let us first consider the capacity mode. When we apply an alternating voltage to the plates, each ion or dipole molecule in the liquid tends to align itself alternately opposite to the voltage, i.e., the ion moves ahead of its ionic atmosphere, thus showing dissymmetry, while the dipole molecules try to orientate themselves and may possibly change their dipole moments; these counter-effects will diminish when the ions and dipole molecules cannot follow the alterations in time at frequencies higher than 1 MHz, resulting in an increase in electric conductance. As the H.F. conductance also increases with increasing concentration of the solute, measurements are not feasible for normalities above 0.01 (e.g. 0.01 M NaCl or 0.001 M HCl) at 2 MHz, nor above 0.2 (e.g. 0.2 M NaCl, 0.1 M $CaCl_2$ or 0.03 M HCl) at 400 MHz; in normal analytical laboratory practice one does not operate above 30 MHz, so that the H.F. technique remains limited to the low concentration regions.

In order to achieve a suitable frequency $f = \omega/2\pi$ and cell performance, one may consider the measuring system as consisting of a glass capacitor (C_g) in series with an ohmic resistance R, the latter being parallel to a liquid capacitor (C_l); for convenience one may consider the system as consisting of an ohmic resistance R_p parallel to an overall capacitor (C_p). By fundamental studies[5] the following relationships for a 1-cm^3 cell have been verified:

$$\text{H.F. conductance} = \frac{1}{R_p} = \frac{\kappa\omega^2 C_g^2}{\kappa^2 + \omega^2(C_g + C_l)^2} \tag{2.22}$$

and

$$\text{Overall capacitance} = C_p = \frac{C_g\kappa^2 + \omega^2(C_g C_l^2 + C_g^2 C_l)}{\kappa^2 + \omega^2(C_g + C_l)^2} \tag{2.23}$$

where κ represents the low-frequency specific conductance of the liquid. The values of R_p, C_p and ω can be inserted in eqn. 2.22 in order to establish for the ac circuit the reciprocal of the impedance $Z = R_p$) at H.F. The capacitance acts as the major factor; modulation of the capacity mode is still possible by introduction of a pure inductance L parallel to the measuring system, so that the impendance will be

$$Z = \sqrt{R^2 + \left(\omega L - \frac{1}{\omega C}\right)^2} \tag{2.24}$$

The introduction of a pure inductance L becomes necessary in a resonance procedure (see oscillometry below).

The dielectric constant ε has a direct influence on C_p and influences κ indirectly, so the presence of non-aqueous solvent, if any, will be of great importance.

In the inductive mode the magnetic permeability of a solution-containing cell placed in the field of an inductance coil represents the measuring principle of the method; however, this property is not very sensitive to differences in solution composition, so that the inductive mode still remains of limited applicability.

For both H.F. modes a choice can be made from a few possibilities of measurement, viz., the observance of voltage and current in the ac circuit at a frequency imposed externally, or the establishment of internal resonance conditions where

$$f = \frac{1}{2\pi\sqrt{LC}} \tag{2.24a}$$

In the latter procedure, which is often called oscillometry[6], one observes the frequency at which resonance occurs or one retunes the oscillator to the original frequency, e.g., with the aid of a calibrated capacitor parallel to the sample cell. Further, in oscillometry it is useful to compare in parallel with a reference frequency unit.

Next the question arises of the extent to which the resonance condition is influenced by changes in the electrolyte. From eqn. 2.22 it follows that the H.F. conductance approaches zero for a very small or very large value of κ, so that by differentating $1/R_p$ with respect to κ one finds a peak position at

$$\kappa_{peak} = \omega(C_g + C_l) = 2\pi f_{peak}(C_g + C_l) \quad (2.25)$$

If C_g becomes small in comparison with C_l and because $C_l = C_0\varepsilon$, where for the flat capacitor filled with air $C_0 = \varepsilon_0 A/4\pi d$ ($A/d = 10^{-2}$m and $\varepsilon_0 = 1/9 \cdot 10^9$), one finds from eqn. 2.25

$$f_{peak} = \frac{1.8 \cdot 10^{12}\kappa}{\varepsilon} \quad (2.26)$$

As f_{peak} represents the most sensitive resonance condition, eqn. 2.26 illustrates once again the great influence of the dielectric constant ε of the solution on the H.F. capacitance mode; on the other hand, ε is almost independent of the electrolyte concentration.

As far as the inductive mode is concerned this technique is less sensitive because the low permittivity of the solution keeps the magnetic flux low.

2.1.2. Conductometric titration

In conductometric titration the reaction is followed by means of conductometry; there is little interest in the complete titration curve, but rather in the portion around the equivalence point in order to establish the titration end-point.

In principle, any type of titration can be carried out conductometrically provided that during the titration a substantial change in conductance takes place before and/or after the equivalence point. This condition can be easily fulfilled in acid–base, precipitation and complex-formation titrations and also the corresponding displacement titrations, e.g., a salt of a weak acid reacting with a strong acid or a metal in a fairly stable complex reacting with an anion to yield a very stable complex. However, for redox titrations such a condition is rarely met.

The concentration of electrolytes which do not take part in the titration reaction should be small in order to keep the background conductance of the solution sufficiently low.

In titration practice it is usual to plot the conductance,

$$L = \frac{\kappa}{\theta} \quad (2.19)$$

against the titration parameter λ or degree of conversion, i.e., the fraction of titrand titrated (+ fractional excess of titrant beyond the end-point). Especially for the conductometric indication procedure, increments of titrant are added well before and after the equivalence point; at the end of each incremental addition the conductance L'_k is measured and finally corrected with respect to the initial solution volume V_i for the total increase of volume at each stage. In this way the corrected conductance

$$L_k = \left[\left(V_i + \sum_{k=0}^{n} V_K\right) \Big/ V_i\right] L'_k \qquad (2.27)$$

offers via 1000 L_k (cf., eqn. 2.3) a direct additional function of the ion equivalent concentrations times their equivalent conductivities, Λ_0. Of course, this holds only for dilute solutions in titrations and if reliable values of Λ_0 are available. For the rest it is advisable to keep the volume corrections low by the use of a concentrated titrant.

As the most straightforward example we take the titration of a strong acid (HCl) with a strong base (NaOH) in a dilute solution (initial concentration C):

$HCl + NaOH \rightarrow NaCl + H_2O$

Well before the equivalence point we have

$$L_k = \frac{C}{1000} [(1 - \lambda)\Lambda_0^{H_3O^+} + \Lambda_0^{Cl^-} + \lambda\Lambda_0^{Na^+}] \qquad (2.28)$$

beyond the equivalence point

$$L_k = \frac{C}{1000} [\Lambda_0^{Cl^-} + (1 + \lambda)\, \Lambda_0^{Na^+} + (\lambda - 1)\,\Lambda_0^{OH^-}] \qquad (2.29)$$

and at the equivalence point

$$L_k = \frac{C}{1000} [\Lambda_0^{Cl^-} + \Lambda_0^{Na^+} + 10^{-7}(\Lambda_0^{H_3O^+} + \Lambda_0^{OH^-})] \qquad (2.30)$$

As a consequence of the high conductivities of H_3O^+ and OH^- in comparison with those of Na^+ and Cl^-, eqns. 2.28 and 2.29 will yield straight lines with negative and positive slopes, respectively. In practice, therefore, both straight lines are drawn through a sufficient number of measuring points for extrapolation to yield an intersection at the equivalence point. According to eqn. 2.30, at A (see Fig. 2.4) some curvature may occur as a consequence of the protolysis

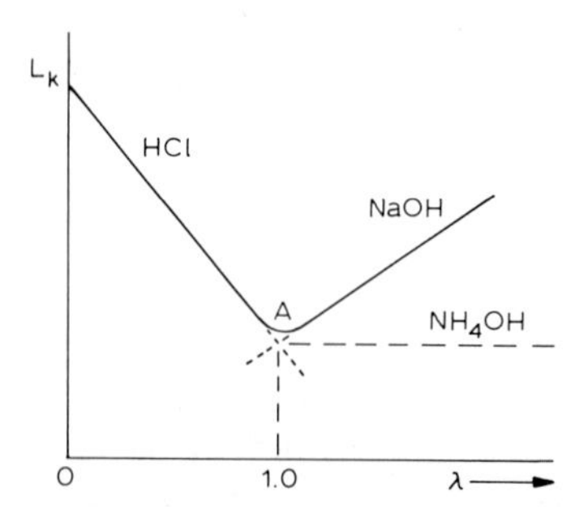

Fig. 2.4. Conductometric acid–base titration.

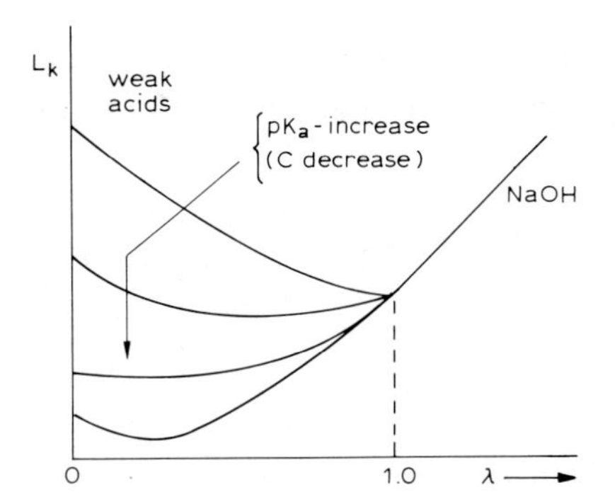

Fig. 2.5. Conductometric titration of weak acids.

of water unless C remains comparatively high (here $10^{-7} = \sqrt{K_W}$ at 25° C). Fig. 2.4 also shows the titration curve of HCl with ammonia.

A different type of curve is obtained when titrating a weak acid with a strong base, as shown in Fig. 2.5; on the acid side the lines have become curved owing to depression of the degree of acid dissociation as soon as the sodium salt is formed. With an increasing pK_a value and a decreasing initial acid concentration the curves will lie lower and start to rise at a smaller λ value; this effect can hamper considerably reliable end-point determinations through graphical extrapolation. Nevertheless, the conductometric titration of a mixture of a strong and a weak acid and of mixtures especially of weak acids such as boric acid and phenol remains of considerable interest because there the potentiometric method may fail.

In Fig. 2.6, curve I illustrates the displacement titration of CH_3COONa with HCl according to

$$(CH_3COO^- + Na^+) + (H_3O^+ + Cl^-) \rightarrow CH_3COOH(H_2O) + Na^+ + Cl^-$$

On the left-hand side the curve is slowly rising because of the displacement of

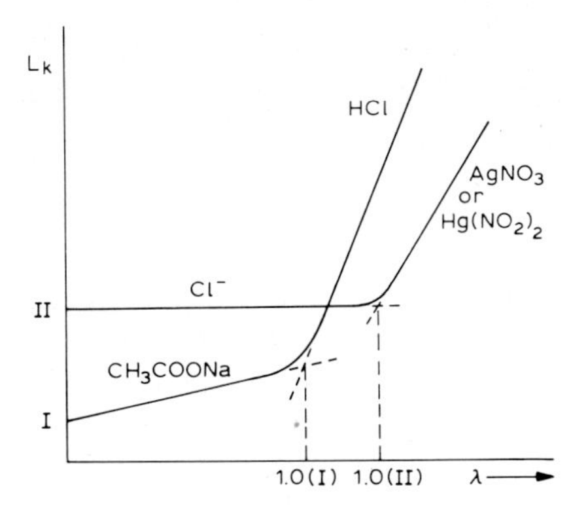

Fig. 2.6. Conductometric titration of basic salt, and of chloride with precipitation or complex formation.

CH_3COO^- (Λ_0 = 40.9) by Cl^- (Λ_0 = 76.34). Curve II illustrates the complexometric titration of Cl^- with $AgNO_3$ or $Hg(NO_3)_2$ respectively, according to

$$Cl^- + (Ag^+ + NO_0^-) \rightarrow AgCl\downarrow + NO_3^- \qquad \text{(argentometric)}$$

or

$$Cl^- + \tfrac{1}{2}(Hg^{2+} + 2NO_3^-) \rightarrow \tfrac{1}{2}[HgCl_2] + NO_3^- \qquad \text{(mercurimetric)}$$

On the left-hand side the curve is almost horizontal because the conductivities of Cl^- (Λ_0 = 76.34) and NO_3^- (Λ_0 = 71.44) are similar. (All values of Λ_0 given above are expressed in $\Omega^{-1}\,cm^2\,equiv.^{-1}$ at 25° C; see Table 2.2.)

An analogous effect on the sharpness at the equivalence point as for an acid–base titration (cf., eqn. 2.30) may not be overlooked in the case of curve II because

$$L_k = \frac{C}{1000}[\Lambda_0^{Na^+} + \Lambda_0^{Cl^-} + \sqrt{S_{AgCl}}\,(\Lambda_0^{Ag^+} + \Lambda_0^{Cl^-})] \qquad (2.31)$$

or

$$L_k = \frac{C}{1000}[\Lambda_0^{Na^+} + \Lambda_0^{Cl^-} + \sqrt[3]{\frac{I_{HgCl_2}}{4}}\,(\Lambda_0^{\frac{1}{2}Hg^{2+}} + \Lambda_0^{Cl^-})] \qquad (2.32)$$

where the solubility product S_{AgCl} is given by

$$S_{AgCl} = a_{Ag^+} \cdot a_{Cl^-} = f_{Ag^+}[Ag^+] \cdot f_{Cl^-}[Cl^-]$$

or for ideal solutions $S_{AgCl}^{25°C} = [Ag^+][Cl^-] = 1.56 \cdot 10^{-10}$ and the ion product I_{HgCl_2} is given by

$$I_{HgCl_2} = a_{Hg^{2+}} \cdot a_{Cl^-}^2 = f_{Hg^{2+}}[Hg^{2+}] \cdot f_{Cl^-}^2[Cl^-]^2$$

or for ideal solutions

$$I_{HgCl_2} = [Hg^{2+}][Cl^-]^2 = \frac{[HgCl_2]}{K} \quad (K_{HgCl_2}^{25°C} \approx 10^{6.5}),$$

where K is the stability or formation constant of the $HgCl_2$ complex. Here again the curvature at the equivalence point will be slight when C remains comparatively high. The assumption that the AgCl dissolved would be completely dissociated to free ions is incorrect; albeit a very small fraction remains present as an ion pair (the associate [AgCl]), which in a sense expresses the Debye–Hückel interaction between ions in strong electrolytes.

For examples of analytical procedures and practical results of conductometric titration, see the selected bibliography (Section 1.1).

2.1.3. High-frequency titration

As the inductive mode of H.F. titration is of limited importance, we shall consider only the capacitive mode. The latter technique, although differing considerably in its indication method, much resembles low-frequency titration in the method of collecting and treating the data; in this instance also one measures well before and beyond the equivalence point while the titrant is added in increments and the measurement at each stage is corrected with respect to the initial volume. Plotting the corrected values of H.F. conductance against the titration parameter one again obtains V-shaped curves with a minimum at the equivalence point and sometimes a reversed type of a curve with a maximum (e.g., for EDTA titrations). In addition to the advantages of an electrodeless method, H.F. titration offers applicability even at high resistivities such as in non-aqueous media.

For practical examples and results see ref. 5 and the selected bibliography (Section 1.1).

2.2. POTENTIOMETRIC ANALYSIS

This method is primarily concerned with the phenomena that occur at electrode surfaces (electrodics) in a solution from which, as an absolute method, through previous calibration a component concentration can be derived. If desirable the technique can be used to follow the progress of a chemical reaction, e.g., in kinetic analysis. Mostly, however, potentiometry is applied to reactions that go to completion (e.g. a titration) merely in order to indicate the end-point (a potentiometric titration in this instance) and so do not need calibration. The overwhelming importance of potentiometry in general and of potentiometric titration in particular is due to the selectivity of its indication, the simplicity of the technique and the ample choice of electrodes.

2.2.1. Potentiometry

2.2.1.1. *Indicator electrodes, reference electrodes and diffusion potentials*

A classification of electrodes has already been given in Section 1.3.1. The function of the indicator electrode is to indicate by means of its potential the concentration of an ion or the ratio of the concentrations of two ions belonging to the same redox system. Under non-faradaic conditions, the relationship between the potential and these concentrations is given by the Nernst or the more extended Nernst–Van't Hoff equation, as explained below. As a single potential between an electrode and a solution cannot be measured in the absolute sense but only in a relative manner, a reference electrode is needed; its function is merely to possess preferably a constant potential or at any rate a known potential under the prevailing experimental conditions. Often both electrodes cannot be placed in the same solution, so that a second solution becomes necessary; however, at the boundary layer between these solutions a diffusion potential, a liquid junction potential, E_j, may occur. First one applies an electrolyte junction through which the solutions cannot mix without hampering the passage of ions, e.g., an agar bridge, an asbestos fibre wick, a porous glass plug, a ground glass plug or a tapered sleeve; second, the junction electrolyte should be nearly equitransferent, i.e., its cation and anion must possess almost equal transference numbers (cf., eqns. 2.9 and 2.10); moreover the electrolyte concentration in the junction should be dominating in comparison with the ion concentrations in the sample solution. In general, nearly saturated KCl solution or (in trace Cl^- determinations) KNO_3 solution is used; where possible a mixture of KCl and KNO_3 of a composition in full accordance with the aim of equitransference is recommended; however, under certain circumstances the build-up of a junction potential must be taken into account, especially in absolute potentiometry. In a few instances E_j can be calculated easily[7], e.g., for a boundary layer between two different concentrations of a same 1:1 electrolyte:

$$E_j = E_{2-1} = (t^+ - t^-)\frac{RT}{F}\ln\left(\frac{a_1}{a_2}\right) \qquad (2.33)$$

and for two different 1:1 electrolytes at the same concentration:

$$E_j = E_{2-1} = \pm\frac{RT}{F}\ln\left(\frac{\Lambda_2}{\Lambda_1}\right) \qquad (2.34)$$

where the positive sign corresponds to a junction with a common cation and the negative sign to a junction with a common anion. However, it should be realized that the above equations are only approximately valid; for instance, in the derivation of eqn. 2.33 the values of t^+ and t^- have been assumed to be equal for both solutions 1 and 2. Table 2.1 shows that the difference is small for 0.01 M HCl ($t^+ = 0.8251$) and 0.1 M HCl ($t^+ = 0.8314$); as both solutions possess nearly the same mean activity factor ($\log f_\pm = -0.509|Z_+Z_-|I^{\frac{1}{2}}$) (cf., eqn. 2.63), we may replace a_1/a_2 by C_1/C_2, so that approximately at 25° C

$$E_j = (0.83 - 0.17)59.1 \log\left(\frac{0.01}{0.1}\right) = -39.1\,\text{mV}$$

This is an appreciable potential compared with the 59.1 mV difference for the hydrogen electrode in these solutions.

The question arises of the extent to which the build-up of an electrode potential may significantly alter the original concentration of the solution in which the electrode is placed. Let us take the example of a silver electrode. Once the electrode has been immersed in an Ag^+ solution, part of the Ag^+ ions will be discharged by precipitation of the corresponding amount of Ag and to an extent such that the Nernst potential has been reached. In fact, a double layer at the electrode/solution interface has been formed whose structure cannot be as precisely described as has appeared from the model proposed by

Helmholtz (1879) or that by Goüy (1910) and Chapman (1913), the latter being modified by Stern (1924)[8].

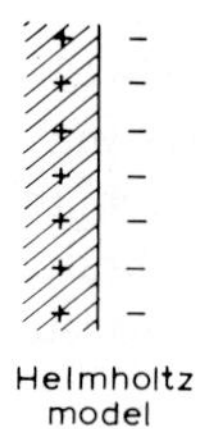

Whatever the most acceptable model may be and as we need only a rough estimate of the amount of ions discharged, we start from the Helmholtz model of a simple parallel-plate capacitor, whose potential difference is

$$\Delta\Phi = \frac{\lambda Q}{\varepsilon_0 \varepsilon A} \tag{2.35}$$

where according to the rationalized (MKSA) system of units $\varepsilon_0 = 10^7/4\pi c^2 = 8.854 \cdot 10^{-12}$, ε (dielectric constant of water) = 80, λ = thickness of layer (metres) and Q/A = surface charge density, so that $\varepsilon_0 \varepsilon/\lambda$ represents the capacitance per unit surface area. The Goüy–Chapman model mentions $\lambda = 10^{-5}$–10^{-7} cm for 10^{-5}–10^{-1} mol dm^{-3}, so that calculated for $\lambda = 10^{-9}$ m we find

$$\frac{8.854 \cdot 10^{-12} \cdot 80}{10^{-9}} \approx 700 \cdot 10^{-3}\,\text{F} \quad \text{or} \quad 700\,\text{mF}\,\text{m}^{-2} = 70\,\mu\text{F}\,\text{cm}^{-2}.$$

So, if we place a silver electrode consisting of a thin sheet of dimensions up to 1 cm^2 (each side) in 10 cm^3 of 0.1 N $AgNO_3$ solution, we will obtain a Nernst potential of 0.80 + 0.0591 log 0.1 = 0.74 V and the charge required for the double layer will be $2 \cdot 70 \cdot 10^{-6} \cdot 0.74 \approx 100 \cdot 10^{-6}$ C. As for total discharge of 10 cm^3 of 0.1 N $AgNO_3$ solution we need 96.5 C, then only 10^{-6} C is required for the electrode charge! Hence one can say that under practical conditions of potentiometric analysis (small electrode surface and with a volume and concentration of solution that are not extremely low), the build-up of the electrode potential does not substantially alter the solution concentration.

From the above and in Fig. 2.1(a) we have seen that the electrochemical cell consists of two half-cells and in its most complete form shows an emf given by $E_{cell} = E_{ind} - E_{ref} + E_j$.

In order to describe its specific structure a shorthand notation is used such as the following:

$^{-}Zn|Zn^{2+}\,\|\,Cu^{2+}|Cu^{+}$
$^{-}Zn|ZnSO_4(1.0\,M)\,\|\,CuSO_4(0.1\,M)|Cu^{+}$
$^{-}Zn|Zn^{2+}, Cl^{-}|AgCl|Ag^{+}$
$^{-}Pt|H_2(1\,atm)|H^{+}, Cl^{-}|AgCl|Ag^{+}$
$^{-}Pt|Fe^{2+}, Fe^{3+}\,\|\,Ce^{4+}, Ce^{3+}|Pt^{+}$

where a single line represents a phase boundary and a double line a salt bridge or boundary with negligible junction potential, and a comma separates two components in the same phase; the components are indicated either in full or by the most essential ions; concentrations or gas pressures are not always mentioned, nor are the signs of the cell necessarily indicated. The latter may be superfluous because by convention the structural sequence is written in such a way that the anodic half-reaction is on the left and the cathodic half-reaction on the right, which automatically places the negative pole to the left and the positive pole to the right.

Apart from the necessity of excluding interferences from any diffusion potential, normal potentiometry requires accurate determination of the emf, i.e., without any perceptible drawing off of current from the cell; therefore, usually one uses the so-called Poggendorff method for exact compensation measurement; the later application of high-resistance glass and other membrane electrodes has led to the modern commercial high-impedance pH and P*I* meters with high amplification in order to detect the emf null point in the balanced system.

On the basis of the type of emf chain, one divides the cells into two groups:

(1) chemical cells, possessing different redox couples in the two half-cells, so that during delivery of energy electrons from the one couple is transferred to the other;

(2) concentration cells, possessing different activities of the participating species within the same redox couple in each of the two half-cells.

Further, the two half-cells in either of these groups may be joined together by a single electrolyte or by different electrolytes, resulting in a liquid junction within the boundary layer; sometimes there may be direct contact between these electrolytes but mostly they are separated by a membrane or a bridge containing a third electrolyte.

2.2.1.2. *Normal electrodes*

In accordance with the classification of electrodes in Section 1.3.1 one can distinguish between redox electrodes (inert type) and electrodes of the 1st, 2nd and 3rd kinds (active type).

2.2.1.2.1. *Redox electrodes*

In view of its inertness Pt is mostly used, but it should be borne in mind that in the presence of strongly reducing agents such as Cr(II), Ti(III) and V(IV) Pt can catalyse the reduction of H^+ ions and then does not satisfy the aim of electrode inertness. Under certain conditions one can use electrodes of Pd, Au, graphite, conducting glass or glassy carbon.

Redox potential (thermodynamic derivation). Suppose we take an electrochemical cell represented by Fig. 2.7. We shall now address the question of both the potential values and the equilibrium state that can be finally attained

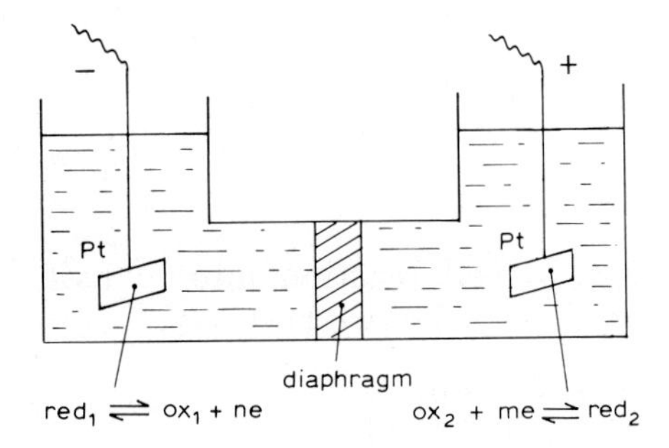

Fig. 2.7. Thermodynamic electrochemical cell.

between the two solutions. Treating the problem thermodynamically we write the total free enthalpy as the sum of the free enthalpy changes of the individual components.

As the chemical potential of a component i in solution is

$$\mu_i \;=\; \mu_i^0 + RT \ln a_i \tag{2.36}$$

we find for a general reaction $aA + bB + \ldots \rightleftharpoons pP + qQ \ldots$ the following equation for the free enthalpy change:

$$\Delta G \;=\; p\mu_P^0 + q\mu_Q^0 + \ldots - a\mu_A^0 - b\mu_B^0 - \ldots + RT \ln\left(\frac{a_P^p \cdot a_Q^q \ldots}{a_A^a \cdot a_B^b \ldots}\right)$$

$$= \;\Delta G^0 + RT \ln\left(\frac{a_P^p \cdot a_Q^q \ldots}{a_A^a \cdot a_B^b \ldots}\right)$$

where ΔG^0 represents the free enthalpy change for the standard conditions chosen.

At equilibrium $\Delta G = 0$, and hence

$$RT \ln\left(\frac{a_P^p \cdot a_Q^q \ldots}{a_A^a \cdot a_B^b \ldots}\right) \;=\; RT \ln K \;=\; -\Delta G^0 \tag{2.37}$$

For a redox reaction in an electrochemical cell the decrease in free enthalpy $(-\Delta G)$ is in accordance with the energy delivered by the transfer of electrons through an external circuit; if this takes place in a reversible way*, i.e., at a rate slow enough to allow complete attainment of equilibrium, the conversion of 1 gram mole will deliver an electrical energy of $-\Delta G = |z|FE$. In total cell reaction $m\,\mathrm{red}_1 + n\,\mathrm{ox}_2 \rightleftharpoons m\,\mathrm{ox}_1 + n\,\mathrm{red}_2$, where $m\delta_1 = n\delta_2$ electrons are transfered (δ_1 and δ_2 represent the respective valence differences of the two redox systems), we have

$$m\delta_1 FE \;=\; -\Delta G^0 - RT \ln\left(\frac{a_{\mathrm{ox}_1}^m \cdot a_{\mathrm{red}_2}^n}{a_{\mathrm{red}_1}^m \cdot a_{\mathrm{ox}_2}^n}\right)$$

*"Reversible" means counteracting the cell emf by means of an external source whose emf is identical so that no current passes through the cell.

where $-\Delta G^0 = RT \ln K = m\delta_1 FE^0$ and hence

$$E = E^0 - \frac{RT}{m\delta_1 F} \ln\left(\frac{a^m_{\mathrm{ox}_1} \cdot a^n_{\mathrm{red}_2}}{a^m_{\mathrm{red}_1} \cdot a^n_{\mathrm{ox}_2}}\right) \qquad (2.38)$$

Analogously to the splitting up of the total cell reaction into to half-reactions (see Fig. 2.7), we can write for the emfs of the two half-cells

$$E_1 = E^0_1 + \frac{RT}{m\delta_1 F} \ln\left(\frac{a^m_{\mathrm{ox}_1}}{a^m_{\mathrm{red}_1}}\right) \rightarrow E_1 = E^0_1 + \frac{RT}{\delta_1 F} \ln\left(\frac{a_{\mathrm{ox}_1}}{a_{\mathrm{red}_1}}\right)$$

and

$$E_2 = E^0_2 + \frac{RT}{n\delta_2 F} \ln\left(\frac{a^n_{\mathrm{ox}_2}}{a^n_{\mathrm{red}_2}}\right) \rightarrow E_2 = E^0_2 + \frac{RT}{\delta_2 F} \ln\left(\frac{a_{\mathrm{ox}_2}}{a_{\mathrm{red}_2}}\right)$$

where $E = E_2 - E_1$ and $E^0 = E^0_2 - E^0_1$. Hence for the redox potential of one half-cell the general Nernst equation is

$$E_{\mathrm{ox}\rightarrow\mathrm{red}} = E^0_{\mathrm{ox}\rightarrow\mathrm{red}} + \frac{RT}{nF} \ln\left(\frac{a_{\mathrm{ox}}}{a_{\mathrm{red}}}\right) \qquad (2.39)$$

where $E^0_{\mathrm{ox}\rightarrow\mathrm{red}}$ represents the standard potential (for $a_{\mathrm{ox}} = a_{\mathrm{red}}$). In order to establish the equilibrium state of two redox couples we take the examples of Ce^{4+}/Ce^{3+} and Fe^{3+}/Fe^{2+}, the standard potentials at 25° C of which are 1.4430 and 0.770 V, respectively. In this instance the cell in Fig. 2.7 can be described by

$$\mathrm{Pt}|Fe^{2+}, Fe^{3+} \| Ce^{4+}, Ce^{3+}|\mathrm{Pt} \qquad (2.40)$$

As each couple requires the transference of only one electron, eqn. 2.38 yields

$$E^0 = 1.443 - 0.770 = 0.673 = \frac{RT}{nF} \ln K = \frac{RT}{nF} \ln\left(\frac{a_{Fe^{3+}} \cdot a_{Ce^{3+}}}{a_{Fe^{2+}} \cdot a_{Ce^{4+}}}\right)$$

Hence $0.05916 \log K = 0.673$ and $K_{25°C} = 10^{11.38}$, which confirms that the titration reaction $Fe^{2+} + Ce^{4+} \rightarrow Fe^{3+} + Ce^{3+}$ proceeds almost to completion.

Redox reaction, ion activities and ion concentrations. In order to express the relationship between cell type 2.40 and the ion concentrations, we can write eqn. 2.38 as follows

$$E = 0.673 + 0.05916 \log\left(\frac{[Ce^{4+}][Fe^{2+}]}{[Ce^{3+}][Fe^{3+}]}\right) + 0.05916 \log\left(\frac{f_{Ce^{4+}} \cdot f_{Fe^{2+}}}{f_{Ce^{3+}} \cdot f_{Fe^{3+}}}\right)$$

In fact, the electrodics of the two half-cells are much more complicated than the current reaction equation illustrates, viz.,

$$[Fe(H_2O)_m]^{2+} + H_2O_{n-m} + M \rightarrow [Fe(H_2O)_n]^{3+} + M(e)$$

and

$$[Ce(H_2O)_p]^{4+} + M(e) \rightarrow [Ce(H_2O)_q]^{3+} + H_2O_{p-q} + M$$

where in addition to the complex ions the M(e) represents the inert metal electrode M after acceptance of an electron. Both "charge transfers" are based on "tunnelling" of an electron from red to M, or from M(e) to ox, with simultaneous rearrangement of the hydration hull of the ions and corresponding changes in their hydration energies[9].

The rapid course of the normal electrodeless titration of Fe^{2+} with Ce^{4+} also proves the occurrence of the direct "tunnelling" of the electron between the ions, notwithstanding the simultaneous hydration rearrangement, viz.,

$$[Fe(H_2O)_m]^{2+} + H_2O_{n-m} + [Ce(H_2O)_p]^{4+} \rightarrow [Fe(H_2O)_n]^{3+} + H_2O_{p-q} + [Ce(H_2O)_q]^{3+}$$

Experimental results have indicated that within the activity coefficients of much resembling redox couples, although their standard redox potentials may differ appreciably, certain compensating effects can be expected, especially in the equilibrium state ($E = 0$), between the equimolecular starting concentrations; hence the logarithmic term of the activity coefficients can be neglected with respect of the final ratio of equilibrium concentrations.

However, under most conditions the activity coefficients cannot be neglected, certainly for a single redox couple where the ox/red concentration ratio cannot be simply calculated from the true standard potential and the potential directly observed. In order to overcome this difficulty the concept of the formal potential was introduced, which represents a formal standard potential E'^0 measured in an actual potentiometric calibration and obeying the Nernst equation, $E = E'^0 + (0.05916/n) \log ([ox]/[red])$ at 25° C, E'^0 must meet the conditions under which the analytical measurements have to be made. Sometimes the formal potential values are decisive for the direction of the reaction between two redox couples even when the E^0 values do not differ markedly[10].

Some special redox electrodes. Within the group of redox electrodes, attention should be paid to the hydrogen and oxygen electrodes, and also to the quinhydrone electrode and its tetrachloro version.

Hydrogen electrode. The classical hydrogen electrode, Pt/H_2(1 atm), represents in combination with a calomel reference electrode the oldest and most straightforward method for determining pH:

$$Pt|H_2(1\,atm)|H^+(x\,M)\,\|\,KCl\ (saturated),\ Hg_2Cl_2,\ Hg^+|Hg \qquad (2.41)$$

The platinum consists of a foil or a wire spiral and should be previously coated with Pt black by means of platinization*.

*The platinization requires great care[11,12].

The method is of limited application because the following components should be absent from the measuring solution: substances that may poison the Pt black, compounds that are catalytically reduced by H_2 (e.g., $KMnO_4$, $FeCl_3$, unsaturated fatty acids) and ions of noble metals, i.e., with a standard potential higher than that of H^+ (e.g., Cu^{2+}, Ag^+). Moreover in practical use the hydrogen electrode does not equal the sturdiness and reliability of pH glass electrodes. However, in theoretical terms it remains of great importance, as we shall now explain in connection with pH measurement.

Application of the Nernst equation (eqn. 2.39) on the basis of the equilibrium reaction $2H^+ + 2e \rightleftharpoons H_2$ yields

$$E_{2H^+ \to H_2} = E^0_{2H^+ \to H_2} + \frac{RT}{2F} \ln \left(\frac{a^2_{H^+}}{p_{H_2}} \right) \tag{2.42}$$

where p_{H_2} atm is the hydrogen gas pressure. In fact, the concentration of the dissolved hydrogen is determinative for the potential, so that the establishment of an equilibrium between H^+ ions, dissolved H_2 and free H_2 gas is required. As the gas is injected at a certain depth h below the solution surface, there must be, notwithstanding the lowering effect of the water vapour pressure, an extra H_2 gas pressure. The latter was measured by Hills and Ives[13], who found $(0.42\,h/13.6)$ mmHg as an empirical value, so that at a barometer pressure p_B, p_{H_2} expressed in mmHg becomes

$$p_{H_2} = p_B - p_{H_2O} + 0.031\,h$$

Fig. 2.8 illustrates the Jackson cell[14], by means of which accurate measurements within 0.1 mV can be made in a few minutes; the sintered-glass disk permits the regular formation of minute H_2 bubbles and the application of the two electrodes allows a check to be made of reproducible results.

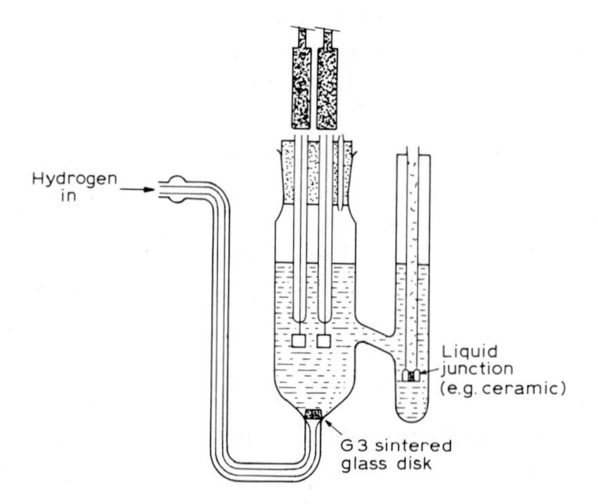

Fig. 2.8. Jackson cell (pH measurement).

In 1909 Sørensen introduced the concept of pH, defined as pH $= -\log [H^+]$. He established a pH scale; however, at a later date this scale was shown to be incorrect with respect to either $-\log [H^+]$ or $-\log a_{H^+}$, because it was simply based on the degree of dissociation according to conductivity measurements. The relation $pH_a = psH + 0.04$ between the activity and the Sørensen scales appears to be a useful approximation for buffer solutions[15].

Now, if a cell functions at exactly 1 atm H_2, the Nernst equation (eqn. 2.42) becomes

$$E_{2H^+ \to H_2(1\,atm)} = E^0_{2H^+ \to H_2(1\,atm)} + \frac{RT}{F} \ln a_{H^+} \qquad (2.43)$$

In the case of a solution with a previously known a_{H^+} (see below), we could determine $E^0_{2H^+ \to H_2(1\,atm)}$, provided that a reference electrode of zero potential is available; however, experiments, especially with the capillary electrometer of Lippmann, did not yield the required confirmation about the realization of such a zero reference electrode[16]. Later attempts to determine a single electrode potential on the basis of a thermodynamic treatment also were not successful[17]. For this reason, the original and most practical proposal by Nernst of assigning to the standard 1 atm hydrogen potential a value of zero at any temperature has been adopted. Thus, for $E_{2H^+ \to H_2(1\,atm)}$ we can write

$$E_{H^+/H_2} = \frac{RT}{F} \ln a_{H^+} \qquad (2.44)$$

From eqn. 2.44 it follows that when using the measuring cell 2.41 one finds

$$pH_a = -\log a_{H^+} = \frac{E - E_{ref}}{2.3026\, RT/F} \quad \text{or at } 25°\text{C} \quad pH_a = \frac{E - E_{ref}}{0.05916} \qquad (2.45)$$

where E represents the cell emf and E_{ref} the constant value of the reference half-cell including a correction for the junction potential, if any.

As this pH_a is the result of direct potentiometry it has now been generally adopted as the definition of pH $= -\log a_{H^+}$, instead of pH $= -\log [H^+]$ as originally introduced by Sørensen.

The question of the relationship between activity and concentration arises. Here the Debye–Hückel theory of activity coefficients, although valid only below 0.01 M, has proved to be most helpful, either for establishing an acid concentration from its H^+ activity or for calculating H^+ activity from its previously known acid concentration.

Referring to more extensive literature[18], we shall limit ourselves to a concise treatment with a view to obtaining some understanding of the possibilities and limitations of the theory in practical use.

In the mathematical derivation we use the following basic concepts:

1. The molality, m = moles per kilogram of solvent, as a temperature-independent concentration unit instead of the molarity, C = moles per litre of solution; this means that the amount of solvent and hence its number of moles is fixed.

2. The mean activity, $a_{\pm}$ = the geometric mean of the individual ion activities; e.g., for an electrolyte $C_{\nu_+}A_{\nu_-}$ dissociating with a total number of ions $\nu = \nu_+ + \nu_-$ according to $C_{\nu_+}A_{\nu_-} \rightarrow \nu_+ C^+ + \nu_- A^-$ we encounter within the thermodynamic treatment of equilibrium the product

$$a = a_+^{\nu_+} \cdot a_-^{\nu_-} = a_{\pm}^{\nu} \qquad (2.46)$$

where $a_{\pm} = (a_+^{\nu_+} \cdot a_-^{\nu_-})^{1/\nu}$, $a_+ = f_+ m_+$ and $a_- = f_- m_-$. Hence for $La_2(SO_4)_3$ we obtain $a = a^2_{La^{3+}} \cdot a^3_{SO_4^{2-}} = a^5_{\pm}$.

3. The ionic strength $I = 1/2 \sum_i m_i z_i^2$; this concept was originally introduced by Lewis and Randall as $\mu = 1/2 \sum_i c_i z_i^2$, where the summation extends over all the ions in the solution. For example,

$$\begin{aligned} 0.01\,M\,\text{NaCl:} \quad \mu &= \tfrac{1}{2}[(0.01)(1)^2 + (0.01)(1)^2] = 0.01 \\ 0.01\,M\,\text{Na}_2\text{SO}_4\text{:} \quad \mu &= \tfrac{1}{2}[(0.02)(1)^2 + (0.01)(2)^2] = 0.03 \\ \hline 0.01\,M\,\text{NaCl} + 0.01\,M\,\text{Na}_2\text{SO}_4\text{:} \quad \mu &= \tfrac{1}{2}[(0.03)(1)^2 + (0.01)(1)^2 + (0.01)(2)^2] = 0.04 \end{aligned}$$

Lewis and Randall stated that in dilute solutions the activity coefficient of a strong electrolyte is the same in all solutions of the same ionic strength; this statement was confirmed in thermodynamic deductions of activity coefficients*. The molality version of I can be applied in a fully analogous way and allows a more straightforward treatment of solution properties. [Conversion of molality into molarity requires the solution densities; e.g., for a solute of molar mass M and a solution of density ϱ we have

$$c = \frac{m}{(1 + mM)/\varrho} = \frac{m\varrho}{1 + mM}$$

in dilute solution and $c \approx \varrho_0 m$ (where ϱ_0 = solvent density), so that

$$I = \frac{1}{2}\sum_i m_i z_i^2 \approx \frac{1}{2\varrho_0}\sum_i c_i z_i^2$$

(cf., ref. 19).]

4. The charge density within a sphere, σ, in relation to the electric potential, U, as a function of r is obtained as follows. As the electric force in a dielectric ε between two charges if $F = q_1 q_2/\varepsilon r^2$ and a unit electric field corresponds to one line of force per cm^2, the electric field strength $\mathscr{E}$ at a distance r from a central point with a charge q is $\mathscr{E} = q/\varepsilon r^2$; the result is the same when the charge is enclosed in a dispersal manner within a sphere with radius r, so that, if we assume a uniform charge density σ, the enclosed charge can be represented by $\frac{4}{3}\pi r^3 \sigma$. As a result, the electric field intensity at r is

$$\mathscr{E} = \frac{\frac{4}{3}\pi r^3 \sigma}{\varepsilon r^2} = \frac{4}{3}\pi r \sigma/\varepsilon$$

*There are limits to the application of this rule, e.g., for uni-univalent electrolytes $\not> 0.2$, for bi-bivalent electrolytes $\not> 0.1$ and for uni-trivalent electrolytes $\not> 0.02$ values for I, although the total value of I may be considered higher than the partial value for the electrolyte concerned.

Multiplying both sides by r^2 and differentiating with respect to r, which takes inhomogeneity of the field within the sphere into account, we obtain

$$\frac{\mathrm{d}(r^2\mathscr{E})}{\mathrm{d}r} = \frac{\mathrm{d}(\frac{4}{3}\pi r^3\sigma/\varepsilon)}{\mathrm{d}r} = 4\pi r^2\sigma/\varepsilon$$

or

$$\frac{1}{r^2}\cdot\frac{\mathrm{d}(r^2\mathscr{E})}{\mathrm{d}r} = 4\pi\sigma/\varepsilon$$

Because the potential energy increases as the unit positive test charge is brought closer to the positive charge q, the potential change, $\mathrm{d}U$, for an infinitesimal distance $\mathrm{d}r$ is $\mathrm{d}U = -\mathscr{E}\mathrm{d}r$, so that $\mathscr{E} = -\mathrm{d}U/\mathrm{d}r$, which by insertion yields

$$\frac{1}{r^2}\cdot\frac{\mathrm{d}}{\mathrm{d}r}\left(r^2\cdot\frac{\mathrm{d}U}{\mathrm{d}r}\right) = -4\pi\sigma/\varepsilon \qquad (2.47)$$

This represents the Poisson differential equation for the electrical potential U as a function of r and a charge density of σ.

The theory of Debye and Hückel* started from the assumption that strong electrolytes are completely dissociated into ions, which results, however, in electrical interactions between the ions in such a manner that a given ion is surrounded by a spherically symmetrical distribution of other ions mainly of opposite charges, the ionic atmosphere. The nearer to the central ions the higher will be the potential U and the charge density; the limit of approach to the central ion is its radius $r = a$.

According to the Boltzmann distribution $N_i' = N_i \cdot \mathrm{e}^{-E/kT}$, which indicates for any solution region the number of ions i in a unit volume (in our case 1 cm^3) whose potential is E above the average; as for a charge Q_i brought to a region of potential U the potential energy is $Q_i U$, we obtain $N_i' = N_i \mathrm{e}^{-QU/kT}$. In order to obtain the charge density in that region we need the summation of all different kinds of ions multiplied by their appropriate charge Q_i, so that

$$\sigma = \sum_i N_i' Q_i = \sum_i N_i Q_i \mathrm{e}^{-Q_i U/kT} \qquad (2.48)$$

As only very dilute solutions are considered, where the ions are rarely close together, the interionic potential energy is small ($Q_i U \ll kT$). Therefore, the exponent factor may be replaced by

$$\mathrm{e}^{-Q_i U/kT} = 1 - \frac{Q_i U}{kT} + \frac{1}{2!}\left(\frac{Q_i U}{kT}\right)^2 - \ldots$$

*As the derivation in the Giorgi system of units (ref. 8: Moore, 5th ed.) is more complicated (e.g., a factor of $1000^{-\frac{1}{2}}$ in b should not be overlooked), we followed the older version (Moore, 4th ed., pp. 352–356) in the electrostatic cgs system because the concepts of molecularity and ionic strength are more easily fitting in.

so that eqn. 2.48 becomes

$$\sigma = \sum_i N_i Q_i - \frac{U}{kT} \sum_i N_i Q_i^2$$

where the first term vanishes because of overall electrical neutrality and $Q_i = z_i e$. Hence

$$\sigma = -\frac{e^2 U}{kT} \sum_i N_i z_i^2 \tag{2.49}$$

By substituting into eqn. 2.47 we obtain the Poisson–Boltzmann equation

$$\frac{1}{r^2} \cdot \frac{\mathrm{d}}{\mathrm{d}r}\left(r^2 \cdot \frac{\mathrm{d}U}{\mathrm{d}r}\right) = \frac{4\pi e^2 U}{\varepsilon kT} \sum_i N_i z_i^2$$

or

$$\frac{\mathrm{d}}{\mathrm{d}r}\left(r^2 \cdot \frac{\mathrm{d}U}{\mathrm{d}r}\right) = b^2 r^2 U \tag{2.50}$$

where

$$b^2 = \frac{4\pi e^2}{\varepsilon kT} \sum_i N_i z_i^2 \tag{2.51}$$

The so-called Debye length, $1/b$, approximately indicates the thickness of the ionic atmosphere; its value depends on the temperature, concentration and salt type, e.g., for 0.1 and 0.01 M NaCl $1/b = 0.96$ and 3.04 nm (at 25° C), respectively, for a radius (or better the closest approach) a of Na^+ 0.1 nm (0.98 Å).

Eqn. 2.50 can easily be solved by substitution of $u = rU$, yielding $\mathrm{d}^2 u/\mathrm{d}r^2 = b^2 u$, so that $u = A\mathrm{e}^{-br} + B\mathrm{e}^{br}$ or $U = (A/r)\,\mathrm{e}^{-br} + (B/r)\,\mathrm{e}^{br}$. As U becomes zero if $r \to \infty$, B must be zero; hence

$$U = \frac{A}{r}\mathrm{e}^{-br} \tag{2.52}$$

According to eqns. 2.47 and 2.50 we find $-4\pi\sigma/\varepsilon = b^2 U$, so that by inserting eqn. 2.52 we obtain

$$\sigma = \frac{-Ab^2\varepsilon}{4\pi r} \cdot \mathrm{e}^{-br}$$

which represents the charge density in their ion cloud as a function of r. The total charge of this cloud must be equal to the charge of the central ion, but opposite in sign, so that

$$\int_a^\infty 4\pi r^2 \sigma \mathrm{d}r = -z_i e$$

and hence

$$Ab^2\varepsilon \int_a^\infty r\mathrm{e}^{-br}\mathrm{d}r = z_i e$$

According to the general integral

$$x\mathrm{e}^{ax}\mathrm{d}x = \frac{\mathrm{e}^{ax}}{a^2}(ax - 1)$$

we obtain

$$A = \frac{z_i e}{\varepsilon} \cdot \frac{e^{ba}}{1 + ba} \quad (2.53)$$

so that

$$U = \frac{z_i e}{\varepsilon} \cdot \frac{\mathrm{e}^{ba}}{1 + ba} \cdot \frac{\mathrm{e}^{-br}}{r} \quad (2.54)$$

From U we can subtract the potential of the central ion, $z_i e/\varepsilon r$, in order to find the potential of the ionic atmosphere,

$$U' = \frac{z_i e}{\varepsilon r}\left(\frac{\mathrm{e}^{ba}}{1 + ba} \cdot \mathrm{e}^{-br} - 1\right) \quad (2.55)$$

The ions of the atmosphere cannot approach the central ion more closely than $r = a$ so that their potential at the site of the central ion is obtained by setting $r = a$ in eqn. 2.55; hence

$$U'_{r=a} = \frac{-z_i e}{\varepsilon}\left(\frac{b}{1 + ba}\right)$$

which becomes

$$U'_{r=a} = \frac{-z_i eb}{\varepsilon} \quad (2.56)$$

because $ba \ll 1$. From this extra potential we can calculate the activity coefficient of the central ion via the extra free enthalpy of the ionic solution as follows. Suppose that the ion is introduced into the solution in an uncharged state and let the charge Q be gradually increased to its final value ze. Then, the electrical energy required per ion will be in accordance with eqn. 2.56:

$$\Delta G = \int_0^{ze} \frac{-bQ}{\varepsilon}\mathrm{d}Q = -\frac{bz^2e^2}{2\varepsilon} \quad (2.57)$$

The extra electric free energy per ion is represented by $kT \ln f_i$ and the chemical potential of ion i is $\mu_i = RT\ln a_i + \mu_i^0 = \mu_i(\text{ideal}) + \mu_i(\text{electric})$. As $\mu_i(\text{ideal}) = RT \ln m_i + \mu_i^0$ and $a_i = f_i m_i$, the extra electric free energy is $\mu_i(\text{electric}) = RT \ln f_i$ and therefore $kT \ln f_i$ per ion, which corresponds to the expression in eqn. 2.57, so that

$$\ln f_i = -\frac{z^2 e^2 b}{2\varepsilon kT} \tag{2.58}$$

or, in the more extended form,

$$\ln f_i = -\frac{z^2 e^2}{2\varepsilon kT}\left(\frac{b}{1 + ba}\right) \tag{2.58a}$$

In eqn. 2.51, N_i represents ions per cm^3 and as the molarity c_i is expressed in $mol\,dm^{-3}$, the relationship between them is $N_i = c_i N_A/1000$, where N_A represents Avogadro's number. Hence eqn. 2.51 can be written as

$$b^2 = \frac{4\pi N_A^2 e^2}{1000\varepsilon RT}\sum_i c_i z_i^2$$

As $c_i = \varrho m_i$ and the ionic strength

$$I = \tfrac{1}{2}\sum_i m_i z_i^2 \approx \frac{1}{2\varrho_0}\sum_i c_i z_i^2$$

we find

$$b = \left(\frac{8\pi N_A^2 e^2 \varrho_0}{1000\varepsilon RT}\right)^{\frac{1}{2}} I^{\frac{1}{2}} = BI^{\frac{1}{2}} \tag{2.59}$$

In contrast with the individual ion activity coefficients f_i, the mean activity coefficient $f_\pm$ can be measured, calculation of which can be achieved through eqn. 2.46 as follows:

$$(\nu_+ + \nu_-)\ln f_\pm = \nu_+ \ln f_+ + \nu_- \ln f_-$$

By substitution in eqn. 2.58 one obtains

$$\ln f_\pm = -\left(\frac{\nu_+ z_+^2 + \nu_- z_-^2}{\nu_+ + \nu_-}\right)\frac{e^2 b}{2\varepsilon kT}$$

and as $|\nu_+ z_+| = |\nu_- z_-|$, we obtain the equation with the valence factor $|z_+ z_-|$:

$$\ln f_\pm = -|z_+ z_-|\frac{e^2 b}{2\varepsilon kT} \tag{2.60}$$

or, in the more extended form,

$$\ln f_\pm = -|z_+ z_-|\frac{e^2}{2\varepsilon kT}\left(\frac{b}{1 + ba}\right) \tag{2.60a}$$

Substitution of b from eqn. 2.59 into eqn. 2.60 and transformation into logarithms to the base 10 leads to

$$\log f_\pm = \frac{-|z_+ z_-|}{2.3026}\,\frac{e^2}{2\varepsilon kT}\left(\frac{8\pi N_A^2 e^2 \varrho_0}{1000\varepsilon RT}\right)^{\frac{1}{2}} I^{\frac{1}{2}}$$

or, by setting apart the data dependent on the type of electrolyte, solvent and temperature,

$$\log f_{\pm} = \frac{-|z_+ z_-|}{2.3026} \cdot \frac{e^3 N_A^2 (2\pi)^{\frac{1}{2}}}{1000^{\frac{1}{2}} \cdot R^{\frac{3}{2}}} \left(\frac{I \varrho_0}{\varepsilon^3 T^3}\right)^{\frac{1}{2}} \tag{2.61}$$

Introducing the values of the universal constants $e = 4.8029 \cdot 10^{-10}$ esu, $R = 8.314 \times 10^7\,\mathrm{erg\,K^{-1}mol^{-1}}$ and $N_A = 6.0222 \cdot 10^{23}$ we obtain the Debye–Hückel limiting law for the activity coefficient:

$$\log f_{\pm} = -1.825 \cdot 10^6 |z_+ z_-| \left(\frac{I \varrho_0}{\varepsilon^3 T^3}\right)^{\frac{1}{2}} = -A|z_+ z_-| I^{\frac{1}{2}} \tag{2.62}$$

In dilute aqueous solution at 25° C, where $\varepsilon = 78.54$ and $\varrho_0 = 0.997$, this becomes

$$\log f_{\pm} = -0.509 |z_+ z_-| I^{\frac{1}{2}} \tag{2.63}$$

or in general form

$$\log f_{\pm} = -A|z_+ z_-| I^{\frac{1}{2}} \tag{2.64}$$

We shall use this result in practice for the following electrochemical cell without a junction:

$Pt|H_2(1\,atm)|HCl\,(xM)|AgCl|Ag$

where the half-cell reactions are

$\frac{1}{2}H_2(g) \rightleftharpoons H^+ + e$

$AgCl(s) + e \rightleftharpoons Ag(s) + Cl^-$

and the total-cell reaction is

$AgCl(s) + \frac{1}{2}H_2(g) \rightleftharpoons H^+ + Cl^- + Ag(s)$

For its emf we can write

$$E = E_{AgCl/Ag} - E_{H^+/H_2} = E^0_{AgCl/Ag} + \frac{RT}{F} \ln \left(\frac{a_{AgCl}}{a_{Ag} \cdot a_{Cl^-}}\right) - \frac{RT}{F} \ln a_{H^+}$$

As we may set the activities of the solid phases to unity we obtain

$$E = E^0_{AgCl/Ag} - \frac{RT}{F} \ln a_{H^+} \cdot a_{Cl^-}$$

which by introduction of the concepts of activity coefficients and molality can be written as

$$E = E^0_{AgCl/Ag} - \frac{RT}{F} \ln f_+ m_+ \cdot f_- m_- = E^0_{AgCl/Ag} - \frac{RT}{F} \ln f_{\pm}^2 \cdot m^2$$

where $f_{\pm} = \sqrt{f_{H_+} \cdot f_{Cl^-}}$ and $m = m_{HCl}$ as the quantity originally added.
Hence we obtain

$$E + \frac{2RT}{F} \ln m = E^0_{AgCl/Ag} - \frac{2RT}{F} \ln f_{\pm}$$

which by introduction of the Debye–Hückel equation for dilute HCl solution,

$$\ln f_{\pm} = -A|z_+ z_-|I^{\frac{1}{2}} = (-A)m^{\frac{1}{2}}$$

yields

$$E + \frac{2RT}{F} \ln m = E^0_{AgCl/Ag} + \frac{2RT}{F} (A)m^{\frac{1}{2}} \qquad (2.65)$$

By plotting the left-hand side of eqn. 2.65 against $m^{\frac{1}{2}}$ and extrapolating back to $m = 0$ the intercept at $m = 0$ gave the value of $E^0_{AgCl/Ag} = 0.2225$ V at 25° C[20] (cf., 0.2223 V[21]). Once E^0 is known, the measurement of E as a function of m provides an important method for determining molal ionic activity coefficients and a paH scale with the above type of cell, which permits a comparison with the Sørensen psH scale (cf., p. 33).

It should be mentioned that the results of the above extrapolation can be improved when its application is extended to concentrations even higher than 0.01 M, by means of an extended Debye–Hückel equation, viz., introduction of the factor $1/(1 + ba)$ (cf., eqns. 2.60 and 2.60a) into the final expression 2.62 for $\log f_{\pm}$ and its general form 2.64 leads to

$$\log f_{\pm} = -A|z_+ z_-|I^{\frac{1}{2}}/(1 + ba)$$

where a is an estimate of closest approach to the central ion by surrounding ions and

$$b = BI^{\frac{1}{2}} = \left(\frac{8\pi N_A^2 e^2 \varrho_0}{1000\varepsilon RT}\right)^{\frac{1}{2}} I^{\frac{1}{2}} \qquad (2.59)$$

Hence

$$\log f_{\pm} = \frac{-A|z_+ z_-|I^{\frac{1}{2}}}{1 + aBI^{\frac{1}{2}}} \qquad (2.66)$$

In relation to this expression a few empirical extensions have been applied, such as that by Davies[22]:

$$\log f_{\pm} = -0.50|z_+ z_-|\left(\frac{I^{\frac{1}{2}}}{1 + I^{\frac{1}{2}}} - 0.20I\right)$$

(in solutions of single electrolytes up to $I = 0.1$ and at 25° C), and one at very high concentrations:

$$\log f_{\pm} = \frac{-A|z_+ z_-|I^{\frac{1}{2}}}{1 + aBI^{\frac{1}{2}}} - CI$$

where C relates empirically to variations of the activity coefficient. The latter equation may be useful with appreciable additions of neutral salt in order to maintain I constant during reactions or other changes within the solution; for these highly concentrated solutions it should be taken into account that if I is expressed as molality, c_i cannot be replaced simply by $\varrho_0 m$ but instead by $\varrho m/(1 + mM)$ (cf., eqns. 2.51 and 2.59 and the definition of I), but causes some complication in establishing the values of A and B.

Eqn. 2.66 is also of interest for the calculation of activity coefficients in other solution media, where as a consequence of considerably different densities (ϱ_0) and dielectric constants (ε) another value for A is calculated, and also the extended equation $\log f_{\pm} = -A|z_+ z_-|I^{\frac{1}{2}}/1 + I^{\frac{1}{2}}$ has been applied[23].

The important question arises of the actual precision of pH measurement in analytical control. In this connection, it has become common practice to standardize pH determinations on standard buffer solutions with pH regions where the pH of the solution under test is to be expected. As currently commercially available pH meters, pH electrodes and buffer solutions are of outstanding quality, the reliability of the pH measurement becomes shifted to the performance of the measuring electrochemical cell*; here as first principle the same cell should be used for the test solution and the standard solution, so that according to the Bates–Guggenheim convention

$$\mathrm{pH} = \mathrm{pH_s} + \frac{Et - Es}{2.3026\,RT/F}$$

where subscript t = test and s = standard. It is hoped that although a junction potential cannot be completely eliminated, its value will not change substantially from one test solution to another, including the standard; even with a KCl bridge one cannot be absolutely certain of this. However, the whole set-up relies on the precision of the pH buffer solution; therefore, users of a commercial buffer must be well informed of the background of its standardisation**. In order to illustrate what order of precision can be expected we shall confine ourselves to the NBS scale, which is mostly used.

In the NBS system the following cell without a liquid junction is applied:

$$\mathrm{Pt}|\mathrm{H_2}(1\,\mathrm{atm})|\mathrm{HCl}\,(xM,\ \text{plus KCl})|\mathrm{AgCl}|\mathrm{Ag}$$

the emf of which can be represented (see p. 39) by

$$E = E^0_{\mathrm{AgCl/Ag}} - \frac{RT}{F}\ln a_{\mathrm{H^+}} \cdot a_{\mathrm{Cl^-}} = E^0_{\mathrm{AgCl/Ag}} - \frac{RT}{F}\ln f_{\mathrm{H^+}} f_{\mathrm{Cl^-}} m_{\mathrm{H^+}} m_{\mathrm{Cl^-}}$$

$$= E^0_{\mathrm{AgCl/Ag}} - 0.000198T\log f_{\mathrm{H^+}} f_{\mathrm{Cl^-}} m_{\mathrm{H^+}} m_{\mathrm{Cl^-}}$$

*Research-type pH meters often show a resolving power ΔpH = 0.001 and normal meters mostly better than pH = 0.01, so that a pH accuracy of 0.01 can be attained for the combination of a pH meter and a pH glass electrode with temperature correction (preferably automatic).

**Well known standards are those from the NBS (National Bureau of Standards), BS (British Standards Institution), etc.

and accordingly the acidity function $p(a_{H^+}f_{Cl^-})$ will be

$$p(a_{H^+}f_{Cl^-}) = -\log f_{H^+}f_{Cl^-}m_{H^+} = \frac{E - E^0_{AgCl/Ag}}{0.000198T} + \log m_{Cl^-} \tag{2.67}$$

The pH_s of the chloride-free buffer solution can be obtained from

$$pH_s = p(a_{H^+}f_{Cl^-})^0 + \log f^0_{Cl^-} \tag{2.68}$$

where $p(a_{H^+}f_{Cl^-})^0$ is evaluated from $p(a_{H^+}f_{Cl^-})$ at several chloride concentrations with extrapolation to zero. According to the suggestion by MacInnes[24] that the activity coefficients of K^+ and Cl^- are equal in pure KCl solution, one can write $f_{K^+} = f_{Cl^-} = f_{\pm KCl}$, so that by means of the Bates–Guggenheim equation one obtains $\log f^0_{Cl^-} = -AI^{\frac{1}{2}}/1 + 1.5I^{\frac{1}{2}}$ (cf., eqn. 2.66). The value thus obtained yields pH_s by substitution in eqn. 2.68. The results for a collection of buffer solutions of different pH ranges and temperatures have been mentioned in the CRC "Handbook of Chemistry and Physics".

It will be clear that once these standards have been generally applied, pH measurements in practice can henceforth be made without any objection by the use of pH glass electrodes instead of the classical hydrogen electrode.

Oxygen electrode. In principle, a classical oxygen electrode in a liquid electrolyte would be possible if an electrode material were known on the surface of which the redox system O_2/OH^- is electrochemically reversible*; however, Luther[26] measured its standard potential from the following cell without a liquid junction:

$Pt|H_2(1\,atm)|NaOH\,(xM)|Ag_2O|Ag$

where the half-cell reactions are

$2H_2(g) \rightleftharpoons 4H^+ + 4e$

$O_2(g) + 2H_2O + 4e \rightleftharpoons 4OH^-$

and the total-cell reaction is

$O_2(g) + 2H_2(g) \rightleftharpoons 2H_2O$ (oxyhydrogen gas reaction)

Further, there is the chemically reversible reaction:

$2Ag_2O(s) \rightleftharpoons 4Ag(s) + O_2(g)$

Hence for the emf of the cell we can write

$$E = E_{O_2/OH^-} - E_{H^+/H_2} = E^0_{O_2/OH^-} + \frac{RT}{4F}\ln\left(\frac{pO_2}{a^4_{OH^-}}\right) - \frac{RT}{4F}\ln a^4_{H^+}$$

At 25° C the value of E was 1.172 V, so that

$$E^0_{O_2/OH^-} + \frac{RT}{4F}\ln\left(\frac{pO_2}{a^4_{H^+}\cdot a^4_{OH^-}}\right) = 1.172\,V \tag{2.69}$$

*For oxygen concentration cells in solid electrolytes see ref. 25.

As

$$\frac{RT}{4F} \ln \left(\frac{1}{a_{H^+}^4 \cdot a_{OH^-}^4} \right) = -\frac{RT}{F} \ln K_w$$

at 25° C we insert the value $-0.000198 \cdot 298 \log 10^{-14} = +0.826$ V. pO_2, the dissociation pressure of Ag_2O at 25° C (298 K), is too small to be determined directly; from the values of 20.5, 32 and 207 atm at 575, 598 and 718 K, respectively, measured by Lewis[27], we calculated via the simplified Clausius–Clapeyron equation, $\log p_2/p_1 = K(1/T_1 - 1/T_2)$, a value of $4.20 \cdot 10^{-4}$ atm at 298 K; hence $(RT/4F) \ln pO_2$ becomes $(0.05916/4)\log(4.20 \cdot 10^{-4}) = -0.050$ V. By substitution in eqn. 2.69 we obtain $E^0_{O_2/OH^-} = 1.72 + 0.050 - 0.826 = 0.396$ V. For the atmospheric oxyhydrogen cell this means $E = 0.396 + 0.826 = 1.222$ V. Brönsted[28], in a comparable experiment with an electrode, obtained a value of 1.231 V, and thermodynamically a value of 1.23 V was found[29].

Quinhydrone electrode. If the equimolar compound formed by (benzo)quinone (Q) and (benzo)hydroquinone (QH_2), the so-called quinhydrone (Q_2H_2), is dissolved in water, it dissociates to the extent of about 90% into the two components. In conjunction with H^+ ions and electrons an equilibrium is established on the basis of a completely reversible redox reaction:

$$Q_{(s)} + 2H^+ + 2e \rightleftharpoons QH_{2(s)}$$

At a white Pt sheet electrode one obtains the following redox potential:

$$E_{(Q_2H_2)} = E^0_{(Q_2H_2)} + \frac{RT}{2F} \ln \left(\frac{a_Q \cdot a_{H^+}^2}{a_{QH}} \right)$$

As an appreciable amount of quinhydrone (50–100 mg) is added and for attainment of the redox potential a relatively small conversion is required, the situation of equimolarity still holds, so that

$$E_{(Q_2H_2)} = E^0_{(Q_2H_2)} + \frac{RT}{F} \ln a_{H^+}$$

where $E^0_{(Q_2H_2)}$ at 25° C = 0.6992 V. The result indicates that the quinhydrone electrode functions as a pH electrode, albeit with a special standard potential. However, there are a few limitations to its use, viz.:

(1) strong reducing or oxidizing agents should be absent as they disturb the Q/QH_2 equimolarity;

(2) the pH of the solution should not exceed 8, where QH_2 starts to react as an acid ($K_1 = 1.7 \cdot 10^{-10}$);

(3) air oxygen within an alkaline solution under test may oxidize both Q and QH_2 to oxyquinone and coloured resinous products[30].

Many substituted Q/QH_2 systems can also be used, e.g., the tetrachloroquinhydrone electrode in a non-aqueous medium[31].

Redox electrodes with more complicated reactions. In many redox systems hydrogen ions take part, which means that the pH also influences the redox voltage. As such a situation often occurs in biological processes, Clark[32] suggested expressing the influence of both the redox pair and the pH by means of a single number, the rH value: rH $= -\log p_{H_2}$ (where p_{H_2} is the partial pressure of hydrogen). When, for instance, a Pt electrode is placed in a system such as

$$HCrO_4^- + 7H^+ + 3e \rightleftharpoons Cr^{3+} + 4H_2O$$

we can calculate the voltage from

$$E_{ox/red} = E^0_{ox/red} + \frac{2.303RT}{3F} \cdot \log \left(\frac{a_{HCrO_4^-} \cdot a^7_{H^+}}{a_{Cr^{3+}} \cdot a^4_{H_2O}} \right)$$

where $E^0_{ox/red} = 1.195$ V, and for dilute solutions $a_{HCrO_4^-}/a_{Cr^{3+}}$ can be taken the ratio of their molarities while a_{H_2O} is unity and pH (measured) $= -\log a_{H^+}$. However, for the Pt hydrogen electrode in this solution we can write

$$E_{2H^+ \to H_2} = E^0_{2H^+ \to H_2} + \frac{2.303RT}{2F} \cdot \log \left(\frac{a^2_{H^+}}{p_{H_2}} \right) \tag{2.42}$$

and as a value of zero has been assigned to $E^0_{2H^+ \to H_2}$ at 1 atm we simply obtain

$$E_{2H^+ \to H_2} = \frac{2.303RT}{2F} \cdot \log \left(\frac{a^2_{H^+}}{p_{H_2}} \right)$$

or

$$\text{rH} = \frac{E_{2H^+ \to H_2}}{0.0992(273.16 + t)} + 2\text{pH}$$

for $E_{2H^+ \to H_2} = E_{ox/red}$ in mV. The rH scale runs from 0 (below free H_2 is evolved) or 42.6 (above free O_2 is evolved); solutions below 15 tend to be reducing and those above 25 tend to be oxidising[32].

2.2.1.2.2. *Electrodes of the 1st, 2nd and 3rd kinds*

An electrode of the first kind consists of a metal (or its amalgam) whose surface shows a reversible equilibrium with its cation (under test) in the contact solution.

Here the pure metal electrode represents the simplest ion-selective electrode, which according to the Nernst–Van't Hoff equation (cf., eqn. 2.39) yields a potential

$$E_{M^{n+}/M} = E'^0_{M^{n+}/M} + \frac{RT}{nF} \ln \left(\frac{a_{M^+}}{a_M} \right)$$

As the activity of any solid phase by convention is taken as unity, we obtain according to the well-known original Nernst equation

$$E_{M^{n+}/M} = E^0_{M^{n+}/M} + \frac{RT}{nF} \ln a_{M^+} \tag{2.70}$$

For a metal amalgam electrode one cannot simplify the equation so that

$$E_{M^{n+}/M_a} = E^0_{M^{n+}/M_a} + \frac{RT}{nF} \ln \left(\frac{a_{M^+}}{a_{M_a}} \right) \tag{2.71}$$

The choice of the amalgam concentration is free, but one always obtains the same standard potential for the condition $a_{M^{n+}} = a_{M_a}$ or approximately for $[M^{n+}]_{solution} = [m]_{amalgam}$; this is the standard reduction potential that is usually mentioned in tables; e.g.,

$Cd^{2+} + 2e \leftrightarrows Cd(Hg)$	E^0	=	-0.3521 V
$Cd^{2+} + 2e \leftrightarrows Cd$	E^0	=	-0.4026 V
$Zn^{2+} + 2e \leftrightarrows Zn(Hg)$	E^0	=	-0.7628 V
$Zn^{2+} + 2e \leftrightarrows Zn$	E^0	=	-0.7628 V

Hence Cd behaves nobler in the amalgam than as the pure metal, whereas there is no such difference for Zn. For proper functioning of metal electrodes such as Cd and Zn (and many others) the prior removal of oxygen from the solution by deaeration is essential.

An electrode of the second kind consists of a metal whose surface is reversibly in contact or coated with one of its compounds (also containing the anion under test) that is sparingly soluble in the contact solution.

The most classical and reliable example of such an electrode is that of a piece of silver dipped into chloride solution that has been saturated with AgCl with the addition of a few drops of $AgNO_3$ solution; then the electrode potential will be

$$E_{Ag^+/Ag} = E^0_{Ag^+/Ag} + \frac{RT}{F} \ln a_{Ag^+} \tag{2.72}$$

The question arises of the relationship between a_{Ag^+} and the chloride concentration under test. Let us write the product $a_{Ag^+} \cdot a_{Cl^-} = f_{Ag^+} m_{Ag^+} \cdot f_{Cl^-} \cdot m_{Cl^-}$; only if the ionic strength I is not higher than 0.01 molal, which in this instance is mainly determined by the chloride concentration, can we write according to eqn. 2.39 $a_{Ag^+} \cdot a_{Cl^-} = f_\pm^2 \cdot m_{Ag^+} \cdot m_{Cl^-}$, where $\log f_\pm = -0.509\, m_{AgCl}^{1/2}$ and, as, m_{AgCl} is very small, $f_\pm = 1 - 0.509\, m_{AgCl}^{1/2}$ and $f_\pm \approx 1$.

Further, the concentration $C = m\varrho/1 + mM$; hence for this dilute solution $c \approx \varrho_0 m$, where for water as a solvent (at 25° C) $\varrho_0 = 0.997$. Hence under these conditions

$$a_{Ag^+} \cdot a_{Cl^-} = f_\pm^2 \cdot m_{Ag^+} \cdot m_{Cl^-} = 0.997^2 [Ag^+][Cl^-] = 0.997^2 S^{25°C}_{0AgCl}$$

where $S^{25°C}_{0AgCl} = 1.56 \cdot 10^{-10}$, as commonly mentioned in tables of solubility products of sparingly soluble salts. Substituting in eqn. 2.72 one obtains

$$E_{Ag^+/Ag} = E^0_{Ag^+/Ag} + \frac{RT}{T} \ln \left(\frac{0.997^2 S_{0AgCl}}{a_{Cl^-}} \right)$$

or

$$E_{AgCl/Ag} = E^0_{AgCl/Ag} - \frac{RT}{F} \ln a_{Cl^-} \qquad (2.73)$$

where

$$E^{0(25°C)}_{AgCl/Ag} = E^{0(25°C)}_{Ag} + 0.05916 \log 0.997^2 S^{(25°C)}_{0\,AgCl}$$

which is in fact the standard potential of a pCl electrode.

In the tables we find $E^0_{Ag^+/Ag} = 0.7996$ V and $E^0_{AgCl/Ag} = 0.2223$ V. From the above it is clear that primarily the silver–silver chloride electrode functions as a pAg electrode, i.e., it measures a_{Ag^+}; at an ionic strength above 0.01 (cf., extended Debye–Hückel expressions) the calculation of $[Ag^+]$ becomes more difficult, and even more so for $[Cl^-]$, where the solubility product value is also involved.

Although not essential, one often uses a previously coated AgCl–Ag electrode or a silver-plated Pt wire coated electrolytically in KCl solution with a thin deposit of AgCl. Such dry AgCl–Ag electrodes are much in favour as reference electrodes (although in the absence of oxidants), in addition to calomel electrodes (Pt wire in contact with Hg, covered with calomel paste in contact with KCl solution), which also belong to the second kind, viz.,

$$E_{Hg_2Cl_2/Hg} = E^0_{Hg_2Cl_2/Hg} - \frac{RT}{F} \ln a_{Cl^-}$$

Their potentials in 0.1 N, 1 molal, 1 N and saturated KCl solutions are 0.3337, 0.2800, 0.2897 and 0.2415 V, respectively. The dilute types reach their equilibrium potentials more quickly and these potentials are less dependent on temperature; the SCE has the advantage of being less sensitive to current flow (electrolysis). The AgCl–Ag electrodes are more compact, do not need a liquid function, which makes them exceedingly attractive for analysis in non-aqueous media, and support high temperatures.

Metal oxide electrodes are also of the second type. A well known example is a rod of antimony coated with Sb_2O_3 (or bismuth with Bi_2O_3), which can function as a pH electrode[33]:

$$Sb_2O_3(s) + 6H^+ + 6e \rightleftharpoons 2Sb(s) + 3H_2O$$

so that

$$E_{Sb_2O_3/Sb} = E'^0_{Sb_2O_3/Sb} + \frac{RT}{6F} \ln \left(\frac{a_{Sb_2O_3(s)} \cdot a^6_{H^+}}{a^2_{Sb(s)}} \right) = E^0_{Sb_2O_3/Sb} + \frac{RT}{F} \ln a_{H^+}$$

where the standard potential $E^0_{Sb_2O_3/Sb} = 0.144$ V. The electrodes, owing to their robustness, are more suitable for industrial control.

An electrode of the third kind consists of a metal whose surface is in reversible contact with a small, fixed dissolved amount of one of its chelates, while the chelating agent Y concerned can also yield in a competitive way a chelate with another metal ion (under test) in the contact solution. The application of such an electrode opens the way to potentiometry of a metal ion for which no electrode of the first kind exists. The most illustrative example is that by Reilley and co-workers[34], who used a static mercury drop electrode. With EDTA as a chelating agent they added a small amount of Hg(II) chelate to the solution of the metal ion (M^{n+}) under test at a certain pH. Then the apparent stability constant of the mercury complex will be

$$K_{Hg_{(app.H)}} = K_{Hg/\alpha H} = \frac{[HgY^{2-}]}{[Hg^{2+}][Y^{4-}]}$$

and the one of the other metal complex

$$K_{M_{(app.H)}} = K_{M/\alpha H} = \frac{[MY^{(n-4)+}]}{[M^{n+}][Y^{4-}]}$$

From these two equations one can eliminate $[Y^{4-}]$ and the pH influence factor α_H, so that

$$\frac{K_M}{K_{Hg}} = \frac{[MY^{(n-4)+}]}{[HgY^{2-}]} \cdot \frac{[Hg^{2+}]}{[M^{n+}]}$$

This equation enables us to calculate the mercury potential according to

$$E_{Hg^{2+}/Hg} = E^0_{Hg^{2+}/Hg} + \frac{0.05916}{2} \log \left([M^{n+}] \frac{[HgY^{2-}]}{[M^{(n-4)+}]} \cdot \frac{K_M}{K_{Hg}} \right)$$

However, as $[HgY^{2-}]$, K_M and K_{Hg} are constants their logarithmic term can be combined with $E^0_{Hg^{2+}/Hg}$ so that

$$E = \left(E'^0 + \frac{0.05916}{2} \log [HgY^{2-}] \right) + \frac{0.05916}{2} \log \left(\frac{[M^{n+}]}{[MY^{(n-4)+}]} \right)$$

or simply

$$E = E^0 + \frac{0.05916}{2} \log \left(\frac{[M^{n+}]}{[MY^{(n-4)+}]} \right) \qquad (2.74)$$

The chelate HgY^{2-} must be considerably more stable than $MY^{(n-4)+}$; if necessary one can mask M^{n+} to a certain desirable extent by means of a selective complexing agent A, so that K_M becomes an apparent stability constant, $K_{M(app.A)} = K_{M/\beta_a}$, which changes E^0 as a consequence of the logarithmic term of β_A.

2.2.1.3. *Membrane electrodes*

Together with active metal electrodes, the membrane electrodes represent the best known ion-selective electrodes (ISEs); however, the membrane type has the advantages of insensitivity to redox agents and surface poisons. As the prototype of membrane electrodes, the glass electrode still remains the most successful and reliable.

In order to explain the working of membrane electrodes, one generally talks of perm-selective membranes, although on this basis not all phenomena can be fully understood. However, some more fundamental remarks on membranes may be useful,

In electroanalysis at least three possibilities are involved in chemically inert membranes:

(1) A permeable membrane, which merely serves to prevent rapid mixing of components within solutions on both sides of the membrane; in principle, no potential occurs unless a diffusion potential occurs.

(2) A Donnan membrane, i.e., a membrane impermeable to certain kinds of ions, which results in the occurrence of the Donnan potential.

(3) A semi-permeable membrane, which is unequally permeable to different components and thus may show a potential difference across the membrane. In case (1), a diffusion potential occurs only if there is a difference in mobility between cation and anion. In case (2), we have to deal with the biologically important Donnan equilibrium; e.g., a cell membrane may be permeable to small inorganic ions but impermeable to ions derived from high-molecular-weight proteins, so that across the membrane an osmotic pressure occurs in addition to a Donnan potential. The values concerned can be approximately calculated from the equations derived by Donnan[35]. In case (3), an intermediate situation, there is a combined effect of diffusion and the Donnan potential, so that its calculation becomes uncertain.

In addition to the above-mentioned inert members there are the electrochemically active membranes such as the perm-selective membranes, the working of which is mainly based on ion-exchange activity. They include the simple ion-exchange membranes which, once placed in a contact solution, still keep a sufficiently large free pore radius and become thoroughly soaked; for instance, through the cation-exchange type anions pass freely whilst cations replace the cation originally situated on the membrane; the reverse takes place with the anion-exchange type.

In coulometry these exchange membranes are often used to prevent the electrolyte around the counter electrode from entering the titration compartment (see coulometry, Section 3.5). However, with membrane electrodes the ion-exchange activity is confined to the membrane surfaces in direct contact with the solutions on both sides, whilst the internal region must remain impermeable to the solution and its ions, which excludes a diffusion potential; nevertheless, the material must facilitate some ionic charge transport internally in order to permit measurement of the total potential across the membrane. The specific way in which all these requirements are fulfilled in practice depends on the type of membrane electrode under consideration.

2.2.1.3.1. *General theory of membrane electrodes*

Consider an ion-exchange membrane between two electrolytes with a

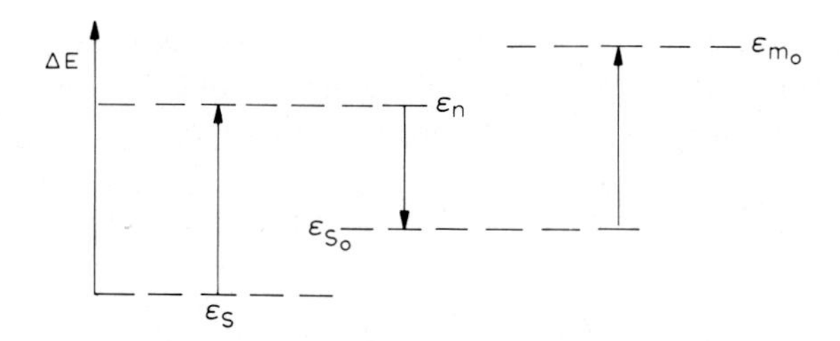

Fig. 2.9. Membrane ISE (potential profile).

common ion i within the following measuring cell:

external reference electrode (of choice)	aqueous solution S with variable ion activity a_i^{S}	ion-exchange membrane n with respect to ion i	aqueous solution S_0 with fixed ion activity $a_i^{S_0}$	internal reference electrode (generally metal electrode of the 2nd kind)
reference half-cell	indicator half-cell (membrane ISE)			

then the potential profile of the membrane ISE can be depicted by Fig. 2.9 and represented by

$$\Delta E \;=\; \varepsilon_{m_0} - \varepsilon_S \;=\; (\varepsilon_{m_0} - \varepsilon_{S_0}) + (\varepsilon_{S_0} - \varepsilon_n) + (\varepsilon_n - \varepsilon_S)$$

where ε_{m_0} is the potential of the internal reference electrode and ε_n is the potential of the membrane.

As the chemical potential of a component i in solution is

$$\mu_i \;=\; \mu_i^0 + RT \ln a_i \tag{2.36}$$

for an ion i, owing to its internal electrostatic or Galvani potential ϕ, there is still a contribution $z_i F\phi$, so that the electrochemical potential becomes

$$\mu_i \;=\; \mu_i^0 + RT \ln a_i + z_i F\phi \tag{2.75}$$

where z_i represents the charge number, including the sign, of ion i.

We assume that between the membrane phase (n) and the solutions (S and S_0) no diffusion potentials occur, i.e., the potential differences $\phi_{(n/S)}$ and $\phi_{(n/S_0)}$ are merely caused by ion-exchange activity. Further, all phases are considered to be homogeneous so that $\mu_{(n)}^0$ is constant within the membrane and also $\mu_S^0 = \mu_{S_0}^0$ within the solutions. Let us at first consider the simple system of selective indication of an univalent anion A^- in the presence of an interfering univalent anion B^-. Here we obtain the following exchange equilibrium:

$$A_{(n)}^- + B_{(S)}^- \rightleftharpoons A_{(S)}^- + B_{(n)}^- \tag{2.76}$$

with an exchange constant K_{A^-/B^-} (see eqn. 2.77). Moreover, one of the two ions will possess a predominating direct ion-exchange affinity for the membrane, giving it a potential difference (negative for anions) versus the solution. Because at the interface concerned the electrochemical potentials of the adjacent phases under equilibrium condition must be equal, we can write

$$\mu^0_{A^-(n)} + RT \ln a_{A^-(n)} - F\phi_{(n)} = \mu^0_{A^-(S)} + RT \ln a_{A^-(S)} - F\phi_{(S)}$$

and

$$\mu^0_{B^-(n)} + RT \ln a_{B^-(n)} - F\phi_{(n)} = \mu^0_{B^-(S)} + RT \ln a_{B^-(S)} - F\phi_{(S)}$$

so that

$$\phi_{(n/S)} = \frac{1}{F}(\mu^0_{A^-(n)} - \mu^0_{A^-(S)}) + \frac{RT}{F} \ln \left[\frac{a_{A^-(n)}}{a_{A^-(S)}}\right]$$

and

$$\phi_{(n/S)} = \frac{1}{F}(\mu^0_{B^-(n)} - \mu^0_{B^-(S)}) + \frac{RT}{F} \ln \left[\frac{a_{B^-(n)}}{a_{B^-(S)}}\right]$$

hence, if

$$\mu^0_{A^-(n)} - \mu^0_{A^-(S)} = A^0_{A^-} \quad \text{and} \quad \mu^0_{B^-(n)} - \mu^0_{B^-(S)} = A^0_{B^-}$$

we find

$$\frac{A^0_{A^-}}{F} + \frac{RT}{F} \ln \left[\frac{a_{A^-(n)}}{a_{A^-(S)}}\right] = \frac{A^0_{B^-}}{F} + \frac{RT}{F} \ln \left[\frac{a_{B^-(n)}}{a_{B^-(S)}}\right]$$

so that the exchange constant is

$$K_{A^-/B^-} = \frac{a_{A^-(S)} \cdot a_{B^-(n)}}{a_{A^-(n)} \cdot a_{B^-(S)}} = \exp\left(\frac{A^0_{A^-} - A^0_{B^-}}{RT}\right) \tag{2.77}$$

According to the potential profile in Fig. 2.9 the potential of the ISE is

$$E = \Delta E = \varepsilon_{m_0} - \varepsilon_S = (\varepsilon_{m_0} - \varepsilon_{S_0}) + (\varepsilon_{S_0} - \varepsilon_n) + (\varepsilon_n - \varepsilon_S)$$

and hence

$$E = E_{\text{int.ref.}} - \phi_{(n/S_0)} + \phi_{(n/S)} = E_{\text{int.ref.}} + \phi_{(n/S)} - \phi_{(n/S_0)}$$

We only have to consider the case of both solutions S_0 and S being aqueous, and for the time being we assume that B^- is absent. Hence

$$\phi_{(n/S)} - \phi_{(n/S_0)} = \left(\frac{A^0_{A^-}}{F} + \frac{RT}{F} \ln \left[\frac{a_{A^-(n)}}{a_{A^-(S)}}\right]\right) - \left(\frac{A^0_{A^-}}{F} + \frac{RT}{F} \ln \left[\frac{a_{A^-(n)}}{a_{A^-(S_0)}}\right]\right)$$

Therefore

$$E = E_{\text{int.ref.}} + \frac{RT}{F} \ln a_{A^-(S_0)} - \frac{RT}{F} \ln a_{A^-(S)} = E^0 - \frac{RT}{F} \ln a_{A^-(S)}$$

Accordingly, the standard potential of the ISE depends on the specific choice of the internal reference electrode (e.g. Ag–AgCl) and also that of the fixed A^- concentration of the filling solution, so that simply

$$E_{ISE} = E^0_{ISE} - \frac{RT}{F} \ln a_{A^-} \tag{2.78}$$

The question arises of how E will be influenced by the presence of an interfering ion B^- within the solution under test. The answer has already been given by Nikolski[36] with his equation

$$E_{ISE} = E^0_{ISE} - \frac{RT}{F} \ln (a_{A^-} + K_{A^-/B^-} \cdot a_{B^-}) \tag{2.79}$$

That this offers a plausibly approximation of the interference by B^- may follow from substitution of eqn. 2.77 in

$$a_{A^-(S)}(\text{apparent}) = a_{A^-(S)} + K_{A^-/B^-} \cdot a_{B^-(S)} = a_{A^-(S)} + \frac{a_{A^-(S)} \cdot a_{B^-(n)}}{a_{A^-(n)} \cdot a_{B^-(S)}} \cdot a_{B^-(S)}$$

or

$$a_{A^-(S)}(\text{apparent}) = a_{A^-(S)} \left[1 + \frac{a_{B^-(n)}}{a_{A^-(n)}} \right] \tag{2.80}$$

In the above derivation we may assume that $a_{A^-(S)} = a_{A^-}$ and $a_{B^-(S)} = a_{B^-}$, because by analogy with the build-up of an electrode potential (see pp. 26–27) the build-up of the ion-exchange potential will not significantly alter the original concentrations of A^- and B^- in the solution under test. Hence in eqn. 2.80 the ratio $a_{B^-(n)}/a_{A^-(n)}$, which reflects the exchange competition of B^- versus A, still depicts the interference ratio of B^- in more straightforward manner than does the so-called selectivity constant (k), usually mentioned by ISE suppliers.

The reader will have realized that eqn. 2.78 has been specifically derived for a univalent anion (cf., $-RT/F$ before the logarithm), which means that for an ion of valence z_i, including the sign of the charge, the more general equation becomes

$$E_{ISE} = E^0_{ISE} + \frac{RT}{z_i F} \ln a_i \tag{2.81}$$

As a consequence, the Nikolski equation will be more complicated, especially if the interfering ion has a different valence. For instance, when, in addition to the ion A with valence z_A, the interfering ion B has a valence z_B, the ion-exchange equilibrium (cf., eqn. 2.76) is

$$|z_B| A^{z_A}_{(n)} + |z_A| B^{z_B}_{(S)} \rightleftharpoons |z_B| A^{z_A}_{(S)} + |z_A| B^{z_B}_{(n)} \tag{2.82}$$

with the exchange constant $K_{A^{z_A}/B^{z_B}}$ (see eqn. 2.83).

If

$$(\mu^0_{A^{z_A}_{(n)}} - \mu^0_{A^{z_A}_{(S)}} = A^0_{A^{z_A}} \quad \text{and} \quad (\mu^0_{B^{z_B}_{(n)}} - \mu^0_{B^{z_B}_{(S)}}) = A^0_{B^{z_B}}$$

we find

$$\frac{A^0_{A^{z_A}}}{z_A F} + \frac{RT}{z_A F} \ln \left[\frac{a_{A^{z_A}_{(n)}}}{a_{A^{z_A}_{(S)}}} \right] = \frac{A^0_{B^{z_B}}}{z_B F} + \frac{RF}{z_B F} \ln \left[\frac{a_{A^{z_A}_{(n)}}}{a_{B^{z_B}_{(S)}}} \right]$$

As z_A and z_B have the same sign, we can multiply by $|z_A| \cdot |z_B|$, and obtain for the exchange constant the equation

$$K_{A^{z_A}/B^{z_B}} = \left(\frac{a_{A^{z_A}_{(S)}}}{a_{A^{z_A}_{(n)}}} \right)^{|z_B|} \cdot \left(\frac{a_{B^{z_B}_{(n)}}}{a_{B^{z_B}_{(S)}}} \right)^{|z_A|} = \exp \frac{|z_B| A^0_{A^{z_A}} - |z_A| A^0_{B^{z_B}}}{RT} \tag{2.83}$$

so that the Nikoski equation now becomes

$$E_{ISE} = E^0_{ISE} + \frac{RT}{z_A F} \ln \left[a_{A^{z_A}} + K_{A^{z_A}/B^{z_B}} \cdot (a_{B^{z_B}})^{|z_A|/|z_B|} \right] \tag{2.84}$$

In an analogous manner it can be extended to the interference of other ions:

$$E_{ISE} = E^0_{ISE} + \frac{RT}{z_A F} \ln \left[a_{A^{z_A}} + K_{A^{z_A}/B^{z_B}} \cdot (a_{B^{z_B}})^{|z_A|/|z_B|} + K_{A^{z_A}/C^{z_C}} \cdot (a_{C^{z_C}})^{|z_A|/|z_C|} + \cdots \right] \tag{2.85}$$

When using the selectivity constant or coefficient (k) mentioned by ISE suppliers, one must be sure that if the ion under test and the interfering ion have different valence the exponent in the activity term according to Nikolski has been taken into account; it has become common practice to mention the interferent concentration that results in a 10% error in the apparent ion concentration; these data facilitate the proper choice of an ISE for a specific analytical problem. Often maximum levels for no interference are indicated.

A more comprehensive notation of eqn. 2.85 often encountered in literature is

$$E = E^0 + \frac{RT}{z_i F} \ln (a_i + \sum_j K_{ij} a_j^{z_i/z_j}) \tag{2.86}$$

where i is the primary ion and j the interfering ions B, C, etc.

Whether interfering ions are present or not, there is still the problem of how to obtain the concentration from an actually measured activity value. In order to achieve this we can choose either a standard calibration method or an incremental (or decremental) method[37].

In the standard calibration technique one adds a few millilitres of a concentrated solution of a non-interfering strength adjuster (ISA) and/or a pH adjuster to both the standard and sample solutions before measurement. Usually ISE suppliers provide a list of ISA/pH adjusters appropriate to their ISEs in order to maintain the activity coefficient, $f_{\pm}$, and/or pH at the most suitable fixed value.

In the incremental or decremental technique, another designation for the standard addition (or subtraction) technique, one adds increments of standard solution to the sample, or vice versa. (In the decremental technique the standard precipitates or complexes the ion under test.) When the sample itself is incrementally added to the standard, the latter may have received a previous addition of ISA and/or pH adjuster, but in the reverse method this addition may be made to the sample. However, for the specific example of a univalent anion we shall show how the normal incremental method works[38] and that in fact the addition of ISA is not necessary.

Consider an anion solution of unknown concentration C_0 with volume V_0. An appropriate ISE together with a reference electrode is introduced and the emf of the cell measured. Next, a series of additions of a standard solution with concentration C and volume V are made, and after each incremental addition the emf is measured.

It is recommended that the concentration of the standard solution is sufficiently high to keep the volume V relatively small; further, it is advisable that by the total amount of standard finally used the solution concentration has not exceeded about 2.5 times the sample concentration.

From the above it follows that, if a junction potential is also taken into account, the potential versus that of the reference electrode will be

$$E_n = E^0 - \frac{RT}{F} \ln \left[f_{a_i} \frac{C_0 V_0 + C \sum_{k=0}^{n} V}{V_0 + \sum_{k=0}^{n} V} \right] + E_j$$

so that

$$-\frac{FE}{2.3026RT} = -\frac{F(E^0 + E_j)}{2.3026RT} + \log f_{a_i} \left(C_0 V_0 + C \sum_{k=0}^{n} V \right) \Big/ \left(V_0 + \sum_{k=0}^{n} V \right)$$

or

$$\left(V_0 + \sum_{k=0}^{n} V \right) \cdot 10^{-\frac{FE}{2.3026RT}} = f_{a_i} \left(C_0 V_0 + C \sum_{k=0}^{n} V \right) \cdot 10^{-\frac{F(E^0 + E_j)}{2.3026RT}} \tag{2.87}$$

If we plot a graph of the values of the left-hand side of eqn. 2.87 on the ordinate against the volumes V on the abscissa, we can draw through the points obtained a straight line that intersects the x-axis at the volume V_e; at

TABLE 2.3

ION-SELECTIVE MEMBRANE ELECTRODES

Working principle	Physical characterization of membrane		Sensor material		Active component (examples)	Electrode designation	Ions detectable and measurable
Direct	Solid state	Homogeneous	Glass	Lattice structure	Special glass	Glass electrode	H^+, Na^+, Li^+, K^+, NH_4^+
			Crystalline	Monocrystalline	LaF_3	Single-crystal (membrane) electrode	La^{3+}, F^-
					Ag_2S,AgX	Homogeneous monocrystalline membrane electrode (compact or pressed)	Ag^+, S^{2-}, $X = Cl^-$, Br^-, I^-, CN^-
				Polycrystalline	$Ag_2S + AgX$	Homogeneous polycrystalline membrane electrode (pressed or sintered)	Ag^+, $X = Cl^-$, Br^-, I^-, CN^-, CNS^-
					$Ag_2S + MS$	As above (pressed or sintered)	Ag^+, $M = Pb^{2+}$, Cu^{2+}, Cd^{2+}
					Ion exchanger (solid)	As above (pressed)	H^+, Na^+, K^+, Ca^{2+}, Zn^{2+}, Ni^{2+}, SO_4^{2-}, Cl^-, OH^-, but of limited success
		Heterogeneous	Inert matrix with emulsified active component	Silicone rubber (Pungor) Polythene, polypropene, paraffin	Sulphides and/or halides, Ca oxalate or stearate,	Heterogeneous precipitate membrane electrode	Ag^+, Cl^-, Br^-, I^-, S^{2-}, CN^-, Ca^{2+}
					Ba sulphate, Bi or Fe(III) phosphate, etc.		Ba^{2+}, SO_4^{2-} PO_4^{3-}, etc.
				Graphite + PTFE (Růžička)	Various ionic compounds		Various ions
				Polystyrene	Ion exchanger (solid)		Various ions, but of limited success

	Liquid	Homogeneous	Inert matrix with emulsified active component	Gel of collodion, PVC, etc.	Solution (with low ε) of ion exchangers or complex in suitable solvent	(apparently dry) "solid-state" membrane electrode	K^+, NH_4^+, Ca^{2+}, BF_4^-, NO_3^-
		Heterogeneous	Emulsified active component trapped in carrier skeleton	Skeleton of porous plastic (PTFE, PVC, etc.), sintered glass, filtering textile (Dacron, glass fibre, etc.)	As above	Liquid membrane electrode	K^+, Ca^{2+}, Me^{2+}*, Cl^-, BF_4^-, NO_3^-, ClO_4^-
Indirect	Liquid	Homogeneous	Inert matrix with emulsified active component occasionally protected by nylon gauze or cellophane foil	Gel of polymer around glass electrode	Solution of enzyme or of enzyme substrate	Enzyme electrode or its substrate electrode	H^+ (or OH^-) as an indirect measure of substrate or enzyme
		Heterogeneous	0.1 M HCO_3^-, HSO_3^- or NH_4^+ solution covered with gas-permeable membrane	Foil of PTFE or polythene	Chemical reagent + glass membrane	Gas-sensing electrode	H^+ (or OH^-) as an indirect measure of NH_3, SO_2, nitrous vapours, H_2S, HCN, $Cl_2(X_2)$, CO_2

*For a few membrane electrodes the ion selectivity is limited, e.g., the Metrohm EA 301-Me^{2+} electrode shows the same sensitivity for various divalent cations; in fact, even non-selectivity may be accepted, e.g., where only one type of cation is present as in aqueous hydrofluoric acid where the pH can be determined with a cation-exchange resin electrode[40].

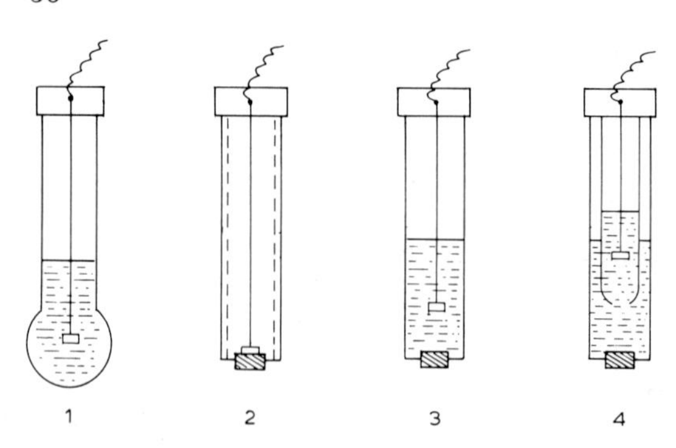

Fig. 2.10. Membrane electrode types. 1: Glass electrode; 2, 3 and 4: crystal membrane electrodes.

at this point, $C_0 V_0 + CV_e$ must be equal to zero, so that finally the original concentration is $C_0 = -CV_e/V_0$.

This result shows that in the incremental method it is not necessary to know the actual values of f_{a_i} and E_j. Apart from this advantage, it may be considered as a disadvantage that the order of magnitude of the sample concentration must be known in advance and that all solution volumes involved should be accurately determined.

2.2.1.3.2. *Types of membrane electrodes*

Table 2.3 shows the various types of membrane electrodes, a few of which are shown schematically in Fig. 2.10.

The general theory given above was treated in more detail for the example of an ion-exchange membrane between two electrolytes for which the validity of the Nikolski equation could be proved. Fortunately, the relationship appeared also to be valid for other types of membrane electrodes, as illustrated by Koryta[39]. Therefore, in the following specific treatment of the more important electrode types we shall confine ourselves to some constructional details and a few theoretical remarks. Further, starting from the division into solid-state and liquid membrane electrodes, we have denoted the glass membrane as solid, although glass represents an undercooled liquid; however, the homogeneous monocrystalline and polycrystalline membrane electrodes and the heterogeneous precipitate electrodes are considered as true solid-state sensors.

2.2.1.3.2.1. *Solid-state membrane electrodes.* All members of this class are direct working.

Glass electrode [see Fig 2.10 (1)]. The pH glass electrode, as the most important representative of the glass electrodes, will be the first subject to be treated, and especially in its application to aqueous solutions. Attached to the stem of high-resistance glass, the electrode proper consists of a pH-sensitive glass bulb that acts as a membrane between an inner reference electrolyte and an outer electrolyte (the sample under test). On both sides of the membrane there will be a superficial swelling resulting in a "dry" glass continuum between two hydrated gel layers; via contact of these layers with the electrolytes ion exchange occurs between Na^+ ions from the glass and H_3O^+ ions from the electrolytes so that the surface layers become more or less acidic. For the pH-sensitive electrode we obtain direct predominant ion exchange (cf., p. 49) of the H_3O^+ ions from the electrolyte towards the gel surface, giving it a positive potential difference versus the electrolyte, in spite of the fact that the still existing ion-exchange equilibrium reaction between H_3O^+ and Na^+ ions may yield some (although negligible) interference (cf., the Nikolski equation, eqn. 2.79); probably the "direct ion exchange" mentioned can be best explained by the postulation of Eisenman et al.[41]: "when a hydrated cation approaches the negatively charged site on the glass gel there will be a competition between the glass negative site and the hydrating water molecule for the bare cation". In this respect it can also be well understood that with an inner and an outer electrolyte of identical compositon normally a slight difference in potential remains between the inside and outside glass surface, the so-called asymmetry potential; such an effect is caused by differing states of strain distortion of the concave and convex surfaces; even with flat surfaces one cannot completely eliminate such slight structural differences. The question arises of the extent to which other cations present in the acidic electrolyte still can interfere; this is only significantly possible at very low a_{H^+} and high metal cation activity a_{M^+}, where

$$a_{M^+} \approx a_{OH^-} = \frac{Kw}{a_{H^+}}$$

so that

$$E = E^0 + \frac{RT}{F} \ln \left(a_{H^+} + K_{a_{H^+/M^+}} \cdot \frac{Kw}{a_{H^+}} \right)$$

The validity of curves obtained from this equation has been verified by Nikolski et al.[42].

Further, there is the question of the stability of the glass electrode. Even with a high resistance against chemical attack we have to deal with acid and alkali errors at the extremes of the pH scale; below pH 1 the measured values are slightly higher than the correct values and above pH 11 they are lower (cf., Fig. 2.11); most electrode suppliers do mention the data concerned. Apart from this phenomenon, chemical attack is not always excluded; this attack depends very much on the composition* of the glass in connection with the conditions required for the specific ion selectivity of the electrode. Here again the theory of Eisenman and co-workers[41,44] has proved to be of most value, but we must confine ourselves to a few remarks about it.

---

*For a review of compositional and constructional aspects of glass electrodes (especially the pH electrode), see Mattock[43].

Today two theories may explain the structure of glass as an undercooled liquid silicate mixture, viz., the crystallites theory (small ordered areas within an unordered amorphous matrix) and the network theory (a three-dimensional network of regular units of SiO_4^{4-} tetrahedra with interstitial charge compensating cations irregularly spread within the material). Although both theories are approaching one another, it is the latter that has proved most useful in explaining the ion sensitivity, electrical conductance and other (e.g., chemical) properties of glass electrodes as a function of their composition. On the basis of the above we can make the following statements:

(1) Quartz glass (silica), pyrex (borosilicates) and other household and laboratory glasses (boroaluminosilicates) owe their high chemical resistance to the (tetrahedral) network forming properties of Si, B, P and Al (cf., the borax and phosphate bead tests in qualitative dry reactions on the salts of numerous metals).

(2) Addition of an alkali metal oxide as a "network modifier" to the "network former" causes pH sensitivity, i.e., small amounts of alkali metal induce superficial gel layer formation as a merely local chemical attack and so with limited alkali error; larger amounts will result in more pronounced dissolving properties of the glass up to complete dissolution, e.g., water-glass with large amounts of sodium oxide. Simultaneous addition of an alkaline earth metal oxide, however, diminishes the dissolution rate. Substitution of lithium for sodium in pH-sensitive glass markedly reduces the alkali error.

(3) According to the theory of Eisenman and co-workers the chemical compositon of the "network former" in itself is of extreme importance in view of ion selectivity; they stressed the influence of increasing availability of free negative sites within the glass as a function of B_2O_3, Al_2O_3 or La_2O_3 added to the silica in that sequence and with increasing ratio of each of them versus alkali metal oxide as a "network modifier"; in that sequence or ratio the negative field strength of the glass increases with the following selectivity response towards the alkali metal ions: with a weakly negative field strength $Cs^+ > Rb^+ > K^+ > Na^+ > Li^+$, and the reverse with a strongly negative field strength. The influence of the boron, aluminium and lanthanum can be ascribed to the fact that their coordination number for oxygen is higher than their oxidation number; this is especially true for lanthanum, whose substitution into the silicate system is not isomorphic (six surrounding atoms for oxygen instead of four). However, whilst an increasing Al/Na ratio in sodium aluminosilicate glass causes a transition from a primarily pH-responsive alkali metal silicate to a primarily alkali-responsive aluminosilicate, the polar bonds between La and O are so strong that the pH response of the electrode prevails over all alkali metal ion responses.

(4) In view of the practical measurement possibilities, the following points are still of importance:

(I) The electrical (apparently ionic) conductance of the glass as a function of:

(a) chemical composition: alkali metal ions increase the conductance to an increasing extent on going from the smaller Li ion (greater coulombic force between conducting ion and lattice) up to the larger Rb and Cs ions, and alkaline earth metal ions give little change (Ba increases the resistance slightly more than Ca, but Be decreases it markedly); however, Li glasses show a considerably better conductance if they contain Pb instead of Ca;

(b) temperature: there is a very marked exponential decrease of conductance with temperature, e.g., for a certain glass electrode a resistance of 200 MΩ at 10° C became 100 MΩ at 20° C.

(II) The conditioning of the glass surface with respect to gel layer formation: in general, the electrodes should preferably be stored in distilled water to prevent them from drying out; manufacturers often recommend pre-soaking in dilute acid or a high pH buffer and at temperatures corresponding to the conditions under which the measurements have to be made; Li within the glass has the advantage of requiring less water for an adequate gel layer, so that the electrode also becomes applicable in partly non-aqueous media.

From the above remarks, it will be clear that the development and manufacture of a properly functioning and sufficiently stable pH glass electrode requires a sophisticated compromise based on research and skill. This may be illustrated as follows:

(1) The high resistance (5–500 MΩ) of the glass membrane sets a maximum on its thickness and a minimum on its surface area and the production of a shockproof membrane sets a minimum on its thickness and a maximum on its surface area. The choice of a compromise between these conflicting requirements also depends on the amount and character of the sample under test, viz., the shape and size of the membrane may be standard (in general a bulb 0.05–0.15 mm thick and about 10 mm in diameter) or special (smaller bulbs with thinner membranes in microelectrodes[25,32], protected bulbs[37] in plant control electrodes, other shapes of tips such as a flat[37], spear or cup form[32,37,45], research electrodes, etc.)

(2) With respect to pH sensitivity and an adequate speed of response (time constant $\tau = RC$ where R is the resistance of the measuring circuit and C the capacitance of the electrode), a certain degree of superficial swelling is needed; however, the gel layer thus formed should remain thin in order to minimize the solubility of the glass and to guarantee sufficient durability of the electrode. In this respect lithium barium silicates offer an attractive compromise[32].

(3) The resistance of the electrode stem must be considerably higher (resistivity $10^{14}\,\Omega$ cm) than that of the membrane attached, so that in a voltage measurement the actual current only passes the membrane; further, to permit measurements without significant errors the input impedance of the voltmeter should be at least 100 times greater than the resistance of the membrane (i.e., voltmeter impedance $> 10^{13}\,\Omega$). Moreover, to avoid any voltage leakage to the inner reference through moisture condensed along the stem, the latter or at least the electrode cap together with the cable needs to be highly insulated. Owing to this insulating material and the capacitance of the electrode, ac fields or charged bodies may influence the pH measurement so that a screen connected to earth directly or through a capacitor is often required.

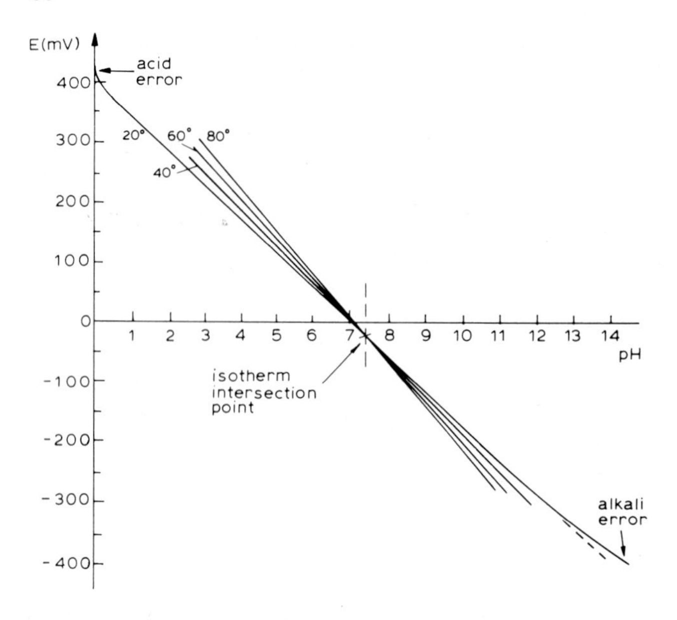

Fig. 2.11. Iso-pH value (courtesy of Metrohm).

It has become fairly common to adopt the manufacture of combinations of internal reference electrode and its inner electrolyte such that the (inner) potential at the glass electrode lead matches the (outer) potential at the external reference electrode if the glass electrode has been placed in an aqueous solution of pH 7. In fact, each pH glass electrode (single or combined) has its own iso-pH value or isotherm intersection point; ideally it equals 0 mV at pH 7 $\pm$ 0.5 according to a DIN standard, as is shown in Fig. 2.11; the asymmetry potential can be easily eliminated by calibration with a pH 7.00 $\pm$ 0.02 (at 25° C) buffer solution.

The advantage of the general adoption of such a standardization system is two-fold: (1) the use of pH electrodes will no longer be strictly bound to pH meters from a specific manufacturer and (2) the intermediate position of the isotherm intersection point on the pH scale means that, on average, temperature corrections remain limited except at extremely low or high pH values. Further, Fig. 2.11 clearly illustrates the Nernstian linear relationship between potential and pH and the effect of acid and alkali errors at extreme pH values; the temperature influence is also in accordance with the Nernst equation. Considering Nernstian behaviour, in general electrode suppliers provide data for approximate slope (mV per decade); for pH glass electrodes the almost theoretical value of 59 at 25° C is mostly attained but may alter slightly during use, so that calibration with buffer solutions from time to time is recommended. As the alteration effect is most pronounced at pH 11, manufacturers supply more chemically resistant pH electrodes for measurements in this range. Considering temperature, it should be realized that standard pH glass electrodes mostly have an internal (saturated) calomel reference and so can only be applied from 0 to 60° C (some suppliers indicate ranges from -10 or -5° C up to 70 or 80° C); for application from -10 to $+110$° C one uses Ag–AgCl–KCl both internally and externally.

The pNa glass electrode is produced by a few manufacturers[32,45] by preparing glasses with a deliberately high alkali error; when applying these electrodes one must realize that they are still sensitive to H^+ and to other univalent cations such as Ag^+, Tl^+ and alkylammonium ions. There are also glass electrodes that are sensitive to cations in general and insensitive to anions.

Single-crystal electrode [see Fig. 2.10 (3)]. By far the commonest example is the Orion fluoride electrode[37] (types 94-09 and 96-09), which contains an LaF_3 crystal doped with Eu^{2+} to lower its electrical resistance (1–5 MΩ) by facilitating the displacement only of F^- ions through the holes within the hexagonal crystal lattice[46]; the internal solution, 0.1 *M* each in NaF and NaCl, controls the respective potentials at the crystal inner side and at the Ag–AgCl wire reference electrode. At the crystal outer side, in contact with the sample, there may be some disturbing effect when stirring unless the electrical double layer remains thin by addition of an indifferent electrolyte. Further, the electrode shows a considerable alkali error above pH 10 and some acid error below pH 5. Nernstian behaviour (a slope of -56 mV per decade at 25° C) is obtained at fluoride concentrations above 10^{-5} *M*; in general, the solubility product at the surface of a single crystal is appreciably lower than that at the surface of the corresponding precipitate.

Homogeneous monocrystalline membrane electrodes [see Fig. 2.10 (3)]. Well known representatives of this group are the silver halide electrodes, which contain membranes of a compact silver halide manufactured by casting or pressing it into a pellet. They are mostly used to determine the halide ion (X^-) identical with that in the pellet; other halides can be measured provided that their solubility products with Ag^+ are higher. Very small amounts of either X^- or Ag^+ can be sensitively determined after addition of a buffer solution of either Ag^+ or X^-, respectively, to the sample solution. The disadvantages of the silver halide electrodes, however, are (1) their high electrical resistance (AgCl) in spite of charge transport by displacement of Ag^+ ions and (2) the occurrence of photoelectric potentials (AgBr). Improvements have been obtained by the addition of Ag_2S (see below). A silver sulphide electrode containing compact monoclinic β-Ag_2S has proved to be one of the most reliable sensors of this kind*.

Homogeneous polycrystalline membrane electrodes [see Fig. 2.10 (3)]. The relatively high electrical conductance of monoclinic β-Ag_2S and its extremely low solubility product led to the development of halide and other metal ISEs with addition of silver sulphide.

*As these electrodes are in fact not produced from precipitates, it seems illogical to classify them as homogeneous precipitate electrodes.

For the halide ISEs on the basis of silver sulphide we may assume, e.g., in the case of chloride, the following exchange reaction:

$$2AgCl + S^{2-} \rightleftharpoons 2Cl^- + Ag_2S$$

the equilibrium constant of which is

$$K_{Cl^-/S^{2-}} = \frac{a^2_{Cl^-} \cdot a_{Ag_2S}}{a^2_{AgCl} \cdot a_{S^{2-}}} = \left(\frac{a_{Cl^-} \cdot a_{Ag^+}}{a_{AgCl}}\right)^2 \left(\frac{a_{Ag_2S}}{a^2_{Ag^+} \cdot a_{S^{2-}}}\right) = \frac{a_{Ag_2S}}{a^2_{AgCl}} \cdot \frac{S^2_{0AgCl}}{S_{0Ag_2S}}$$

so that

$$a^2_{Cl^-} = \frac{S^2_{0AgCl}}{S_{0Ag_2S}} \cdot a_{S^{2-}} \tag{2.88}$$

If we substitute eqn. 2.88 in the Nernst equation, it must necessarily lead to the same potential value whether we consider the electrode as a Cl^- or an S^{2-} ISE, so

$$E_{ISE_{Cl^-}} = E^0_{ISE_{Cl^-}} - \frac{RT}{F} \ln a_{Cl^-} = E^0_{ISE_{Cl^-}} - \frac{RT}{2F} \ln \left(\frac{S^2_{0AgCl}}{S_{0Ag_2S}}\right) \cdot a_{S^{2-}}$$

$$= \left[E^0_{ISE_{Cl^-}} + \frac{RT}{2F} \ln \left(\frac{S_{0Ag_2S}}{S^2_{0AgCl}}\right)\right] - \frac{RT}{2F} \ln a_{S^{2-}} \tag{2.89}$$

and

$$E_{ISE_{S^{2-}}} = E^0_{ISE_{S^{2-}}} - \frac{RT}{2F} \ln a_{S^{2-}} = E^0_{ISE_{S^{2-}}} - \frac{RT}{2F} \ln \left(\frac{S_{0Ag_2S}}{S^2_{0AgCl}}\right) \cdot a^2_{Cl^-}$$

$$= \left[E^0_{ISE_{S^{2-}}} + \frac{RT}{2F} \ln \left(\frac{S^2_{0AgCl}}{S_{0Ag_2S}}\right)\right] - \frac{RT}{F} \ln a_{Cl^-} \tag{2.90}$$

Now, according to eqn. 2.89, we find

$$E^0_{ISE_{S^{2-}}} = E^0_{ISE_{Cl^-}} + \frac{RT}{2F} \ln \frac{S_{0Ag_2S}}{S^2_{0AgCl}}$$

which substituted in eqn. 2.90 shows that $E_{ISE_{S^{2-}}} = E_{ISE_{Cl^-}}$.

The above equations have been derived under the condition that the mixture of halide sulphide is really polycrystalline, i.e., that no mutual compound has been formed between them, otherwise a_{Cl^-} and $a_{S^{2-}}$ (and so S_{0AgCl} and S_{0Ag_2S}) would reflect considerable interaction. Further, the equations can be extended to any halide X instead of chloride. The aim that, in spite of the silver sulphide present, we can in fact deal with halide electrodes, can be fulfilled by the following conditions:

(1) S_{0AgX} should be appreciably higher than S_{0Ag_2S}, so that the exchange reaction goes almost to completion;

(2) S_{0AgX} must nevertheless be relatively low in order to provide the electrodes with sufficient sensitivity to lower halide concentrations; and

(3) Manufacture of the electrode should be directed to the attainment of a low response time.

According to the foregoing theory, any halide electrode can measure Ag^+, but not a halide whose solubility product is lower than that of the halide in the membrane; so, although an AgI electrode measures I^-, Br^- and Cl^-, an AgCl electrode measures only Cl^-.

Further, recommendations by suppliers for avoiding interferences must be followed, e.g., high concentrations of non-alkali metal ions, sulphide and complexing agents such as cyanide may be harmful.

For metal ISEs based on silver sulphide we can assume for other divalent metals analogously the exchange reaction

$$MS + 2Ag^+ \rightleftharpoons M^{2+} + Ag_2S$$

the equilibrium constant of which is

$$K_{M^{2+}/Ag^+} = \frac{a_{M^{2+}} \cdot a_{Ag_2S}}{a_{MS} \cdot a^2_{Ag^+}} = \left(\frac{a_{M^{2+}} \cdot a_{S^{2-}}}{a_{MS}}\right)\left(\frac{a_{Ag_2S}}{a^2_{Ag^+} \cdot a_{S^{2-}}}\right) = \frac{a_{Ag_2S}}{a_{MS}} \cdot \frac{S_{oMS}}{S_{0Ag_2S}}$$

so that

$$a_{M^{2+}} = \frac{S_{0MS}}{S_{0Ag_2S}} \cdot a^2_{Ag^+} \tag{2.91}$$

Substitution of eqn. 2.91 leads to the following Nernst equation:

$$E_{ISE_{M^{2+}}} = E^0_{ISE_{M^{2+}}} + \frac{RT}{2F} \ln a_{M^{2+}} = E^0_{ISE_{M^{2+}}} + \frac{RT}{2F} \ln \left(\frac{S_{oMS}}{S_{0Ag_2S}}\right) \cdot a^2_{Ag^+}$$

$$= \left[E^0_{ISE_{M^{2+}}} + \frac{RT}{2F} \ln \left(\frac{S_{0MS}}{S_{0Ag_2S}}\right)\right] + \frac{RT}{F} \ln a_{Ag^+} \tag{2.92}$$

Further, the same reasoning as for the halide electrodes applies to these metal ISEs, including the following specific conditions for their proper working:

(1) S_{0MS} should be appreciably higher than S_{0Ag_2S} so that the exchange reaction goes almost to completion;

(2) S_{0MS} must nevertheless be relatively low to provide the electrode with sufficient sensitivity to lower metal concentrations (see above).

Good results are obtained for electrodes with sulphides of Pb, Cd and Cu(II), but with certain other sulphides the response time is unsatisfactory. Interference occurs in highly acidic solutions (H_2S formation) and in alkaline solutions (at pH > 11); other metal ions sometimes disturb determinations with the metal ISE; also, anions may cause difficulties, e.g., in a Cu(II) determination at a Cu(II) ISE if Cu^{2+} and Cl^- are simultaneously present in the solution ($Ag_2S + Cu^{2+} + Cl^- \rightleftharpoons 2AgCl + CuS$) or in a Pb(II) determination at a Pb(II) ISE if Pb^{2+} and SO_4^{2-} are simultaneously present in the solution ($Pb^{2+} + SO_4^{2-} \rightleftharpoons PbSO_4$). At any rate, the recommendations by suppliers should be carefully consulted.

Heterogeneous precipitate membrane electrodes [see Fig. 2.10 (3)]. In many instances it appeared difficult to obtain satisfactory ISEs from crystalline materials as such, which led to the development by Pungor[47] of the oldest type of heterogeneous precipitate membrane electrode composed of a silver halide precipitate in a silicone-rubber matrix. In its preparation a mixture of precipitate and polysiloxane is homogenized and then a cross-linking agent (silane derivative) and a catalyst are added; by rolling of this material a cylinder of suitable diameter is obtained, from which sufficiently cured disks of 0.3–0.5 mm thickness are cut, which by means of silicone-rubber glue are fixed on to the end of a glass tube; within this tube a suitable connection of electric conductance to the membrane must be provided.

For the Pungor-type electrode not only many variants of the matrix have been proposed, e.g., thermoplastic polymer, collodion, paraffin or alkyd resin, but also provisions have been made for achieving rapid renewal of the membrane either as a whole or only its surface, which concerns mainly carbon rod and carbon paste electrodes. In this connection, mention should be made of the so-called Radiometer Selectrodes[48], originally proposed by Růžička and Lamm based on mixtures of halide and sulphide with graphite hydrophobized with PTFE[49]. Radiometer now supply for a greater number of ions ion-sensitive carbon rods as well as Selectrode powders for refilling together with kits for assembling, abrasing or refilling. Other investigators[50], however, envisaged a different method of removal of the membrane surface by using a plastic body in which the ion-sensitive carbon paste can be quickly replaced (type 1) or periodically pressed out by screwing (type 2) and superficially cleaned with smoothing paper.

Many other heterogeneous electrodes have been developed based on, e.g., calcium oxalate or stearate in paraffin, barium sulphate in paraffin or silicone-rubber, bismuth phosphate or iron(III) phosphate in silicone-rubber, caesium dodecamolybdophosphate in silicone-rubber and amminenickel nitrate in phenol–formaldehyde resin[39]; these permit the determination, respectively, of Ca and oxalate, Ba and sulphate, Bi or Fe(III) and phosphate, Cs, Ni and nitrate, etc.

It can be concluded that in principle the heterogeneous precipitate membrane electrodes act in the same way as the corresponding homogeneous electrodes, but often they are slower in response; in practice however, they still offer manufacturing possibilities where suitable pellets of the pure crystalline material cannot be obtained.

2.2.1.3.2.2. *Liquid membrane electrodes.* The members of this class can be divided into direct or indirect working. In terms of ion selectivity, the direct working group are completely comparable to the solid-state sensors, so that we simply talk about homogeneous or heterogeneous liquid membrane electrodes. Within the indirect working group the more specific enzyme (normally homogeneous) and gas sensing (heterogeneous) electrodes are in common use.

Homogeneous liquid membrane electrodes. This type, which is in limited use, is sometimes considered as a solid ion-exchange electrode as the electroactive species, e.g., calcium dioctylphosphate, after being dissolved in an ethanol–diethyl ether solution of collodion, is left to "dry" and can function as an ion-selective pellet in an electrode tip. Orion[37] use these electrodes with PVC-gelled membranes for Ca^{2+}, K^{+}, BF_4^- and NO_3^-.

Heterogeneous liquid membrane electrodes. This type, which has become of considerable practical importance, consists of a liquid ion-exchange layer or a complex-forming layer within a hydrophobic porous membrane of plastic (PTFE, PVC, etc.), sintered glass or filtering textile (glass-fibre, etc.). The construction of such an electrode is depicted in Fig. 2.12.

In a Ca^{2+} ISE the inner compartment contains the Ag–AgCl reference electrode in a solution with fixed concentrations of Ca^{2+} and Cl^-; this solution is in contact with the membrane, whose pores (100 μm) are kept filled with, e.g., calcium bis(2-ethylhexyl)phosphate, $Ca(D2EHP)_2$, dissolved in a straight-chain alcohol, or calcium didecylphosphate dissolved in di-*n*-octylphenyl phosphonate. The active compound and its solvent must be insoluble in water and stable, and the solution should be sufficiently viscous to prevent rapid passage through the membrane. These requirements, together with the necessity to allow considerable association of ions within the membrane, mean that the solvent must have a low dielectric constant and a high molecular weight with a low vapour pressure. A slight overpressure may if necessary prevent intrusion of the sample solution into the membrane; on the other hand, with intensive use the lifetime of these electrodes is limited as a consequence of diffusion of active compound into the sample solution; therefore, the construction permits easy replacement of the same or another ion-selective membrane.

Sometimes the choice of the solvent can exert a great influence on ion selectivity, e.g., whereas calcium dialkylphosphate in dioctylphenyl phosphonate

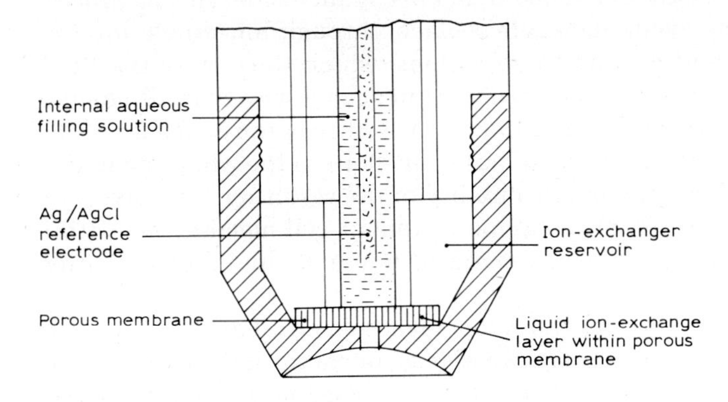

Fig. 2.12. Heterogeneous liquid membrane electrode (Courtesy of Orion).

yields a membrane selective to Ca^{2+} with practically no interference from Mg^{2+}, replacement of the solvent with 1-decanol makes the selectivities towards Ca^{2+} and Mg^{2+} roughly identical with a considerable influence of other divalent metal ions; in other words, one obtains a divalent cation electrode, suitable for the determination of water hardness.

Of course, the choice of the electroactive species, being a neutral salt of the ion under test and the ion of the ion exchanger or complex former, has a decisive influence on the ion selectivity, as is illustrated in Table 2.4.

TABLE 2.4

LIQUID MEMBRANE ELECTRODES

See data on interfering ions, selectivity constants and references in the original Table XV by Koryta[39] and also data given by electrode suppliers[32,37].

Ion under test	Ion-exchanger ion or complex-former ion	Solvent
Ca^{2+}	$(RO)_2PO_2^-$ (R = C_8H_{17} to $C_{16}H_{33}$)	Dioctylphenylphosphonate
NO_3^-	Ni(*o*-phenanthroline)$_3^{2+}$	Not given
	Methyltricaprylammonium ion	1-Decanol
ClO_4^-	Fe(*o*-phenanthroline)$_3^{2+}$	Not given
	Methyltricaprylammonium ion	1-Decanol
Cl^-	Distearyldimethylammonium ion	Not given
	Methyltricaprylammonium ion	1-Decanol
BF_4^-	Ni(*o*-phenanthroline)$_3^{2+}$	Not given
CNS^-	Methyltricaprylammonium ion	1-Decanol

There have been developments in liquid membranes with electroneutral carriers: these are oxygen-containing macrocyclic compounds known from biochemical activity where under electrically uncharged conditions, i.e., in an organic solvent of low dielectric constant such as decane, they can form 1:1 complexes with alkali metal and ammonium ions by inclusion. In this process the ring oxygens of the cyclic molecule replace, through ion–dipole interactions, the hydration shell around these cations, which thus enter the lipid-soluble molecule while changing its structure in such a way that all popular groups become directed towards the cation in the centre of the molecule whilst the non-polar lipophilic groups form an outer envelope to the whole structure. This phenomenon explains the high solubility of the resulting complexed ion in non-polar solvents, e.g., in phospholipid layers of cell membranes; within these layers the charged complexes remain mobile and thus provide cation permeation across them[39].

The best known groups of the cyclic compounds concerned are the depsipeptides (e.g., valinomycin), the macrotetrolides (e.g., nonactin and monactin) and the synthetic polyethers (crown ethers); they are used at concentrations of 10^{-4}–10^{-7} *M*, e.g., in decane. Valinomycin membranes show a K^+ selectivity of about 3800 to 1 with respect to Na^+ and of 18,000 to 1 with respect to H^+; the NH_4^+ selectivity of an actin membrane is 4 with respect to K^+.

The crown ether here was named by its decoverer Pedersen[51] dicyclohexyl-18-crown-6 (18 = number of atoms in the heterocyclic ring, 6 = number of oxygen atoms in the ring); its membrane shows an appreciably higher K^+ selectivity with respect to the other alkali metal ions. There is still much research being carried out on the synthesis and practical use of crown ethers.

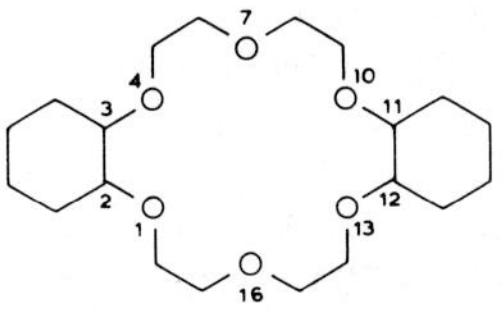

2,3,11,12-dicyclohexyl-1,4,7,10,13,16-hexaoxacyclo-octadecene

Enzyme electrodes. Guilbault[52] was the first to introduce enzyme electrodes. The bulb of a glass electrode was covered with a homogeneous enzyme-containing gel-like layer (e.g., urease in polyacrylamide) and the layer was protected with nylon gauze or Cellophane foil; when placed in a substrate solution (e.g., urea) an enzymatic conversion took place via diffusion of substrate into the enzymatic layer.

There are many examples of enzymatic reactions that yield free ammonium ion and as a consequence an alteration of pH around the glass electrode, e.g.,

$$\text{urea} \xrightarrow{\text{urease}} NH_4^+ \ (+CO_2)$$

$$\text{L- and D-amino acids} \xrightarrow[\text{oxidase}]{\text{L- and D-amino acid}} NH_4^+$$

Most suitable would be the use of a perfectly NH_4^+ ion-selective glass electrode; however, a disadvantage of this type of enzyme electrode is the time required for the establishment of equilibrium (several minutes); moreover, the normal Nernst response of 59 mV per decade (at 25° C) is practically never reached. Nevertheless, in biochemical investigations these electrodes offer special possibilities, especially because they can also be used in the reverse way as an enzyme-sensing electrode, i.e., by testing an enzyme with a substrate layer around the bulb of the glass electrode.

Gas-sensing electrodes. A gas-sensing electrode consists of a combination electrode that is normally used to detect a gas in its solution by immersion. The sensor contains the inner sensing element, usually a glass electrode or another ISE, and around this a layer of a 0.1 *M* electrolyte, surrounded by a gas-permeable membrane. On immersion of the sensor this membrane contacts the solution of the gas which diffuses through it until an overall equilibrium is established, i.e., the partial pressure of the gas attains an equilibrium between sample solution and membrane and between membrane and sensor electrolyte. For a better understanding of the interaction between this electrolyte and the

References pp. 94–96

TABLE 2.5

GAS-SENSING ELECTRODES

Component to be determined	Diffusing agent	Equilibrium in sensor electrolyte	ISE
NH_3 or NH_4^+	NH_3	$NH_3 + H_2O \rightleftharpoons NH_4^+ + OH^-$ $xNH_3 + M^{n+} \rightleftharpoons M(NH_3)_x^{n+}$	H^+ $M = Ag^+, Cd^{2+}, Cu^{2+}$
SO_2, H_2SO_3, SO_3^{2-}	SO_2	$SO_2 + H_2O \rightleftharpoons H^+ + HSO_3^-$	H^+
NO_2^-, NO_2	$NO_2 + NO$	$2NO_2 + H_2O \rightleftharpoons 2H^+ + NO_3^- + NO_2^-$	H^+, NO_3^-
S^{2-}, HS^-, H_2S	H_2S	$H_2S + H_2O \rightleftharpoons HS^- + (H^+ + H_2O)$	S^{2-}
CN, HCN	HCN	$Ag^+ + 2CN \rightleftharpoons Ag(CN)_2^-$	Ag^+
Cl_2, ClO^- Cl^-	Cl_2	$Cl_2 + H_2O \rightleftharpoons 2H^+ + ClO^- + Cl^-$	H^+, Cl^-
X_2, XO^-, X^-	X_2	$X_2 + H_2O \rightleftharpoons 2H^+ + XO^- + X^-$	$X = I^-, Br^-$
CO_2, H_2CO_3, HCl_3^-, CO_3^{2-}	CO_2	$CO_2 + H_2O \rightleftharpoons H^+ + HCO_3^-$	H^+

gas under test, let us first contemplate the data in Table 2.5, originally given by Orion[53].

The interaction mostly takes place via the solvent H_2O and more specifically with either H_3O^+ or OH^- so that the detection concerns OH^- or H^+, or the counter ion M^+ or A^-, respectively. For instance, in the measurement of SO_2 gas we can detect H^+; hence

$$E_{\text{sensor}} = \frac{2.3026\, RT}{F} \log a_{H^+}$$

For the equilibrium reaction

$$K(G)_{SO_2} = \frac{a_{H^+} \cdot a_{HSO_3^-}}{p_{SO_2} \cdot a_{H_2O}}$$

As $a_{H_2O} = 1$ and one uses in the sensor electrolyte the relatively high fixed concentration of 0.1 M $NaHSO_3$, we find

$$a_{H^+} = \frac{K(G)_{SO_2}}{a_{HSO_3^-}} \cdot p_{SO_2}$$

From Henry's law, $p_{SO_2} = k_H[SO_2]$, we obtain as a Nernst equation

$$E_{\text{sensor}} = \frac{2.3026RT}{F} \cdot \log\left(\frac{K(G)_{SO_2}}{HSO_3^-}\right) \cdot k_H[SO_2]$$

so that

$$E_{\text{sensor}} = E^0_{\text{sensor}} + \frac{23026RT}{F} \cdot \log[SO_2] \tag{2.93}$$

where

$$E^0_{\text{sensor}} = \frac{23026RT}{F} \cdot \log\left(\frac{K(\mathrm{G})_{\mathrm{SO_2}}}{a_{\mathrm{HSO_3^-}}}\right) k_{\mathrm{H}}$$

(Orion Model 95-64). In practice, one simply determines E^0_{sensor} by calibration with a standard solution without the necessity of knowing the various constants mentioned. The SO_2 electrode allows the determination of concentrations down to $10^{-6}\,M$ with a response time of a few minutes. From the above it appears that the gas-sensing electrodes show Nernstian behaviour provided that the concentrations to be measured are not high; there is little or no interference by other components in the sample solution.

2.2.2. Practice of potentiometry

As in normal potentiometry one uses and indicator electrode versus a reference electrode, the electrodes should, especially in pH measurements, be those recommended by the supplier of the pH meter in order to obtain a direct reading of the pH value displayed. In redox or other potential measurements any suitable reference electrode of known potential can be applied. However, a reference electrode is only suitable if a junction potential is excluded, e.g., an Ag–AgCl electrode in a solution of fixed Ag^+ concentration or a calomel electrode in a saturated KCl solution as a junction; in many instances a direct contact of Cl^- with the solution under test (possibly causing precipitation therein) is not allowed, so that an extra or so-called double junction with KNO_3 solution is required. Sometimes micro-electrodes or other adaptations of the surface are required.

Most manufacturers not only have taken the above requirements and possibilities into account, but also deliver for convenience combined electrodes that contain both the indicator and reference electrodes; this is the case for glass electrodes (see Fig. 2.13) and other ISEs as well as for redox electrodes.

The cell voltage measurement in itself represents a point of decisive significance, where factors such as temperature of the measurement, and Nernstian behaviour and asymmetry of the electrode play a role together with the reliability and flexibility of the pH/mV meter. Such a meter consists of a null-point or a direct-reading meter.

In the null-point instruments use is made of the well-known compensating method according to Poggendorf, by which the emf of the cell under test is compared with that of a standard cell. The circuit diagram of such a method[54] is illustrated in Fig. 2.14.

The procedure is as follows. In switch position 1 and while repeatedly depressing tap key K, the variable resistor R_1 is adjusted once for each measurement to zero current through galvanometer G, so that the emf of the standard cell C_{st} (Weston: 1.01832 V) becomes accurately compensated over the constant resistor R_2. Next, in switch position 2 the unknown emf of cell C_x is

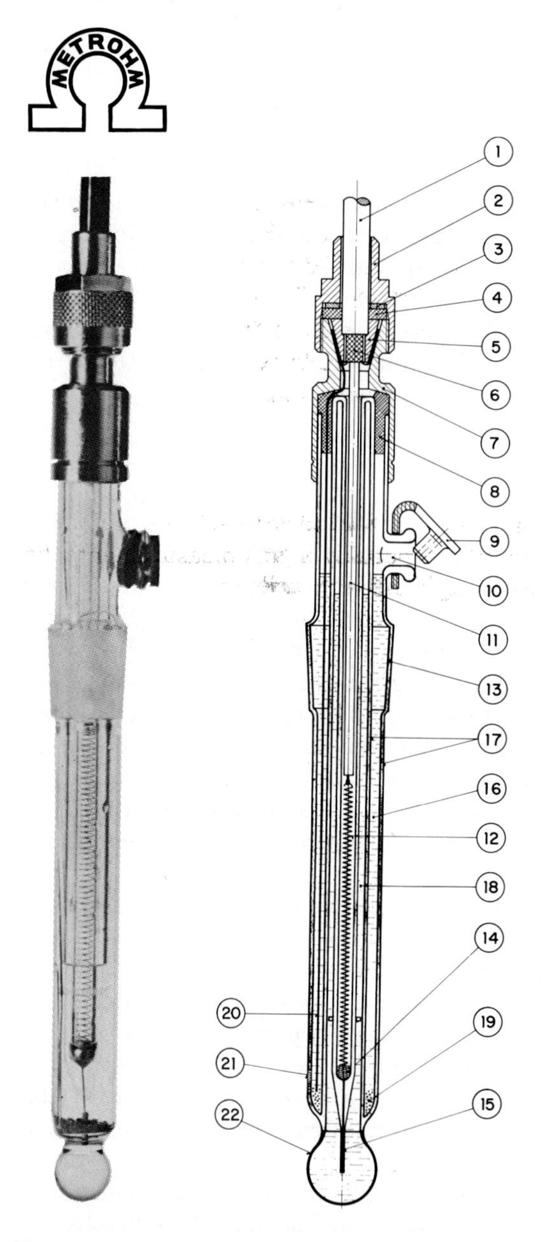

1 Screened, highly insulated electrode cable
2 Pinch nut
3 Washer
4 Sealing
5 Pinch cone
6 External conductor = screen (reference electrode)
7 Socket
8 Mastic sealing
9 Rubber stopper (opened for measurements)
10 Orifice for filling with KCl
11 Highly insulated inner conductor (indicator electrode)
12 Contact spring
13 Standard ground joint
14 Solder
15 Internal reference electrode
16 Reference electrode electrolyte (3 mol l^{-1} KCl)
17 Shaft, double wall
18 Internal reference liquid (buffer)
19 Silver chloride powder
20 Reference electrode (silver wire)
21 Porous ceramic diaphragm
22 pH-sensitive glass membrane

Fig. 2.13. Combined pH glass electrode Metrohm 121 (Courtesy of Metrohm).

carefully compensated in the same way by adjusting the coarse-control resistor R_3 and the fine-control resistor R_4. Suppose, for instance, we choose R_3 and R_4 such that $R_3/R_4 = 10$ and $(R_3 + R_4)/R_2 = 1000.0/1018.32$, then the voltage over R_3 will be 900 mV and over R_4 100 mV, which now permits a direct reading of the

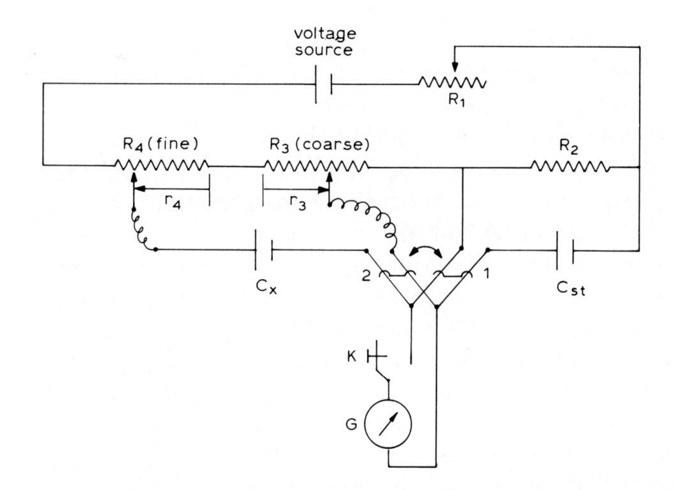

Fig. 2.14. EMF measurement (Poggendorf compensating method).

emf C_x from these dial resistors, being the combination of shifts $(r_3 + r_4)$ mV; alternatively, this millivolt reading may be a direct display in pH units.

From the above, it is clear that the precision of the measurement depends considerably on the sensitivity by which the off-balance can be controlled on the galvanometer; here two factors play a role: (1) the internal resistance of the cell, which must allow an albeit limited current withdrawal without markedly affecting the emf and (2) the sensitivity of the galvanometer itself, which should be high. With regard to these factors the situation has been considerably improved in recent years, first via the development of better electrodes, especially the glass electrodes (normal types showing an impedance of 100–400 Ω at 20° C[32]), and second by the introduction of the more sensitive and reliable transistorized amplifiers, especially those with feedback such as the operational amplifiers. Today only in very special instances, e.g., in non-aqueous media of extremely low conductance, may amplifiers based on the vibrating capacitor principle[55] be required.

In the direct-reading instruments the emf of the cell is led through an (operational) amplifier across a standard high resistor yielding a current that is measured by a milliammeter calibrated to be read in pH units or millovolts. So, while the null-point system provides a truly potentiometric (non-faradaic) measurement where the off-balance adjustment remains limited to an interrupted temporary current draw-off, the direct-reading system represents an amperometric measurement where a continuous steady-state current draw-off takes place as long as the meter is switched on. In fact, the latter is a deflection method as a pointer indicates the pH units or millivolts by its deflection on the meter scale.

In principle, a standard cell is not required as the standard high resistor, provided that its resistance is stable, has taken over the calibration role. However, it is clear that the current draw-off will be larger in both its instantaneous and its cumulative value; hence it is of great importance that, the combination of electrodes and pH/mV meter chosen on the recommendation of the suppliers should guarantee a relatively insignificant current draw-off with regard to the truly non-faradaic emf of the cell.

To summarize*, one may say that in principle the null-point instruments yield more precise results; however, the direct-reading instruments not only provide still adequate results in practice, but also are much more convenient in use for a number of reasons: reliable functioning with any type of energy supply (battery-operated portable apparatus in field use, and dc or ac mains-operated laboratory instruments), straightforward and easy presentation of results by analog or digital display (on laboratory instruments), direct access to automatic temperature correction, and not least continuous indication of the measurement result, which can be readily adapted to automatic recording and is desirable when acting as a sensor in an automatic process control circuit. As a consequence, the direct-reading instruments have become the pH/mV meters of choice in normal potentiometry.

In addition to the desired performance of the electrodes and required working of the potentiometer itself, some other points important to the potentiometric measurement on the electric cell must be considered. For instance, external conditions such as the energy source to the instrument and the temperature may change during even a single measurement. Considering the energy source, variations in voltage have an influence, especially with null-point meters, but the effect can be virtually eliminated by using an oversize battery and by frequently adjusting on the standard cell; when a.c. mains feed is applied, a mains stabiliser may prevent the cumbersome repeated adjustment on the standard cell. With direct-reading meters, there is no such problem with the introduction of the proper functioning operational amplifiers. However, temperature-control remains necessary with all types of instruments because the temperature directly influences the electrodics and ionics in the cell itself. So, let us consider the Nernst equation for the emf, with a practical example for a pH glass electrode (cf., pp. 56–61) in particular.

Here the potential equation is

$$E_{\mathrm{GE}} = E^0_{\mathrm{GE}} + \frac{RT}{F} \ln a_{\mathrm{H}^+}$$

and as pH $=$ $-\log a_{\mathrm{H}^+}$ we obtain

$$E_{\mathrm{GE}} = E^0_{\mathrm{GE}} - \frac{2.3026RT}{F} \cdot \mathrm{pH}$$

If the inner reference of the electrode is such that the iso-pH value (isotherm intersection point) lies at pH $= 7 \pm 0.5$ (according to a DIN standard), which may agree with the E_{GE} vs. pH plot in Fig. 2.11, then the overall potential of this

*For a more detailed discussion of pH-measuring instruments see Taylor[56].

glass electrode will be positive at pH < 7 and negative at pH > 7; it also means that, as the potential E at pH 7 is zero, the standard potential is $E^0_{GE} = (2.3026RT/F) \cdot 7$, which at 25° C equals 0.4141 V. In order to allow a pH measurement, the glass electrode is accompanied by a reference electrode; thus in the Metrohm Type EA 121 combined pH electrode (see Fig. 2.13) the outer reference consists of Ag–AgCl–KCl (3 mol l^{-1}), so that on the basis of $E^0_{Ag^+} = 0.7996$ V and $S_{0AgCl} = 1.56 \cdot 10^{-10}$, both at 25° C, we calculate a fixed potential of 0.1912 V; hence the outer reference represents the mostly positive side of the combined system with a potential range of approximately -0.21 to 0.59 V for pH 0–14.

Hence we can write for the emf of the combined electrode

$$E_{CE} = E^0_{CE} + \frac{2.3026RT}{F} \cdot \text{pH} = (E^0_{Ref.} + E_j)$$

$$- (E^0_{GE} + E_{as.}) + (0.0019842T)\text{pH} \qquad (2.94)$$

where E_j = junction potential and $E_{as.}$ = asymmetry potential.

In order to judge the influence of temperature on the measurement we shall consider successively the various factors mentioned.

The standard-potential, E^0, shows a temperature dependence called the "zero shift", according to its direct relationship with the free enthalpy for the standard conditions chosen, $-\Delta G^0 = RT \ln K$ (eqn. 2.37), and the Arrhenius equation for the reaction rate,

$$\ln K = -\frac{E_a}{RT} + \ln A \text{ (or } K = A e^{-E_a/RT})$$

so that $-\Delta G^0 = -E_a + RT \ln A$; hence the zero shift depends on the logarithmic term with the pre-exponential factor A. In fact it concerns empirical data and, as we do not know the precise composition of the inner solution of the Metrohm EA 121 electrode, we shall take as an example a similar system treated by Mattock[57]:

Hg | Hg_2Cl_2 | KCl(satd.) ‖ test soln. (outer) | glass | 0.1 M HCl soln. (inner) | AgCl | Ag

Reference electrode	Junction	Glass electrode
Reference half-cell		Indicator half-cell

The above notation indicates, according to convention (cf., pp. 27–28), that the indicator electrode is cathodic, which is only so for pH < 5.8, whilst the reference electrode becomes cathodic for pH > 5.8 (cf., comparable situation in Fig. 2.13). We shall consider the latter situation (all potentials being indicated vs. solutions), and hence the emf is

$$E = E_{ref} - E_{ind} = (E^0_{ref} + E_j) - (E_{\substack{m/soln.\\(outer)}} - E_{\substack{m/soln.\\(inner)}} + E_{as}) - E_{\substack{Ag/AgCl/soln.\\inner}}$$

$$= \underbrace{E^0_{outer}}_{} + \frac{2.3026RT}{F} \cdot pH_{\substack{soln.\\(outer)}} - \underbrace{(E^0_{\substack{m/soln.\\(outer)}} - E^0_{\substack{m/sol.\\(inner)}} + E_{as})}$$

E_j negligibly sensitive to small temperature variations

Cancel if correction for E is applied

$$- \frac{2.3026RT}{F} pH_{\substack{soln.\\(inner)}} - E^0_{inner}$$

Hence the zero emf can be represented by

$$E^0 = E^0_{outer} - E^0_{inner} - \frac{2.3026RT}{F} \cdot pH_{inner} \tag{2.95}$$

The general equation

$$E = E^0 + \frac{2.3026RT}{F} \cdot pH \tag{2.96}$$

(cf., eqn. 2.94) is used when testing a solution with previous calibration on a pH buffer solution at the same temperature, so that together with correction for E_{as}, E^0 also cancels out; hence

$$\Delta E = E_t - E_b = \frac{2.3026RT}{F} (pH_t - pH_b)$$

By calibration on pH_b at T_c the pH meter scale expresses the voltage in pH units at that temperature, so difficulties may arise when the test solution is measured at a deviating temperature T. If we assume for the present that the true pH value is not significantly altered by temperature variation (see later) and that the error in the pH read from the scale bears a linear relationship to the relative temperature difference we can correct pH_t by

$$\Delta pH = pH_t - pH'_t = \frac{T_c - T}{T} (pH'_t - pH_b) \tag{2.97}$$

Whether the assumption about this linear relationship can be used for the zero shift as such is doubtful; the situation becomes more reliable if the internal and external reference electrodes are equal so that E^0_{inner} and E^0_{outer} cancel, hence eqn. 2.95 becomes $E^0 = (-2.3026RT/F) \cdot pH_{inner}$. Therefore, the zero shift can be eliminated instrumentally by setting the mechanical zero of the pH meter to pH_{inner} (if previously known). With a non-combined glass electrode the external reference electrode can be kept at the calibration temperature so that the temperature variation effect remains restricted to the glass electrode itself.

For the foregoing example, Mattock[57] mentioned for calibration at 20° C and measurement at 25° C a zero shift error of 2.5 mV, which is about 0.04 pH unit (0.008 pH unit $°C^{-1}$); hence, for accurate determinations the temperature difference between calibration and measurement should be zero or less than 1° C unless a reliable (instrumental) Nernstian slope compensation can be applied; larger temperature fluctuations may be harmful owing to "hysteresis" effects whereas an internal reference electrode, especially the calomel electrode calibration and measurement should be zero or less than 1° C unless a reliable (instrumental) Nernstian slope compensation can be applied; large temperature fluctuations may be harmful owing to "hysteresis" effects whereas an internal reference electrode, especially the calomel electrode, appears sluggish unlike the preferred Ag–AgCl electrode. There is little drift of the asymmetry potential in the intermediate pH range even for changes of 10–20° C.

When appreciable temperature differences between calibration and measurement unavoidably occur, a correction must be applied either by calculation or most practically by instrumental means. It is in this instrumental method that the concept of the isopotential suggested by Jackson* plays a basic role, and can be explained as follows.

The emf of a half-cell as a function of temperature may be generally written as $E^0 = E^0_0 + AT + BT^2 + CT^3 + \cdots$; hence the emf of a pH cell at a deviating temperature T can be expressed as

$$E_T = E^0_{T_0} + \alpha(T - T_0) + \beta(T - T_0)^2 + \frac{2.3026RT}{F} \cdot \text{pH}$$

($T - T_0 = t$° C, while A and power terms above 2 are neglected).

$$\left(\frac{\partial E_T}{\partial T}\right)_{\text{pH}_x} = \alpha + 2\beta(T - T_0) + \frac{2.3026R}{F} \cdot \text{pH}_x$$

(at a constant pH $=$ pH_x) and for $(\partial E/\partial T)_{\text{pH}_x} = 0$ the pH_x value is given by

$$\text{pH}_i = -\frac{F[\alpha + 2\beta(T - T_0)]}{2.3026R} \tag{2.98}$$

(β is generally very small compared with α). Jackson called pH_i the isopotential pH of the cell, representing the pH at which the cell emf is invariant with temperature for the range where β may be neglected (i.e., over a range $\not> 20$° C).

(Note: from the definition is follows that the pH of the isotherm intersection point in Fig. 2.13 represents the isopotential pH_i of the Metrohm EA 121 combined electrode.)

*For a more or less self-supporting shorthand explanation, we have tried to quote the most essential elements and the nomenclature given by Mattock[57], as the concerning text was checked by Jackson himself; for a full explanation, the original text is recommended.

By substituting α (from eqn. 2.98) in the general equation for E_T one obtains

$$E_T = E_{T_0}^{0'} + \frac{2.3026RT}{F}(\mathrm{pH} - \mathrm{pH}_i) \quad (2.99)$$

where (by neglecting β)

$$E_{T_0}^{0'} = (E_{T_0}^{0} - \alpha T_0)$$

or

$$E_{T_0}^{0'} = E_{T_0}^{0} + \frac{2.3026RT_0}{F} \cdot \mathrm{pH}_i$$

Hence, by applying a bias emf corresponding to $2.3026RT/F \cdot \mathrm{pH}_i$ or by offsetting the mechanical zero of the potentiometer by pH_i units, the zero shift is compensated for any temperature within the range (about 20° C) over which pH_i is fairly constant. Modern instruments often include the possibility of varying the pH_i bias (to allow the use of different electrode systems) by incorporating the isopotential bias into the meter circuits.

An interesting situation is obtained when pH_b coincides with pH_i so that eqn. 2.97 becomes

$$\Delta\mathrm{pH} = \mathrm{pH}_t - \mathrm{pH}_t' = \frac{T_c - T}{T}(\mathrm{pH}_t' - \mathrm{pH}_b) = \frac{T_c - T}{T}(\mathrm{pH}_t' - \mathrm{pH}_i)$$

Hence, if by instrumental means only Nernstian slope correction is applied, zero shift compensation is automatically achieved without additional incorporation of pH_i into the meter circuit; the example illustrates that the choice of electrode systems having $\mathrm{pH}_i = 7$* or any other intermediate pH_i value offers an attractive possibility of using less sophisticated potentiometers, i.e., without setting a pH_i or $\mathrm{p}I_i$ value. Under these experimental conditions it is to be expected that the broader the pH range, the narrower the temperature range allowed will be, e.g., for an error not exceeding 0.01 pH unit the conditions are within the pH range 4–10 a $T - T_c$ value of less than 10° C and within the pH range 0–15 a $T - T_c$ value of less than 4° C.

In the foregoing derivations we have assumed that the true pH value would be invariant with temperature, which in fact is incorrect (cf., eqn. 2.58 of the Debye–Hückel theory of the ion activity coefficient). Therefore, this contribution of the solution to the temperature dependence has still to be taken into account. Doing so by differentiating E_T with respect to T at a variable pH we obtain in $\mathrm{d}E/\mathrm{d}T$ the additional term $(2.3026RT/F) \cdot \mathrm{dpH}/\mathrm{d}T$, which if β (cf., eqn. 2.98) is neglected and when $\mathrm{d}E/\mathrm{d}T = 0$ for the whole system yields

$$\mathrm{pH}_{i(\mathrm{overall})} = \mathrm{pH}_i - T \cdot \frac{\mathrm{dpH}}{\mathrm{d}T} \quad (2.100)$$

*Most electrode suppliers have adopted a pH_i of about 7, although some manufacturers use a pH_i of 0.

The setting of $pH_{i(overall)}$ on the potentiometer instead of pH_i, the latter virtually representing $pH_{i(electrodes)}$, is of special importance with alkaline solutions where dpH/dT has a significant negative value (-0.01 to 0.04 pH unit $°C^{-1}$; cf., ref. 12, Table 5).

In practice, electrode suppliers can provide the relevant isopotential data and a few dpH/dT coefficients of electrolytes are known from the literature or can be obtained from measurements at two different temperatures in the area of practical interest. Nevertheless, it may be necessary to carry out some separate determinations of isopotential values[57].

Automatic zero shift compensation by previously setting the isopotential values of $pH_{i(electrodes)}$ or even $pH_{i(overall)}$ may be especially attractive for on-line control of process streams.

The pH (or pI) term of the Nernst equation contains the electrode slope factor as a linear temperature relationship. This means that a pH determination requires the instantaneous input, either manual or automatic, of the prevailing temperature value into the potentiometer. In the manual procedure the temperature compensation knob is previously set on the actual value. In the automatic procedure the adjustment is permanently achieved in direct connection with a temperature probe immersed in the solution close to the indicator electrode; the probe usually consists of a Pt or Ni resistance thermometer or a thermistor normally based on an NTC resistor. An interesting development in 1980 was the Orion Model 611 pH meter, in which the pH electrode itself is used to sense the solution temperature (see below).

The theoretical Nernstian slope amounts to, e.g., $0.05916/n$ mV at 25° C, where n is the ion charge number. However, in many electrodes, although showing a linear relationship between potential and pH (or pI), the actual value does not reach the theoretical potential and can even diminish during use. Naturally, it is an appreciation if the difference is only slight and so electrode suppliers generally provide the actual slope values. However, today most potentiometers for this reason have a separate dial for percentage slope (regularly checked by calibration buffers), in addition to the normal temperature compensation knob, and if desired a separate means of setting an isopotential; the overall performance of the instrumental has been arranged in such a way that the single temperature compensation knob serves for both the zero shift and the pH (or pI) term.

Also, in modern, accurate apparatus there is a general trend to have a digital display not only of pH units or millivolts but often also of isopotential setting and solution temperature; in the more sophisticated instruments all this can be actuated by a push-button via a microprocessor.

Finally, in order to illustrate the above explanation of the working of potentiometers Fig. 2.15 shows a few representative commercial instruments*,

*Other manufacturers have equivalent instruments, e.g. Philips, Radiometer, Tacussel, etc.

Instruments specially constructed for automatic titration or autoanalysis in general will be treated in Part C.

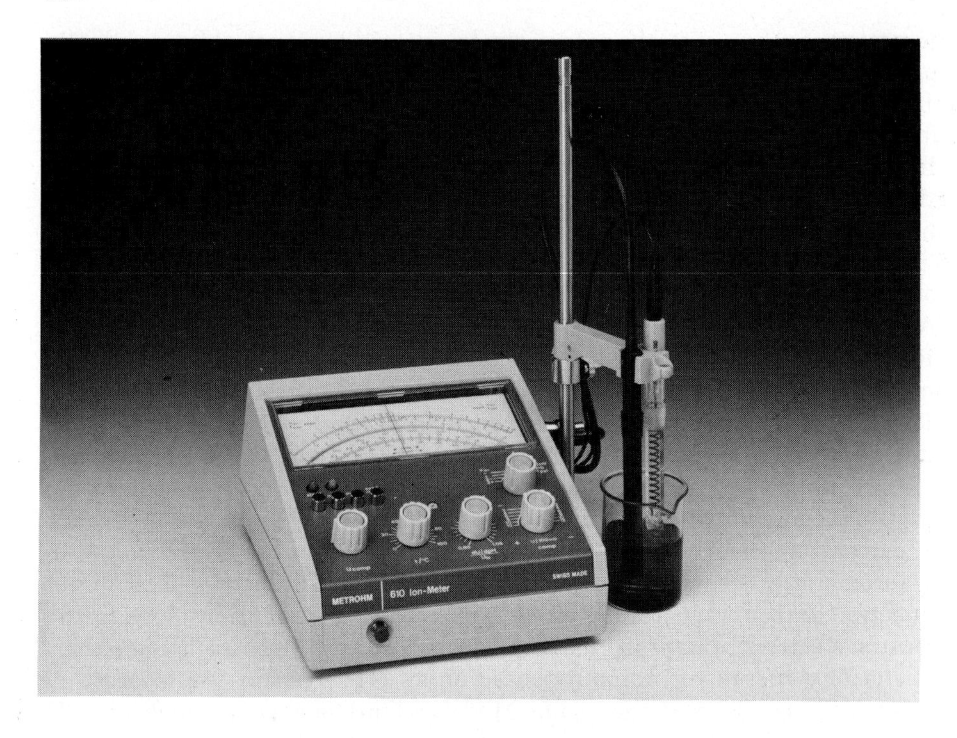
METROHM
610 Ion-Meter
SWISS MADE

(a)

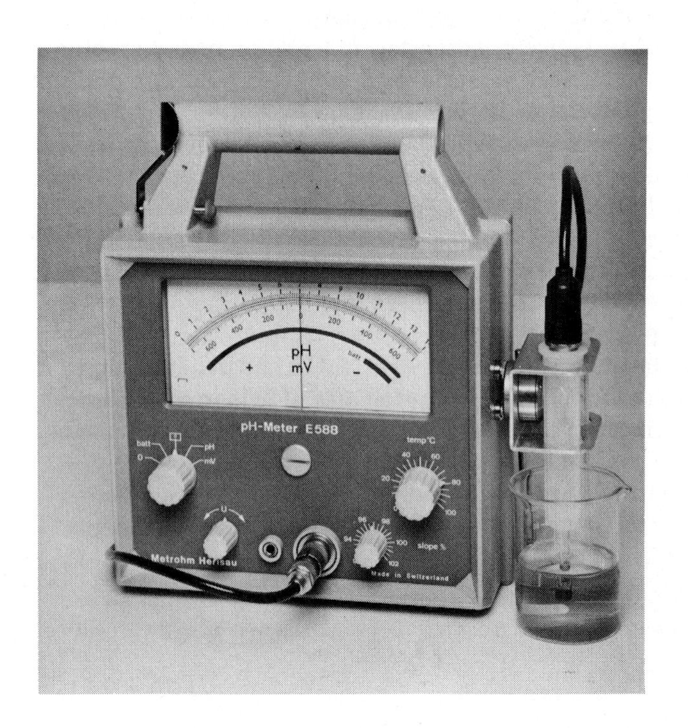
pH
mV
pH-Meter E588
temp°C
slope %
Metrohm Herisau
Made in Switzerland

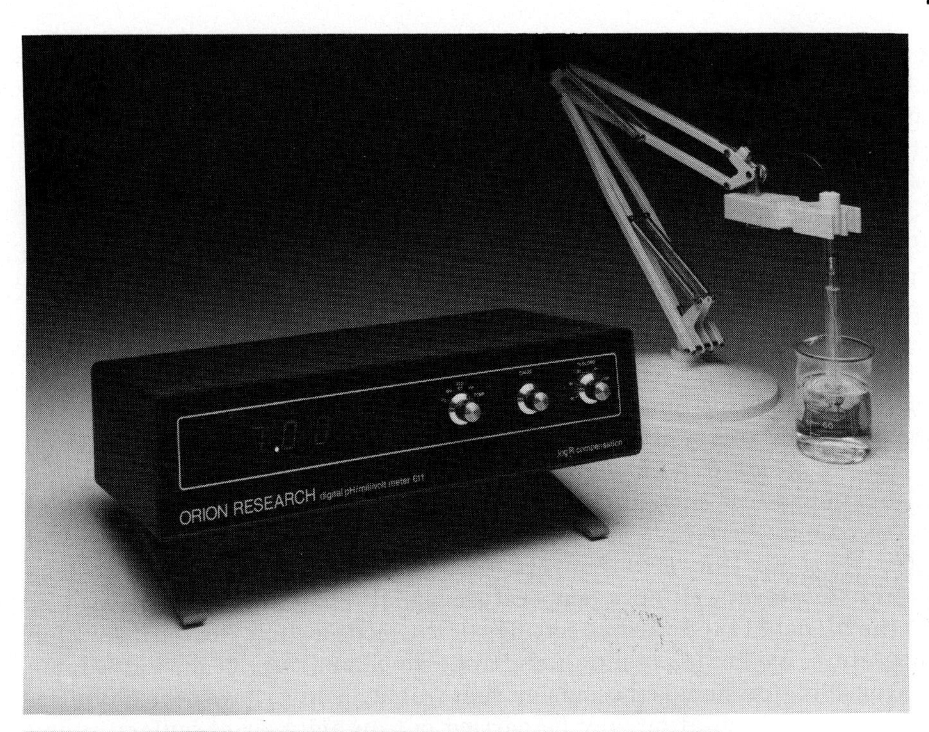

(b)

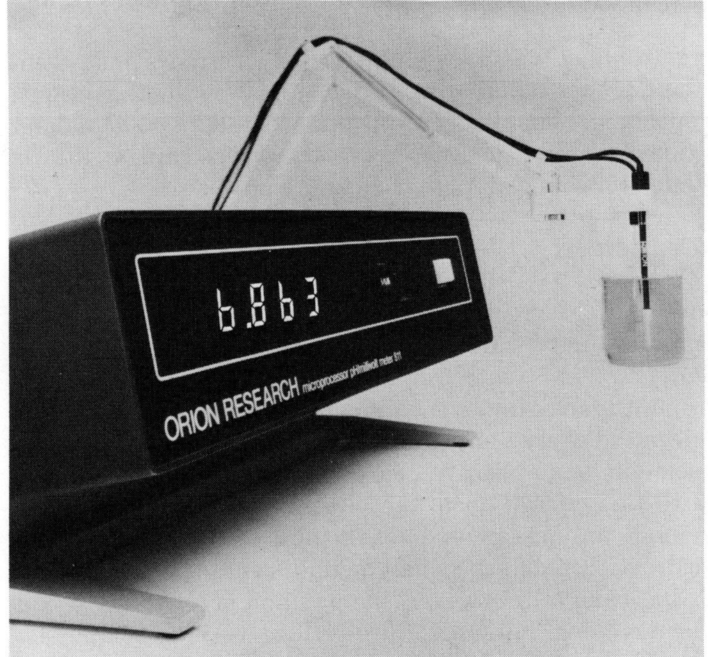

Fig. 2.15. Commercial potentiometers. (a) Metrohm ion activity meter 610 and pH-meter E 588; (b) Orion Model 611 pH meter and Model 811 microprocessor pH meter.

e.g., the Metrohm 610 ion activity meter and E 588 pH meter and the Orion Model 611 pH meter and Model 811 microprocessor pH meter. It can be seen that the Metrohm 610 is very suitable for laboratory use (resolving power ΔpH = 0.001 and ΔU = 0.1 mV) and the E 588 is a robust field instrument (resolving power ΔpH = 0.05 and ΔU = 5 mV). Both Orion instruments, although not intended to give the highest accuracy (Model 611, pH accuracy ± 0.01; Model 811, relative accuracy ± 0.003 pH unit or ± 0.2% of reading or 0.1 mV), offer some interesting new possibilities for easy manipulation and rapid display of results.

As indicated above, the Model 611 does not require a separate temperature probe and so it has no temperature knob to be operated; its circuits instead perform the following functions (abbreviated as in the Orion specification): (1) induce ac signal across pH probe; (2) measure average dc potential of probe; (3) convert amplitude of ac signal to dc potential (V); (4) calculate log V; (5) measure in-phase ac current through probe; (6) convert current to dc potential proportional to current (I); (7) calculate log I; (8) calculate log R (resistance of probe) = log V − log I; (9) convert log R into signal proportional to temperature (displayed); (10) use temperature signal to correct pH, to be read.

In the Model 811 the various settings of the mode switch yield pH, solution temperature, electrode potential, electrode slope and time of day; an illuminated push-button signals the moment that stable readings have been obtained. Two-point standardization may be carried out on any pair of buffers in any order.

Finally, one must realize that, whatever sophisticated electrodes and potentiometers one uses, their proper functioning may be disturbed by the adsorption of adventitious substances such as proteins or surface-active agents and by precipitation or film formation on the electrode surface, unless this is prevented by supersonic vibration or another cleaning procedure.

2.2.3. ISFETs in potentiometry

An ISFET (ion-sensitive field effect transistor) represents a special n–p–n type of transistor. In the ordinary transistor there is a thin p slice (with free holes) between two large n regions (with free electrons) and barrier potentials are built up at both n|p interfaces; under the influence of a positive charge on collector c versus a negative charge on emitter e the upper barrier potential can still not be overcome unless a positive charge is applied on base b; in an amount proportional to this charge on b, the electrons, attracted from the lower n region, overshoot the barrier and reach the collector (current I_c), whereas only a minor fraction reach the small metal b contact (current I_b).

In the FETs [Fig. 2.16(b) and (c)], however, the p region is relatively broad in comparison with the n regions; moreover, the p surface has been completely covered by an insulating layer preferably of SiO_2, so that the transistor steering can only be achieved by a field effect. In FET (b) the insulating layer is in touch with a metal plate, the so-called "gate", and the area between both n

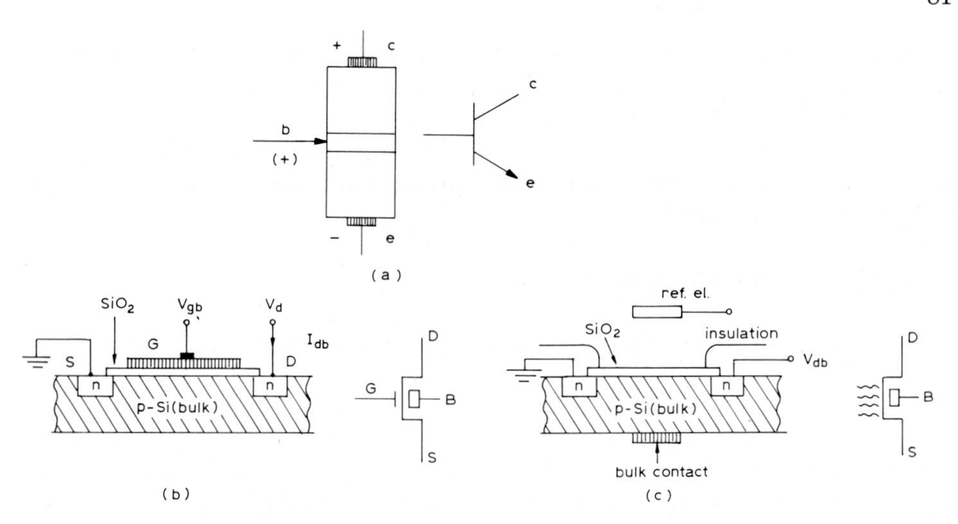

Fig. 2.16. Ion-sensitive field effect transistor (intersection and symbol), (a) n–p–n transistor, (b) IGFET (MOSFET), (c) ISFET.

regions ("source" S and "drain" D) represents the inversion layer. This insulated-gate field effect transistor (IGFET) or metal oxide semiconductor field effect transistor (MOSFET), also abbreviated to MOST, yields a drain current

$$I_d = \beta(V_{gb} - V_T)V_{db} \tag{2.101}$$

where V_T is a threshold voltage and β depends on the geometry of the gate and other constructional items. The metal gate can be omitted (OSFET), e.g., when direct contact with a locally occurring potential is desired as in muscular tissue. The remaining insulation layer of SiO_2 can also be hydrated by treating it with aluminium vapour and subsequently etching the aluminium off, followed by hydration in water for several hours*; the FET (c) has thus become an ion-sensitive field effect transistor (ISFET), especially with respect to hydrogen ion activity. In the build-up of potential at the hydrated surface layer it is comparable to a membrane electrode, so Nernstian behaviour can be expected, although this is not true for the entire pH range, but as a pH sensor the ISFET shows a much shorter response time than do the pH glass electrodes[59].

For the drain current I_d of the ISFET, again eqn. 2.101 is valid; there is a freedom of choice as to the drain-bulk voltage V_{db}, but once its value has been chosen one has to calibrate with buffers. In fact, a reference electrode is not essential, but it contributes to more stable results.

Apart from the advantage of very rapid response, ISFETs can be easily constructed as real microprobes, although their construction requires considerable skill. At present the ISFET as a pH sensor is less selective than the

*The metal gate may also be replaced with silicon nitride or polymeric pH-selective membranes[58].

glass electrode; Na^+, K^+ and to a lesser extent even Ca^{2+} ions interfere markedly, but the functioning of the ISFET as a whole can still be described by a V_{gb}, based on the Nikolski equation (eqn. 2.85).

By covering the SiO_2 insulating layer with a thin film of ion-selective carbon paste one would expect interesting possibilities for determining other ions such as halides and sulphide*.

2.2.4. Potentiometric titration

In potentiometric titration the reaction is pursued by means of potentiometry; interest is sometimes taken in the complete titration curve, but mostly in the part around the equivalence point in order to establish the titration end-point.

In fact, any type of titration can be carried out potentiometrically provided that an indicator electrode is applied whose potential changes markedly at the equivalence point. As the potential is a selective property of both reactants (titrand and titrant), notwithstanding an appreciable influence by the titration medium [aqueous or non-aqueous, with or without an ISA (ionic strength adjuster) or pH buffer, etc.] on that property, potentiometric titration is far more important than conductometric titration. Moreover, the potentiometric method has greater applicability because it is used not only for acid–base, precipitation, complex-formation and displacement titrations, but also for redox titrations.

In the practice of potentiometric titration there are two aspects to be dealt with: first the shape of the titration curve, i.e., its qualitative aspect, and second the titration end-point, i.e., its quantitative aspect. In relation to these aspects, an answer should also be given to the questions of analogy and/or mutual differences between the potentiometric curves of the acid–base, precipitation, complex-formation and redox reactions during titration. Excellent guidance is given by the Nernst equation, while the acid–base titration may serve as a basic model. Further, for convenience we start from the following fairly approximate assumptions: (1) as titrations usually take place in dilute (0.1 M) solutions we use ion concentrations in the Nernst equation, etc., instead of ion activities and (2) during titration the volume of the reaction solution is considered to remain constant.

2.2.4.1. *Acid–base titrations*

We shall distinguish three possibilities:

(A) titration of a strong acid with a strong base or vice versa;

(B) titration of a weak acid with a strong base or acid, respectively or vice versa;

---

*Comparable is the CHEMFET (Chemical Field Effect Transistor), a chemical sensor on a FET, e.g., for H^+, Na^+, K^+ and Ca^{2+} in blood, four CHEMFETs had been mounted on one plate [Clin. Chem., 30 (1984) 136].

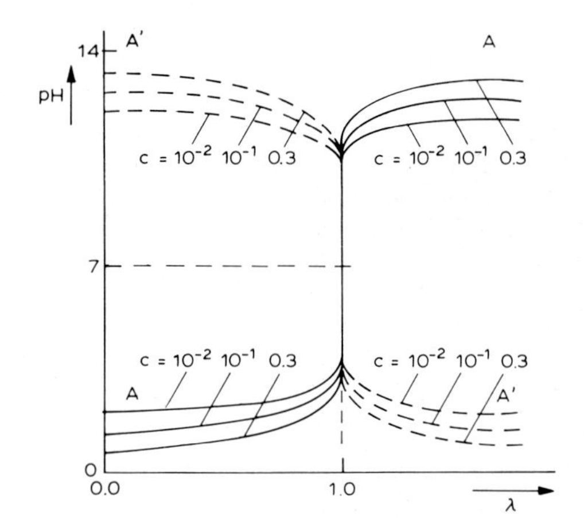

Fig. 2.17. Potentiometric titration of strong acids and bases.

(C) titration of a weak acid with a weak base or vice versa. In all cases we consider the pH as a function of the titration parameter λ (or degree of conversion) and at 25° C, so that $K_w = 10^{14}$ or pH + pOH = 14.

Case A is depicted in Fig. 2.17. The pH curves were obtained by assuming full dissociation of the acid and the base and by calculating the pH at each λ from the H^+ concentration remaining, i.e., on the basis of the part $(1 - \lambda)$ not yet titrated. Considering the shape and vertical position of the curves and the pH value of the equivalence point, we can list the following characteristics:

(1) an initially flat course with increasing λ (e.g., at half-neutralization of the acid the pH has increased by only log 2 = 0.3 with respect to the starting-point);

(2) λ mainly determines the shape of the curve and the initial concentration (at $\lambda = 0.0$) influences its pH height;

(3) the curves are symmetrical and the equivalence point, at the same time being a true inflection point, lies at pH = 7.0 according to pH = pOH (compare curves AA and A′A′).

Case B is illustrated in Fig. 2.18 for a few different weak acids and bases at 25° C. The pH curves were obtained on the basis of the relevant dissociation constants. For instance, for acid HA we simply write

$$K_a = \frac{[H^+][A^-]}{[HA]}$$

Suppose a total amount of acid dissolved such that without dissociation a concentration c would occur, but that after dissociation concentrations $[H^+]$ and $[A^-]$, both equal to x, are actually present; then equilibrium situation corresponds to

$$K_a = \frac{x^2}{c - x}$$

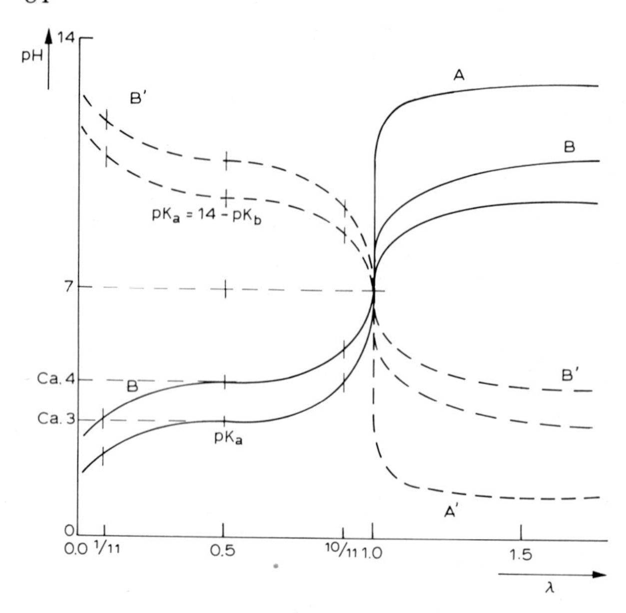

Fig. 2.18. Potentiometric titration of weak acids and bases.

Depending on the degree of dissociation of the acids (or bases) we distinguish three cases:

(a) For a very weak acid x can be neglected compared with c, so that K_a would be equal to x^2c; however, this is not correct because under at electroneutrality $x = [H^+]$ is not simply equal to $b = [A^-]$, as within the negative ion concentration there is a still perceptible contribution of $y = [OH^-]$ from water dissociation; hence $x = b + y$, while $K_w = xy$ and $K_ac = xb$, so that $x = \sqrt{K_ac + K_w}$.

(b) For a weak acid the influence of y can be neglected so that K_a simply becomes equal to x^2/c and hence $x = \sqrt{K_ac}$.

(c) For a fairly strong acid the influence of x versus c cannot be neglected, so that $K_a = x^2/(c - x)$ and hence

$$x = -\frac{K_a}{2} + \sqrt{\left(\frac{K_a}{2}\right)^2 + K_ac}$$

In titrations we normally have to deal mainly with weak to fairly strong acids (or bases), so that for acids we can use the equation $K_a = [H^+][A^-]/[HA]$; hence $[H^+] = K_a \cdot [HA]/[A^-]$. When only a part λ of the acids has been titrated, we find $[H^+] = K_a \cdot (1 - \lambda)/\lambda$; this equation is approximately valid, because the salt formed is fully dissociated, whereas the dissociation of the remaining acid has been almost completely driven back. Hence for the pH curve we obtain the Henderson equation for acid titration:

$$pH = pK_a + \log\left(\frac{\lambda}{1 - \lambda}\right) \tag{2.102}$$

It follows that at half-neutralization pH = pK_a, while at $\lambda = 1/11$ the pH = $pK_a - 1$ and at $\lambda = 10/11$ the pH = $pK_a + 1$; in fact this means that the whole titration takes place within 2 pH units, which agrees with the maximum pH range of acid–base colour indicators.

The above arguments can be given in an analogous form for bases where $K_b = [M^+][OH^-]/[MOH]$; hence $[OH^-] = K_b \cdot [MOH]/[M^+]$, so that the Henderson equation for base titration becomes

$$\mathrm{pOH} = \mathrm{p}K_b + \log\left(\frac{\lambda}{1-\lambda}\right)$$

Hence,

$$\mathrm{pH} = (\mathrm{p}K_w - K_b) - \log\left(\frac{\lambda}{1-\lambda}\right)$$

or

$$\mathrm{pH} = \mathrm{p}K_a - \log\left(\frac{\lambda}{1-\lambda}\right) \qquad (2.103)$$

This equation corresponds to today's general convention of expressing base strength also be means of pK_a, where K_a is considered in the sense of the Brönsted acid–base theory as a protolysis constant of the following protolytic reactions:
for acids:

$HA + H_2O \rightleftharpoons A^- + H_3O^+$ (acid protolysis constant K_a)

for bases:

$B + H_2O \rightleftharpoons BH^+ + OH^-$ (base protolysis constant $K_{a'}$)

Both equilibria are shifted to the left for both the weak acid and the weak base; however, BH^+ is dissociated much less than HA, which is strikingly expressed by its high $pK_{a'}$ value.

On the basis of the Henderson equation for titration of acid or base one can prove mathematically that the half-neutralization point represents a true inflection point and that as the titration end-point $\mathrm{dpH}/\mathrm{d}\lambda$ is maximal or minimal, respectively (the latter is only strictly true for titration of a weak acid with a weak base and vice versa).

Considering the pH curves BA and B′A′ in Fig. 2.18 as a whole we can list the following characteristics of shape and vertical position:

(1) a symmetrical shape with respect to the half-neutralization point pH = pK_a, a true inflection point;

(2) the increasing λ determines the shape (without a concentration influence) and pK_a the vertical position;

(3) the original concentration c determines both the initial point and the titration end-point.

At the initial point the pH will be $-\frac{1}{2}\log(K_a c + K_w)$ for a very weak acid, $\frac{1}{2}pK_a - \frac{1}{2}\log c$ for a weak acid and

$$-\log\left[-\frac{K_a}{2} + \sqrt{\left(\frac{K_a}{2}\right)^2 + K_a c}\right]$$

for a fairly strong acid; for the corresponding bases analogous formulae for pOH are obtained, K_a being replaced with K_b; e.g., for a weak base pOH $= \frac{1}{2}pK_b - \frac{1}{2}\log c$ or pH $= pK_w - \frac{1}{2}pK_b + \frac{1}{2}\log c$ and if we replace pK_b with $pK_w - pK_{a'}$ we obtain pH $= \frac{1}{2}pK_w + \frac{1}{2}pK_{a'} + \frac{1}{2}\log c$.

Considering the titration end-point or equivalence point, we in fact have to deal with the salt of a weak acid and a strong base or the reverse. Such a salt undergoes hydrolysis, e.g., for NaA, $A^- + H_2O \rightleftharpoons HA + OH^-$, so that the hydrolysis constant can be written as

$$K_h = \frac{[HA][OH^-]}{[A^-]} = \frac{[HA]}{[A^-]} \cdot \frac{[OH^-][H^+]}{[H^+]} = \frac{K_w}{K_a}$$

As $[HA] = [OH^-]$ and for a slight degree of hydrolysis $[A^-] \approx c$, we obtain

$$\frac{[OH^-]^2}{c} = \frac{K_w}{K_a}$$

so that

$$[OH^-] = \sqrt{\frac{K_w c}{K_a}}$$

and hence

$$[H^+] = \frac{K_w}{\sqrt{\frac{K_w c}{K_a}}} \quad \text{or pH} = \tfrac{1}{2}pK_w + \tfrac{1}{2}pK_a + \tfrac{1}{2}\log c$$

Analogously for the hydrolysis of a salt of a weak base and a strong acid, e.g. MX, $M^+ + H_2O \rightleftharpoons MOH + H^+$, we obtain

$$K_h = \frac{[MOH][H^+]}{[M^+]} = \frac{[MOH]}{[M^+]} \cdot \frac{[H^+][OH^-]}{[OH^-]} = \frac{K_w}{K_b}$$

which leads to

$$[H^+] = \sqrt{\frac{K_w c}{K_b}} = \sqrt{K_{a'} c} \quad \text{or pH} = \tfrac{1}{2}pK_{a'} - \tfrac{1}{2}\log c$$

In connection with the above, we shall still consider the pH curves of the displacement titrations, because in fact they represent a back-titration of an alkaline reacting salt (e.g., NaA) with a strong acid or of an acidic reacting salt (e.g., MX) with a strong base, so that in Fig. 2.19 the foregoing data of equivalence point and initial point are of direct application.

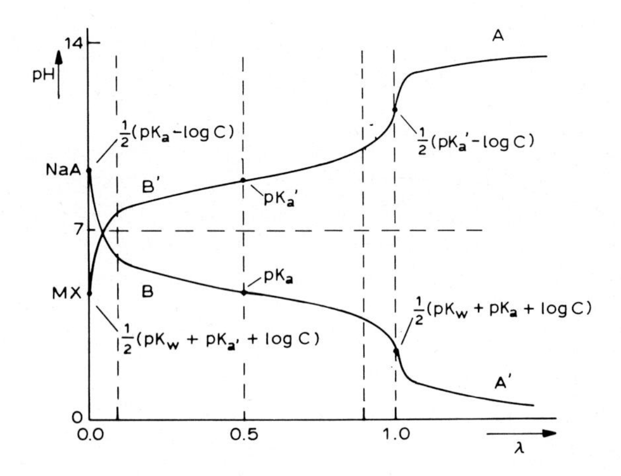

Fig. 2.19. Potentiometric titration of alkaline and acidic reacting salts.

Case C, the titration of a weak acid with a weak base and vice versa, has in fact already been illustrated in Fig. 2.18 by the curves BB and B′B′ are fully valid and for characteristic (3) the initial point is still dependent on the original concentration c; however; for the further main part of the curve we see a clean symmetry versus the equivalence point, which has become a true inflection point, independent of the concentration and simply determined by the mean value of pK_a and pK_b, i.e., $(pK_a + pK_b)/2$ or $(pK_a + pK_w - pK_{a'})/2$. It also means that in the simultaneous titration of a polyvalent acid or a series of weak acids of different strength with a strong base and vice versa, (1) the stronger the acid the earlier it is titrated within the series, (2) the initial point and the final end-point of the series are still influenced by the concentration, but (3) the intermediate steps are only determined at the pH of the inflection point by the mean value of the pK_as of the subsequent acids and in its steepness by the difference between these pK_as. Therefore, consultation of pK_a tables provides the most suitable way of predicting the results of such simultaneous titrations.

2.2.4.2. *Precipitation titrations*

As an example we take the titration of $AgNO_3$ with NaCl by means of an Ag electrode as an indicator and a double junction calomel electrode (the external junction being filled with a KNO_3 solution) as a reference.

The titration course can be illustrated (see Fig. 2.20) by a pAg curve whose values are obtained from the silver potential $E_{Ag} = E^0_{Ag} + RT/F \ln a_{Ag^+}$ or at 25° C $E_{Ag} = E^0_{Ag} - 0.05916$ pAg. As $AgNO_3$ as a salt is fully dissociated into ions, the initial point of the curve is determined by the original concentration: the later part of the curve up to the titration end-point can be obtained in the same way because the Ag^+ concentration undergoes a simple reduction as a consequence of the withdrawal of Ag^+ into the AgCl precipitate. At the

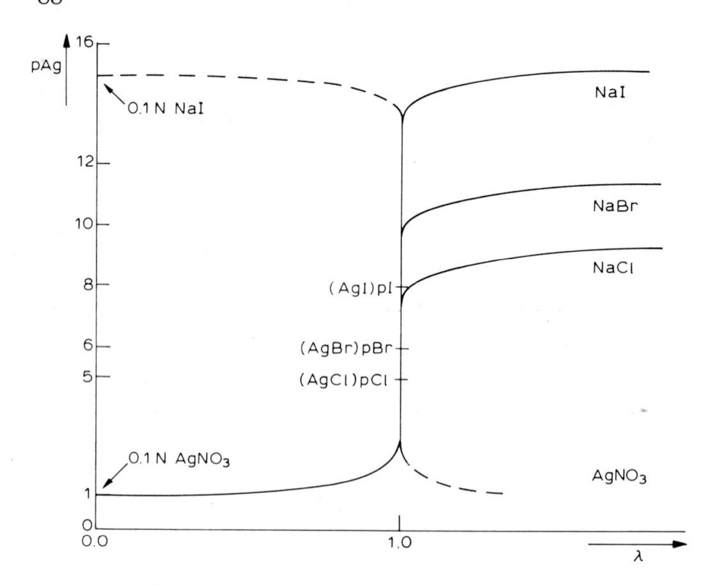

Fig. 2.20. Potentiometric precipitation titrations.

equivalence point we have to deal merely with a saturated AgCl solution, although with an equivalent amount of $NaNO_3$ dissolved. Therefore, as at 25° C the solubility product $S_{0AgCl} \approx 10^{-10}$, we obtain pAg = pCl = 5.

In the later part of the curve the solubility product remains the governing factor so that the whole pAg curve lies within a range of 10 units. This situation is in all respects analogous to that of titrating a strong acid with a strong base depicted in Fig. 2.17; there at 25° C the pH curves lie within a range of 14 pH units while for all points pH + pOH = 14 and the equivalence point is 7; here pAg + pCl = 10 and the equivalence point is 5. Fig. 2.20 also illustrates the pAg curves for bromide and iodide, whose solubility products are about 10^{-12} and 10^{-16}, respectively. Also shown is the curve of the reverse titration of iodide with $AgNO_3$; in a way it may be considered as a displacement titration. For a mixture of halides a simultaneous titration with $AgNO_3$ is possible in the sequence iodide, bromide and chloride, the intermediate inflection points lying at pAg 8, 6 and 5, respectively.

2.2.4.3. *Complex-formation titrations*

For a complex-forming metal ion detectable by its own metal electrode, e.g. in the titration of Cu^{2+} with EDTA by means of a Cu electeode and a double junction calomel electrode, p*I* curves are obtained of a nature comparable to those in Fig. 2.20 and with a Cu range of about 20 (cf., stability constant $K_{CuEDTA} = 10^{18.7}$ at pH 4.7, e.g., in a sodium acetate–acetic acid buffer and at 25° C).

For a metal ion for which its own metal electrode is not available, e.g., Ca^{2+}, Reilley et al.[34] recommended a static mercury electrode, as already mentioned

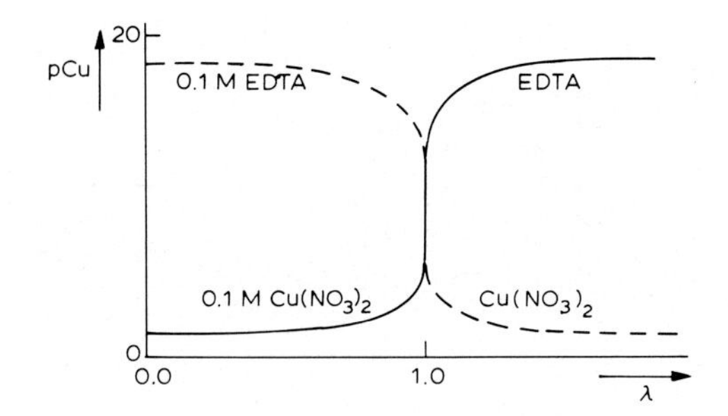

Fig. 2.21. Potentiometric complex-formation titration (under conditions favouring the overall reaction: $Cu^{2+} + Y^{4-} \rightarrow CuY^{2-}$).

on p. 47. From eqn. 2.74 at 25° C and its derivation, it follows that the standard potential depends on the amount of Hg(II) chelate added and on the stability constants of both chelates; further, the potential alters during titration as a function of the concentration ratio of metal ion to metal chelate. As a consequence, the voltage curve[60] resembles those of the redox titrations (see below) where the standard potential is the potential found half way through the titration (see Fig. 2.21).

2.2.4.4. *Redox titrations*

Let us first take the example of titrating Fe^{2+} with Ce^{4+} and vice versa. There are two redox couples, Fe^{3+}/Fe^{2+} (1) and Ce^{4+}/Ce^{3+} (2), for each of which the Nernst equation

$$E_{ox \rightarrow red} = E^0_{ox \rightarrow red} + \frac{RT}{nF} \cdot \ln\left(\frac{a_{ox}}{a_{red}}\right) \tag{2.39}$$

can be written. For 0.1 *M* solutions we shall write concentrations instead of activities.

The titration is represented in Fig. 2.22 by plotting the Pt electrode potential versus the titration parameter λ. BB is the voltage curve for titration of Fe^{2+} with Ce^{4+} and B′B′ that for titration of Ce^{4+} with Fe^{2+}; they correspond exactly to the pH curves BB and B′B′ in Fig. 2.18, with the exception that the initial point in Fig. 2.22 would theoretically have an infinitely negative and an infinitely positive potential, respectively. In practice this is impossible, because even in the absence of any other type of redox potential there will be always a trace of Fe^{3+} in addition to Fe^{2+} and of Ce^{3+} in addition to Ce^{4+} present. Further, half way through the oxidation or reduction the voltage corresponds to the standard reduction potentials of the respective redox couples; it also follows that the equivalence point is represented by the mean value of both standard potentials:

$$E_{eq.} = \frac{E^0_{Fe^{3+} \rightarrow Fe^{2+}} + F^0_{Ce^{4+} \rightarrow Ce^{3+}}}{2}$$

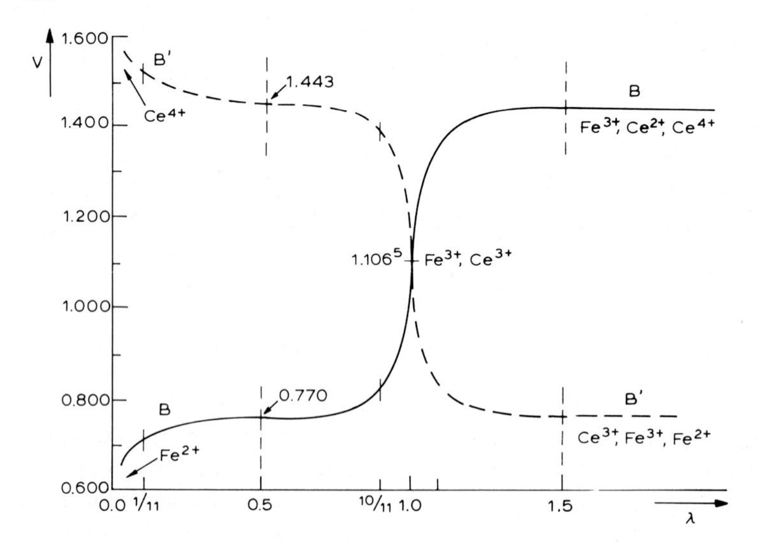

Fig. 2.22. Potentiometric redox titrations.

However, the above conclusions are not so simple when the redox couples require different numbers of electrons for reduction, especially if, in addition to the ions whose valence changes, other ions such as H^+ are involved.

A striking example of the latter is the oxidation of Fe^{2+} with permanganate in acidic solution:

1st couple: $5[Fe^{2+} \rightarrow Fe^{3+} + e]$ $(E_1^0 = 0.770\,V)$

2nd couple: $[MnO_4^- + 8H^+ + 5e \rightarrow Mn^{2+} + 4H_2O]$ $(E_2^0 = 1.491\,V)$

Total reaction: $MnO_4^- + 5Fe^{2+} + 8H^+ \rightarrow Mn^{2+} + 5Fe^{3+} + 4H_2O$

This means that with attainment of the Pt electrode potential the permanganate couple has a contribution five times greater than that of the iron couple, and as for each point of the voltage curve the equilibrium potential is $E = E_1 = E_2$ we can write

$$E = \frac{1}{6}E_1 + \frac{5}{6}E_2 = \frac{E_1^0 + 5E_2^0}{6} + \frac{RT}{6F} \cdot \ln\left(\frac{[Fe^{3+}]}{[Fe^{2+}]}\right)$$

$$+ \frac{5}{6} \cdot \frac{RT}{5F} \cdot \ln\left(\frac{[MnO_4^-][H^+]^8}{[Mn^{2+}]}\right)$$

$$= \frac{E_1^0 + 5E_2^0}{6} + \frac{RT}{6F} \cdot \ln\left(\frac{[Fe^{3+}][MnO_4^-]}{[Fe^{2+}][Mn^{2+}]}\right) + \frac{RT}{6F}\ln\,[H^+]^8$$

During the whole titration $5[Fe^{3+}] = [Mn^{2+}]$ and at the equivalence point all Fe^{2+} has vanished without any MnO_4^- in excess; if at this point the reaction were to go backwards then equivalent amounts of $5[Fe^{2+}]$ and $[MnO_4^-]$ should be formed. Hence it follows that at the equivalence point the second term becomes zero, and moreover if $[H^+]$ is kept at a value of 1, $E_{eq} = (E_1^0 + 5E_2^0)/6 = 1.371$ V; hence the voltage curve is high and far from symmetrical. If a glass pH electrode is used as the reference electrode the determination of the equivalence point becomes even more reproducible.

2.2.4.5. *Experimental titration curves and end-point detection*

In normal manual titration practice, increments of titrant are added while the pH, p*I* or voltage values are measured after each incremental addition and registered as a function of λ. The increments are taken smaller and added more frequently in the area where the end-point is expected and should be accurately detected.

With today's titrimeters* the titration can be programmed so that not only the curve is directly registered but also its first derivative and often even its second derivative. Once the empirical curve has been obtained, a method of end-point detection must be applied, and this should be such that the end-point detected agrees with the true equivalence point.

For a completely symmetrical curve, the end-point can be easily established as the inflection point through which a tangent can be drawn; here for convenience the "rings" method (Fig. 2.23) can be used, where the inflection point is obtained by intersection of the titration curve with the line joining centres of fitting circles (marked on a thin sheet of transparant plastic; see ref. 61).

Another procedure consists in finding the point on the curve where the first derivative, $\Delta E/\Delta$ ml, or ΔpH$/\Delta$ ml, is maximal and its second derative, $\Delta^2 E/\Delta \text{ml}^2$ or $\Delta^2 \text{pH}/\Delta \text{ml}^2$, equals zero, which can be verified graphically or instrumentally for volume increments (ml) around the end-point sought.

For an asymmetrical curve, the end-point can possibly be pre-calculated from the numbers of equivalents required by the reaction equation and from

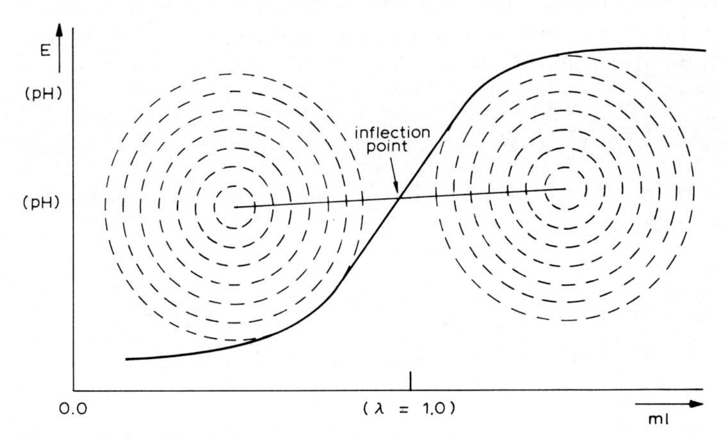

Fig. 2.23. Graphic titration end-point determination.

*For automatic titrimeters see Part C.

both standard reduction potentials involved, which may be given in tables and/or verified approximately by the experimental half-titration potentials (see the previous example of the Fe^{2+}/MnO_4^- titration). However, with a very small jump around the equivalence point such that even for a symmetrical curve the titration end-point cannot be adequately distinguished, one may increase the jump by changing the titration medium, e.g., applying a partly or wholly non-aqueous solvent. When this is impossible, e.g., in a plant stream, one must intensify the evaluation of the titration curve, especially the parts before and beyond the presumed end-point, as these parts represent mainly the qualitative aspects of the titrand and titrant, respectively, revealing themselves in the standard reduction potentials. Moreover, if the incidental effects of activity coefficients become controlled by an ISA (ion strength adjuster) and more exact relationships are applied than the approximate Henderson equation, true standard reduction potentials become available; these give the equilibrium constants such as K_a and $K_{a'}$, which together with the experimental curve shape permit a more correct prediction of the intermediate titration end-point. It was along these lines and with the aid of computerization that Bos[62] could still determine weak acids in addition to strong acids under conditions where visual detection of inflection points was impossible; the method has also been applied successfully to weak bases[63].

Another well known procedure for titration end-point detection is Gran's method (the Gran plot). In 1950 already Gran[64] proposed the plotting of ΔV (volume increments of titrant) versus the resulting ΔE or ΔpH, which yields the differential curves $\Delta V/\Delta E$ or $\Delta V/\Delta$pH, respectively, consisting of two branches intersecting at the equivalence point [Sørensen[65] in 1951 reported a similar graphical method, plotting antilog($-$pH) versus titrant volume]. Gran worked out his method for various types of titrations and demonstrated its advantages; we shall explain it with two examples.

(1) Gran plot for titration of a strong acid with a strong base. If to V_0 ml of a strong acid (concentration C_A) V ml of a strong base (concentration C_B) have been added, C_{H^+} will be given by

$$C_{H^+} = C_A \cdot \frac{V_0}{V_0 + V} - C_B \cdot \frac{V}{V_0 + V}$$

At the equivalence point, V_e ml of base have been added, which means $C_A V_0 = C_B V_e$ and so by substitution

$$C_{H^+} = C_B \cdot \frac{V_e - V}{V_0 + V}$$

As antilog ($-$pH) $= 10^{-\text{pH}} = a_{H^+} = f_{H^+} \cdot C_{H^+}$, we now obtain

$$10^{-\text{pH}} = f_{H^+} \cdot C_B \cdot \frac{V_e - V}{V_0 + V}$$

and hence $(V_0 + V) \cdot 10^{-\text{pH}} = f_{H^+} \cdot C_B(V_e - V)$. Here f_{H^+} and C_B are constants so that expressing their influence as k_1 and k_2, respectively, we can write

$$(V_0 + V) \cdot 10^{k_1 - \text{pH}} = k_2(V_e - V) \quad (2.104)$$

(this equation is valid before the equivalence point, viz., $V < V_e$). Beyond the equivalence point we obtain analogously:

$$C_{\text{OH}^-} = C_B \cdot \frac{V - V_e}{V_0 + V}$$

and

$$10^{\text{pH}} = 1/a_{\text{H}^+} = 1/f_{\text{H}^+} \cdot C_{\text{H}^+} = C_{\text{OH}^-}/f_{\text{H}^+} \cdot K_w$$

so that

$$10^{\text{pH}} = \frac{C_B}{f_{\text{H}^+} \cdot K_w} \cdot \frac{V - V_e}{V_0 + V}$$

and accordingly

$$(V_0 + V)10^{\text{pH} - k_3} = k_4(V - V_e) \quad (2.105)$$

Plotting the left-hand side of the equations against V we find according to eqn. 2.104 a titration line descending to approximately zero and according to eqn. 2.205 a titration line ascending from approximately zero. If corrected for the changing volume $(V_0 + V)$ (cf., such a correction in conductometric titration), these are straight lines, each intersecting the abscissa at the same titrant volume V_e (see Fig. 2.24; the curved lines are those by Sørensen[65]).

(2) Gran plot for titration of a weak acid with a strong base. Under comparable conditions and as for a weak acid $a_{\text{H}^+} = K_a \cdot C_{\text{HA}}/C_{\text{A}^-}$ (activity coefficients have been taken into account in K_a) we obtain before the equivalence point $C_{\text{HA}} = C_B \cdot (V_e - V)/(V_0 + V)$ and $C_{\text{A}^-} = C_B \cdot V/(V_0 + V)$, so that $a_{\text{H}^+} = K_a \cdot (V_e - V)/V = 10^{-\text{pH}}$; so analogously to eqn. 2.104 we find

$$V \cdot 10^{k_5 - \text{pH}} = k_6(V_e - V) \quad (2.106)$$

Beyond the equivalence point we can apply eqn. 2.105 again.

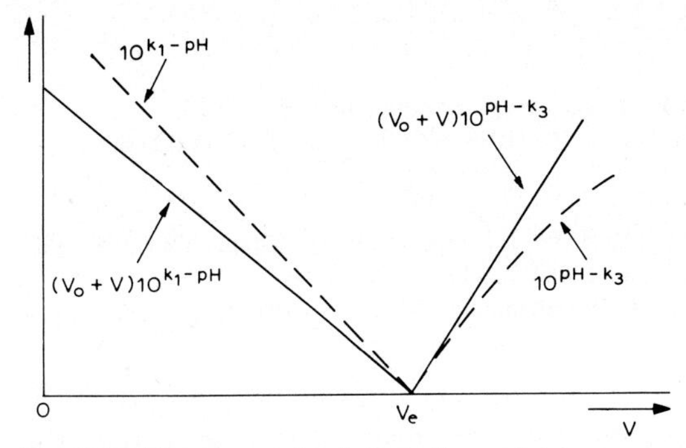

Fig. 2.24. Titration end-point via Gran plot.

(3) Gran plots for other types of titrations. Gran[64] gave the equations for dibasic acid titration and for precipication, complex-formation and redox titrations; especially for the precipitation and complex-formation titrations the equations are complicated.

Final remarks on end-point detection. In addition to our remarks above on the types of titration curves and the Henderson equation or more extended relationships, we can state that in Gran's method activity coefficients are taken into account; however, these were assumed to be constant, which is incorrect, and therefore the addition of an ISA (ion strength adjuster) must be recommended (for errors of the Gran method see ref.[66]).

It remains possible to check the correctness of the end-point detection by calibration on samples of known composition under the same measurement conditions; a similar procedure consists in the differential titrations introduced by Pinkhof and Treadwell, who used a reference electrode, identical with the indicator electrode, but dipped it into a solution buffered to the end-point potential value[67].

A special advantage of the Gran plot is that it lends itself extremely well to computerized automation and data treatment.

REFERENCES

1 O. Masson, Z. Phys. Chem., 29 (1899) 501.
2 D. A. MacInnes, The Principles of Electrochemistry, Reinhold, New York, 1961, pp. 85 and 322.
3 C. Buchböck, Z. Phys. Chem., 55 (1906) 563.
4 A. I. Vogel, A Textbook of Quantitative Inorganic Analysis, Longmans, London, 3rd ed., 1961, p. 972.
5 K. Cruse and R. Huber, Hochfrequenz Fitration, Verlag Chemie, Weinheim, 1957; C. N. Reilley and W. W. McCurdy, Jr., Anal. Chem., 25 (1953) 86.
6 E. Pungor, Oscillometry and Conductometry, Pergamon Press, Oxford, 1965.
7 A. J. Bard and L. R. Faulkner, Electrochemical Methods, Fundamentals and Applications, Wiley, New York, 1980, pp. 68–71.
8 W. J. Moore, Physical Chemistry, Longmans, London, 5th ed., 1972, pp. 510–513.
9 J. O. M. Bockris and A. K. N. Reddy, Modern Electrochemistry, Vol. 2, Plenum Press, New York, 1970, p. 982.
10 Ref. 4, pp. 97–98.
11 Ref. 4, p. 912.
12 G. Mattock, pH Measurement and Titration, Heywood, London, 1961, pp. 79–81.
13 G. J. Hills and D. J. G. Ives, Nature (London), 163 (1949) 997; J. Chem. Soc., (1951) 305.
14 Ref. 12, p. 77, Fig. 10(b).
15 Ref. 12, pp. 25–29 and 45.
16 W. Palmaer, Z. Phys. Chem., 59 (1907) 129; S. W. J. Smith, Z. Phys. Chem., 32 (1900) 433; K. Bennewitz and J. Schulz, Z. Phys. Chem., 124 (1926) 115.
17 H. S. Rossotti, Chemical Applications of Potentiometry, Van Nostrand, London, 1969, pp. 30–32.
18 Ref. 8, pp. 449–456.
19 Ref. 8, p. 443.
20 A. S. Brown and D. A. MacInnes, J. Amer. Chem. Soc., 57 (1935) 1356; D. A. MacInnes, ref. 2, pp. 184–187.

21 R. C. Weast (Editor), Handbook of Chemistry and Physics, CRC, Cleveland, OH, 61st ed., 1980–81, p. D-155.
22 C. W. Davies, J. Chem. Soc., (1938) 2093.
23 M. Bos and W. Lengton, Anal. Chim. Acta, 76 (1975) 149.
24 D. A. MacInnes, J. Amer. Chem. Soc., 41 (1919) 1086; ref. 2, (1939 ed.), Ch. 13.
25 Ref. 17, pp. 200–203.
26 R. Luther, Z. Anorg. Allg. Chem., 57 (1908) 290.
27 G. N. Lewis, Z. Phys. Chem., 55 (1906) 449.
28 J. N. Brönsted, Z. Phys. Chem., 65 (1909) 84 and 744.
29 W. Nernst and H. von Wartenberg, Z. Phys. Chem., 56 (1906) 544.
30 P. Dingemans, Thesis, Delft, 1928.
31 H. B. van der Heijde, Chem. Weekbl., 51 (1955) 823.
32 Electrodes, Metrohm, Switzerland, Herisau, 1977, pp. 25–26.
33 J. T. Stock, W. C. Purdy and L. M. Garcia, Chem. Rev., 58 (1958) 611.
34 C. N. Reilley and R. W. Schmid, Anal. Chem., 30 (1958) 947; C. N. Reilley, R. W. Schmid and D. W. Lamson, Anal. Chem., 30 (1958) 953.
35 Ref. 8, pp. 544–546.
36 B. P. Nikolski, J. Phys. Chem. (USSR), 10 (1937) 495; G. Gavach, Bull. Soc. Chim. Fr., 9 (1971) 3395.
37 Analytical Methods Guide, Orion Research, Cambridge, MA, 9th ed., 1978.
38 G. Horvai, K. Tóth and E. Pungor, Anal. Chim. Acta, 82 (1976) 45.
39 J. Koryta, Anal. Chim. Acta, 61 (1972) 329.
40 A. V. Gordievski and E. L. Filippov, Zh. Fiz. Khim., 36 (1962) 2280.
41 G. Eisenman, D. O. Rudin and J. K. Casby, Science, 126 (1957) 831.
42 B. P. Nikolski, M. M. Shulz and N. V. Peshekonova, J. Phys. Chem. (USSR), 32 (1958) 19 and 362; 33 (1959) 1922; B. P. Nikolski and A. Materova, J. Phys. Chem. (USSR), 25 (1951) 1335; B. P. Nikolski, J. Phys. Chem. (USSR), 27 (1953) 724.
43 Ref. 12, pp. 93–137.
44 G. Eisenman, in R. A. Durst (Editor), Ion-Selective Electrodes, Publication No. 314, National Bureau of Standards, Washington, DC, 1969.
45 Everything for pH, Fisher Scientific, Pittsburgh, PA, Bull. No. 252, November 1978.
46 M. S. Frant and J. W. Ross, Science, 154 (1966) 1553.
47 E. Pungor, Anal. Chem., 39 (13) (1967) 28A.
48 Analytical Electrode Guide, Radiometer, Copenhagen, May 1982.
49 J. Růžička and C. G. Lamm, Anal. Chim. Acta, 54 (1971) 1; 59 (1972) 403.
50 S. Mesarić and E. A. M. F. Dahmen, Anal. Chim. Acta, 64 (1973) 431.
51 C. J. Pedersen, J. Amer. Chem. Soc., 89 (1967) 7017.
52 G. G. Guilbault, Enzymatic Methods of Analysis, Pergamon Press, Oxford, 1970.
53 Orion Res. Newsl., 5 (1973) No. 2; J. W. Ross, J. H. Riseman and J. A. Krueger, Pure Appl. Chem., 36 (1973) 473.
54 Ref. 17, p. 20.
55 Ref. 12, pp. 227–230.
56 G. R. Taylor, in G. Mattock, pH Measurement and Titration, Heywood, London, 1961, Ch. 10, pp. 198–246.
57 Ref. 12, Ch. 9.
58 P. T. McBride, J. Janata, P. A. Comte, S. D. Moss and C. C. Johnson, Anal. Chim. Acta, 101 (1978) 239.
59 M. Bos, P. Bergveld and A. M. W. van Veen-Blaauw, Anal. Chim. Acta, 109 (1979) 145; N. F. de Rooy, The ISFET in Electrochemistry, Thesis, Technical University, Twente, 1978.
60 Ref. 4, pp. 959–963.
61 Ref. 12, pp. 328–329, fig. 74.
62 M. Bos, Anal. Chim. Acta, 90 (1977) 61.
63 M. Bos, Anal. Chim. Acta, 112 (1979) 65; W. B. Roolvink and M. Bos, Anal. Chim. Acta, 122 (1980) 81.

64 G. Gran, Acta Chem. Scand., 4 (1950) 559; Analyst (London), 77 (1952) 661.
65 P. Sørensen, Kem. Maanedsbl. Nord. Handelsbl. Kem. Ind., 32 (1951) 73.
66 J. Buffle, N. Parthasarathy and D. Monnier, Anal. Chim. Acta, 59 (1972) 427; J. Buffle, Anal. Chim. Acta, 59 (1972) 439; N. Parthasarathy, J. Buffle and D. Monnier, Anal. Chim. Acta, 59 (1972) 447.
67 Ref. 12, pp. 329–330.

Chapter 3

Faradaic methods of electrochemical analysis

Here the most important techniques are voltammetry, electrogravimetry and coulometry.

In voltammetry (abbreviation of voltamperometry), a current–potential curve of a suitably chosen electrochemical cell is determined, from which qualitative and/or quantitative analytical data can be obtained.

In electrogravimetry, also called electrodeposition, an element, e.g., a metal such as copper, is completely precipitated from its ionic solution on an inert cathode, e.g., platinum gauze, via electrolysis and the amount of precipitate is established gravimetrically; in the newer and more selective methods one applies slow electrolysis (without stirring) or rapid electrolysis (with stirring), both procedures either with a controlled potential or with a constant current. Often such a method is preceded by an electrolytic separation using a stirred cathodic mercury pool, by means of which elements such as Fe, Ni, Co, Cu, Zn and Cd are quantitatively taken up from an acidic solution whilst other elements remain in solution.

In coulometry, one measures the number of coulombs required to convert the analyte specifically and completely by means of direct or indirect electrolysis.

Before treating specific faradaic electroanalytical techniques in detail, we shall consider the theory of electrolysis more generally and along two different lines, viz., (a) a pragmatic, quasi-static treatment, based on the establishment of reversible electrode processes, which thermodynamically find expression in the Nernst equation, and (b) a kinetic, more dynamic treatment, starting from passage of a current, so that both reversible and non-reversible processes are taken into account.

3.1. PRAGMATIC TREATMENT OF THE THEORY OF ELECTROLYSIS

A description of an electrolytic cell has already been given under cell features (Section 1.3.2, Fig. 1.1c). Another example is the cell with static inert electrodes (Pt) shown in Fig. 3.1 where an applied voltage ($E_{\text{appl.}}$) allows a current to pass that causes the evolution of Cl_2 gas at the anode and the precipitation of Zn metal on the cathode. As a consequence, a galvanic cell, $^{-}(\text{Pt})\text{Zn}\,|_2\,\text{ZnCl}_2|\text{Cl}_2|_1\text{Pt}^{+}$, occurs whose emf counteracts the voltage applied; this counter- or back-emf can be calculated with the Nernst equation to be

$$E_{\text{pol.}} = \varepsilon_1 - \varepsilon_2 = \left\{\varepsilon_{0_{\text{Cl}_2 \to 2\text{Cl}^-}} + \frac{RT}{2F} \ln\left(\frac{p_{\text{Cl}_2}}{[\text{Cl}^-]^2}\right)\right\} - \left\{\varepsilon_{0_{\text{Zn}^{2+} \to \text{Zn}}} + \frac{RT}{2F} \ln\left(\frac{[\text{Zn}^{2+}]}{[\text{Zn}]}\right)\right\}$$

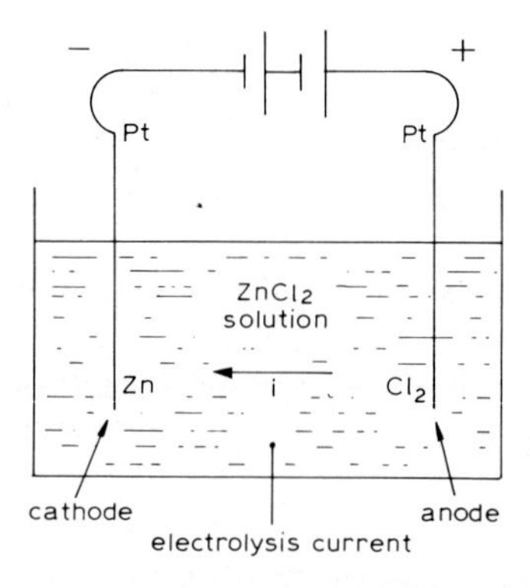

Fig. 3.1. Cell for electrolysis of $ZnCl_2$.

or

$$E_{pol.} = (\varepsilon_{0_{Cl_2 \to 2Cl^-}} - \varepsilon_{0_{Zn^{2+} \to Zn}}) + \frac{RT}{2F} \ln \left(\frac{p_{Cl_2}[Zn]}{[Cl^-]^2[Zn^{2+}]} \right)$$

Theoretically, at a low $E_{appl.}$ the counteraction would be expected to result in full polarization of the electrodes, i.e., $E_{pol.}$ would become equal to $E_{appl.}$, so no current will be passed; however, the actual p_{Cl_2} at the electrode surface is continuously diminished by diffusion of the Cl_2 gas into the solution and so there results a residual current, $i = (E_{appl.} - E_{pol.})/R$. The amount of the latter increases more or less gradually with increasing $E_{appl.}$, because the actual p_{Cl_2} increases until it finally becomes 1 atm, where Cl_2 gas starts to escape from the solution. In the meantime, the anode has been completely covered with Zn metal, so that [Zn] has become unity. In fact, $E_{pol.}$ has now attained a constant maximum value, the so-called decomposition potential, where electrolysis really breaks through. Any further increase in $E_{appl.}$ would, according to first expectations, cause a linear current increase, $i = (E_{appl.} - E_{decomp.})/R$. However, Fig. 3.2 shows that the experimental current curve deviates more and

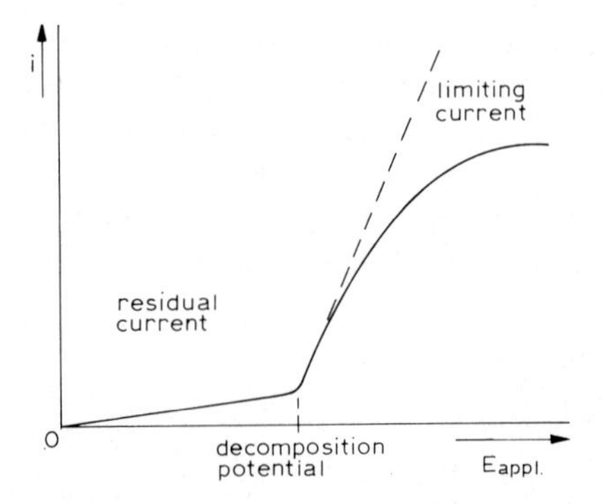

Fig. 3.2. Electrolytic current vs. potential applied.

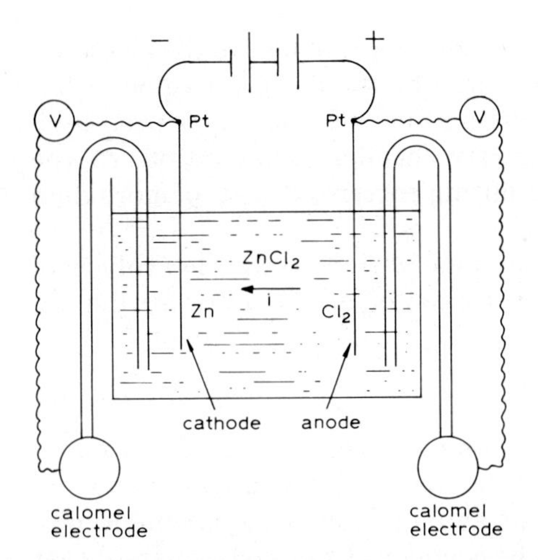

Fig. 3.3. Measurement of cathodic and anodic potential in $ZnCl_2$ electrolysis.

more from linearity with increasing $E_{appl.}$ values and finally ends in a limiting current. This phenomenon is explained by the decomposition rate increasing at higher potentials to such an extent that finally dynamic equilibria occur between the decomposition at the electrodes and the supply of the ions concerned from the bulk of the solution to the electrodes.

In Fig. 3.4 the electrolysis current is plotted against the cathodic and anodic potentials as measured in the experiment illustrated in Fig. 3.3.

In fact, the occurrence of $i_{residual}$ represents an electrochemical polarization and that of $i_{limiting}$ a concentration polarization; the term depolarization should be used only if a polarizing agent occurring at an electrode is eliminated, e.g., Cl_2 (and/or O_2) at an anode is reduced by a reductant, or Zn (and/or H_2) at a cathode oxidized by an oxidant.

It appears from Fig. 3.2 that the experimental curve has been rounded off at $E_{decomp.}$, as there is no such sharp transition between Cl_2 gas at 1 atm diffusing

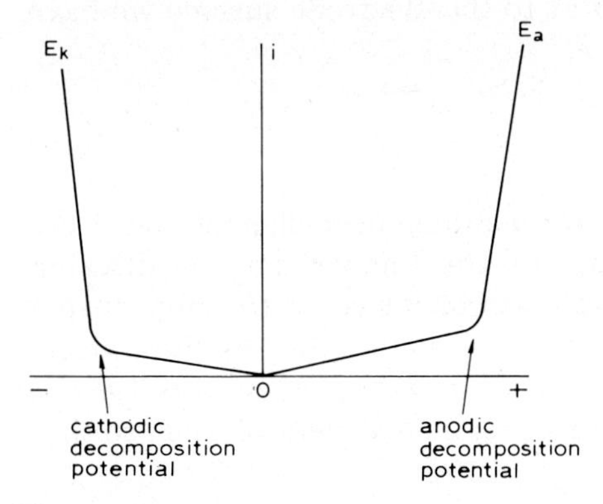

Fig. 3.4. Electrical current vs. cathodic and anodic potential.

into and escaping from the solution. Further, the supply of ions to the electrode surface at high $E_{appl.}$ may be hindered not only by the diffusion velocity but occasionally also by the slowness of dissociation into ions as a preceding process, which can be considered as a purely chemical polarization. At the same time this shows that sometimes the limiting current consists of more than the concepts of a diffusion current.

Let us consider an example of a chemical polarization. In the electrolysis of $KCu(CN)_2$ solution in excess of KCN by means of an AC current and Cu electrodes,

$$KCu(CN)_2 \rightleftharpoons (K^+)\,Cu(CN)_2^- \rightleftharpoons Cu^+ + 2CN^-$$

evolution of H_2 at the cathode and dissolution of Cu to Cu^+, which is rapidly converted into $Cu(CN)_2$ at the anode, occur, both electrodes alternatively acting as cathode and anode; with a high ac frequency an appreciable proportion of the Cu^+ ions generated cannot be complexed quickly enough so that precipitation occurs at the next current inversion; as a result, the dissolution of copper is less than that corresponding to the amount of current passed through.

In an earlier note (p. 9) we mentioned the occurrence of overvoltage in an electrolytic cell (and overpotentials at single electrodes), which means that often the breakthrough of current requires an $E_{appl.} = E_{decomp.}\ \eta$ V higher than E_{back} calculated by the Nernst equation; as this phenomenon is connected with activation energy and/or sluggishness of diffusion we shall treat the subject under the kinetic treatment of the theory of electrolysis (Section 3.2).

If we consider the limiting current (i_l) to be confined to a merely diffusion-limited current (i_d), we can consider its value as follows. As an example we take the cathodic reduction of a Zn^{2+} solution with a considerable amount of KCl. We chose an $E_{appl.}$ value greater than $E_{decomp.}$ of Zn^{2+} and less than $E_{decomp.}$ of K^+, so that only Zn^{2+} is reduced. The transport of electricity is completely provided for by the excess of K^+ and Cl^- ions and hence Zn^{2+} ions can reach the cathode only by diffusion. Suppose $[Zn^{2+}]$ in the bulk of the solution is equal to C and at the cathode surface is equal to c; the latter therefore determines the electrode potential. For diffusion perpendicular to the electrode surface we have Fick's first law:

$$\frac{dN}{dt} = -DA \cdot \frac{dc}{dx} \tag{3.1}$$

where N represents the number of moles, D the diffusion coefficient and A the electrode surface area (cm^2). Hence D indicates the number of moles diffusing through unit surface area at a concentration gradient equal to unity during unit time. The current passed, i, due to diffusion is determined by the concentration gradient at the electrode surface, i.e., $(\delta c/\delta x)_{x=0}$, which (as for the concentration gradients at all other distances) is independent of time, so that

$$i = -nF \cdot \frac{dN}{dt} = nFDA\left(\frac{\delta c}{\delta x}\right)_{x=0}$$

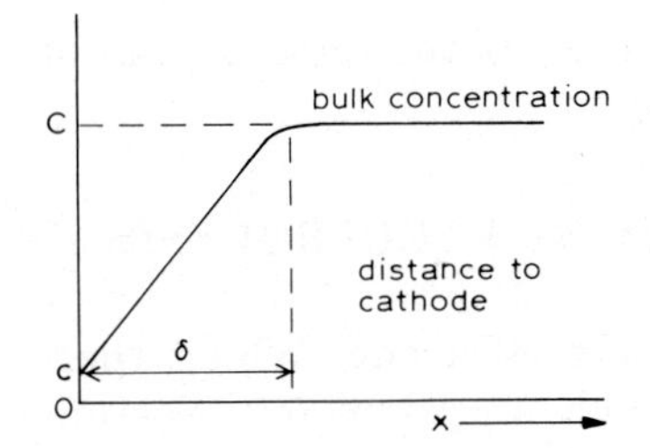

Fig. 3.5. Concentration gradient vs. distance to electrode surface.

This concentration gradient can be calculated via Fick's second law:

$$\frac{\delta c}{\delta t} = D \cdot \frac{\delta^2 c}{\delta x^2} \tag{3.2}$$ [1]

If one stirs intensely, the diffusion layer becomes sharply defined and relatively thin; under these circumstances, the concentration in the diffusion layer approaches a linear function of x (see Fig. 3.5)[2], i.e., $\mathrm{d}c/\mathrm{d}x = (C - c)/\delta$, and hence

$$i = nFDA \cdot \frac{C - c}{\delta} \tag{3.3}$$

Hence i attains a maximum for $c = 0$ (c negligible in comparison with C) which is called the limiting current (i_l), and as in this instance we have to deal with diffusion only it is the diffusion current:

$$i_d = nFDAC/\delta \tag{3.4}$$

In Fig. 3.6 it is shown, that according to expectation, i_d increases (for the same C) with (i) intense stirring (δ then will be smaller) and (ii) increasing temperature (D increases).

Two other concepts are of importance in electrolysis, viz., the current density, i.e., the current per unit surface area, $j = i/A$, and the current efficiency, i.e., the amount of ion converted as a fraction of that which, according to

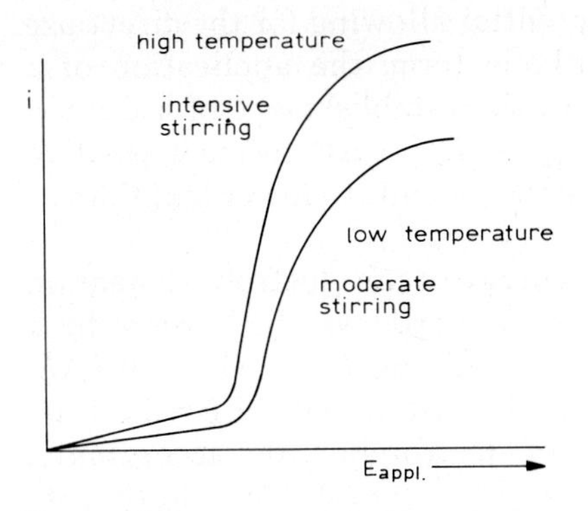

Fig. 3.6. Temperature and stirring intensity influence on electrolytic current vs. potential applied.

Faraday's law, would have been converted on the basis of the total amount of electricity passed through.

3.2. KINETIC TREATMENT OF THE THEORY OF ELECTROLYSIS

Let us consider a redox system at a static inert electrode. Whilst thermodynamics only describe the equilibrium of such a system (cf., Section 2.2.1.2.1), kinetics deal with an approach to equilibrium and assume a dynamic maintenance of that state. For that purpose the equilibrium reaction

$$\text{ox} + n\text{e} \underset{k_b}{\overset{k_f}{\rightleftharpoons}} \text{red}$$

is split up into a forward and a backward reaction, viz.,

$$\text{cathodic reaction:} \quad \text{ox} + n\text{e} \xrightarrow{k_c} \text{red}$$

$$\text{anodic reaction:} \quad \text{red} \xrightarrow{k_a} \text{ox} + n\text{e}$$

According to the Arrhenius equation for the reaction rate constant, $k = A\text{e}^{-E_A/RT}$, where A is the frequency factor and the exponential factor contains the activation energy, E_A, we can write for the respective rate constants

$$\text{cathodic:} \quad k_c = A_c\text{e}^{-E_{A_c}/RT}$$

and

$$\text{anodic:} \quad k_a = A_a\text{e}^{-E_{A_a}/RT}$$

According to the Nernst equation, $E = E^{0\prime} + (RT/F)\ln([\text{ox}]/[\text{red}])$ (cf., p. 29), where $E^{0\prime}$ represents the formal standard potential allowing for the direct use of concentration values within the logarithmic term, the application of a specific potential E to the electrode leads to the establishment of a certain equilibrium [ox]/[red]. It is clear, that if $E_{\text{appl.}} - E_{\text{eq}} = \Delta E$ becomes positive there will be a shift to greater [ox] and if negative a shift to lower [ox] than at the original ratio [ox]/[red].

As a change in $-\Delta E$ changes the relative energy of the electron resident on the electrode by $-nF\Delta E$, the anodic activation energy will be lowered by a fraction $1 - \alpha$ of that total energy, so that E_{A_a} becomes $E_{A_a} - (1 - \alpha)nF\Delta E$, where α represents the transfer coefficient, indicating in particular the fraction of the potential that influences the rate of electroreduction; accordingly, one obtains for the cathodic activation energy $E_{A_c} + \alpha nF\Delta E$ (ref. 3). By substitution in the equations for the rate constants, we obtain cathodic: $k_c = A_c e^{-E_{A_c}/RT} \cdot e^{-\alpha nF\Delta E/RT}$

and

anodic: $k_a = A_a e^{-E_{A_a}/RT} \cdot e^{(1-\alpha)nF\Delta E/RT}$

The first two factors in both equations are independent of the potential and remain the same also for $\Delta E = 0$; hence we can write

$$k_c = k_c^0 e^{-\alpha nF\Delta E/RT}$$

and

$$k_a = k_a^0 e^{(1-\alpha)nF\Delta E/RT}$$

This means that k_c^0 and k_a^0 represent the rate constants at $E_{appl.} = E_{eq}$. In the special case when $E_{eq} = E^{0\prime}$, where [ox] = [red], in addition to the rule at any equilibrium that the reaction rates $r_c = k_c^{0\prime}[ox]$ and $r_a = k_a^{0\prime}[red]$ are equal, it implies that $k_c^{0\prime} = k_a^{0\prime} = k^{0\prime}$, so that

$$k_c = k^{0\prime} e^{-\alpha nF(E_{appl.} - E^{0\prime})/RT}$$

and

$$k_a = k^{(0\prime} e^{1-\alpha)nF(E_{appl.} - E^{0\prime})/RT}$$

This, together with the inferences derived, is known as the Butler–Volmer[4] formulation of electrode kinetics. Apart from writing E instead of $E_{appl.}$, one prefers k^0 instead of $k^{0\prime}$, which in fact means that one has to take into acount ion activities instead of their concentrations and thus E^0 instead of $E^{0\prime}$ to be completely correct.

Hence the above equations are generally written as

$$k_c = k^0 e^{-\alpha nF(E-E^0)/RT} \tag{3.5}$$

and

$$k_a = k^0 e^{(1-\alpha)nF(E-E^0)/RT} \tag{3.6}$$

In electrochemical literature the standard rate constant k^0 is often designated as $k_{s,h}$ or k_s, called the specific heterogeneous rate constant or the intrinsic rate constant. According to eqns. 3.5 and 3.6, we have

$$k_c e^{\alpha nF(E-E^0)/RT} = k_a e^{-(1-\alpha)nF(E-E^0)/RT} = k_{s,h} \tag{3.7}$$

The importance of the Butler–Volmer formulation lies in the possibility of setting up the relationship between an electrolytic current i and $E_{appl.}$ on the electrode. In view of this, we shall first address the question of whether the concept of dynamic equilibria at the electrode interface is realistically correct; if so, at $E_{eq.}$ the forward current i_c must be equal to the backward current i_a and they will compensate one another, so that the net current $i_c - i_a = i = 0$.

In fact, it means that there exists* a so-called exchange current, $i_0 = i_c = i_a$, which, in accordance with Faraday's law, can be written as

$$i_c = nFAk_c[\text{ox}] = nFAk_{s,h}[\text{ox}]e^{-\alpha nF(E_{eq.} - E0)/RT}$$

or

$$i_a = nFAk_a[\text{red}] = nFAk_{s,h}[\text{red}]e^{(1-x)nF(E_{eq.} - E0)/RT}$$

Hence

$$E_{eq.} = E^0 + \frac{RT}{nF}\ln\left(\frac{[\text{ox}]}{[\text{red}]}\right)$$

which of course represents the Nernst equation. In those instances where $E_{appl.} \neq E_{eq.}$ the resulting current i will be $i_c - i_a \neq 0$ and hence

$$i = nFAk_{s,h}[\text{ox}]e^{-\alpha nF(E-E0)/RT} - nFAk_{s,h}[\text{red}]e^{(1-\alpha)nF(E-E0)/RT} \quad (3.8)$$

Writing this equation for the current density $j = i/A$ and introducing the exchange current density $j^0 = nFk_{s,h}$, we obtain

$$j = j^0[[\text{ox}]e^{-\alpha nF(E-E0)/RT} - [\text{red}]e^{(1-\alpha)nF(E-E0)/RT}] \quad (3.9)$$

3.2.1. Current–potential curves at fixed redox concentrations

The electron transfer reaction at the electrode may be (a) rapid or (b) slow. For case (a) it is of interest to distinguish between two possibilities, viz., an infinitely rapid (a_1) and a reasonably rapid electron transfer (a_2).

(a_1) An infinitely rapid electron transfer means that $k_{s,h}$ (and therefore j^0) is infinitely great. Hence the term in brackets in eqn. 3.9 will be virtually zero, so that

$$[\text{ox}]e^{-\alpha nF(E-E0)/RT} \approx [\text{red}]e^{(1-\alpha)nF(E-E0)/RT}$$

or

$$E \approx E^0 + \frac{RT}{nF}\ln\left(\frac{[\text{ox}]}{[\text{red}]}\right)$$

so with passage of any current the potential always agrees with the Nernst equation. Further, at $E > E^0$ oxidation and at $E < E^0$ reduction occurs [see Fig. 3.7 (a_1)]**; there is a sharp and separated transition between oxidation and reduction.

---

*On the basis of exchange of radioactive tracers between electrode and solution, estimates of the exchange current could be made[5].

**By convention the reduction current is indicated as positive and the oxidation current as negative.

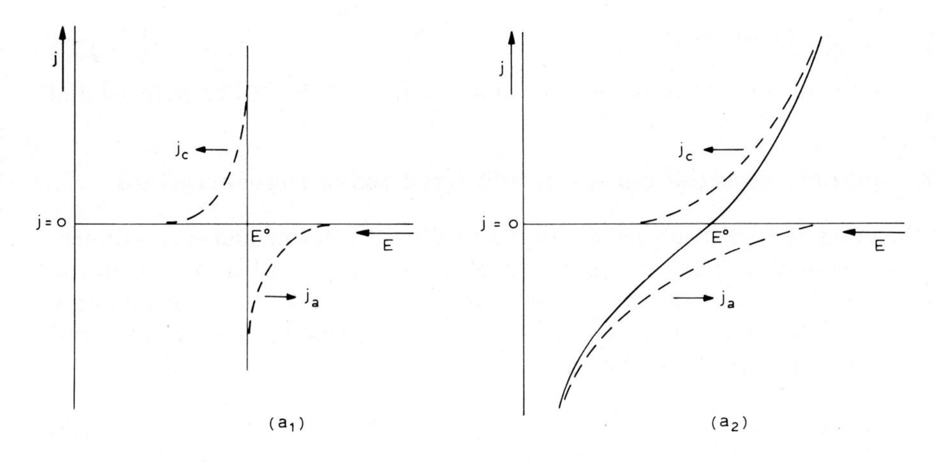

Fig. 3.7. Current density vs. redox potential, (a_1) at infinitly rapid electron transfer, (a_2) at reasonably rapid electron transfer.

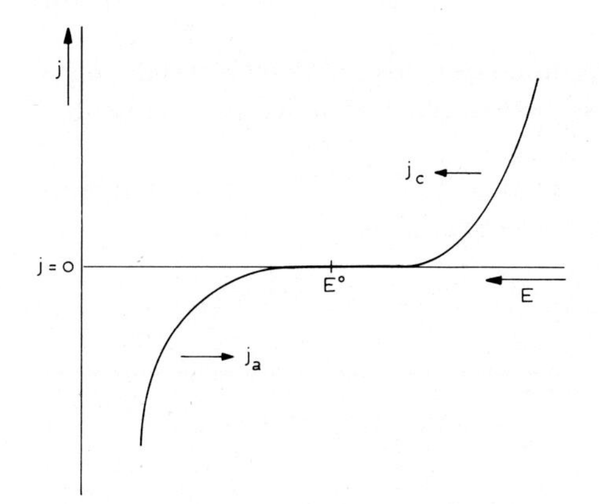

Fig. 3.8. Current density vs. redox potential at slow electron transfer.

(a_2) A reasonably rapid electron transfer means that $k_{s,h}$ (and therefore j^0) is great; with passage of a current the potential deviates from the Nernst equation under non-faradaic conditions as illustrated in Fig. 3.7(a_2), where at $E > E^0$ oxidation and at $E < E^0$ reduction predominates without a net separation between them.

(b) A slow electron transfer means that $k_{s,h}$ (and therefore j^0) is small, so both reduction and oxidation processes are hindered. From Fig. 3.7(a_2) this means that the j_c curve shifts to the right and the j_a curve to the left; hence one can arrive at the situation where the two curves no longer overlap, as shown in Fig. 3.8. For the separate parts of the combined curve we can write, according to eqn. 3.9, the following relationships:

$$j_c = j^0[\text{ox}]e^{-\alpha nF(E-E0)/RT} \tag{3.10}$$

and

$$j_{\mathrm{a}} = j^{0}[\mathrm{red}]\mathrm{e}^{(1-\alpha)nF(E-E0)/RT} \tag{3.11}$$

From eqns. 3.10 and 3.11 it can be seen that for $E \ll E^{0}$ ox will be reduced and for $E \gg E^{0}$ red will be oxidized.

3.2.2. Current–potential curves at non-fixed redox concentrations

In the practice of electrolysis one mostly deals with altering and even exhausting redox concentrations at the electrode interface, so-called concentration polarization; this has been considered already on pp. 100–102 for exhaustion counteracted by mere diffusion. The equations given for partial and full exhaustion (eqns. 3.3 and 3.4) can be extended to the current densities:

$$j = i/A = nFD \cdot \frac{C - c}{\delta} \tag{3.12}$$

and

$$j_{\mathrm{d}} = i_{\mathrm{d}}/A = nFD \cdot \frac{C}{\delta} \tag{3.13}$$

Now, assuming again the concepts of dynamic equilibria at the electrode, e.g., the cathode (cf., p. 104), applying a steady-state treatment we may consider the following:

(1) the steady state for ox, being the current supply of ox by diffusion, the anodic input $(-i_{\mathrm{a}})$ and the cathodic output i_{c}, resulting in

$$i_{\mathrm{cath.}} = nFD_{\mathrm{ox}}A \cdot \frac{C_{\mathrm{ox}} - c_{\mathrm{ox}}}{\delta} = -i_{\mathrm{a}} + i_{\mathrm{c}}$$

and for the diffusion current

$$i_{\mathrm{cath.d}} = nFD_{\mathrm{ox}}A \cdot \frac{C_{\mathrm{ox}}}{\delta}$$

(2) the steady state for red, being the current supply of red by diffusion, the cathodic input i_{c} and the anodic output $(-i_{\mathrm{a}})$, resulting in

$$i_{\mathrm{an.}} = -FD_{\mathrm{red}}A \cdot \frac{C_{\mathrm{red}} - c_{\mathrm{red}}}{\delta} = -i_{\mathrm{a}} + i_{\mathrm{c}}$$

and for the diffusion current

$$i_{\mathrm{an.d}} = -nFD_{\mathrm{red}}A \cdot \frac{C_{\mathrm{red}}}{\delta}$$

From state (1) we obtain

$$\frac{i_{\mathrm{cath.d}} - i_{\mathrm{cath.}}}{nFD_{\mathrm{ox}}A} \cdot \delta = c_{\mathrm{ox}} = [\mathrm{ox}]$$

and from state (2)

$$\frac{i_{\mathrm{an.}} - i_{\mathrm{an.d}}}{nFD_{\mathrm{red}}A} \cdot \delta = c_{\mathrm{red}} = [\mathrm{red}]$$

As both states concern the same electrode, $i_{cath.} = i_{an.} = i$; further, for $i_{cath.d}$ and $i_{an.d}$ we shall simply write $i_{c,d}$ and $i_{a,d}$ as diffusion currents.

With rapid electron transfer we can substitute the above [ox] and [red] relationships into the Nernst redox equation, from which n, F, A and δ then disappear, so that

$$E = E^0 + \frac{RT}{nF}\ln\left(\frac{D_{red}}{D_{ox}}\right) + \frac{RT}{nF}\ln\left(\frac{i_{c,d} - i}{i - i_{a,d}}\right) \tag{3.14}$$

For a solution with only ox we obtain

$$E = E^0 + \frac{RT}{nF}\ln\left(\frac{D_{red}}{d_{ox}}\right) + \frac{RT}{nF}\ln\left(\frac{i_{c,d} - i}{i}\right)$$

and for $i = \frac{1}{2} i_{c,d}$ the cathodic half-wave potential

$$E_{\frac{1}{2},c} = E^0 + \frac{RT}{nF}\ln\left(\frac{D_{red}}{D_{ox}}\right)$$

Hence

$$E = E_{\frac{1}{2},c} + \frac{RT}{nF}\ln\left(\frac{i_{c,d} - i}{i}\right) \tag{3.15}$$

or

$$\mathrm{E} = E_{\frac{1}{2}(ox)} + \frac{RT}{nF}\ln\left[\frac{i_{d(ox)} - i}{i}\right]$$

Analogously we find for a solution with only red and so for the anodic wave

$$E = E_{\frac{1}{2},a} + \frac{RT}{nF}\ln\left(\frac{i}{i_{a,d} - i}\right) \tag{3.16}$$

or

$$E = E_{\frac{1}{2}(red)} + \frac{RT}{nF}\ln\left[\frac{i}{i_{d(red)} - i}\right]$$

The term "half-wave potential" relates to the symmetrical wave that one obtains when E is plotted against i in eqn. 3.15 (cathodic wave) and eqn. 3.16

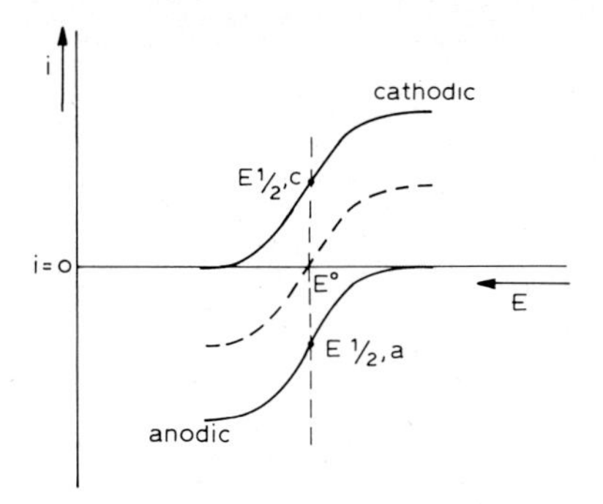

Fig. 3.9. Cathodic and anodic electrolysis waves at very rapid electron transfer.

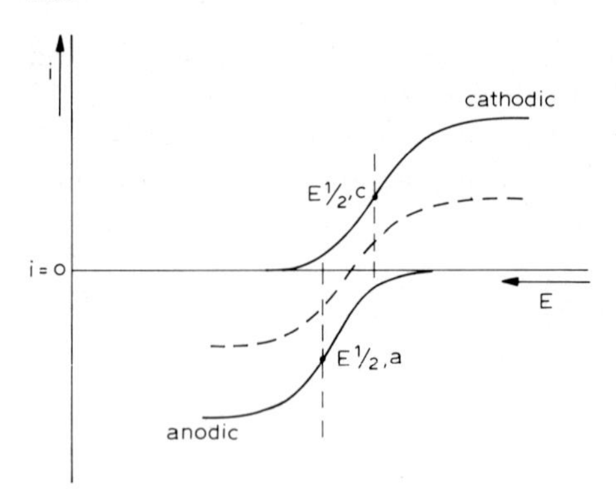

Fig. 3.10. Cathodic and anodic electrolysis waves at rather rapid electron transfer.

(anodic wave). For very rapid electron transfer both waves are depicted in Fig. 3.9 together with that (dashed line) obtained for equal concentrations of ox and red in the solution. For such a redox couple the half-wave potentials $E_{\frac{1}{2},c}$ and $E_{\frac{1}{2},a}$ will be virtually equal; moreover, $E_{\frac{1}{2}}$ will deviate only slightly from E^0, as for such systems generally $D_{red} \approx D_{ox}$.
In Fig. 3.10 the corresponding waves are depicted for rapid electron transfer; as expected [cf., Fig. 3.7(a_2)], $E_{\frac{1}{2},c}$ has shifted to the right and $E_{\frac{1}{2},a}$ to the left, while the position of E^0 must be somewhere between them.

With slow electron transfer (cf., Fig. 3.8) we must substitute the above relationships for [ox] and [red] in the equations for the separate parts of the curve:

$$i_c = nFAk_{s,h}[\text{ox}]e^{-\alpha nF(E-E0)/RT} \qquad \text{(cf., eqn. 3.10)}$$

and

$$i_a = nFAk_{s,h}[\text{red}]e^{(1-\alpha)nF(E-E0)/RT} \qquad \text{(cf., eqn. 3.11)}$$

For the cathodic wave this leads to

$$E = E^0 + \frac{RT}{\alpha nF}\ln\left(\frac{k_{s,h}\delta}{D}\right) + \frac{RT}{\alpha nF}\ln\left(\frac{i_{c,d} - i_c}{i_c}\right)$$

or

$$E = E^0 + \frac{RT}{\alpha nF}\left(\ln\frac{k_{s,h}\delta}{D_{ox}}\right) + \frac{RT}{\alpha nF}\ln\left[\frac{i_{d(ox)} - i}{i}\right] \qquad (3.17)$$

and for the anodic wave

$$E = E^0 - \frac{RT}{(1-\alpha)nF}\ln\left(\frac{k_{s,h}\delta}{D_{red}}\right) + \frac{RT}{(1-\alpha)nF}\ln\left(\frac{i_a}{i_a - i_{a,d}}\right)$$

or

$$E = E^0 - \frac{RT}{(1-\alpha)nF}\ln\left(\frac{k_{s,h}\delta}{D_{red}}\right) + \frac{RT}{(1-\alpha)nF}\ln\left[\frac{i}{i - i_{d(red)}}\right] \qquad (3.18)$$

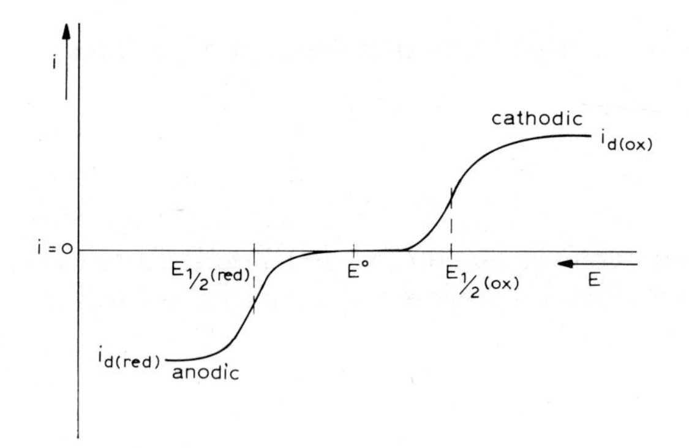

Fig. 3.11. Cathodic and anodic electrolysis waves at slow electron transfer.

In Fig. 3.11 both waves are depicted; they do not influence one another, so that it does not matter whether or not ox and red are simultaneously present in the solution.

3.2.3. The concept of overpotential

In order to obtain a definite breakthrough of current across an electrode, a potential in excess of its equilibrium potential must be applied; any such excess potential is called an overpotential. If it concerns an ideal polarizable electrode, i.e., an electrode whose surface acts as an ideal catalyst in the electrolytic process, then the overpotential can be considered merely as a diffusion overpotential (n_{D}) and yields (cf., Section 3.1) a real diffusion current. Often, however, the electrode surface is not ideal, which means that the purely chemical reaction concerned has a free enthalpy barrier; especially at low current density, where the ion diffusion control of the electrolytic conversion becomes less pronounced, the thermal activation energy (ΔG^0) plays an appreciable role, so that, once the activated complex is reached at the maximum of the enthalpy barrier, only a fraction α (the transfer coefficient) of the electrical energy difference $nF(E_{\mathrm{appl.}} - E_{\mathrm{eq.}}) = nF\eta_{\mathrm{t}}$ is used for conversion.

If ΔG_{c}^0 and ΔG_{a}^0 represent the thermal free enthalpy barriers for cathodic and anodic conversions, respectively, eqns. 3.10 and 3.11 become

$$j_{\mathrm{c}} = j^0[\mathrm{ox}]\mathrm{e}^{[-\Delta G_{\mathrm{c}}^0 - \alpha nF(E_{\mathrm{eq.}} - E0)]/RT}$$

and

$$j_{\mathrm{a}} = j^0[\mathrm{red}]\mathrm{e}^{[\Delta G_{\mathrm{a}}^0 + (1-\alpha)nF(E_{\mathrm{eq.}} - E0)]/RT}$$

hence for $[\mathrm{ox}] = [\mathrm{red}]$ and $\eta_{\mathrm{t}} = E_{\mathrm{appl.}} - E_{\mathrm{eq.}}$, eqn. 3.9 will become

$$j = j_{\mathrm{c}} - j_{\mathrm{a}} = j^0[\mathrm{e}^{\alpha nF\eta_{\mathrm{t}}/RT} - \mathrm{e}^{(1-\alpha)nF\eta_{\mathrm{t}}/RT}] \quad (3.19)$$

This Butler–Volmer[6] equation allows the current density to be plotted against $nF\eta_{\mathrm{t}}/RT$ for different values of the transfer coefficient α; the slope of the curves at $\eta_t = 0$ gives the exchange current density j^0, where it does not depend on α, as

$$\left(\frac{\partial j}{\delta \eta_t}\right)_{\eta_t \to 0} = -\frac{nF}{RT} \cdot j^0$$

For large negative or positive overpotentials, i.e., for $|\eta_t| \gg RT/nF$, either the cathodic or the anodic partial current density predominates, so that according to eqn. 3.19

$$j_e = j^0 e^{-\alpha nF\eta_t/RT}$$

and hence

$$\ln j_c = \ln j^0 - \alpha nF\eta_t/RT \qquad (3.20)$$

or

$$j_a = j^0 e^{(1-\alpha)nF\eta_t/RT}$$

and hence

$$\ln j_a = \ln j^0 + (1 - \alpha)nF\eta_t/RT \qquad (3.21)$$

These are of the type of equations found empirically by Tafel[7] in 1905. For any η_t value and for the j and j^0 concerned, eqn. 3.20 or 3.21 permits the calculation of α, which is generally in the neighbourhood of 0.5.

Tafel obtained his logarithmic relationship between current density and overpotential from a study of the reduction of H_3O^+ ions on various metal electrodes. For such a system η cannot be simply considered solely as a combination (η_t) of diffusion overpotential (η_D) and activation overpotential (η_A), as after diffusion of an H_3O^+ ion towards the electrode and its electrolytic reduction to an H atom absorbed on the metal (MeH), either a second H_3O^+ ion reacts under electrolytic reduction with the MeH to give H_2 (Heyrovský reaction) or two MeHs react with one another to give H_2 (Tafel reaction). Whatever the true reaction mechanism may be, it means that the overall electrolytic reduction of H^+ to H_2 gas is hindered by a sequential chemical reaction, so that one can talk of an additional reaction overpotential (η_R). In other electrolytic reductions a crystallization overpotential (η_c) may also be considered. All this indicates that in electrolytic processes we may often have to deal with a sequence of various types of reactions, which on the suggestion of Testa and Reinmuth[8] can be symbolized by $E_{heterogeneous}$ (electron transfer step), $C_{homogeneous}$ (chemical step), C′ (homogeneous chemical step regenerating the starting material), etc., so denominations such as CE, EC, EC′, ECE, etc., briefly describe the character of reaction sequence of the overall electrolytic process. Modern polarography (see later) offers the best possibilities for elucidating such complex mechanisms.

For the cathodic reduction of H_3O^+ ion we can write for the electrode potential (cf., eqn. 2.42)

$$E = E^{0\prime} + \frac{RT}{2F} \ln \left(\frac{[H_3O^+]^2}{p_{H_2}} \right) + \eta_{H_2}$$

where η_{H_2} is negative, and for the anodic oxidation of OH^- ion (cf., eqn. 2.69)

$$E = E^{0\prime} + \frac{RT}{4F} \ln \left(\frac{p_{O_2}}{[OH^-]^4} \right) + \eta_{O_2}$$

where η_{O_2} is positive. The occurrence of these overpotentials is of great practical significance, e.g., not only does it still permit zinc plating in acidic zinc solution without development of appreciable amounts of H_2 gas, but also the polarography of ions of metals less noble than hydrogen is normally feasible in acidic solution.

3.3. VOLTAMMETRY

In voltammetry as an analytical method based on measurement of the voltage–current curve we can distinguish between techniques with non-stationary and with stationary electrodes. Within the first group the technique at the dropping mercury electrode (dme), the so-called polarography, is by far the most important; within the second group it is of particular significance to state whether and when the analyte is stirred.

3.3.1. Voltammetry at a non-stationary electrode, in particular at the dme (polarography)

By a non-stationary electrode we mean an electrode whose surface is periodically and/or continuously renewed, so that any influence of its contamination becomes virtually negligible; in fact it represents a dynamic electrode. Here the dme, with its periodically growing and renewing mercury drop, characteristic of polarography and mainly diffusion controlled under the usual conditions, is the most important and attractive electrode. However, the less important streaming mercury electrodes are also strongly convection controlled and can be more usefully classified with the so-called hydrodynamic electrodes such as rotating disc and solid tubular electrodes[9].

3.3.1.1. *Conventional (DC) polarography*

Polarography as an analytical technique was invented in 1920 by Prof. Jaroslav Heyrovský of the Charles University, Prague, who was awarded the Nobel Prize for Chemistry in 1959. Basically it involves the application of a slowly increasing DC voltage across a dme and a mercury pool as a reference while registering the passing direct current as a function of the voltage on photographic paper by means of a sensitive mirror galvanometer; this also explains the still existing name "polarography". This technique represents the conventional form of DC polarography.

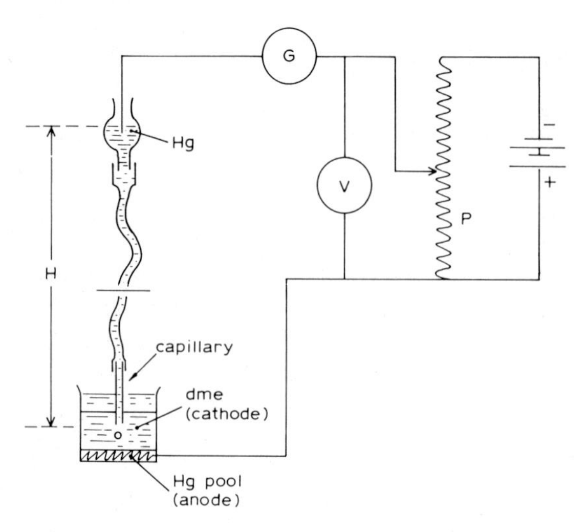

Fig. 3.12. Scheme of conventional polarography.

3.3.1.1.1. *Principles of conventional polarography*

Polarography in its simplest form can be represented by Fig. 3.12. The height (H) of the mercury column is adjusted to yield one drop in 2-4 s for a glass capillary of diameter 0.03–0.04 mm, delivering drops of $\phi \approx 0.8$ mm, $A \approx 2\,\text{mm}^2$ and $V \approx 0.3\,\text{mm}^3$ at the maximum. By means of the slide-wire potentiometer (P) the cathodic potential (versus the anode) is slowly increased, during which period the current (in mA or μA) indicated by the rapidly responding galvanometer (G) is continuously registered on a recorder as a function of the voltage on the voltmeter (V). As a reference any other non-polarizable electrode can be used, e.g., a calomel electrode instead of the Hg pool.

Owing to the considerable mercury overpotential for hydrogen, reductive polarography even in acidic media is frequently used (potentials varying from 0 to 2.2 V), whereas oxidative polarography, owing to possible dissolution of Hg metal, remains limited (potentials varying from 0 to +0.6 V, cf., Fig. 3.26).

In order to illustrate the character of the polarographic curves we can consider Fig. 3.13 (a_1), (a_2) and (b) for the reductive polarography of Cd^{2+} and Zn^{2+} in 0.1 N KCl.

In agreement with the theory of electrolysis, treated in Sections 3.1 and 3.2, the parts of the residual current and the limiting current are clearly shown by the nature of the polarographic waves; because for the cathodic reduction of Cd^{2+} and Zn^{2+} at the dme we have to deal with rapid electron transfer and limited diffusion of the cations from the solution towards the electrode surface and of the metal amalgam formed thereon towards the inside of the Hg drop, we may conclude that the half-wave potential, $E_{\frac{1}{2}}$, is constant [cf., Fig. 3.13 (a_1)] and agrees with the redox potential of the amalgam, i.e., −0.3521 V for $Cd^{2+} + 2e \rightarrow Cd(Hg)$ and −0.7628 V for $Zn^{2+} + 2e \rightarrow Zn(Hg)$ (ref. 10). The Nernst equation is

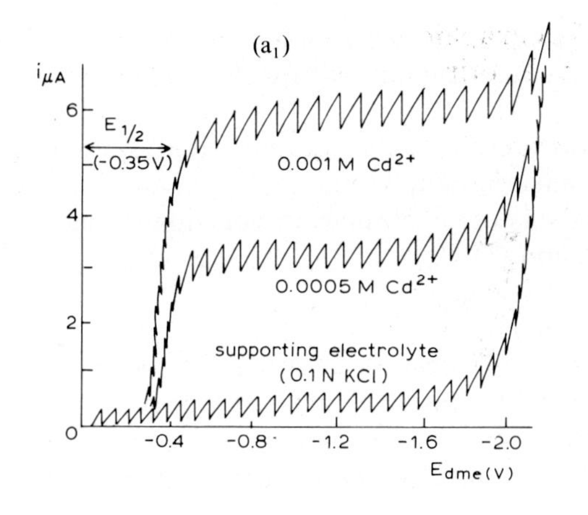

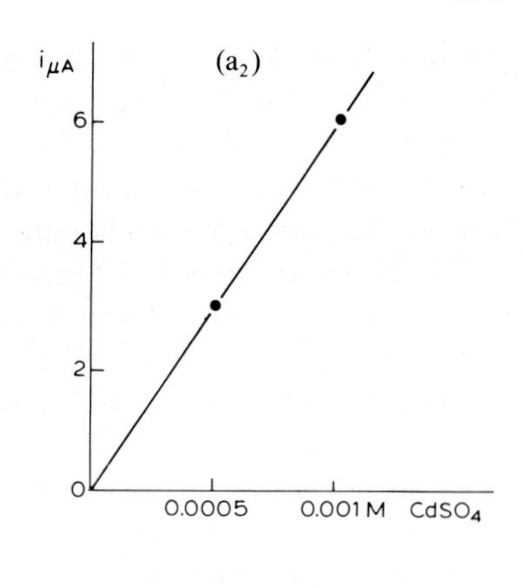

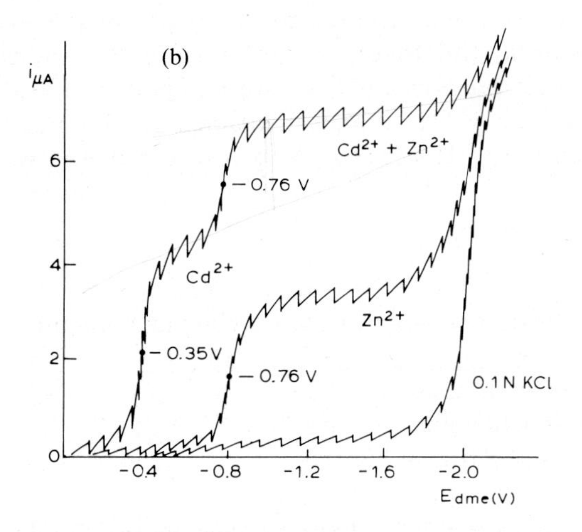

Fig. 3.13. Reductive polarography of Cd^{2+}, (a_1) undamped waves, (a_2) limited current vs. Cd^{2+} concentration, (b) undamped waves in presence of Zn^{2+}.

$$E = E^0 + \frac{RT}{nF} \ln\left(\frac{C_M}{C_a}\right)$$

where C_M is the concentration of the metal ion in solution and C_a is the concentration of the metal in the amalgam. For Cd^{2+} as a representative example, Fig. 3.13 (a_2) shows that the limiting current is linearly dependent on the concentration, as expected.

So far, we have intentionally omitted the fine structure of the classic polarographic wave, which is caused by the repeating effect of the Hg drop growing and falling off. However, even if this is taken into account the above remarks remain valid, as will be shown by the theory in the next section.

Finally, Fig. 3.13(b) illustrates the polarographic wave of Zn^{2+} and also that for a mixture of Cd^{2+} and Zn^{2+}; here the limiting current for Zn^{2+} is obtained by subtraction of that for Cd^{2+}.

On the basis of the above, the advantages of polarographic analysis over other analytical methods can be summarized as follows:

(1) the measurements allow the analysis of substances in very low concentrations with a fairly good precision (1–2%);
(2) the determination itself is rapid (a few minutes);
(3) the method is qualitative and quantitative;
(4) simultaneous determination of several substances is possible with one polarogram, provided that the $E_{\frac{1}{2}}$ values differ by at least 0.2 mV;
(5) the method can be applied to many types of compounds or ions such as inorganic and organic molecules or reducible and oxidizable ions.

3.3.1.1.2. *Theory of conventional polarography*

As a consequence of the particular geometry of the dme and its periodically altering surface area, the theory of polarography necessarily requires an amplification of what has already been stated about the theory of electrolysis in Sections 3.1 and 3.2. Such an amplification was given in 1934 by Ilkovič in close collaboration with Heyrovský, and resulted in the well known Ilkovič equation:

$$i_t(\text{amps}) = 0.734\, nFD^{\frac{1}{2}}\, Cm^{\frac{2}{3}} t^{\frac{1}{6}} \tag{3.22}*$$

where

i_t = limiting (diffusion-controlled) current at a certain time t during the drop life;
nF = number of coulombs per mole;
D = diffusion coefficients in $cm^2\, s^{-1}$;
C = bulk concentration in $mol\, cm^{-3}$;
m = flow-rate of mercury in $g\, s^{-1}$;
t = time in s passed since previous drop has fallen off.

Derivation of Ilkovič equation[11]. Ilkovič started from the simplification of a dme behaving as a flat electrode; during drop growth the original differential equation for diffusion will alter as the concentration gradient is influenced by the relative rate at which the drop grows against the diffusion direction and by the decrease in the thickness of the diffusion layer. If one considers the picture (see Fig. 3.14) of the concentration distribution near the electrode surface during simultaneous movement of the liquid at a relative velocity v with respect to the surface, then the synchronous change in the concentration by this convection can be expressed by

$$\left(\frac{\partial c}{\partial t}\right) \text{conv.} = -v\left(\frac{\partial c}{\partial x}\right)$$

*The factor 0.734 has been adopted from the latest figures for the constants concerned.

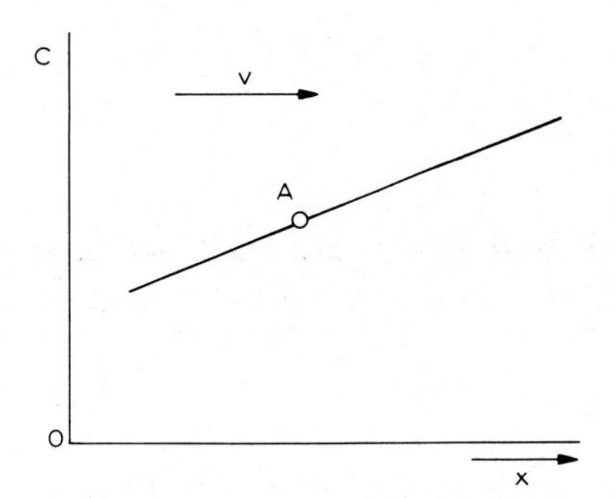

Fig. 3.14. Concentration distribution near Hg drop surface.

so that the total synchronous change will be

$$\frac{\partial c}{\partial t} = \left(\frac{\partial c}{\partial t}\right)_{\text{diff.}} + \left(\frac{\partial c}{\partial t}\right)_{\text{conv.}}$$

Hence by substitution of the convection term and from Fick's second law of diffusion (eqn. 3.2), we obtain

$$\frac{\partial c}{\partial t} = D \cdot \frac{\partial^2 c}{\partial x^2} - v \cdot \frac{\partial c}{\partial x} \tag{3.23}$$

In order to obtain the velocity v, we consider two concentric spheres, the first representing the drop surface with radius r_1 and the second one having a fictive radius r_2; as the liquid is not compressible, we can choose r_2 (although altering) in such a way that the volume difference ΔV of the spheres is constant, i.e. $\Delta V = V_2 - V_1 = \frac{4}{3}\pi(r_2^3 - r_1^3)$, so that for a very slight difference $r_2 - r_1 = x$ we find

$$\Delta V = \tfrac{4}{3}\pi[(r_1 + x)^3 - r_1^3] \approx 4\pi r_1^2 x = qx = \text{constant}$$

By differentiating we obtain

$$q \cdot \frac{\mathrm{d}x}{\mathrm{d}t} + x \cdot \frac{\mathrm{d}q}{\mathrm{d}t} = 0$$

so that

$$v = \frac{\mathrm{d}x}{\mathrm{d}t} = -\frac{x}{q} \cdot \frac{\mathrm{d}q}{\mathrm{d}t}$$

As at a drop lifetime t the drop volume is

$$V = \frac{mt}{d_{25°\mathrm{C(Hg)}}} = \tfrac{4}{3}\pi r_1^3 \tag{3.24}$$

its radius is

$$r_1 = \left(\frac{3mt}{4\pi d_{25°\mathrm{C(Hg)}}}\right)^{1/3}$$

Hence

$$q = 4\pi r_1^2 = 4\pi\left(\frac{3mt}{4\pi \cdot 13.534}\right)^{2/3} = 0.8515\, m^{2/3} t^{2/3}$$

By substituting this result into eqn. 3.24 we find $v = -2/3 \cdot x/t$, so that eqn. 3.23 becomes

$$\frac{\partial c}{\partial t} = D \cdot \frac{\partial^2 c}{\partial x^2} + \frac{2}{3} \cdot \frac{x}{t} \cdot \frac{\partial c}{\partial x} \qquad (3.25)$$

Resolving this equation for c at the electrode surface, i.e., at $x = 0$, Ilkovič assumed the following initial and final conditions: $c = C$ for $t = 0$ and $c = c_0$, the constant value obtained at $t > 0$, dependent on the electrode potential; thus he obtained the expression

$$\left(\frac{\partial c}{\partial x}\right)_{x=0} = \frac{C - c_0}{\sqrt{\frac{3}{7}\pi D t}} \qquad (3.26)$$

If we compare this with the approximately linear eqn. 3.3, there is a difference only in the denominator, so that for the instantaneous current at the dme we can write

$$i = nFqD \cdot \frac{C - c_0}{\sqrt{\frac{3}{7}\pi D t}} \qquad (3.26a)$$

which on substitution of $q = 0.8515\, m^{2/3} t^{2/3}$ yields the general Ilkovič equation at 25° C:

$$i(\text{amps}) = 0.734 nF(C - c_0)D^{1/2} m^{2/3} t^{1/6} \qquad (3.27)$$

As at the limiting (diffusion-controlled) current $c_0 \rightarrow 0$, the Ilkovič equation becomes

$$i_{t_d}(\text{amps}) = 0.734 nFCD^{1/2} m^{2/3} t^{1/6} \qquad (3.22)$$

Many extensions have been derived for the Ilkovič equation from the consideration that the dme does not behave as a flat electrode but in fact shows a spherical growth. For instance, Fick's second law of diffusion (cf., eqn. 3.2) becomes[12]

$$\frac{\partial c}{\partial t} = D\left(\frac{\delta^2 c}{\partial r^2} + \frac{2}{r} \cdot \frac{\partial c}{\partial r}\right)$$

Derivations based on the introduction of this modified equation lead to a small correction factor within the Ilkovič equation, viz.,

$$i_{t_d} = 0.734 nFCD^{1/2} m^{2/3} t^{1/6}(1 + KD^{1/2} m^{-1/3} t^{1/6})$$

where K has been evaluated as 17–39 by several investigators[12]. By expressing in eqn. 3.22 i_{t_d}, C and m in the more usual units μA, mmol l^{-1} and mg s^{-1}, we obtain (with $F = 96{,}487$ C included in the factor)

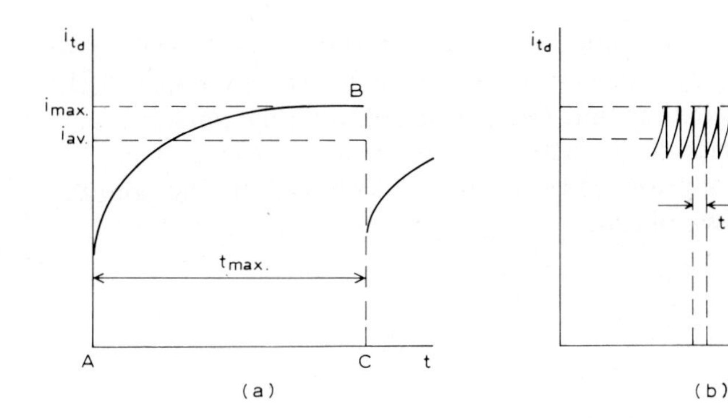

Fig. 3.15. Diffusion-controlled current, (a) vs. Hg drop growth, (b) average value.

$$i_{t_d}(\mu\text{A}) = 708nD^{1/2}Cm^{2/3}t^{1/6} \tag{3.28}$$

As i_{t_d} is proportional to $t^{2/3}$ (drop volume is proportional to t, so its surface area is proportional to $t^{2/3}$) and is also proportional to $t^{-1/2}$ (because of the concentration gradient decreasing with x in addition to D expressed in $\text{cm}^2\,\text{s}^{-1}$*, so that $D^{1/2}$ occurs in the equation), we obtain the relationship $i_{t_d} = kt^{1/6}$.

Fig. 3.15(a) shows this relationship, where $k = 708nD^{1/2}Cm^{2/3}$; in Fig. 3.15(b) the zigzag line shows that the galvanometer can only partly follow the rapid changes of the current ($t_{max.} \approx 3$ s).

By integration one obtains the enclosed surface ABC:

$$S = \int_0^{t_{max.}} i_{t_d}\,\text{d}t = \int_0^{t_{max.}} kt^{1/6}\,\text{d}t = \tfrac{6}{7}kt_{max.}^{7/6}$$

and hence i_d on the average is

$$\bar{i}_d = \frac{1}{t_{max.}} \cdot \tfrac{6}{7}kt_{max.}^{7/6} = \tfrac{6}{7}kt_{max.}^{1/6} = \tfrac{6}{7}i_{max.}$$

so that the average limiting (diffusion-controlled) current at 25°C obeys the Ilkovič equations

$$\bar{i}_d(\text{A}) = 0.629nCD^{1/2}m^{2/3}t^{1/6} \tag{3.29}$$

and

$$\bar{i}_d(\mu\text{A}) = 607nD^{1/2}Cm^{2/3}t^{1/6} \tag{3.30}$$

for C and m expressed in $\text{mmol}\,\text{l}^{-1}$ and $\text{mg}\,\text{s}^{-1}$, respectively.

Influence of mercury column height. In the above Ilkovič equations for the average limiting current m and $t_{max.}$ require some further consideration; m

*Cf., Section 3.1, p. 100: D is expressed in $\text{mol}\,\text{cm}^{-2}\,\text{s}^{-1}$ per concentration gradient ($\text{mol}\,\text{cm}^{-3}\,\text{cm}^{-1}$), so that $D = \frac{\text{mol}}{\text{cm}^2\,\text{s}} \cdot \frac{\text{cm}^3\,\text{cm}}{\text{mol}} = \text{cm}^2\,\text{s}^{-1}$.

depends considerably on the inner diameter of the capillary* and the effective mercury column height ($h_{eff.}$), $t_{max.}$ depends on m and the surface tension of Hg (σ) in the solution to such an extent that $mt_{max.}$ is approximately constant with constant σ and $t_{max.}$ is proportional to σ with constant m. To obtain $h_{eff.}$ (ref. 13), one must correct for the back-pressure from the solution and for the counter effect exerted by σ_{Hg}; hence we obtain

$$h_{eff.} = h_{Hg} - h_{soln.}\left(\frac{d_{soln.}}{d_{Hg}}\right) - \frac{2\sigma}{r_0 d_{Hg} g}$$

where

h_{Hg} = the actual Hg column height in cm, kept virtually constant by the large area of the Hg reservoir;
$h_{soln.}$ = depth of immersion of the tip in cm;
$d_{soln.}$ and d_{Hg} = densities of solution and Hg, respectively;
r_0 = radius of the mercury drop in cm;
g = gravitational acceleration in cm s^{-2};
σ = surface tension of Hg in dyne cm^{-1}; $h_\sigma = 2\sigma/r_0 d_{Hg} g$.

For h_{Hg} between 20 and 80 cm and a glass capillary of i.d. 0.03–0.04 mm with $t_{max.}$ between 3 and 6 s, $h_{soln.} \approx 0.1$ cmHg and can be neglected, but $h_\sigma \approx 1.5$ cmHg and therefore is not always negligible; although at the beginning of drop growth r_0 is so small that h_σ may be considerably higher, this does not have a practical influence on the i_d measurement. Another point is the fact that the potential alters σ (and so also $t_{max.}$), as can be seen from the electrocapillary curve of Hg (see later $\sigma_{max.}$ at -0.52 V). Above it was stated that $m = k' h_{eff.}$ and $t_{max.} = k'' \cdot 1/h_{eff.}$, so that

$$m^{2/3} t_{max.}^{1/6} = k''' h_{eff.}^{1/2} \tag{3.31}$$

This means that i_d (diffusion-controlled) is proportional to the square root of $h_{eff.}$ and its linear plot passes through the origin, which property is often used as a check on the diffusion-controlled electron-transfer reaction[14].

Shape of the polarographic curve. The kinetic theory of electrolysis (Section 3.2) for a redox system at a static inert electrode for partial and full exhaustion at the electrode under merely diffusion-controlled conditions leads, for ox + $ne \rightarrow$ red, to the relationship

$$E = E^0 + \frac{RT}{nF} \ln \left(\frac{i_{c,d} - i}{i - i_{a,d}}\right) \cdot \frac{D_{red}}{D_{ox}}$$

(cf., eqn. 3.14) and hence for a solution with only ox to

$$E = E^0 + \frac{RT}{nF} \ln \left(\frac{i_d - i}{i}\right) \cdot \frac{D_{red}}{D_{ox}}$$

*Cf., Poiseuille's equation, according to which the fluid velocity through a capillary is linearly dependent on the pressure gradient along it.

The question now is what these relationships will be at the dme. For reductive polarography (current positive according to convention)

$$E = E^0 + \frac{RT}{nF} \ln \left\{ \frac{[\mathrm{ox}]_{(x=0)}}{[\mathrm{red}]_{(x=0)}} \right\} \tag{3.31a}$$

where ox/red is a redox couple between ions or between an ion and its metal (amalgam). During the cathodic reduction ox diffuses towards the dme, so that according to the Ilkovič eqns. 3.27 and 3.22

$$i_\mathrm{c} = 0.734nF\{[\mathrm{ox}] - [\mathrm{ox}]_{(x=0)}\} D_\mathrm{ox}^{1/2} m^{2/3} t^{1/6}$$

$$= i_\mathrm{c,d} - 0.734nF[\mathrm{ox}]_{(x=0)} D_\mathrm{ox}^{1/2} m^{2/3} t^{1/6}$$

and thus

$$[\mathrm{ox}]_{(x=0)} = \frac{i_\mathrm{c,d} - i_\mathrm{c}}{0.734nFD_\mathrm{ox}^{1/2} m^{2/3} t^{1/6}} \tag{3.32}$$

The red produced by electrolysis diffuses either into the bulk solution (for an ion) or into the mercury (for an amalgamating metal), so that

$$i_\mathrm{a} = -0.734nF\{[\mathrm{red}] - [\mathrm{red}]_{(x=0)}\} D_\mathrm{red}^{1/2} m^{2/3} t^{1/6}$$

$$= i_\mathrm{a,d} + 0.734nF[\mathrm{red}]_{(x=0)} D_\mathrm{red}^{1/2} m^{2/3} t^{1/6}$$

and thus

$$[\mathrm{red}]_{(x=0)} = \frac{i_\mathrm{a} - i_\mathrm{a,d}}{0.734nFD_\mathrm{red}^{1/2} m^{2/3} t^{1/6}} \tag{3.33}$$

As both steady states concern the same electrode, $i_\mathrm{c} = i_\mathrm{a} = i$. Hence, by substituting eqns. 3.32 and 3.33 into eqn. 3.31a we obtain

$$E = E^0 + \frac{RT}{nF} \ln \left(\frac{i_\mathrm{c,d} - i}{i - i_\mathrm{a,d}}\right)\left(\frac{D_\mathrm{red}}{D_\mathrm{ox}}\right)^{1/2} \tag{3.34}$$

and for a solution with only ox

$$E = E^0 + \frac{RT}{nF} \ln \left(\frac{i_\mathrm{c,d} - i}{i}\right)\left(\frac{D_\mathrm{red}}{D_\mathrm{ox}}\right)^{1/2} \tag{3.35}$$

In comparison with eqns. 3.14 and 3.15 for a static inert electrode, eqns. 3.34 and 3.35 differ only in the power of $D_\mathrm{red}/D_\mathrm{ox}$; as for the usual reversible diffusion-controlled redox couples D_red and D_ox are approximately equal, $(D_\mathrm{red}/D_\mathrm{ox})^{1/2}$ will even be closer to unity, so that we can simply write

$$E = E^0 + \frac{RT}{nF} \left(\ln \frac{i_\mathrm{c,d} - i}{i - i_\mathrm{a,d}}\right) \tag{3.36}$$

and in case of ox only

$$E = E^0 + \frac{RT}{nF} \ln \left(\frac{i_\mathrm{c,d} - i}{i}\right) \tag{3.37}$$

Eqn. 3.37 for the reduction curve was first given by Heyrovský and Ilkovič[11] and for this reason is often referred to as the Heyrovský–Ilkovič equation; it shows that by plotting i as a function of E the polarographic curve obtained is S-shaped, i.e. a sigmoidal-shaped wave (see Fig. 3.13), and that for $i = \frac{1}{2} i_{c,d}$ the potential $E = E_{\frac{1}{2}} = E^0$, the so-called half-wave potential, which is independent of the concentration. In the more general situation (cf., eqn. 3.35), $E_{\frac{1}{2}}$ is no longer equal to the standard redox potential, but becomes

$$E_{\frac{1}{2}} = E^0 + \frac{RT}{2nF} \ln \left(\frac{D_{\text{red}}}{D_{\text{ox}}} \right) \tag{3.37a}$$

By plotting E against $\log (i_{c,d} - i)/i$ one can assess the reversibility of the electrode process, because a straight line with slope $2.303RT/nF$ should then be obtained; from the slope one can also find n; many workers* determine the slope by simply calculating $E_{\frac{1}{4}} - E_{\frac{3}{4}}$, which yields

$$\left(E_{\frac{1}{2}} + 2.3026 \frac{RT}{nF} \log 3 \right) - \left(E_{\frac{1}{2}} + 2.3026 \frac{RT}{nF} \log \tfrac{1}{3} \right)$$

$$= 2.3026 \frac{RT}{nF} \log 9 = 0.954 \left(2.3026 \frac{RT}{nF} \right)$$

In view of the possibility of resolution between subsequent waves, it is of interest to estimate the voltage range of a complete wave, e.g.,

$$E_{0.003} - E_{0.997} = 2.3026 \frac{RT}{nF} \log \left(\frac{0.997}{0.003} \right)^2 = \frac{2 \cdot 0.059 \cdot 2.52}{n} = 0.30/n \text{ V}$$

Hence the ranges for the reduction waves of mono-, di- and trivalent metal ions will be 0.30, 0.15 and 0.10 V, respectively.

A different situation occurs for the reductive polarography of an ion whose metal does not amalgamate; then the metal precipitates on the dme surface without further diffusion, so that

$$E = E^0 + \frac{RT}{nF} \ln [\text{Me}^{n+}]_{(x=0)}$$

Here we can only substitute eqn. 3.32 and thus

$$E = E^0 + \frac{RT}{nF} \cdot \frac{i_{c,d} - i}{0.734nFD_{\text{ox}}^{1/2} m^{2/3} t^{1/6}}$$

where for $E_{\frac{1}{2}}$ we can replace the term $i_{c,d} - i$ by $\frac{1}{2} i_{c,d}$, which according to $i_{c,d} = nFA_{\text{Hg}} D_{\text{ox}} C/\delta_1$ and $A_{\text{Hg}} = 0.8515 m^{2/3} t_{\text{max.}}^{2/3}$ yields

$$E_{\frac{1}{2}} = E^0 + \frac{RT}{nF} \ln \left[\frac{\frac{1}{2} nFD_{\text{ox}} \left(\frac{C}{\delta_1} \right) \cdot 0.8515 m^{2/3} t_{\text{max.}}^{2/3}}{0.734 mFD_{\text{ox}}^{1/2} m^{2/3} t^{1/6}} \right]$$

*Criterion of reversibility suggested by Tomeš[15].

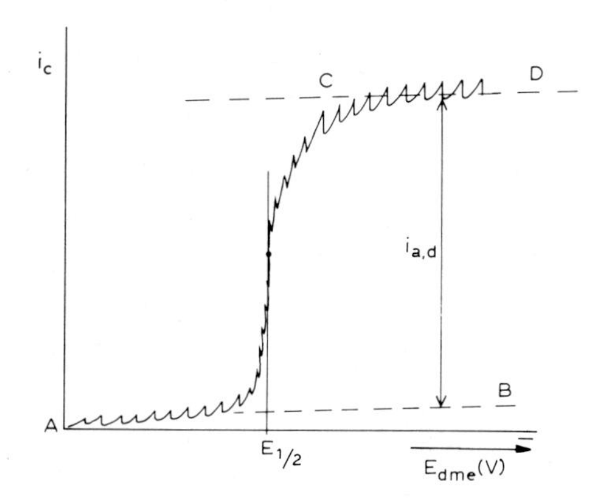

Fig. 3.16. Reductive polarographic wave of a non-amalgamating metal.

If we consider it for $t = t_{max.}$ only, we obtain

$$E_{\frac{1}{2}} = \left[E^0 + \frac{RT}{nF}\ln\left(\frac{0.8515}{0.734}\right)\cdot\frac{D_{ox}^{1/2}}{\delta_1}\cdot t_{max.}^{1/6}\right] + \frac{RT}{nF}\ln\left(\frac{C}{2}\right) \tag{3.38}$$

Hence, for the non-amalgamating metals such as Fe, Cr, Mo, W and V, even if all other conditions included $t_{max.}$ are kept constant, the half-wave potential is not constant but still shifts to more positive values with increasing concentration. The precise value of E^0 depends to a great extent on the supporting electrolyte, e.g., in a KCl electrolyte it may be approximately the potential of the calomel electrode concerned. In contrast with both previous reversible redox systems with diffusion of both ox and red, here the polarographics wave is not symmetrical versus $E_{\frac{1}{2}}$ but starts almost vertically and rises steeply at the beginning (see Fig. 3.16). The same applies to reversible redox couples with non-amalgamating metals, where either ox is completed[16] or red is becoming complexed[17] to give an insoluble complex.

Residual current in polarography. In the pragmatic treatment of the theory of electrolysis (Section 3.1) we have explained the occurrence of a residual current on the basis of back-diffusion of the electrolysis product obtained. In conventional polarography the wave shows clearly the phenomenon of a residual current by a slow rise of the curve before the decomposition potential as well as beyond the potential where the limiting current has been reached. In order to establish the value one generally corrects the total current measured for the current of the blank solution in the manner illustrated in Fig. 3.16 (vertical distance between the two parallel lines CD and AB). However, this is an unreliable procedure especially in polarography because, apart from the troublesome saw-tooth character of the i versus E curve, the residual current exists not only with a faradaic part, which is caused by reduction (or oxidation)

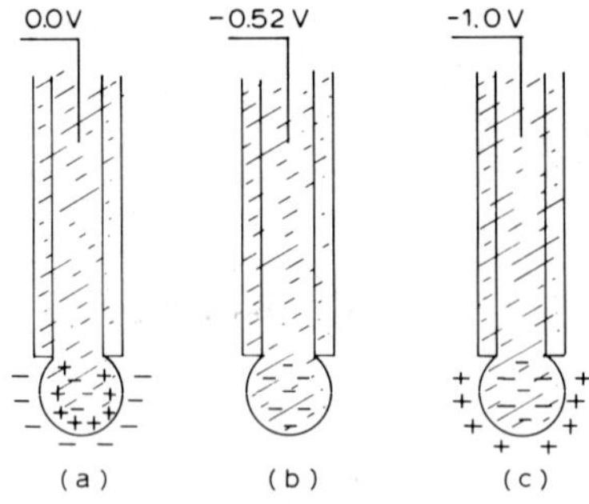

Fig. 3.17. Electric double layer at Hg droop vs. potential applied.

of traces of impurities in the supporting electrolyte or in incidental components of the analyte (such as heavy metals, oxygen or chlorine), but also with a non-faradaic part; the latter is known as the charging, capacitive or condenser current and its intensity is considerable in comparison with the faradaic current to be measured. In order to explain the charging current in polarography we shall consider a mercury drop attached to the orifice of the capillary (see Fig. 3.17).

Certain negative ions such as Cl^-, Br^-, CNS^-, NO_3^- and SO_4^{2-} show an adsorption affinity to the mercury surface; so in case (a), where the overall potential of the dme is zero, the anions transfer the electrons from the Hg surface towards the inside of the drop, so that the resulting positive charges along the surface will form an electric double layer with the anions adsorbed from the solution. Because according to Coulomb's law similar charges repel one another, a repulsive force results that counteracts the Hg surface tension, so that the apparent σ_{Hg} value is lowered.

In situation (b) the anion adsorption is compensated by the negative overall potential of the dme. In situation (c), with a further increase in the negative potential, an electric double layer will now be formed with cations from the solution, so that the apparent σ_{Hg} is lowered again. Hence σ_{Hg} as a function of the negative dme potential, yielding the so-called electrocapillary curve, shows a maximum at about -0.52 V (see Fig. 3.18).

Further, as σ_{Hg} has a more direct effect on $t_{max.}$ but only a marginal influence on $h_{eff.}$ and accordingly on m (see pp. 117–118), $t_{max.}$ follows the course of σ_{Hg}, although to a less extent for $t^{1/6}$; the latter, therefore, proves the minor effect of σ_{Hg} on i_t according to $m^{2/3}t^{1/6}$ in the Ilkovič equation (eqn. 3.22).

The dme in the analyte solution acts as a spherical condenser with a periodically renewed growing surface $\pi r_1^2 = q = 0.8515\, m^{2/3} t^{1/6}$ (cf., p. 116) and a double-layer thickness δ_{dl}, which is negligibly small compared with r_1; hence the capacity of the condenser is

$$C = \frac{\varepsilon_0 r_1^2}{\delta_{dl}} = \frac{\varepsilon_0}{\pi \delta_{dl}} \cdot \pi r_1^2 = \frac{\varepsilon_0}{\pi \delta_{dl}} \cdot q$$

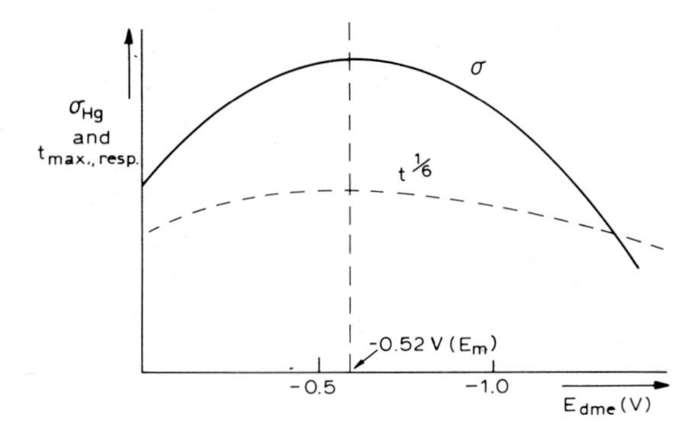

Fig. 3.18. Electrocapillary curve of Hg drop.

Now we can write the equation for the charging current, i_C, during the growth of the drop as

$$i_C = \frac{dQ}{dt} = V \cdot \frac{dC}{dt} = \frac{V\varepsilon_0}{\pi\delta_{dl}} \cdot \tfrac{2}{3} \cdot 0.8515 m^{2/3} t^{-1/3} = kVt^{-1/3} \tag{3.40}$$

This variation of i_C with $t^{-1/3}$ is only correct if the potential V during the drop's lifetime may be considered as constant and δ_{dl} as nearly constant, which in practice is justified. In Fig. 3.19, the plot of the asymptotic decrease in the (non-faradaic) charging current i_C and the increase in the (faradaic) diffusion current i_F (cf., i_{t_d} in Fig. 3.15) versus the drop's lifetime illustrates the following important feature: at $t = 0$, i_C is high and i_F is low, whereas at $t \rightarrow t_{max.}$, i_C is low and i_F is high, i.e., approaching a minimal increase. This means that in polarographic analysis, where the analyte must be determined from i_F, a measurement nearest to $t_{max.}$ offers the most attractive conditions; realization of the latter in conventional polarography will be considered in Section 3.3.1.1.3, and also plays a major role in the choice of more modern polarographic techiques. As far as the above derivation is concerned only an Hg drop of a potential at which a double layer can actually occur was considered. Fig. 3.20 shows the

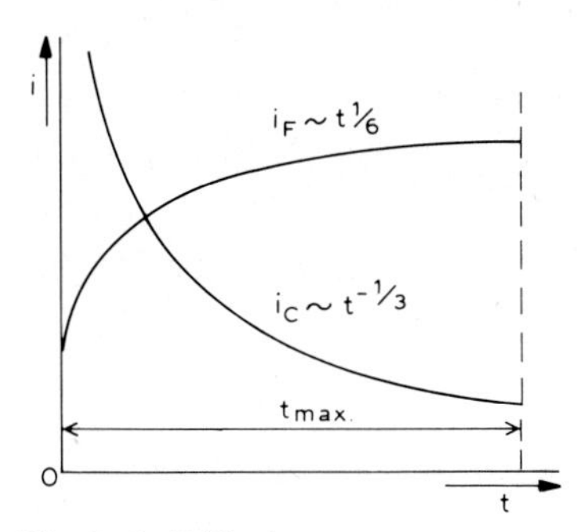

Fig. 3.19. Diffusion current i_F and charging current i_C during Hg drop growth.

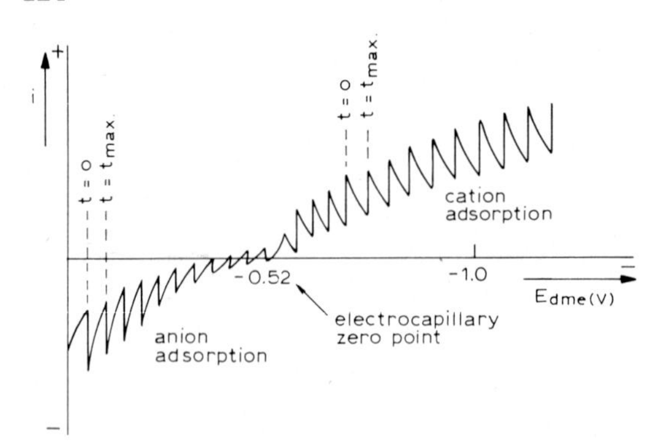

Fig. 3.20. Saw-tooth character of i_C vs. dme potential.

saw-tooth picture of i_C as obtained during a slowly increasing negative dme potential; according to expectation, i_C is virtually zero at the capillary zero point but becomes negative (counteraction) below it and positive (coaction) beyond it (in comparison with the direction of i_F)*.

Irreversibility versus reversibility in polarography. Previously in this chapter we dealt only with reversible redox systems, i.e., with truly Nernstian behaviour and merely diffusion control. This also applies to combined processess of electron transfer and chemical reaction (e.g., complexation) provided that both take place instantly. For instance, in EC such as

$$\mathrm{ox} + n\mathrm{e} \rightarrow \mathrm{red}' \underset{k_2}{\overset{k_1}{\rightleftharpoons}} \mathrm{red}$$

the Nernst equation is

$$E = E^0 + \frac{RT}{nF} \ln \left(\frac{[\mathrm{ox}]}{[\mathrm{red}']} \right)$$

and the chemical reaction constant $K = [\mathrm{red}]/[\mathrm{red}'] = k_1/k_2$, so that

$$E = \underbrace{E^0 + \frac{RT}{nF} \ln K}_{E_{\frac{1}{2}}} + \frac{RT}{nF} \ln \left(\frac{[\mathrm{ox}]}{[\mathrm{red}]} \right)$$

and according to eqn. 3.34

$$E = \underbrace{E^0 + \frac{RT}{nF} \ln K}_{E_{\frac{1}{2}}} + \frac{RT}{nF} \ln \left(\frac{i_{\mathrm{c,d}} - i}{i - i_{\mathrm{a,d}}} \right) \left(\frac{D_{\mathrm{red}}}{D_{\mathrm{ox}}} \right)^{1/2}$$

*In general the capacitance of the double layer with anions exceeds that with cations[18].

Hence, for $(D_{\text{red}}/D_{\text{ox}})^{\frac{1}{2}} \approx 1$ and for a solution of ox or red only:

$$E = \underbrace{E^0 + \frac{RT}{nF}\ln K}_{E_{\frac{1}{2}}} + \frac{RT}{nF}\ln\left(\frac{i_{\text{c,d}} - i}{i}\right) \qquad \text{(3.41, reductive)}$$

or

$$E = \underbrace{E^0 + \frac{RT}{nF}\ln K}_{E_{\frac{1}{2}}} + \frac{RT}{nF}\ln\left(\frac{i}{i - i_{\text{a,d}}}\right) \qquad \text{(3.41, oxidative)}$$

In CE such as

$$\text{ox} \underset{k_2}{\overset{k_1}{\rightleftharpoons}} \text{ox}' + ne \rightarrow \text{red}$$

where $K = [\text{ox}']/[\text{ox}] = k_1/k_2$, an analogous treatment leads to the same eqns. 3.41. Considering the results for both reversible systems we conclude that owing to the chemical reaction equilibrium neither the S-shape of the polarographic curve nor its slope will change; the curve only shifts with $(RT/nF)\ln K$, i.e., if K is high (>1) to a less negative potential (reduction becomes easier but oxidation more difficult), and if K is low (<1) to a more negative potential (reduction becomes more difficult, but oxidation easier).

Concerning the above requirement "instantly" in relation to "reversibility", Delahay[19] has given the following criterion: if $k_{\text{s,h}} \geqslant 2 \cdot 10^{-2}\,\text{cm}\,\text{s}^{-1}$ (see eqns. 3.8 and 3.9) a dc electrode process (at a drop time of ca. 3 s) can be considered as reversible; however, in ac polarography with moderate frequency reversibility requires $k_{\text{s,h}} \geqslant 1\,\text{cm}\,\text{s}^{-1}$; hence "reversibility" in practice is a relative concept, as an electrode process that appears reversible in the dc sense may appear irreversible in the ac sense. Sometimes electrode processes in dc polarography (with a drop time of ca. 3 s) are called quasi-reversible if at comparable rates of the forward and backward reactions (k_{f} and k_{b}) $k_{\text{s,h}}$ lies within the range $5 \cdot 10^{-5}$–$2 \cdot 10^{-2}\,\text{cm}\,\text{s}^{-1}$.

If $k_{\text{s,h}} \leqslant 5 \cdot 10^{-5}\,\text{cm}\,\text{s}^{-1}$ in dc polarography we arrive at a totally irreversible electrode process, where the backward reaction can be neglected; we shall treat such a situation for a reduction process as the forward reaction.

As $k_{\text{s,h}}$ in this instance is very small, then according to the Butler–Volmer formulation (eqn. 3.5) the reaction rate of the forward reaction, $k_{\text{c}} = k_{\text{s,h}}\text{e}^{-\alpha nF(E-E^0)/RT}$, even at $E = E^0$, is also very low. Hence $E_{\text{appl.}}$ must be appreciably more negative to reach the half-wave situation than for a reversible electrode process. Therefore, in the case of irreversibility, the polarographic curve is not only shifted to a more negative potential, but also the value of its slope is considerably less than in the case of reversibility (see Fig. 3.21). In fact, the wave character of the curve is rate determining, i.e., kinetically controlled, but the limiting current is controlled partly kinetically and partly by diffusion ($i_l = i_k + i_d$). There are naturally the extreme situations $i_l = i_k$ or $i_l = i_d$, and the ratio i_k/i_d can be judged from Koutecký's expression[20] for the chemical reaction in a CE mechanism:

$$\frac{i_k}{i_d - i_k} = 0.886 \left(\frac{k_1}{k_2^{1/2}}\right) t^{1/2} \tag{3.42}$$

(In an EC mechanism the ratio of the forward and backward reaction rates is decisive for i_k/i_d in i_l; the chemical follow-up reaction has no influence here, so that for a sufficiently rapid electron transfer step the limiting current remains diffusion controlled.)

For the CE mechanism equations for i_k and i_d have been derived[20]; that for i_k contains the term $m^{2/3}t^{2/3}$, which means that, as $mt_{max.}$ is approximately constant (see p. 118), i_k is independent of $h_{eff.}$ whereas i_d is proportional to the square root of $h_{eff.}$ (see eqn. 3.31); this is sometimes an even better criterion for distinguishing between an irreversible and a reversible electrode process than the value of $E_{1/4} - E_{3/4}$ used by Tomeš (see p. 120). With regard to the latter method the following equation in relation to Koutecký's expression 3.42 has been derived for the i versus E curve of the irreversible reduction provided that the limiting current is still diffusion controlled*:

$$E = \underbrace{E^0 + \frac{RT}{\alpha nF} \ln 0.886\, k_{s,h} \left(\frac{t}{D_{ox}}\right)^{1/2}}_{E_{\frac{1}{2}}} + \frac{RT}{\alpha nF} \ln\left(\frac{i_d - i}{i}\right) \tag{3.43}$$

This equation shows a real distortion of the waves, i.e., a shift of $E_{\frac{1}{2}}$ to more negative potentials and a lower slope, which means a higher potential range ($E_{\frac{1}{4}} - E_{\frac{3}{4}}$) than the reversible electrode process could have; for this reason, workers often use this as a distinguishing criterion, although it may sometimes be less reliable[20] as a consequence of alternative stoichiometry in the charge-transfer step. The above considerations with respect to irreversibility are analogously applicable to the electrode process of the oxidation.

In Fig. 3.21 this is illustrated for the same redox couple in the case of reversibility and of irreversibility; in the latter situation $E_{\frac{1}{2}(red)}$ and $E_{\frac{1}{2}(ox)}$ are so different that both the reduction and the oxidation waves can be separately determined. In fact, this is in agreement with the picture in Fig. 3.11 for irreversibility at a static inert electrode.

Again returning to the diffusion-controlled limiting current, we often meet a considerable influence on its height by catalysis, adsorption or other surface phenomena, so that we have to deal with irreversible electrode processes. For instance, when to a polarographic system with a diffusion-controlled limiting

*If α is independent of the potential the logarithmic plot is still a straight line; however, the wave position always shows a dependence on drop time.

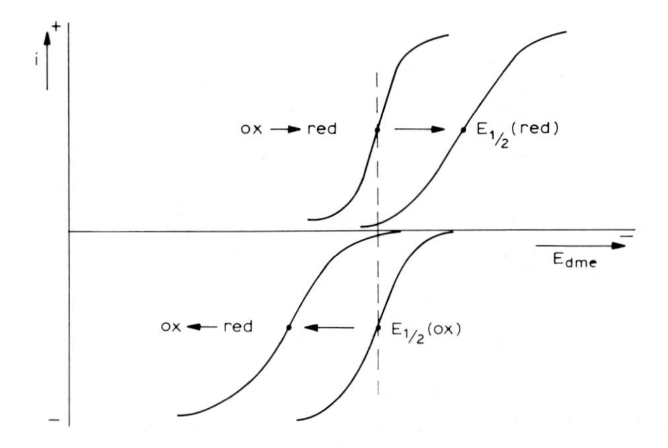

Fig. 3.21. Potential shift in case of irreversibility.

current ($i_l = i_d$) an excess of reagent (reacting with the product of the charge-transfer step) is added, i_l is markedly increased and is actually converted into a more kinetically controlled limiting current ($i_l = i_k \gg i_d$) and so non-varying with $h_{eff.}$; we may then talk of a catalytic wave. With adsorption phenomena at the dme not only the non-faradaic charging currents change but also the faradaic currents alter, mostly in an undesirable way (wave deformation and/or splitting); usually a distinction is made between adsorption of the product of the charge-transfer step, yielding a so-called adsorption wave, or adsorption of another component[21].

Without going into further detail, we have shown above the great importance of distinguishing between reversible and irreversible electrode processes in order to understand the theory of polarography and its implications.

3.3.1.1.3. *Practice of conventional polarography*

For the principle of the apparatus, see Section 3.3.1.1.1; usually the adjusted height of the glass Hg reservoir, and more especially its Hg surface, can be read in order to favour reproducibility of the drop time. Thermostatting of the analyte solution at 25.0 ± 0.5° C is of importance owing to the influence of temperature on the diffusion constants. Before the start of the measurement traces of oxygen should be removed from the solution by flushing with nitrogen or argon (passed through a wash-bottle containing blank supporting electrolyte); during the measurement, of course, the nitrogen or argon flushing is stopped and preferably the solution is kept under a nitrogen cover.

If the measurement itself is made by means of an undamped rapidly responding galvanometer the recorded oscillations are impracticably large; for this reason, conventional polarography was usually made with a damped galvanoter (damping adjustment by a variable resistance in an RC feedback circuit of the galvanometer), which as shown in Fig. 3.22 (cf., also Fig. 3.15) records well the average current $\bar{i} = \frac{6}{7} i_{max.}$ (cf., eqns. 3.29 and 3.30).

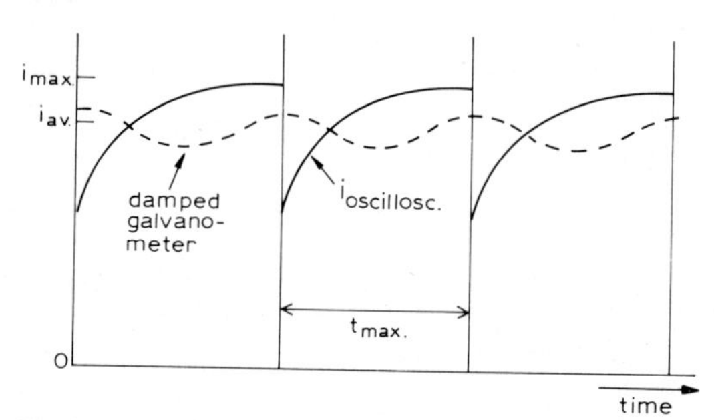

Fig. 3.22. Polarography with damped galvanometer vs. undamped measurement.

The most attractive type of polarogram would be that illustrated in Fig. 3.23 (curve a), where line AB (of the average limiting curve) is parallel to line CD (of the residual curve), so that their distance apart, i_d, yields the corrected limiting current. Most polarographs have a regulation knob labelled "*I* comp.", which permits a current offset proportional to the increase in potential; if such an increase is linear, as is the case for both the residual and limiting currents of curve a, the offset can be adjusted to the situation of curve a′ where i'_d, being simply the distance between A′B′ and the abscissa, is equal to i_d. This offset procedure can still be used when line AB is not parallel to line CD, provided that the deviation is not excessive; however, it requires strict calibration, i.e., i'_d must be measured under standard experimental conditions and at a fixed E value close to the potential where the limiting current can be considered to be just reached.

Unfortunately, AB and CD are often far from parallel as a consequence of the complicated nature of the residual current in polarography (cf., pp. 118–121), where one has to deal with at least two or three factors, as follows.

(1) The true residual current (faradaic) caused by reduction (or oxidation) of traces of impurities in the supporting electrolyte or in incidental components of the analyte. This current can be reasonably well compensated by the current offset "*I* comp." of the polarograph.

(2) A charging current (non-faradaic) due to the formation of an electric double layer on the surface of the growing drop; most polarographs permit a

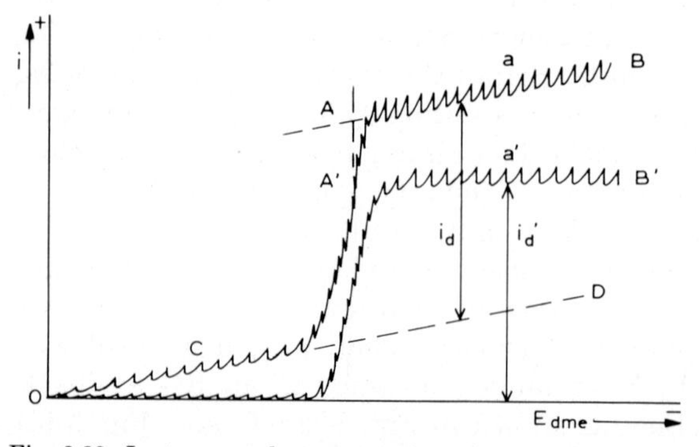

Fig. 3.23. Instrumental compensation of residual current in polarography.

current correction "I_c comp." using the linear dc potential ramp as a source. However, Fig. 3.20 shows that the correction must be zero at − 0.52 V, negative to the more negative side and positive to the less negative side. Moreover, the correction slope on either side of the electrocapillary zero point is different. Altogether this makes the I_c compensation procedure cumbersome and often unreliable, so that modern polarographic modes have been developed with the main objective of actual discrimination against the charging current.

(3) A migration current caused by insufficient concentration of the supporting electrolyte, so that part of the ionic conductance has to be provided by the reacting species itself; in reductive polarography this means for ox in a cation a positive migration current (which contributes to the diffusion current) and for ox in an anion a negative migration current (which detracts from the diffusion current); 50–100-fold concentrations of supporting electrolyte versus the reactant generally suffice to eliminate a migration current (it is useful to determine previously the minimal concentration of supporting electrolyte required for each specific determination).

The addition of supporting (inert) electrolyte is especially desirable as a voltage drop in the solution must be prevented; apart from the distortion of the polarographic wave that this voltage drop will cause, it may also induce the development of heat and hence troublesome convection at the dme surface. In aqueous solutions this *IR* effect is normally negligible, but in non-aqueous media of low conductance the wave distortion must be remedied by applying either *IR* compensation as function of *I* (separately or integrally provided with most polarographs) or a three-electrode system. The latter represents the most effective and fundamentally correct procedure, i.e., the potential across the working electrode (dme) and a reference electrode a short distance away is measured at zero current, and the current through the dme and a non-polarizable auxiliary electrode is determined separately; hence the current can be recorded as a function of the true dme potential.

Fig. 3.24 shows the beneficial use of three electrodes instead of two; in the latter instance there is an apparent wave distortion (shift of *iR* to a more

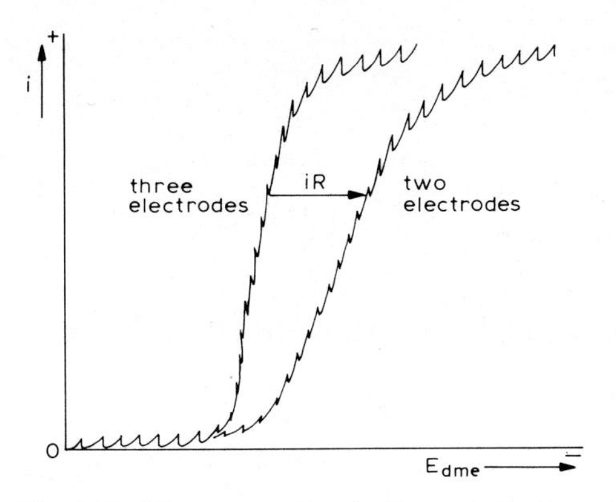

Fig. 3.24. *iR* compensation in three-electrode system.

negative potential) that resembles the wave of an irreversible electrode process, but there (see Fig. 3.21) is real distortion. The Metrohm Polarecord E506, apart from possibilities in addition to other more modern polarographic modes, possesses "I comp" and "I_c comp" adjustment knobs plus an extremely high ohmic potentiostat (iR compensator) integrated within the apparatus; the Polarecord 626 is a three-electrode polarograph with only an "I comp" adjustment knob. The Tacussel PRG polarograph has adjustment knobs for current offset and iR compensation; for measurements at high sensitivity in low-conductivity media the shielded cell type CTP 3B, with a Faraday cage, protects against external electrostatic and electromagnetic interference; the PRG3 polarograph is especially intended for the three-electrode procedure. It can be said that to-day the three-electrode method is the most commonly adopted in current practice.

Notwithstanding all previous precautions taken, some difficulties may still remain. For instance, in the reductive polarography of monovalent metal ions the half-wave potentials should differ preferably by at least 0.30 V (see p. 120) in view of the net wave separation. Simply in order to detect the presence of a second metal a difference of at least 0.1 V is required, but interferences soon arise at low concentration; although derivative polarography yields some improvement, these can best be overcome by complexation of one of the metals, so that its half-wave potential shifts to the more negative side. If we take the complexation of a metal M^{n+} as an example, e.g. with X^{b-} as the complexing ion, then

$$M^{n+} + pX^{b-} \rightleftharpoons MX_p^{(n-pb)}$$

and the stabilization constant is

$$K_{\text{stab.}} = \frac{[MX_p^{(n-pb)}]}{[M^{n+}][X^{b-}]^p}$$

Hence for the Nernst potential

$$E = E^0_{M^{n+}\rightarrow M(Hg)} + \frac{RT}{nF}\ln\left\{\frac{[M^{n+}]}{[M(Hg)]}\right\}$$

we can write

$$E = \underbrace{E^0_{M^{n+}M(Hg)} - \frac{RT}{nF}\ln K_{\text{stab.}} - \frac{pRT}{nF}\ln[X^{b-}]}_{E_{\frac{1}{2}}} + \frac{RT}{nF}\ln\left\{\frac{[MX_p^{(n-pb)}]}{[M(Hg)]}\right\} \tag{3.44}$$

Considering situations at the dme surface and for a certain drop life t, we find according to eqn. 3.3

$$i_t = \frac{nFA[C_{MX_p^{(n-pb)}} - c_{MX_p^{(n-pb)}}]}{\delta_{MX_p^{(n-pb)}}} \tag{3.45}$$

and

$$i_t = \frac{nFAC_{M(Hg)}}{\delta_{M(Hg)}} \tag{3.46}$$

and further

$$i_{t_d} = \frac{nFAc_{MX_p^{(n-pb)}}}{\delta_{MX_p^{(n-pb)}}} \tag{3.47}$$

From eqns 3.45 and 3.47 we obtain

$$c_{MX_p^{(n-pb)}} = \frac{i_{t_d} - i_t}{nF\delta_{MX_p^{(n-pb)}}}$$

and from eqn. 3.46

$$C_{M(Hg)} = \frac{i_t}{nF\delta_{M(Hg)}}$$

so that by substitution in eqn. 3.44 we obtain

$$E = \underbrace{E^0_{M^{n+}\to M(Hg)} - \frac{RT}{nF}\ln K_{stab.} - \frac{pRT}{nF}\ln[X^{b-}] + \frac{RT}{nF}\ln\left[\frac{\delta_{M(Hg)}}{\delta_{MX_p^{(n-pb)}}}\right]}_{E_{\frac{1}{2}}}$$

$$+ \frac{RT}{nF}\ln\left(\frac{i_{t_d} - i_t}{i_t}\right) \tag{3.48}$$

This equation clearly shows that according to expectation the $E_{\frac{1}{2}}$ shift to the negative side is greater the higher is $K_{stab.}$ and the greater is the amount of complexing agent.

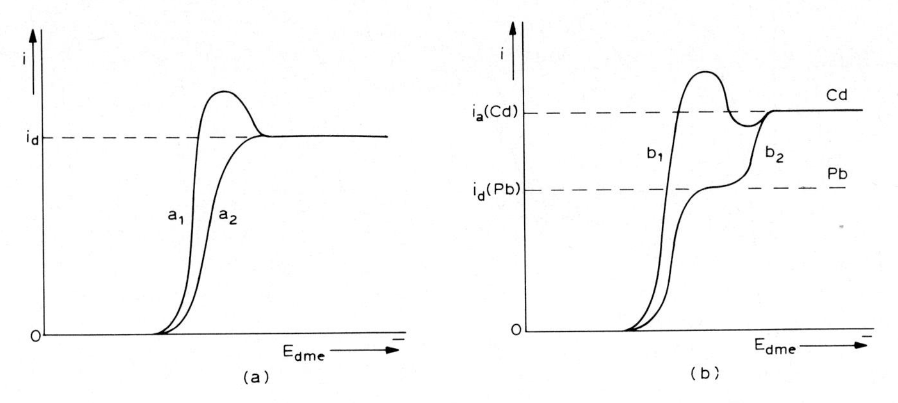

Fig. 3.25. Interference by maxima in polarographic waves of (a) a single analyte, (b) two analytes.

Another type of interference is the occurrence of maxima (see Fig. 3.25) in both cathodic and anodic curves; the steeply ascending curves (a_1 and b_1) do not possess an inflection point, but obey more or less the Ohmic law $i = E_{appl.} - E_{decomp.})/R_{electrolyte}$ (cf., Section 3.1), which suggests that the usually occurring concentration polarization is prevented, probably owing to the effect of solution streaming (convection) along the mercury surface; the shape of the curve remains the same whether it is measured in the forward or backward direction. In case (a) i_d can still be determined, but in case (b) the maximum in b_1 completely obscures the Pb wave. Fortunately these maxima can be mostly suppressed by adding surface-active agents or colloids, and sometimes by changing the solution experimental conditions in different manners. Well known additives are Triton X-100 (0.002%), gelatin (0.01%) and methyl red (0.001%); the concentrations mentioned should not be exceeded in view of unfavourable effects (viscosity increase by gelatin, methyl red reduction, etc.).

In oxidative polarography there is still the difficulty of a considerably limited potential range owing to dissolution of the mercury itself with a direct dependence on the electrolyte composition; this is well illustrated in Fig. 3.26 for the following electrode reactions of Hg:

curve A in 0.1 N KNO_3:

$$Hg \rightarrow Hg^{2+} + 2e \quad \text{or} \quad 2Hg \rightarrow Hg_2^{2+} + 2e \text{ (beyond } +0.2\text{ V)}$$

curve B in 0.1 N KCl:

$$2Hg + 2Cl^- \rightarrow Hg_2Cl_2\downarrow + 2e \text{ (beyond } +0.1\text{ V)}$$

curve C in 0.1 N NaOH:

$$Hg + 2OH^- \rightarrow Hg(OH)_2\downarrow + 2e \text{ (beyond } -0.2\text{ V)}$$

curve D in 0.1 N NaOH with 0.001 M Na_2S, which represents the anodic wave in Na_2S:

$$2Hg + S^{2-} \rightarrow Hg_2S\downarrow + 2e \text{ (beyond } -0.8\text{ V)}$$

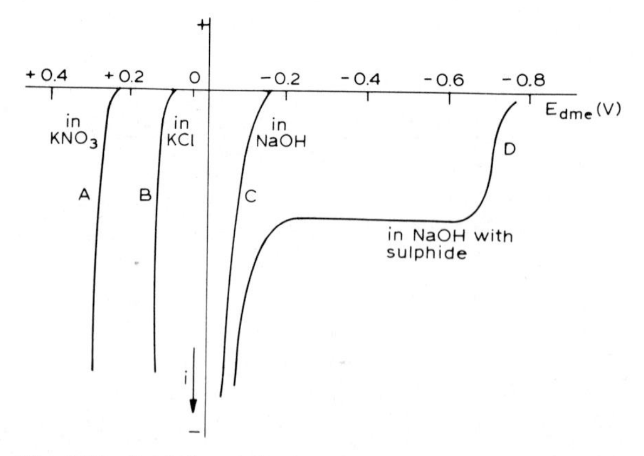

Fig. 3.26. Oxidative polarography in different supporting electrolytes.

and the anodic wave in 0.1 N NaOH of curve C (beyond -0.2 V). (Note that the anodic currents are negative according to convention!)

3.3.1.2. *Newer methods of polarography*

The difficulties in conventional polarography as mentioned in Section 3.3.1.1, especially the interference due to the charging current, have led to a series of most interesting developments by means of which these problems can be solved in various ways and to different extents. The newer methods concerned can be divided into controlled-potential techniques and controlled-current techniques. A more striking and practical division is the distinction between advanced DC polarography and AC polarography. These divisions and their further classification are illustrated in Table 3.1. In treating the different classes we have not applied a net separation between their principles, theory and practice, because these aspects are far too interrelated within each class.

3.3.1.2.1. *Advanced DC polarography*

1. *Rapid DC polarography*. The natural drop time of 2–10 s in conventional polarography has two major disadvantages, viz., (1) in view of the precision of measurement the potential scan rate must be comparatively slow, leading to a long duration of analysis, and (2) the drop time shows some dependence on potential, especially at the very negative side. The latter disadvantage can be eliminated by control of the drop time through a mechanical knock-off of the dme at a fixed time interval. This procedure, once adopted, led as a matter of course to experiments aimed at avoiding the former disadvantage by means of short controlled drop times; notwithstanding the earlier recommendations of long drop times in polarographic work, investigators such as Cover and Connery[22] used drop times as low as milli-seconds with an acceptable correlation with the existing classical theory; moreover, as an additional advantage of high scan rates (up to several hundred millivolts per second), maxima do not appear and catalytic and kinetic waves are minimized or eliminated, so that ill-defined curves in the conventional technique are now simplified and have only a faint zigzag character in the rapid technique[23]. Further, the concentration-exhausting effect remains more limited. On the other hand there is the disadvantage of a poorer limit of detection in the rapid technique for two reasons: the influence of the charging current is not eliminated and only the average limiting current is registered (eqns. 3.29 and 3.30).

Although modern polarographs also permit the choice of short drop times (usually down to 0.5 s), the most attractive apparatus for rapid polarography is that supplied by Metrohm (Herisau, Switzerland); Wolf[24], who was one of the first promoters of the technique, used the Polarecord E 261 R together with the E 354 S polarography stand and its drop time controller (from 0.32 to 0.16 s in five steps) (see Fig. 3.27). Comparison of the normal and rapid (drop time 0.25 s)

TABLE 3.1

NEWER METHODS (NON-CONVENTIONAL) OF POLAROGRAPHY (MOSTLY WITH A CONTROLLED DROP TIME)

I. Controlled-potential techniques				II. Controlled-current techniques*
Advanced DC polarography			AC polarography	
A. Stationary (mostly pseudo-stationary) methods → static measurement		B. Non-stationary methods → dynamic measurement		
1. Rapid DC polarography 2. Current-sampled DC (or Tast) polarography 3. Subtractive DC polarography 4. Derivative DC polarography (1st and 2nd): a. True derivative b. Pseudo-derivative	These principles can be and often are used in any DC technique of Class B as well as in AC polarography	5. Linear-sweep voltammetry: a. Impulse method ("single-sweep") b. Kipp method ("multi-sweep") 6. Pulse polarography: a. Normal pulse polarography b. Differential pulse polarography c. Pseudo-derivative pulse polarography	1. Polarography with superimposed AC signal: a. Sinusoidal AC polarography including tensammetry and AC bridge polarography b. Square-wave polarography and high-frequency polarography c. Kalousek polarography	2. Oscillographic: AC polarography

*For controlled-current DC polarography, especially its current density mode, see under Chronopotentiometry at a dme (p. 172). For charge-step polarography, i.e., a controlled charge of coulostatic technique, see ref. 9, pp. 424–429, and ref. 3, pp. 270–276.

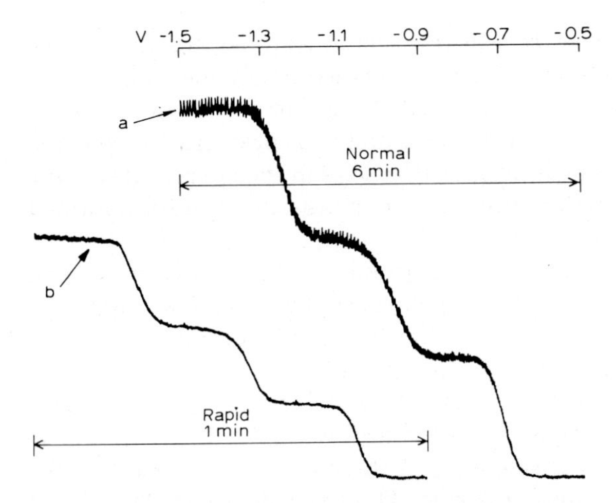

Fig. 3.27. Conventional and rapid polarography of solution: $5 \cdot 10^{-4}$ mol Cd^{2+}, Ni^{2+}, Zn^{2+} in 0.25 M NH_3/NH_4Cl, 0.02% gelatine; curve a normal (strong damping $\rightarrow \tau = 3.9$ s) and curve b rapid (weak damping $\rightarrow \tau = 0.6$ s) (Courtesy of Metrohm).

polarograms clearly shows the simplified shape and considerable reduction in analysis time for the rapid technique.

In the newer Metrohm Polarecord E 506, which permits the application of almost the complete range of modern polarographic methods (see later) including the rapid technique, the drop time controller (from 6 to 0.4 s in eleven steps) has been built in.

2. *Current-sampled DC polarography (Tast polarography).* From Fig. 3.19, we have seen that the faradaic-to-charging current ratio is most favourable at the end of the drop life; one can take advantage of this condition by means of current-sampled DC polarography (originally and sometimes still called Strobe or Tast polarography).

Apart from minor differences between the various instruments the current sampling works a follows. During the continuous voltage scan on the X-axis and in connection with a mechanical drop time controller the current is sampled and registered on the Y-axis with the galvanometer undamped and only during a fixed period of 5–20 ms prior to the drop fall. To ensure proper functioning there is an electronic sample-and-hold circuit, which not only regulates the sampling but also holds the current value until the current of the next-drop sampling period appears. Hence the polarographic curve shows a series of steps, the rising part of each of which corresponding to the sampling period and the horizontal part to the holding interval. A short drop time or a slow voltage scan rate leads to a smooth curve with fine serrations; the opposite situation of a long drop time or sampling period versus a high scan rate makes the rising part of the steps deviate from vertical; drop times longer than 5 s should also be avoided in view of the occurrence of maxima and/or adsorption problems.

For the PAR Model 170 signal processor (Princeton Applied Research; later EG & G Princeton Applied Research Corp.) it was explicitly mentioned that the sample-and-hold system includes an averaging circuitry that averages the current sampled over the entire aperture period and thus prevents unexpected discrepancies produced by sudden slight perturbations in the current flow. One can deduce from the foregoing that this system makes the results obtained easily accessible to digital conversion.

In contrast with normal rapid DC polarography, which only gives the average limiting current, current-sampled polarography gives the maximum limiting current. Hence it is not surprising that the rapid method method has also been combined with the current-sampling procedure[25].

We may consider the question of the extent to which the current-sampling procedure really prevents interference by the charging current. The answer is that a slight effect remains as a consequence of the voltage scan rate and the drop growth, both during the sampling period. The first effect can be avoided by staircase voltammetry, as suggested by Barker[26], i.e., the use of a stepwise increase in the voltage after each drop fall (cf., Ferrier et al.[27]). The second effect is prevented by the excellent method involving the PARC static mercury drop electrode (SMDE), introduced in 1979[28]; here the drop is dispensed rapidly and then caused to hang stationary at the capillary tip until finally the measurement is made (i.e. current sampling only during a fixed short time before the drop fall); the drop is dislodged by an adjustable mechanical knocker.

Next (see Fig. 3.28 for mechanical details), when the solenoid is activated the plunger is lifted, allowing mercury to flow through the capillary (I.D. 0.006 in.), which yields a small-, medium- or large-sized drop corresponding to a pre-set solenoid activation time of 50, 100 and 200 ms, respectively. Then, on deactivation of the solenoid the compensation spring closes the capillary, so that after the end of drop growth both the charging current and the faradaic current decay, the former quickly falling almost to zero ($i_C = k_2 e^{-k_3 t}$) and the latter approaching the diffusion-limited current ($i_F = k_1 t^{-\frac{1}{2}}$)[29]. In the final stage (before drop fall), current sampling delivers the value of the faradaic current only. The operating sequence is then repeated. The above procedure, including the current sampling, is operated by means of PARC Models 174A, 364, 374 and 384 Polarographic Analyzers (see later).

Finally, it can be remarked that the SMDE method with current sampling represents a practically stationary method, which would become truly stationary if also combined with staircase voltammetry. All other DC polarographic techniques, so-called stationary, are in fact pseudo-stationary methods. There are still some less common current-sampling polarographic techniques, e.g. the integrated DC Tast polarography according to Metrohm (Polarecord 626) where the current of a 200-ms controlled final drop time is integrated, and the so-called current-averaged DC polarography, where through low-pass filter systems a voltage signal directly proportional to the average limiting current[30] is obtained as a suitable input for an X–Y recorder or for a computer in derivative polarography (see below).

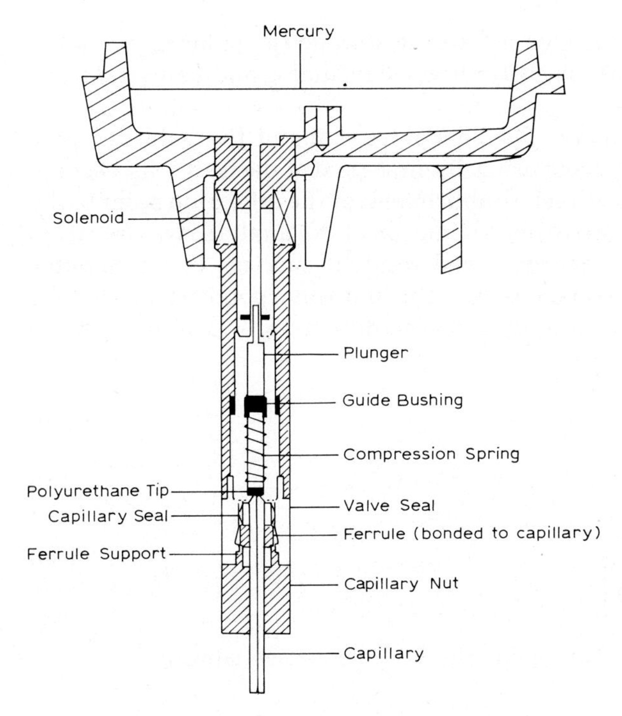

Fig. 3.28. Static mercury drop electrode (SMDE) (Courtesy of PARC).

3. *Subtractive DC polarography.* In the current-sampling technique the interference of the charging current is prevented fairly well, but the residual faradaic current of the blank solution still remains. However, the latter can be suitably eliminated by subtractive DC polarography, i.e., polarography of the analyte and blank supporting electrolyte under the same conditions in order to effect subtraction of the blank value. The method was introduced in 1942 by Semerano and Riccoboni[31] under the name "differences polarography", carried out in a dualcell with two identical dmes and two interconnected reference electrodes; the latter may be a mercury pool in each cell and in that form they may alternatively function as auxiliary electrodes in a three-electrode system, if one separate (calomel) reference electrode communicates with both the analyte and blank via a double salt bridge. To arrive at closely identical working of the two dmes one uses not only identical capillaries and identical mercury flow-rates, but also mechanical drop time control, preferably with current sampling prior to the drop fall; further, the cells should be physically the same with respect to their shape, mutual positioning of the electrodes, solution volume, etc. Nevertheless, with the previous precautions subtractive polarography in twin cells remains difficult and cumbersome. Today, however, with computerized polarographs available, it is easy to record from two different runs in the same apparatus the polarograms of the blank and the sample

solution separately, so that through electronic storage the polarogram of the analyte corrected for the blank is neatly obtained in analog and/or digital form.

4. *Derivative DC polarography*. The sigmoidal shape of the polarographic wave makes it attractive for measuring its first (and if desirable its second) derivative curve, as it enables direct establishment of the inflection point of the original wave and a more sensitive determination of the analyte concentration with less interference by the charging and residual current. In a reversible reaction the inflection point coincides with the half-wave potential so that for the example of reductive polarography and acccrding to eqns. 3.35 and 3.37 the Heysovský–Ilkovič equation

$$E = E_{\frac{1}{2}} + \frac{RT}{nF} \ln \left(\frac{i_{c,d} - i}{i} \right) \tag{3.49}$$

is valid for ox only; hence

$$i = \frac{i_{c,d}}{\left[1 + \exp \frac{nF}{RT} (E - E_{\frac{1}{2}}) \right]} \tag{3.50}$$

Next, differentration of eqn. 3.50 yields the first derivative equation

$$I = \frac{di}{dE} = -i_d \left[\exp \frac{nF}{RT} (E - E_{\frac{1}{2}}) \right] \left[1 + \exp \frac{nF}{RT} (E - E_{\frac{1}{2}}) \right]^{-2} \cdot \frac{nF}{RT} \tag{3.51}$$

from which the peak current $I_p = (di/dE)_{max.}$ is found by the condition

$$\frac{d^2 i}{dE^2} = i_d \left(\frac{nF}{RT} \right)^2 \frac{\left[\exp \frac{nF}{RT} (E - E_{\frac{1}{2}}) \right] \left[-1 + \exp \frac{nF}{RT} (E - E_{\frac{1}{2}}) \right]}{\left[1 + \exp \frac{nF}{RT} (E - E_{\frac{1}{2}}) \right]^3} = 0$$

and so for $E = E_{\frac{1}{2}}$; hence according to eqn. 3.51

$$I_p = -\frac{i_d}{4} \cdot \frac{nF}{RT} \tag{3.52}$$

So, corresponding to i_d for a reversible electrode reaction, I_p is a linear function of concentration; the greater sensitivity of the latter permits determinations down to 10^{-7} *M* (instead of 10^{-6} *M* for i_d). Apart from this advantage, the second derivative curve, by means of the difference between its maximum and minimum as a function of concentration, offers an even better check on reaction reversibility[32] than the straight-line plot of E (according to eqn. 3.49) against $\log(i_{c,d} - i)/i$ (see also p. 120), especially because I_p, as a property at the half-wave potential, is more sensitive to the occurrence of irreversibility (cf., pp. 124–127).

In practice, derivative polarography must be carried out on a smooth curve such as is obtained in current-sampled, current-averaged or rapid DC

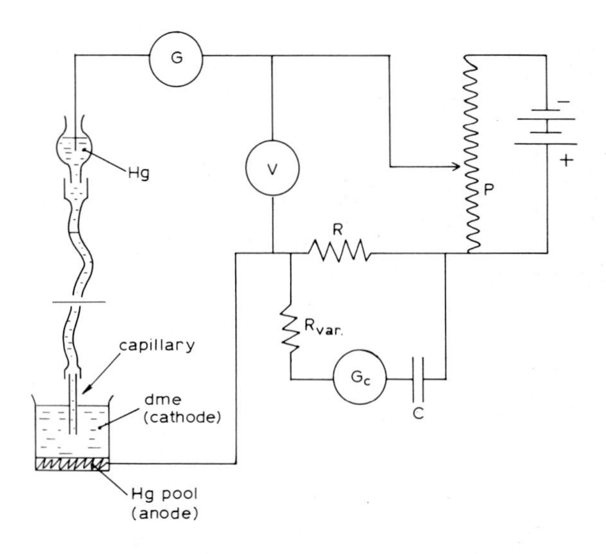

Fig. 3.29. RC circuit for derivative DC polarography.

polarography; it can be done in an indirect way, i.e., graphical or computer-aided, or in a direct way, i.e., electrical; in both instances one may talk of true derivative DC polarography. For the electrical method use is made of instantly registering the charging current of a parallel RC circuit as illustrated in Fig. 3.29 (cf. also Fig. 3.12); any change of i through the dme causes a voltage change over the measuring resistance R, which consequently yields a charging current in the RC circuit, registered by the galvanometer G_c. The method suffers either from noise at a low time constant or from asymmetric distortion at a high time constant of the RC circuit.

Another approach is the so-called pseudo-derivative DC polarography, where by comparative measurement at two separate Hg drops, differing in voltage by a constant ΔE, values of $\Delta I/\Delta E$ are obtained; in a way the technique resembles that of subtractive polarography (cf., pp. 142–143), but instead of comparing the i values of the analyte and blank at the same E, one now compares for separate drops in the analyte solution itself the i value at E with the i value at $(E + \Delta E)$. Originally the method was proposed by Heyrovský (ΔE being 10 mV) for two identical synchronized dmes[33]; today computers facilitate the pseudo-derivative technique by periodically increasing E by a constant amount ΔE (e.g., 0.5 mV) after each drop in the same analyte solution and measuring the current differences between consecutive drops[25]. A similar method is the pseudo-derivative pulse polarography (see later).

5. *Linear sweep voltammetry at the dme.* In linear sweep voltammetry (LSV) at the dme a continuously changing rapid voltage sweep (single or multiple) of the entire potential range to be covered is applied in one Hg drop. Originally the rapidity of the sweep (about 100 mV s^{-1}) required the use of an oscilloscope,

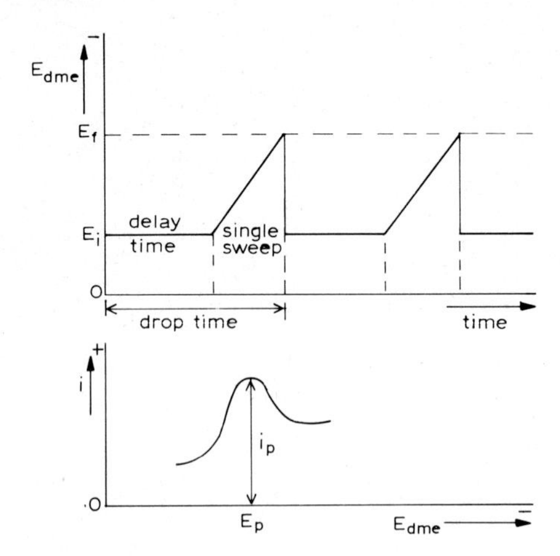

Fig. 3.30. LSV at the dme (single sweep).

so that the method was called oscillographic or cathode-ray DC polarography; however, as less fast sweeps (e.g., 20 mV s^{-1}) can be detected by fast-response X–Y recorders or digital display, the term LSV has been introduced, especially because it is also applied to other types of electrodes (see later); further, the former term fast-sweep polarography is not appropriate in view of the technique of rapid DC polarography.

In LSV usually a single-sweep procedure, the so-called impulse method, is applied with the result illustrated in Fig. 3.30. In the multi-sweep procedure, formerly called Kipp method, Fig. 3.31 is obtained, which shows the saw-tooth character of the sweep and a series of peak curves of increasing height caused by the growing drop surface. Exceptionally, use is made of a triangular sweep in the impulse method; this variant of cyclovoltammetry is depicted in Fig. 3.32

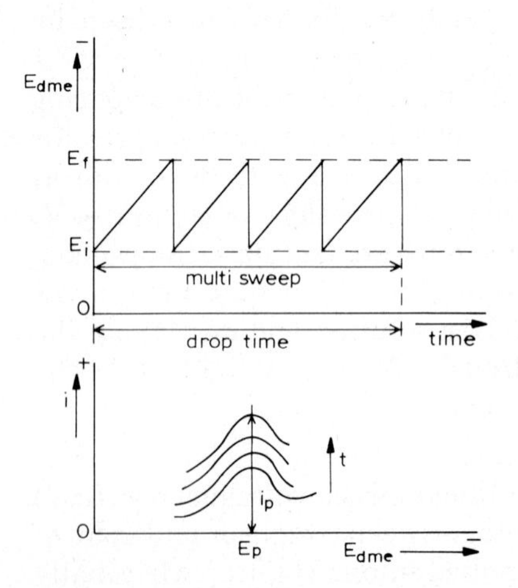

Fig. 3.31. LSV at the dme (multi-sweep).

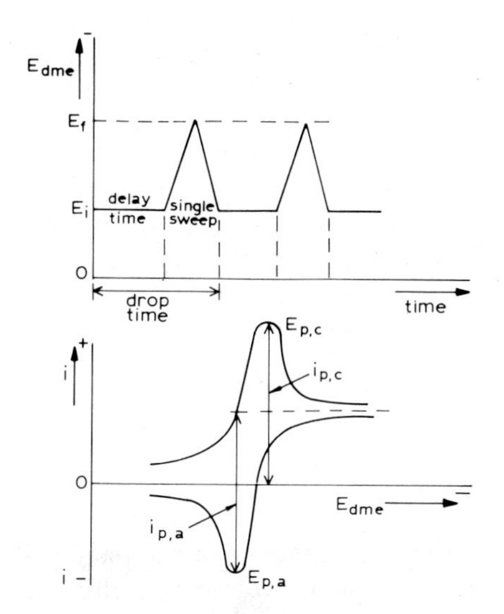

Fig. 3.32. LSV at the dme (triangular sweep).

and again for reductive polarography; the positive peak $i_{p,c}$ concerns analyte reduction to an amount of reductant which in turn by oxidation yields the negative peak $i_{p,a}$.

In fact, both the cyclovoltammetric mode and multi-sweep LSV are mainly of interest in the elucidation of reaction mechanisms, where irreversibility is indicated by a shift and drawing out of the ascending part of the wave (cf., pp. 124–127 and Fig. 3.21) when the scan rate is increased. However, single-sweep LSV remains of importance in chemical analysis, especially from a quantitative point of view[34]; the method is extremely rapid and fairly sensitive (down to 10^{-7} M), provided that precautions are taken to minimize interference by the charging current. In this connection, and as an electronic condenser charge is generally represented by $Q = CV$, we find for a dme of increasing charge E a charging current

$$i_C = \frac{dQ}{dt} = \left(\frac{\delta Q}{\delta E}\right)_t \cdot \frac{dE}{dt} + \left(\frac{\delta Q}{\delta t}\right)_E \tag{3.53}$$

In normal DC polarography where E is assumed to be virtually constant for each Hg drop

$$i_{C_{(DC)}} = \left(\frac{\delta Q}{\delta t}\right)_E \approx kEt^{-1/3} = \frac{E\varepsilon_0}{\pi\delta} \cdot \tfrac{2}{3} \cdot 0.8515\, m^{2/3} t^{-1/3}$$

(cf., eqn. 3.40). In LSV both terms have an influence, so that the faradaic-to-charging current ratio can be unfavourable, but this can be counteracted to a considerable extent by applying several of the above methods, viz., current sampling by means of the sweep, only in the later stages of the drop lifetime, limits the influence of the second term; subtractive LSV compensates not only

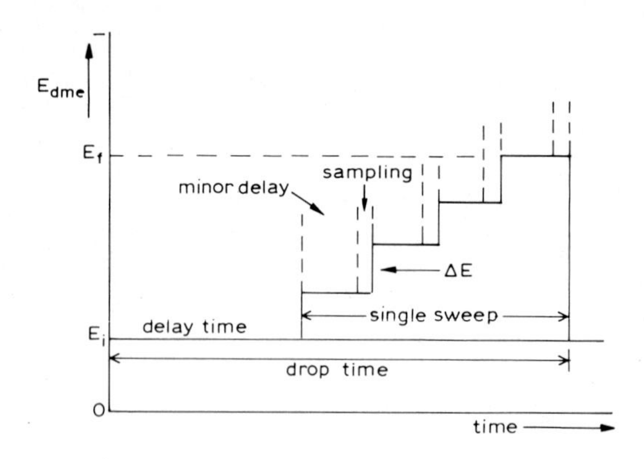

Fig. 3.33. LSV at the dme (stair-case single sweep).

the effect of both terms to a larger extent but also the influence of the residual current; derivative LSV has a similar although less favourable result; finally, stair-case voltammetry[27] and current sampling, both within the sweep (see Fig. 3.33), eliminate the influence of the first term on the charging current.

It is clear, that the various modes of LSV at the dme require an integrated and coherent regulation of sweep time, current sampling and drop knocking, preferably by an electronic device and on with computer guidance. A disadvantage of LSV at the dme, in contrast to normal DC polarography, is that for mixtures of components the latter yields a simple evaluation by curve extrapolation on the basis of additivity [see Fig. 3.34(a)], whereas the former suffers from an uncertain evaluation [see Fig. 3.34(b)].

A partial solution to this problem can be obtained by so-called interrupted LSV, which means that on passing the first peak [see Fig. 3.35(a)] the voltage ramp is held sufficiently long at the value concerned before being continued, and so on with the same procedure for subsequent peaks; the peak splitting in the interrupted technique is impressive [see Fig. 3.35(b)] in contrast to the normal technique.

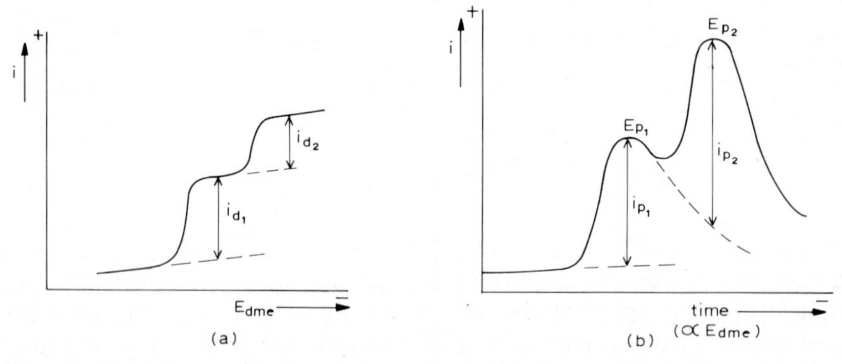

Fig. 3.34. DC polarography of component mixture, (a) normal polarography, (b) LSV at the dme.

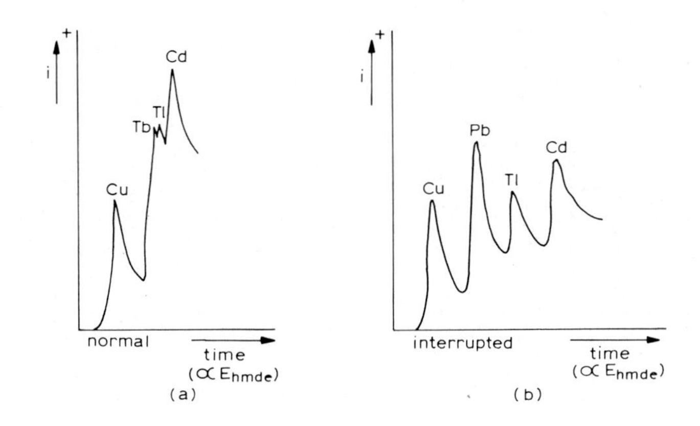

Fig. 3.35. Component mixture at the dme, (a) normal LSV, (b) interrupted LSV.

Finally, a remark should be made on the effect of the scan rate; an increase in the scan rate, e.g., from 50 through 100 to 200 mV s^{-1}, causes a sharper and apprecially higher peak, as expected. If the electrode reaction is reversible, the half-wave potential, $E_{p/2}$, remains nearly unaltered, otherwise there is a shift to the right (more negative in reductive LSV). It should be borne in mind that in a follow-up reaction such as the system EC (see p. 124) an increase in scan rate may cause a transition from irreversibility to apparent reversibility if the charge-transfer reaction E becomes predominant.

6. *Pulse polarography.* Recently there has been increasing interest in the technique of pulse polarography, which is characterized by the application of a single pulse per mercury drop on the background of a constant or a linearly growing potential. As the pulse has a constant value, the technique differs from the impulse method (single-sweep LSV) and from the AC square-wave method (according to Barker[51]). In pulse polarography one distinguishes between normal pulse (NPP), differential pulse (DPP) and pseudo-derivative pulse polarography.

(a) Normal pulse polarography is based on current sampling (5–20 ms) for a sequence of pulses of increasing potential and of constant time duration (chosen within the range 0.5–5 s) superimposed on a constant potential (E_{basic}), as illustrated in Fig. 3.36*. According to Fig. 3.36(b) there is no faradaic reaction at E_{basic} except for the ordinary dc residual current; however, as soon as a pulse with a height beyond the decomposition potential of the analyte is applied a charging current, rapidly decreasing, will occur together with an appreciable faradaic current from the analyte.

Fig. 3.37 illustrates that i_F starts much higher than in Tast DC polarography, because at the onset of the pulse we still have the analyte bulk concentration at the dme surface and rather near to the end of the drop life. Therefore, even if current sampling took place on an averaged basis over the whole pulse time,

*Experimental data obtained with a PARC Model 174A polarograph.

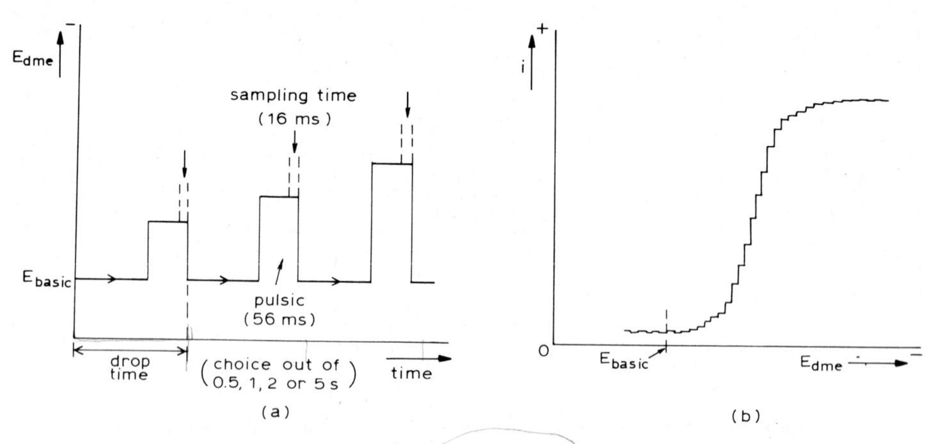

Fig. 3.36. Normal pulse polarography, (a) sampling scheme, (b) current sampled.

which is sometimes done in so-called integral pulse polarography, then discriminative determination of i_F versus i_C can still be realized with high sensitivity.

In most instances, however, current-sampling is performed before the end of the pulse, where i_C is fairly well eliminated; the measurement may be instantaneous or averaged over that sampling time. Nevertheless, some effect of the dc residual current may remain and can be compensated for to a great extent by subtraction of the current before the pulse rise.

The considerable gain in sensitivity with NPP compared with sampled DC polarography can be shown still more clearly on a calculative basis. Thus for a reversible process A + $ne \rightleftharpoons$ B and for a solution with ox only, we obtain from eqn. 3.50 by substituting $P = \exp(nF/RT)(E - E_{\frac{1}{2}})$ the equation

$$i = \frac{i_{c,d}}{1 + P} \tag{3.54}$$

As the current sampling takes place at the end of drop life, the dme then shows such a large surface area that we may consider it as a planar microelectrode;

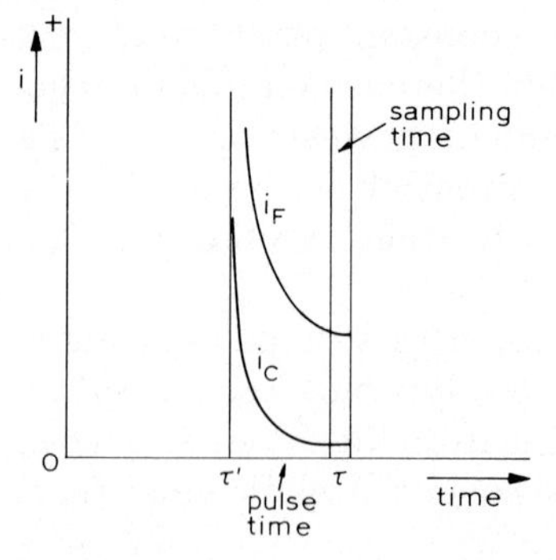

Fig. 3.37. NPP, i_F and i_C vs. pulse-time course.

so the special term in eqn. 3.25 has virtually disappeared, so that the latter becomes

$$\frac{\delta c}{\delta t} = D \cdot \frac{\delta^2 c}{\delta x^2} \tag{3.55}$$

the solution[36] of which leads to the well known Cottrell equation:

$$i_{d(t)} = \frac{nFCAD^{1/2}}{\pi^{1/2} t^{1/2}} \tag{3.56}$$

Hence, for NPP eqn. 3.54 can be written as

$$i_{(pulse)} = \frac{nFCAD^{1/2}}{\pi^{1/2}(\tau - \tau^1)^{1/2}} \cdot \frac{1}{1 + P} \tag{3.57}$$

(cf., Fig. 3.37). If $(E + \Delta E)_{pulse}$ becomes more negative than $E_{\frac{1}{2}}$, P approaches zero, so that the limiting current $i_{d(pulse)}$ agrees with its Cottrell equation

$$i_{d(pulse)} = \frac{nFCAD^{1/2}}{\pi^{1/2}(\tau - \tau^1)^{1/2}} \tag{3.58}$$

According to eqn. 3.26(a) (p. 116), the limiting current in sampled DC polarography can be expressed by

$$i_{d(DC\ Tast)} = \frac{nFCAD^{1/2}}{\pi^{1/2} t_d^{1/2}} \sqrt{7/3} \tag{3.59}$$

Hence

$$\frac{i_{d(pulse)}}{i_{d(DC\ Tast)}} = \left(\frac{t_d}{\tau - \tau^1}\right)^{1/2} \sqrt{3/7} \tag{3.60}$$

If we realize that the drop life (t_d) is of the order of 4 s with a 50 s pulse time, it shows that NPP yields a signal about six times higher than that of DC Tast polarography.

(b) Differential pulse polarography (DPP) consists in a current sampling just before the drop fall versus a sampling prior to the pulse, which here has a constant value (ΔE = 10–100 mV) and is imposed on a linearly growing potential (see Fig. 3.38)*; the latter changes either from drop to drop in small increments (a) or by way of a potential ramp (b), as is usually done.

The DPP measurement per drop (Δi) yields a picture similar to that of NPP (i) (see Fig. 3.37), but there are important differences in the overall result, viz.:

(1) the Δi versus E_{dme} graph shows a differential peak (see Fig. 3.39);

(2) the peak maximum occurs near $E_{\frac{1}{2}}$ if the pulse height is sufficiently small;

(3) the residual current interference (dc effect) is much less, as subtraction always takes place for an E_{basic} (prior to the pulse) that differs only slightly from

*The data in Fig. 3.38 were obtained with a Metrohm E 506 polarograph (a) and a PARC Model 174 A polarograph (b).

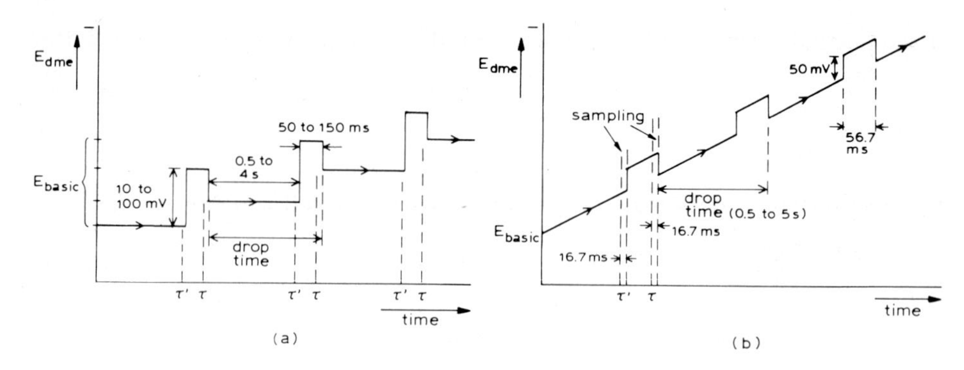

Fig. 3.38. Differential pulse polarography with E_{basic} growing (a) incrementally or (b) linearly.

$E + \Delta E$ at the instant just before the drop fall; this has such an overwhelming effect that, although E_{basic} acquires values with a considerable pre-electrolysis component, Δi remains sufficiently large to reach high analytical sensitivity; limits of detection of 10^{-8} M[35] by DPP compared with 10^{-6}–10^{-7} M by NPP have been obtained. Further, where by DPP in fact $\Delta i/\Delta E$ per drop is determined, $\Delta^2 i/\Delta E^2$ can be established for consecutive drops.

The Metrohm method may have the advantage that the residual current (i_F effect) remains constant during the sampling steps, but the drop still grows (i_C effect); both are miniscule effects. The PARC method can be combined with the PARC SMDE technique (see p. 136), which excludes i_C alterations due to drop growth during sampling.

Sometimes besides direct pulses, reverse pulses are also applied (Tacussel PRG 5 polarograph), because by tracing out the polarogram using first positive-, then negative-going pulses, one may study the reversibility of reactions.

Fig. 3.40[37] illustrates well the character of the curves and the gain in sensitivity on going from conventional DC via sampled DC and normal pulse to differential pulse polarography. It should be realized that, although the DPP curve ($\Delta i/\Delta E$) might seem an approximation of a derivative curve, we cannot speak of derivative polarography.

(c) Pseudo-derivative pulse polarography is based on normal pulse polarography. As has been said with respect to derivative DC polarography

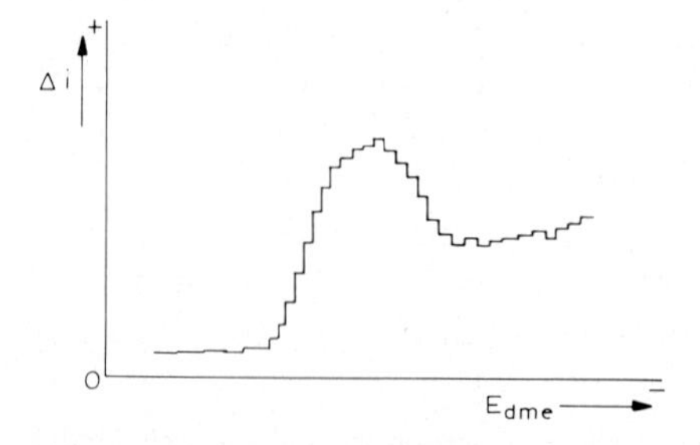

Fig. 3.39. Δi vs. E_{dme} in DPP.

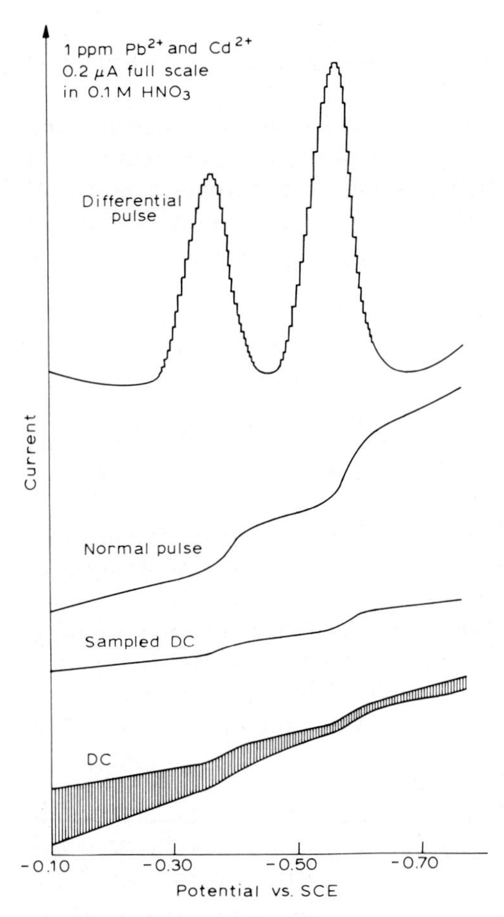

Fig. 3.40. i vs. E_{dme} compared between DC, sampled DC, normal pulse and differential pulse polarography (Courtesy of PARC).

(pp. 138–139), derivative polarography must be carried out on a smooth curve; so in fact a true derivative pulse polarogram can be obtained in principle when it is derived from an NPP curve. Accordingly, eqns. 3.51 for $\mathrm{d}i/\mathrm{d}E$ and 3.52 for $\mathrm{d}^2 i/\mathrm{d}E^2$ are valid, in addition to what has been mentioned for $I_p = (\mathrm{d}i/\mathrm{d}E)_{max.}$ (eqn. 3.52) as a linear function of concentration.

However, the pseudo-derivative approach, already described for DC polarography and based on a comparative measurement at two separate Hg drops differing by a constant ΔE, is most indicated in pulse polarography, i.e., the current of a preceeding pulse is subtracted from that of the next pulse and the resulting difference is plotted against the potential. The method allows very short drop times and is superior to the DPP method[25] as a consequence of the elimination of interfering adsorption phenomena.

3.3.1.2.2. *AC polarography*

The techniques in this field can be divided into two groups, viz., (I) controlled potential and (II) controlled current techniques (cf., Table 3.1).

In the methods of type I an alternating potential signal of a certain well chosen character and with a frequency higher than the Hg drop frequency is applied on a background of a DC potential ramp or sometimes a potential growing from drop to drop in small increments (or even a constant potential in the case of a pulse height growing by increments) and the resulting current is evaluated.

In the methods of type II an alternating current signal of an appreciable frequency is forced through an electric circuit containing a high impedance in series with the polarographic cell and the resulting cell voltage is measured.

Apart from the many variants of applied techniques, the theory is complex, so that we must confine ourselves to the more practical side of those methods which are of direct importance in chemical analysis.

1. *Polarography with superimposed AC signal.* As examples of chemical analytical importance we shall consider (a) sinusoidal AC, (b) AC bridge, (c) square-wave and (d) Kalousek polarography.

(a) Sinusoidal AC polarography is based on the evaluation of an alternating current from a polarographic cell wherein a sine wave potential is superimposed on a DC potential ramp. We shall mainly consider the occurrence of the reversible electrodic reaction: ox + $ne \rightleftharpoons$ red. Then the total potential E applied to the cell can best be described by

$$E = E_{dc} - \Delta E \sin \omega t \tag{3.61}$$

where E_{dc} is the potential ramp as a function of time t and ΔE is half the peak-to-peak value of the superimposed potential with angular frequency ω rad s^{-1}. As both the slope of the DC potential ramp and the alteration of the dme surface are slight in comparison with the AC frequency, E_{dc} and the differential capacity of the double layer, C_{dl}, can be considered to be nearly constant within the time of a sine wave, so that the AC charging current is at first sight represented by

$$\frac{dQ}{dt} = i_C^{\sim} = AC_{dl}\Delta E\omega \cos \omega t \tag{3.62}$$

Further, if within the electrical circuit the ohmic resistance R can be neglected, the $i_C^{\sim}$ wave leads to the potential by 90°, as is known, which means that $i_C^{\sim}$ shows a positive $\pi/2$ phase angle shift (φ) versus E_{ac}; any perceptible R, however, has φ between $\pi/2$ and zero. Our main objective in AC polarography, however, is the faradaic current, so a separating condenser is placed between the amplifier and normal resistor in order to filter out the d.c. current and to evaluate the ac current component. As we want to understand the relationship between $i_{dc}(i^{=})$ and $i_{ac}(i^{\sim})$ as a function of E_{dc} and E_{ac} applied, we may consider Fig. 3.41(a) and (b).

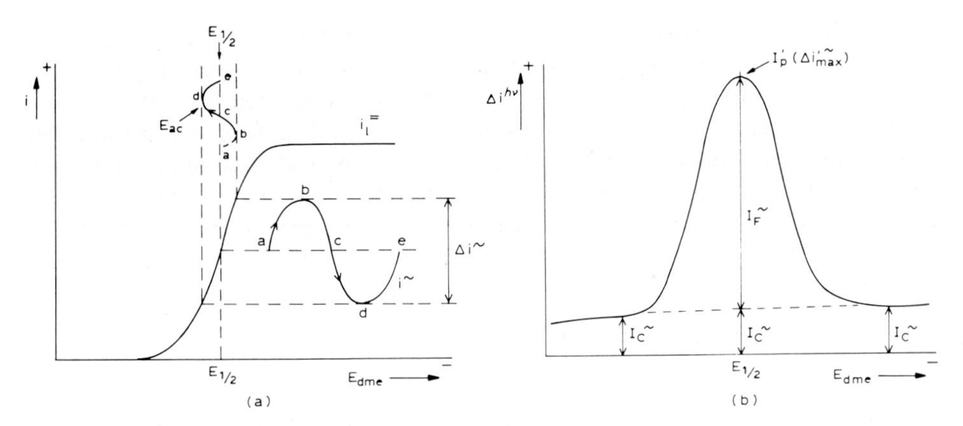

Fig. 3.41. Sinusoidal AC polarography. (a) measuring principle, (b) fundamental harmonic ac polarogram ($i_C^{\sim}$ included).

Fig. 3.41(b) shows the total so-called fundamental harmonic ac polarogram (i.e., including charging current), and Fig. 3.41(a) shows that $\Delta i'^{\sim}_{max.}$ is obtained if E_{ac} is applied at $E_{dc} = E_{\frac{1}{2}}$; $\Delta i'^{\sim}_{max.}$ will be approximately sinusoidal, but $\Delta i'^{\sim}$ below $E_{\frac{1}{2}}$ (i.e., less negative) is impressed at the low side and above $E_{\frac{1}{2}}$ at the high side.

Fig. 3.42 represents the symmetric bell shape curve of I, i.e., the genuine fundamental harmonic ac polarogram, which means the curve of only $I_F^{\sim}$ discriminated for $I_C^{\sim}$, e.g., by means of phase-selective ac polarography. The term "fundamental" is related to the character of the polarographic cell as a non-linearized network whose response is not purely sinusoidal but consists of the sum of a series of sinusoidal signals at ω, 2ω, 3ω, etc.; in this signal collection "fundamental" indicates the first harmonic (ω) response, besides that of the second harmonic (2ω), the third harmonic (3ω), etc.

By the use of tuned or lock-in amplifiers these various harmonics can be detected; the results remain advantageously confined to the faradaic current

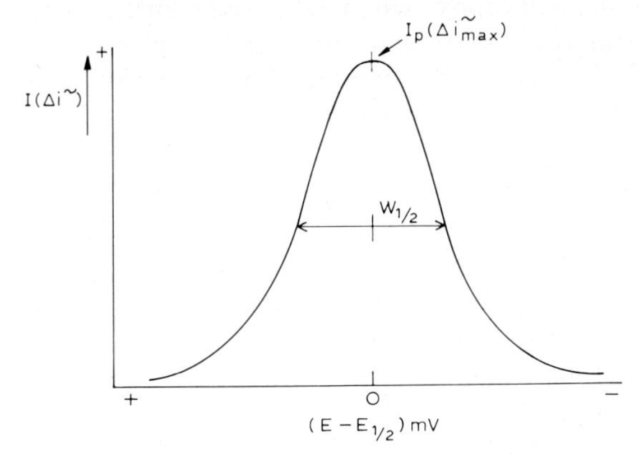

Fig. 3.42. Genuine fundamental ac polarogram (only $i_F^{\sim}$).

only, because the charging current behaves as a much more linear network; however, as most commercial polarographs do not allow this tuning selection, for the higher harmonics one should consult the specialized literature[38].

Returning to the fundamental ac harmonic in Fig. 3.42, we wish to establish the relationship between I and the faradaic impedance Z_f; instead of considering a combination of a series resistance R_s and a pseudo-capacity C_s, the alternative is to separate a pure resistance of charge transfer R_{ct} and a kind of resistance to mass transfer Z_w, the Warburg impedance; the derivation of the polarogram[39] then (for $\Delta E_{ac} \leqslant 8/n\,\text{mV}$) leads to the equation

$$I = \frac{\Delta E_{ac}}{Z_f} = \frac{n^2 F^2 C A \omega^{1/2} D^{1/2} \Delta E}{4RT \cosh^2(a/2)} \sin\left(\omega t + \frac{\pi}{4}\right) \tag{3.63}$$

where $a = (nF/RT)(E_{dc} - E_{\frac{1}{2}})$, so that the peak current comes at $E_{dc} = E_{\frac{1}{2}}$ or $\cosh(0) = 1$, and hence

$$I_p = \frac{n^2 F^2 C A \omega^{1/2} D^{1/2} \Delta E}{4RT} \sin\left(\omega t + \frac{\pi}{4}\right) \tag{3.64}$$

By substituting I_p in eqn. 3.63 one obtains $I = I_p/\cosh^2(a/2)$ and next by solving $a/2$ the real solution

$$E_{dc} = E_{\frac{1}{2}} + \frac{2RT}{nF} \ln\left[\left(\frac{I_p}{I}\right)^{1/2} + \left(\frac{I_p - I}{I}\right)^{1/2}\right] \tag{3.65}$$

to which the shape of the ac polarogram adheres.

In fact, Fig. 3.41 illustrates that, if ΔE_{ac} is sufficiently small, the ac polarogram represents the first derivative of the dc polarogram; this is especially evident because the same eqn. 3.65 is obtained when we substitute eqn. 3.52 (value of I_p) in eqn. 3.51 (value of I), both equations obtained by direct differentation of the d.c. polarographic curve. However, it should be kept in mind that for the above considerations a reversible electrodic process always was assumed; what in conventional DC polarography still shows up as reversible may become irreversible in AC polarography, when ω is sufficiently high. However, eqn. 3.65 provides checks on the reversibility, e.g., the plot of E_{dc} against $\log\{(I_p/I)^{1/2} - [(I_p - I)/I]^{1/2}\}$ must yield a straight line[40], whereas the ac wave should be symmetrical with a calculated half-width of

$$W_{\frac{1}{2}} = \frac{2RT}{nF} \ln \frac{\sqrt{2}+1}{\sqrt{2}-1} = \frac{4RT}{nF} \ln(\sqrt{2}+1) = 1.531\left(2.3026 \cdot \frac{RT}{nF}\right)$$

[so at 25°C $W_{\frac{1}{2}}$ will be $(59.16/n) \cdot 1.531 \approx 90/n\,\text{mV}$].

It has been shown that the charging current $i_C^{\sim}$ leads to the potential E_{ac} by 90° (cf., eqns. 3.62 and 3.61) and the faraday current $i_F^{\sim}$ by 45° (cf., eqn. 3.63); these phenomena can be used to discriminate $i_F^{\sim}$ from $i_C^{\sim}$ and vice versa by means of phase-sensitive AC polarography.

The principle of its measurement is illustrated in Fig. 3.43; I_F can be selectively determined at 0 and 180° and I_C at 135 and 315°, the signals obtained being

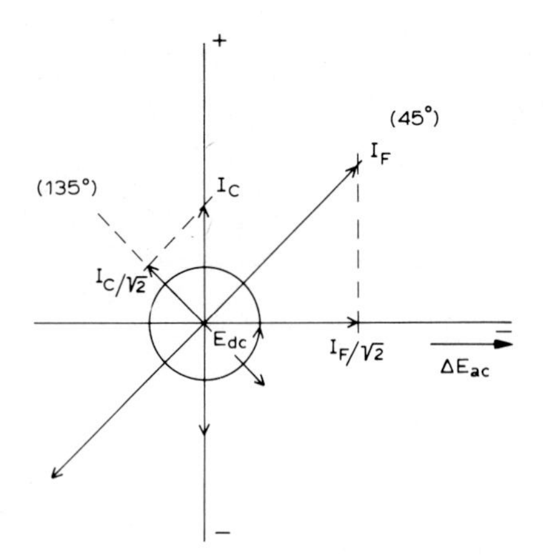

Fig. 3.43. Discrimination of $i_F^{\sim}$ from $i_C^{\sim}$ via phase-sensitive AC polarography.

$I_F/\sqrt{2}$ and $I_C/\sqrt{2}$, respectively. As already mentioned for I_C on p. 148, the phase-angle relationships are subject to interference by ohmic IR drop effects in the measuring circuit; this can be substantially overcome by a three-electrode system (cf., Section 3.3.1.1.3, p. 129) and whether this system is applied or not there may still be substantial improvements with the use of shorter drop times, e.g., improved linearity of calibration graphs[41] (see also Section 3.3.1.2.1, rapid dc polarography). Without use of the phase-sensitive mode one must avoid a low I_F/I_C ratio; for instance, high-frequency ac is now unfavourable, because i_C increases linearly with increasing ω (cf., eqn. 3.62), but i_F with $\omega^{1/2}$ (cf., eqn. 3.64); therefore, moderate frequencies of 20–100 Hz are usually applied with the additional analytical advantage that the electrode process may remain at least quasi-reversible. I_F and I_C show the same dependence on the ac amplitude; hence E can be high and its value should be well above the instrument noise, e.g. 5–25 mV.

Current-sampled AC polarography (at the end of the drop life) is an effective technique again for achieving almost complete elimination of i_C interference; in combination with it, subtractive AC polarography can also eliminate the residual current (cf., the corresponding DC techniques in Section 3.3.1.2.1). In Table 3.1 it was mentioned that the principles of the stationary methods A can be and often are used in the non-stationary methods B; this has been illustrated already for the principles of 1–4, but can also be shown for the principles of 5 and 6 in Table 3.1 in AC polarography. For instance, when we impose a pulsed ac potential on the dc voltage ramp [see Fig. 3.44 in comparison with Fig. 3.38(b)] we obtain differential pulse AC polarography. The term "differential" clearly indicates that the polarogram obtained with a maximum and minimum peak now represents the derivative of the fundamental harmonic ac polarogram (see Fig. 3.42); very flat baselines are found with a substantial decrease in charging current, especially in the phase-selective mode, even at

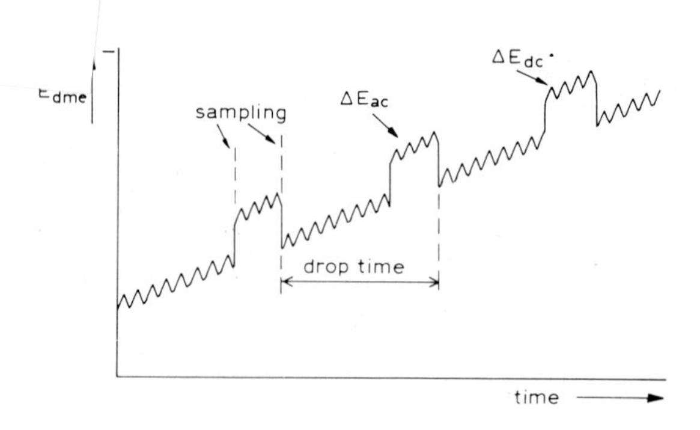

Fig. 3.44. Differential pulse AC polarography.

high frequency (1000 Hz), where in conventional ac polarography i_C would be extremely high.

Bond and O'Halloran[42] claim for the combination of phase-selective detection and differential pulse ac polarography a "virtually complete rejection of i_C, even at high frequencies and low concentrations".

In all previously described ac polarographic techniques the aim was to measure faraday currents without interference of charging currents; however, tensammetry represents an AC polarographic method of opposite character, i.e., with the aim of measuring charging currents without interference from faraday currents; it has appeared to be of increasing interest for the analysis of surface (tension)-active agents, which explains the name tensammetry. Its principle is illustrated in Fig. 3.45 by the differential capacitance–potential curve obtained by Breyer and Bauer[43].

The dotted line represents the base current of the supporting electrolyte only; adsorption at the dme of any other component (e.g., a detergent) than the solvent or ions of that electrolyte causes a lowering of the capacitance; specific adsorption occurs within the lower capacity region delimited by the anodic (E_{T+}) and the cathodic (E_{T-}) adsorption/desorption peaks. A more specific

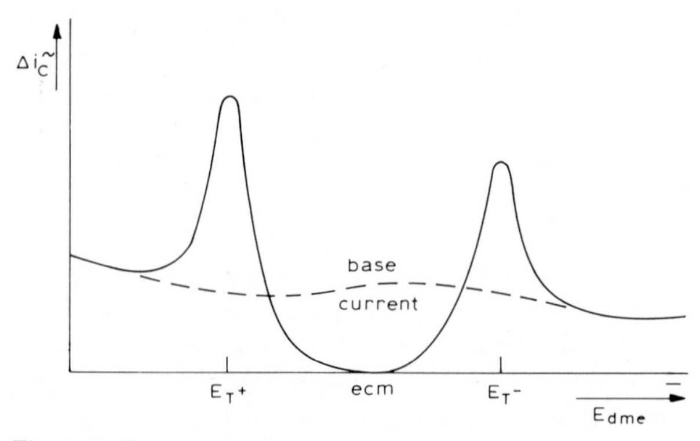

Fig. 3.45. Tensammetry.

elucidation of the character of these peaks was provided by Booth and Fleet[44] by means of cyclic voltammetry at the dme of methylcarbamates (agricultural insecticides); a cathodic scan for Butocarb (3,5-di-*tert.*-butyl N-methylcarbamate) delivered a cathodic peak by desorption of the surfactant, but an anodic scan gave an anodic peak by adsorption replacing the cations of the supporting electrolyte. However, for Aldecarb (2-methyl-2-methylthiopropionaldehyde O-methylcarbamoyl oxime) the cathodic peak corresponded to adsorption and the anodic peak to desorption.

Within the low-capacity region between the adsorption/desorption peaks around the electrocapillary maximum (ecm; see Fig. 3.18), the depression of the base current is greatest because of maximum adsorption of the surfactant in that area.

On the basis of the above, illustrated in Fig. 3.45, we may conclude that tensammetric* analysis can be effected by measuring either (a) the height of the peaks or (b) the base current depression; (a) is normally more sensitive and selective, but sometimes peaks do not show up clearly, so that (b) becomes necessary. However, neither of these methods yields a linear relationship with concentration and, as a consequence, tensammetric titration[45] is preferred, owing also to the high sensitivity of the tensammetric end-point indication. As to the measurement itself, which in fact yields the fundamental AC harmonic polarogram of the charging current, there are a few possibilities, e.g., in the absence of a faraday current of significance a direct measurement, but as a better method a phase-angle selective measurement (see Fig. 3.43) or resonant frequency measurement (cf., oscillometry, p. 20) as an alternative[47], which at any significant faraday current becomes imperative.

Some other less important types of AC polarography may also be considered as sinusoidal ac techniques, as their theoretical treatment can be based on signals from a complex Fourier function; in this context we confine ourselves to mentioning sawtooth or triangular wave[48] superimposed on the dc ramp. Square-wave polarography is also of that type, but in view of its greater importance we shall treat it separately.

(b) Finally, the AC bridge polarography described by Takahashi and Niki[49] merits some attention; they state that "the ac voltage superimposed on the dc voltage at the dme surface may change the amplitude and phase of the ac current in accordance with the gradually changing dc voltage (more ac current flow, resulting in a depression of the ac voltage)". They prevented this depression by means of an ac bridge with internal self-regulation. If their conclusion is substantially correct, even the application of a three-electrode system would not improve the situation significantly. However, we wonder whether the sophisticated potentiostats in today's commercial research polarographs do not evade this trouble as well (cf., the Tacussel three-electrode DC and superimposed AC polarography system of Type PRG 3).

*Tensammetry, although originally a sinusoidal ac polarography, is not limited to that technique, but has been applied also to the Kalousek-type wave form[45] and to square waves[46].

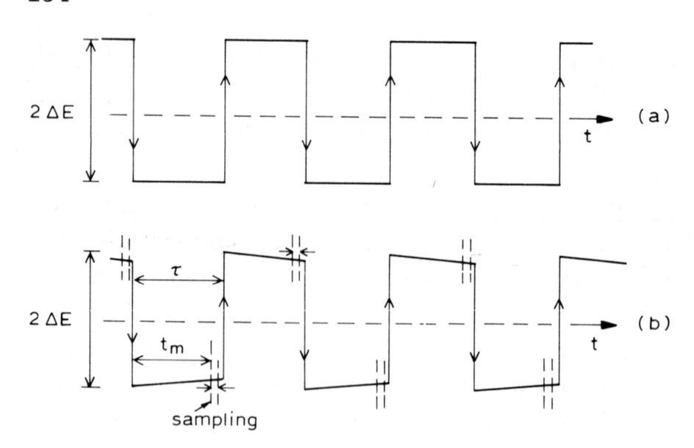

Fig. 3.46. Alternating square-wave potential, (a) rigid wave, (b) tilted wave.

(c) Square-wave polarography and high-frequency polarography are techniques originally introduced by Barker and co-workers. In square-wave polarography an alternating square-wave potential of the type as depicted in Fig. 3.46 (a meander function) is superimposed on a DC potential ramp and the resulting AC current is evaluated; Barker and Jenkins[48] applied instead of the rigid square wave (a) a tilted square wave (b) in order to compensate for a residual capacity current; they used an amplitude ΔE of 10 mV at a frequency of 225 Hz; τ is the time between the steps and t_m the decay time of the charging current before sampling.

Fig. 3.47 is comparable to Fig. 3.41 for sinusoidal ac polarography; if the tilted shape provides a net compensation of the charging current one obtains a symmetric bell-shaped curve of I in the square-wave polarogram, similar to that depicted in Fig. 3.42. In fact, virtually all of the statements made before on the sinusoidal technique are valid for the square-wave mode except for the rigid shape of its wave; this conclusion is according to expectation, especially as Fourier analysis reveals the square wave to be a summation of a series of only

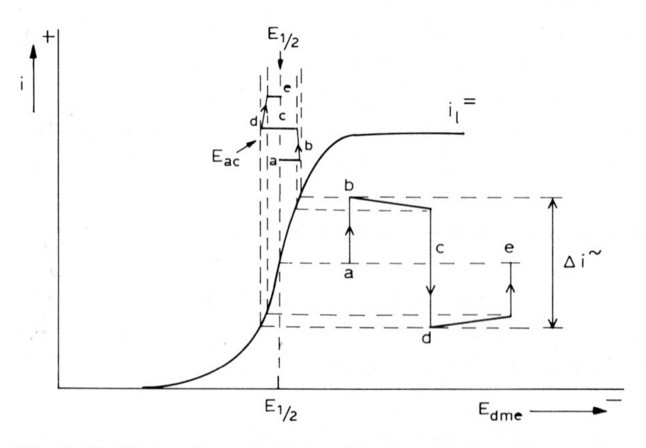

Fig. 3.47. Measuring principle of square-wave polarography.

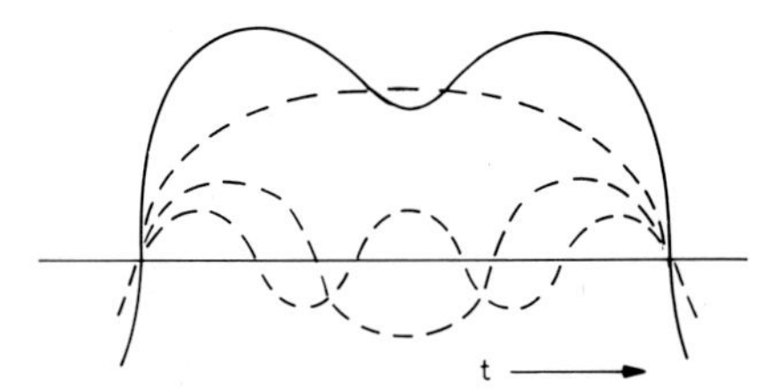

Fig. 3.48. Square-wave as Fourier summation of odd harmonics.

the odd harmonics of the square-wave frequency, which is roughly illustrated in Fig. 3.48 on the basis of the fundamental, the third and the fifth sinusoidal harmonics. Hence there exists a narrow relationship between eqns. 3.63 and 3.64 on the one hand and such equations for $I_{sq.w}$ (i.e., the faraday current only), e.g., eqn. 3.66 given by Barker* for the amplitude limit $\Delta E \ll 2.3026\, RT/nF$ and a fixed square-wave frequency (usually 225 Hz).

$$I_p = \frac{k^1 n^2 F^2 CAD^{1/2} \Delta E}{RT} \tag{3.66}$$

The above considerations concern a reversible electrodic process, ox + $ne \rightleftharpoons$ red; as instead of 20–100 hz in the sinusoidal technique a fixed frequency of 225 Hz is normally used in the square-wave mode, the chance of irreversibility in the latter becomes greater, which then appears as asymmetry of the bell-shaped I curve. Such a phenomenon may occur more especially when the complete i versus E curve is recorded on a single drop, a technique which has appeared useful[51] in cases of sufficient reversibility.

In view of the successful compensation of the charging current by means of the tilted shape of the square-wave, current sampling near the end of the drop life is again advisable; moreover, the ohmic resistance of the supporting electrolyte should if possible be low because, as the dc voltage and the drop size can be considered to be constant during the sampling time**, the charging current decay is a first-order process, which means that via its time constant $\tau = RC_{dl}$ the double layer has become nearly completely discharged within a time delay of 5τ, i.e., the potential U_{c_t} after a time $t = 5\tau$ has decreased according to $U_{c_t} = U_{c_0} e^{-t/\tau}$ down to about 0.67% of its original value U_{c_0}.

Apart from the above requirement of a low ohmic resistance, it is nevertheless recommended to use a three-electrode system in view of the more precise establishment of the dme potential.

In connection with the square-wave technique, mention can be made of high-frequency polarography, also called radiofrequency polarography and developed by Barker[53], in which a sinusoidal radio-frequency ω_1 (100 kHz to 6.4 MHz) square-wave modulated at ω_2 (225 Hz) is superimposed on to the dc potential ramp; as the wave form includes (apart from additional higher

*In fact, this version of the equation of Barker[50] mentions $k'A$ instead of k.

**This is not justified in normal DC polarography (cf., eqn. 3.40).

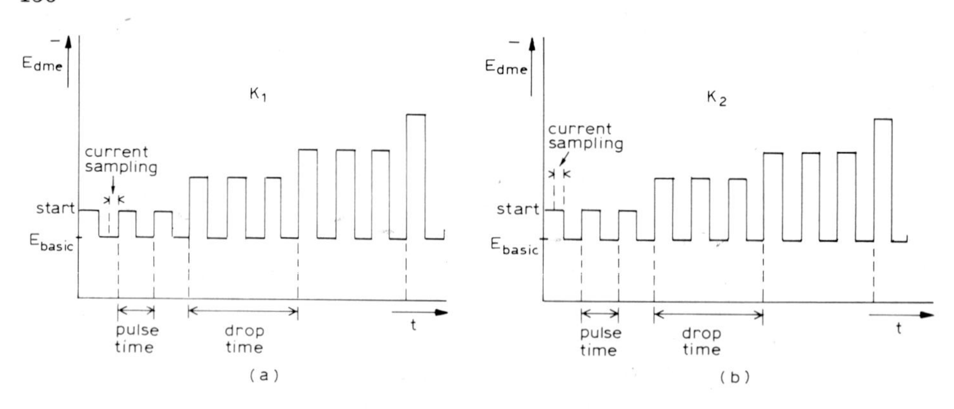

Fig. 3.49. Kalousek polarography on a constant potential background, (a) K_1, (b) K_2.

harmonics) the frequency components ω_1, $\omega_1 - \omega_2$ and $\omega_1 + \omega_2$, the current response measured at 225 Hz by the square-wave technique can be considered as a faradaic intermodulation of ω_1 with two side-band frequencies (cf., intermodulation polarography according to Reinmuth[52]). Barker[53] used 72 kHz modulated at 255 Hz and employed the normal square-wave circuits to measure the high-frequency polarogram, which in its form and character resembles the normal square-wave polarogram.

(d) Kalousek polarography is characterized by the application of several pulses per mercury drop on the background of a constant or incrementally growing potential; as in pulse polarography, where, however, only one pulse per mercury drop is applied (cf., Figs. 3.36 and 3.38), the pulses within one drop-time have a constant potential value. There are four variants of the Kalousek technique, as illustrated in Figs. 3.49 and 3.50 with the indications K_1, K_2, K_3 and K_4 originally introduced by Heyrovský[11].

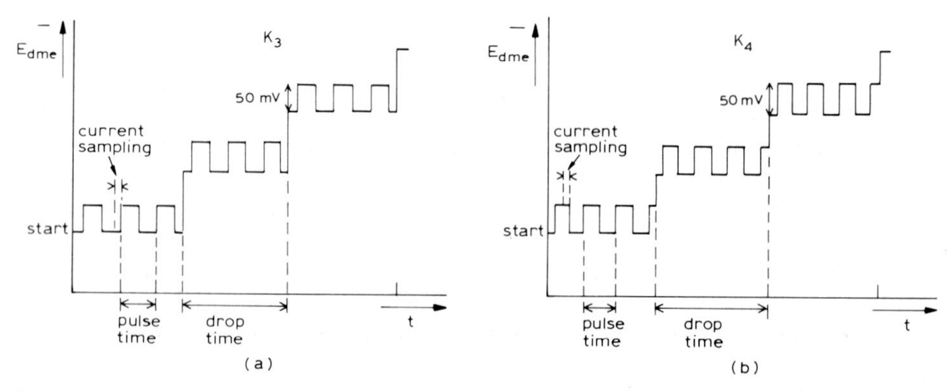

Fig. 3.50. Kalousek polarography on an incrementally growing potential background, (a) K_3, (b) K_4.

In K_1 and K_2 the pulses are superimposed on to a constant potential, but in K_3 and K_4 on to a growing potential; in this connection there is a resemblance (especially for K_2) with NPP [c.f., Fig. 3.36(a)] and (especially for K_4) with DPP [cf., Fig. 3.38(a)].

In order to understand the current evaluation, let us consider again a reversible process, ox + ne $\rightleftharpoons$ red in a dilute solution, so that

$$E = E_{\frac{1}{2}} + \frac{RT}{nF} \ln \left(\frac{[\text{ox}]}{[\text{red}]} \right)$$

For the methods K_1 and K_2 one chooses E_{basic} well below, i.e., less negative than, $E_{\frac{1}{2}}$ ($\approx E^0$). Hence in K_2 one measures at the pulse top versus E_{basic} a cathodic additional current, but in K_1 at the pulse base (= E_{basic}) versus its top an anodic additional current. As a function of the increasing pulse height one obtains a polarogram with an ordinary sigmoidal wave, determined either cathodically (K_2) or anodically (K_1); the wave height depends on the amounts and ratio of [ox] and [red], while $E_{\frac{1}{2}} = E^0$ (cf., Fig. 3.9).

For method K_3 and K_4 where again E_{start} is chosen to be less negative than $E_{\frac{1}{2}}$, we must consider the pulse technique in relation to the polarographic wave in the same ways as in Figs. 3.41(a) and 3.47; however, the measurement result is different as we sample the current either at the pulse top (K_4) versus its base or at the pulse base (K_3) versus its top (see Fig. 3.51).

This method K_4 yields a reduction or cathodic additional current owing to the pulse increase with a current maximum around $E_{\frac{1}{2}}$; analogously, K_3 yields an oxidation or anodic additional current owing to the pulse decrease with a current minimum around $E_{\frac{1}{2}}$. Both additional currents are depicted in Fig. 3.52; for a completely reversible redox system the cathodic maximum and the anodic

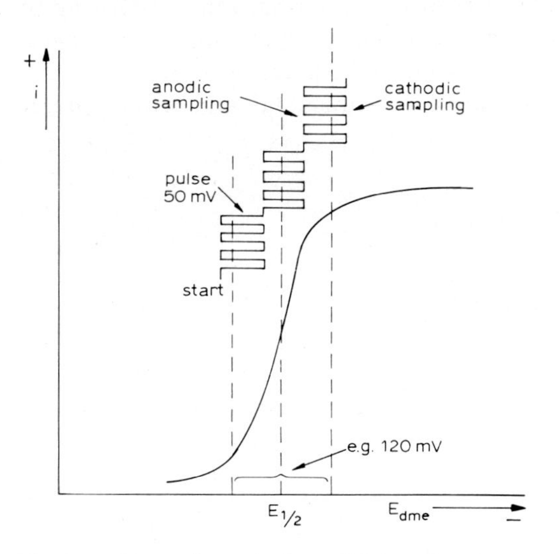

Fig. 3.51. Measuring principle of Kalousek polarography K_4 and K_3.

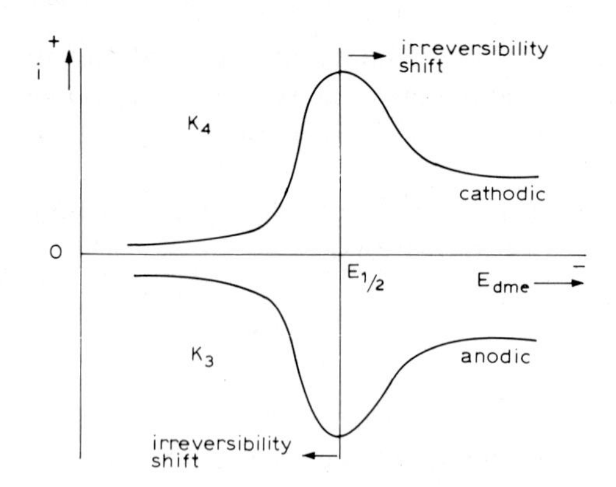

Fig. 3.52. Cathodic (K_4) and anodic (K_3) Kalousek polagrams.

minimum will show the same $E_{\frac{1}{2}}$ value; in the case of irreversibility the cathodic maximum shifts to a more negative potential and the anodic minimum to a more positive potential (cf., Fig. 3.21) and the same shift effects are then obtained for the sigmoidal polarographic waves of K_2 and K_1. The above illustrates that the Kalousek methods are most suitable for studying the electrodics of a redox system; their sensitivity is high because, as a consequence of the rapid succession of the pulse increases and decreases with corresponding reduction and oxidation, the diffusion layer remains thin, so that high limiting currents per concentration unit occur. Among the commercial polarographs the Metrohm Polarecord 506 can be used for Kalousek polarography.

2. *Oscillographic AC polarography.* In recognition of its original inventors the method is often referred to as "oscillographic polarography according to Heyrovský and Forejt"; a comprehensive treatment of the technique has been given by Kalvoda[54]. Basically it is controlled-current polarography, where the ac current may be triangular, square-wave or sinusoidal and where the E

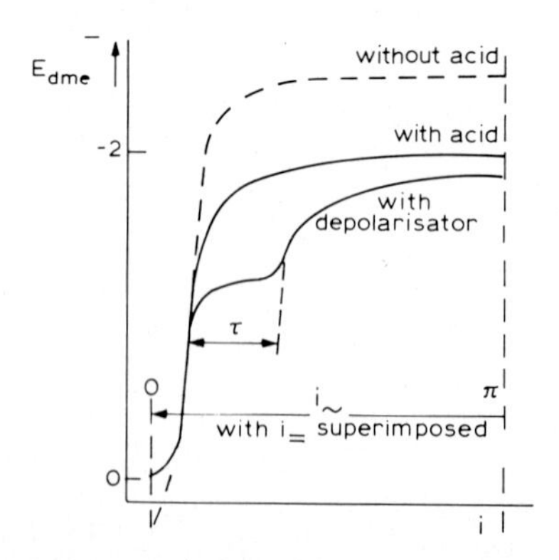

Fig. 3.53. Measuring principle of oscillographic AC polarography.

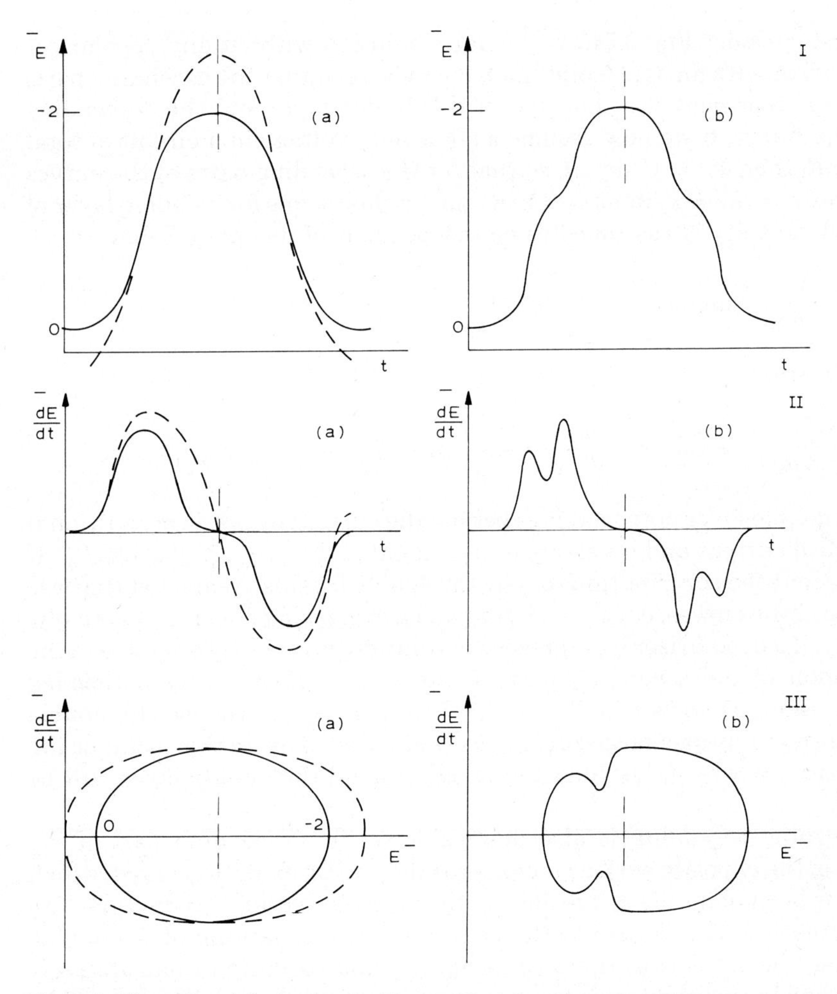

Fig. 3.54. Normal and derivative oscillographic AC polarograms, (a) without and with acid, (b) with depolarisator.

versus t curve is traced oscillographically or more recently oscilloscopically, so that there is a relationship with chronopotentiometry.

In the ac circuit of the polarographic cell there is such an external ohmic resistance that via the alternating voltage (300 V) together with a superimposed dc the voltage over the cell alternates from 0 to -2 V vs. an SCE; within these limits oxidation of Hg and reduction of Na^+ (electrolyte) to Na(Hg) remains sufficiently restricted.

Fig. 3.53 shows the effect of an electroactive species such as an acid or a more active depolarizer that undergoes cathodic reduction in one ac half-period and anodic oxidation in the next ac half-period (see also Fig. 3.54); τ is the so-called transition time, well known from chronopotentiometry (see later), i.e., in Fig. 3.53 the transition time of reduction.

If we first consider Fig. 3.54(a) I, i.e. for a solution without any depolarizer (broken line) or with an acid (solid line), it can be seen that the ascending parts of the curves represent the charging of the double layer and the descending parts its discharge. If we now assume a sinusoidal voltage in a circuit of total resistance R, then $V_t = V \sin \omega t$, so that for the ascending parts of the curves we can write $i_t = i_C = i \sin \omega t = (V/R) \sin \omega t$; this means for a double layer of differential capacity c_{dl} (assumed to be independent of its charge) that

$$\frac{dE}{dt} = -\frac{V}{c_{dl} AR} \cdot \sin \omega t$$

or by integration

$$E = \frac{V}{\omega C_{dl} AR} \cdot \cos \omega t = \frac{i_C}{\omega C_{dl} A} \cdot \sin \left(\omega t - \frac{\pi}{2} \right)$$

Hence the double layer potential lags behind the current by $\pi/2$, is proportional to the applied current and inversely proportional to its frequency ω and C_{dl}. It also means that the contribution to E by the double layer is greatest at the base and the top of the curves, causing a rising and a depressing effect, respectively; fortunately, if a depolarizer ox is present in solution, it is mostly effective in the middle region of the ascending part of the curve, which causes a time-lag (transition time) as shown in Fig. 3.54(b) I around $E_{\frac{1}{2}(ox)}$, because the double layer is considerably discharged during the reduction of ox; in this instance the total current $i = i_C + i_F$; beyond the middle region the double layer will be recharged.

The foregoing reasoning is also applicable to the descending part of the curve, but in the opposite way, as it concerns the anodic oxidation of red, which possibly has been obtained at the dme by the previous cathodic reduction. The peculiar situation with regard to the charging and discharging of the double layer makes the transition times of reduction and oxidation quantitatively unreliable and insensitive; some improvement in sensitivity can be achieved by recording dE/dt as a function of t or E (as shown in Fig. 3.54 by II and III, respectively), viz., one measures the indentations in II(b) or III(b) as being proportional to concentration; moreover, sampling at the end of drop life is highly advisable.

The above considerations concerned a reversible process, ox + $ne \rightleftharpoons$ red, where the E values of the inflection points of curve I(b) agree fairly well with the same $E_{\frac{1}{2}}$; if there are several electroactive species with sufficiently different $E_{\frac{1}{2}}$ values, they show up separately with their own inflection points. In the case of irreversibility the inflection point of ox will shift to a more negative potential and that of red to a more positive potential, or one or both may be absent; time-lags will also appear by sudden changes of the double-layer capacity owing to adsorption or desorption of surface-active agents. It is more in the detection of these kinds of phenomena that oscillographic AC polarography has been shown to be most useful.

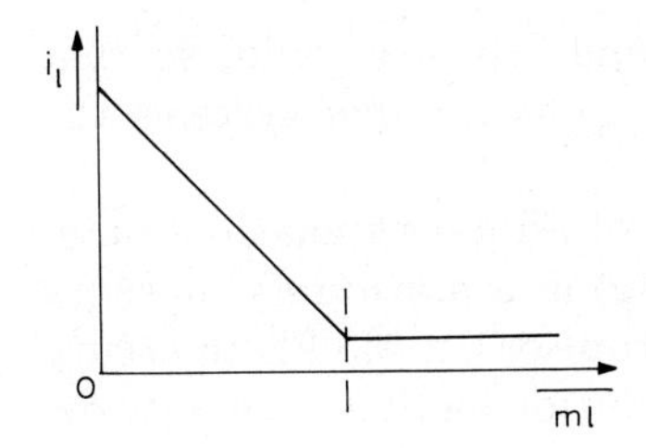

Fig. 3.55. Polarographic complex-formation or precipitation titration.

3.3.1.3. *Polarographic titration*

Majer[55] in 1936 proposed measuring, instead of the entire polarographic curve, only the limiting current at a potential sufficiently high for that purpose; if under these conditions one titrates metal ions such as Zn^{2+}, Cd^{2+}, Pb^{2+}, Ni^{2+}, Fe^{3+} and Bi^{3+} with EDTA[56], one obtains a titration as depicted in Fig. 3.55; i_l decreases to a very low value, in agreement with the stability constant of the EDTA–metal complex and the titration end-point is established by the intersection of the i_l curves before and after that point; correction of the i_l values for alteration of the solution volume by the titrant increments as in conductometric titration is recommended.

Dieker et al.[57] used a similar method but applied a dropping amalgam electrode (DAE) and followed amperometrically by means of pulse polarography the anodic dissolution wave of mercury in the presence of an excess of ligand; by appropriate choices of pH and titrant they achieved selective determinations of metal ions at low concentrations.

Polarographic titration seems of limited application owing to the discontinuity of the dme; moreover, the addition of the titrant increments requires stirring whereas the polarographic measurements require an unstirred solution.

3.3.2. Voltammetry at stationary and hydrodynamic electrodes

At the beginning of Section 3.3 a distinction was made between voltammetric techniques with non-stationary and stationary electrodes; the first group consists of voltammetry at the dme or polarography, already treated, and voltammetry at hydrodynamic electrodes, a later subject in this section; however, we shall now first consider voltammetry at stationary electrodes, where it is of significance to state whether and when the analyte is stirred.

3.3.2.1. *Voltammetry at stationary electrodes*

In general we work in an analyte solution without stirring, as in polarography, unless mentioned otherwise. The simplest situation is that where a reversible redox process such as ox + $ne \rightleftharpoons$ red takes place at an inert electrode such as Pt, Pd, Ir or Rh (and sometimes Au or Ag) and where both ox and red remain soluble; in this instance eqns. 3.14, 3.15 and 3.16 are valid, so that $E_{\frac{1}{2}(ox)} = E_{\frac{1}{2}(red)}$ as shown in Fig. 3.9 and $E_{\frac{1}{2}} \approx E^0_{(Nernst)}$, as in such systems the diffusion coefficients D_{red} and D_{ox} are nearly equal.

The above considerations also apply to the ion of an amalgamating metal with the reversible equilibrium $M^{n+} + ne \rightleftharpoons M(Hg)$ at a stationary mercury electrode such as an HMDE (hanging mercury drop) or an MTFE (mercury thin-film) with the restriction, however, that the solution can contain only ox, so that merely the cathodic wave (cf., eqn. 3.15) represents a direct dependence of the analyte concentration, whilst the reverse anodic wave concerns only the clean-back of amalgam formed by the previous cathodic amplitude. When one or both of the electrodic reactions is or becomes (in the case of a rapid potential sweep) irreversible, the cathodic wave shifts to a more negative potential and the anodic wave to a more positive potential (cf., Fig. 3.10); this may even result in a complete separation of the cathodic and anodic waves (cf., Fig. 3.11).

Irreversibility can be avoided by the use of a metal electrode with sufficient catalytic activity; as to a mercury electrode in connection with an amalgamating analyte metal, its diffusion to and from the mercury surface takes place more easily at the MFTE by the large surface and thin layer of Hg than at the HMDE, where irreversibility may occur more readily.

Serious difficulties may arise when as a consequence of the electrode process, a precipitate on the surface of the inert metal electrode disturbs its catalytic activity. Normally at a mercury electrode, owing to a potential sweep occurring, both charging currents and adsorption or desorption effect have to be taken into account; non-amalgamating analyte metals, apart from their disturbing precipitate formation, also cause the difficulty of an $E_{\frac{1}{2}}$ altering with varying analyte concentration (cf., Fig. 3.16). AC voltammetry offers also for stationary electrodes the best way of pre-establishing[58] the kinetics of an electrode process.

From the previous treatment of newer methods of polarography (see Table 3.1) and from the above remarks, it follows that corresponding measuring techniques (see Table 3.2) can be applied in voltammetry at stationary electrodes.

The methods will now be briefly treated in the sequence indicated by Table 3.2.

1. *Linear sweep voltammetry* (*chronoamperometry*)

The application of this technique (even in its various modes such as cyclic voltammetry) to other electrodes has already been mentioned in the description of LSV at the dme [Section 3.3.1.2.1(5)]. Especially with stationary electrodes LSV becomes fairly simple, under the conditions of sufficient solubility of ox and red, because of the constant and undisturbed electrode surface; at an inert electrode the residual faraday current can be adequately eliminated by means of "I compensation" (cf., Fig. 3.23) or by subtractive [cf., Section 3.3.1.2.1(3)] and derivative[59] [cf., Section 3.3.1.2.1(4)] voltammetry; at a stationary mercury electrode (e.g., HMDE), in addition to the residual faradaic current,

TABLE 3.2
VOLTAMMETRY AT STATIONARY ELECTRODES (INERT TYPES, HMDE, MTFE, ETC.)

II. Controlled-current techniques	I. Controlled-potential techniques	
DC voltammetry		AC voltammetry
4. Chronopotentiometry (cf., also current density mode at a dme)	1. Linear sweep voltammetry (chronoamperometry) 2. Pulse voltammetry a. Normal pulse voltammetry b. Differential pulse voltammetry c. Differential double-pulse voltammetry	3. Voltammetry with superimposed AC signal a. Sinusoidal AC voltammetry b. Square-wave voltammetry
5. Stripping voltammetry (SV)		

a charging current occurs as a consequence of the potential sweep (cf., eqn. 3.53, first term), which can be effectively eliminated by staircase voltammetry with current sampling (see Fig. 3.33). Interrupted sweep voltammetry also shows effective peak splitting at stationary electrodes (cf., Fig. 3.35)[60].

2. *Pulse voltammetry*

Since in this technique, i.e., normal pulse voltammetry (NPV), as well as differential pulse voltammetry (DPV), current sampling takes place at the end of the pulse, such as in pulse polarography where the sampling coincides with the end of the drop life, i.e., at the largest and most slowly growing surface, the phenomena in pulse voltammetry follow essentially the same theory and with corresponding sensitivity results; however, at stationary electrodes, even in pure supporting electrolytes, considerable residual currents, although markedly decreased at the end of the pulse, have been reported[61], but could not be fully explained.

(a) In normal pulse voltammetry and for the example of only ox of a redox couple, we preferably take E_{basic} (cf., Fig. 3.36) just before the cathodic decomposition potential; on application of the pulse an i_F (cathodic) starts at a high value, subsequently slowing down to the equilibrium value i_d (cf., Fig. 3.37, with i_C only at an Hg electrode) which is sampled at the pulse end and which agrees with a certain red/ox ratio at the electrode surface. However, after the pulse E has fallen back to E_{basic} all the red will be reoxidized if the delay time between the pulses is sufficient; this means that even insoluble red on the electrode surface is cleaned off before the next pulse occurs, so that again the NPV yields excellent results. The reversibility of the electrode process can be checked from the E vs. $\log [(i_{c,d} - i)/i]$ by considering the slope value, $2.303\ RT/nF$ (see p. 120); with irreversibility, such as for Au(III)/Au(0) at a graphite electrode,

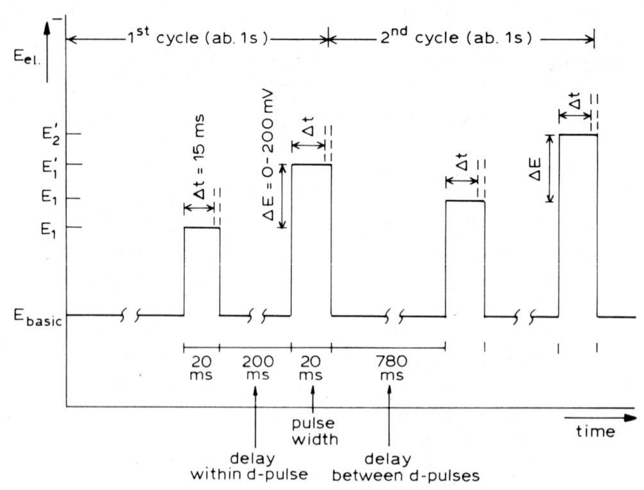

Fig. 3.56. Sampling scheme of differential double-pulse voltammetry.

even a long delay time between the pulses cannot prevent a persistent depletion of Au(III) at the electrode; however, this could be overcome by stirring[62].

(b) Using differential pulse voltammetry at stationary electrodes, excellent results can be obtained provided that ox and red are soluble, or with a mercury electrode if the resulting metal (if any) amalgamates; the voltammetric method can often be more rapid than the corresponding polarographic mode with its dependence on the drop time, provided that the delay time between pulses is not less than twice the pulse width (to avoid transient noise disturbances) and that the scan rate is not too fast (to limit dc distortion). When either ox or red is insoluble and so can precipitate on the electrode surface, NPV is to be preferred to DPV.

(c) Differential double-pulse voltammetry[63]* is another very interesting variant of pulse voltammetry at stationary electrodes, as it combines (see Fig. 3.56) both the advantages of NPV (electrode cleaning at fallback of pulses; cf., Fig. 3.36 for NPP) and those of DPV [an extreme, although not complete, elimination of residual current, especially when $\Delta i/\Delta E$ is based on the lower $\Delta E = (E_2 - E_1)$ and $\Delta i/\Delta E$ on $\Delta E' = (E'_2 - E'_1) = \Delta E$, respectively (cf., Fig. 3.39); moreover, for the stationary mercury electrode, owing to current sampling at the end of the pulse there is complete elimination of the charging current]. An increase in sensitivity can still be obtained by the determination of $\Delta^2 i/\Delta E^2$ on the basis of $(\Delta i' - \Delta i)$; computerization in the double-pulse technique is indicated, or course. It must be realized that the above differential values in fact represent pseudo-derivatives (cf., under pulse polarography).

*A variant of double-pulse voltammetry is the DNPV or differential normal pulse voltammetry, where at the end of each pulse an additional small constant pulse is imposed[63a].

3. *Voltammetry with superimposed AC signal*

At the beginning of this section we explained the attraction of AC voltammetry for pre-establishing the kinetics of electrode process. In principle it does not matter whether the AC signal is sinusoidal or square-wave; it is more interesting to consider that at a stationary electrode such as the HMDE within the reduction/oxidation cycle of an amalgamating metal the anodic wave (reverse sweep) often becomes more enhanced (revealing a higher peak)[64] than the cathodic wave; this is one of the reasons why anodic stripping analysis (see later) has attained so much importance as a sensitive analytical method.

4. *Chronopotentiometry*

In contrast to the previous voltammetric methods at stationary electrodes, chronopotentiometry, which is based on interpretation of E–t curves, represents a controlled current method.

Preferably a three-electrode circuit is used with a galvanostatic system across working and auxiliary electrodes (WE and AE) and a potentiometric system between WE and a reference electrode (RE). Ordinarily a constant current is applied and, as the solution is unstirred, the ox of a reversible redox couple becomes, in cathodic chronopotentiometry, rapidly exhausted at the WE surface whilst passing the situation where [ox] = [red], and so $E_{el} = E^0_{ox/red}$. Fig. 3.57 shows a chronopotentiogram for an acidic Fe^{3+}/Fe^{2+} solution at a Pt electrode. In a way it represents the curve of a local titration at the electrode surface where i acts as a titrant and the so-called transition time τ has, according to expectation, a certain relationship with the analyte concentration. Although chronopotentiometry as an analytical method was discovered much later by Gierst and Juliard[65] in 1950 and the term chronopotentiometry was first used by Delahay and Mamantov[66] as late as in

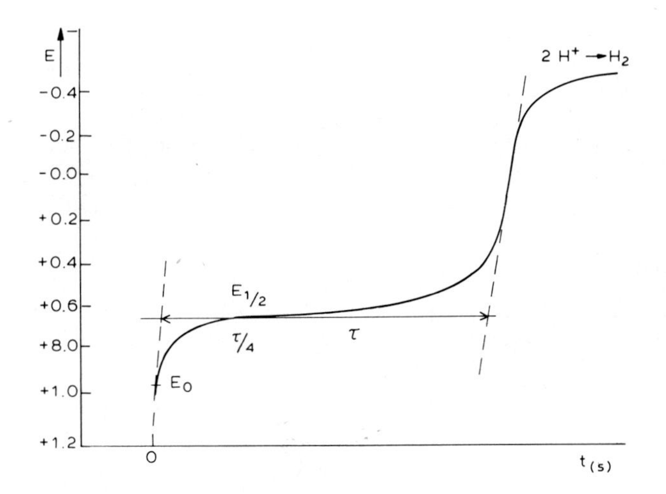

Fig. 3.57. Chronopotentiogram (at Pt) of acid Fe^{3+}/Fe^{2+} solution.

1955, it was Sand[67] in 1901 who originally established the relationship between transition time and concentration as

$$\tau^{1/2} = \frac{\pi^{1/2} nFCAD^{1/2}}{2i} \tag{3.67}$$

This equation can be derived by means of Laplace transformation of Fick's second law for a planar microelectrode:

$$\frac{\delta c}{\delta t} = D \cdot \frac{\delta^2 c}{\delta x^2} \tag{3.55}$$

However, in contrast to the condition of a constant potential and current sampling in pulse polarography, which yields[36] the Cottrell equation:

$$i_{d(t)} = \frac{nFCAD^{1/2}}{\pi^{1/2} t^{1/2}} \tag{3.56}$$

we now obtain under the conditions of a constant current[68] the Sand equation (eqn. 3.67) for the transition time, i.e., the time until the potential inflection point (see Fig. 3.57). The equation shows most importantly, with respect to solution analysis, that $\tau^{1/2}$ is proportional to concentration whatever the electrode kinetics may be; another feature is that for a stationary electrode in a certain solution the product $i\tau^{1/2}$ is constant and that in general in $2 \cdot 10^{-4}$–10^{-3} *M* solutions $i\tau^{1/2}/AC = \frac{1}{2}\pi^{1/2} nFD^{1/2} = k$ shows a nearly constant value within a precision of 1%[69].

If we consider Fig. 3.57 for the redox couple Fe^{3+}/Fe^{2+} at a Pt electrode it can be seen that the curve of this completely reversible system follows the Nernst equation, which means that starting from a solution with only Fe^{3+} of concentration C_{ox} we shall obtain after time t and with a constant current i:
for Fe^{3+} remaining:

$$C_{ox}(0, t) = C_{ox} - \frac{2\, it^{1/2}}{\pi^{1/2} nFAD_{ox}^{1/2}} \tag{3.68}$$

and for Fe^{2+} obtained:

$$C_{red}(0, t) = \frac{2\, it^{1/2}}{\pi^{1/2} nFAD_{red}^{1/2}} \tag{3.69}$$

so that according to Nernst

$$E = E^0 + \frac{RT}{nF} \ln \left[\frac{C_{ox} - t^{1/2}(2i/\pi^{1/2} nFAD_{ox}^{1/2})}{t^{1/2}(2i/\pi^{1/2} nFAD_{red}^{1/2})} \right] \tag{3.70}$$

Because after $t = \tau$ according to Sand (eqn. 3.67) all c_{ox} would have disappeared,

$$C_{ox} = \tau^{1/2}(2i/\pi^{1/2} nFAD_{ox}^{1/2}) \tag{3.68a}$$

so that substitution into eqn. 3.70 leads to

$$E = \left[E^0 + \frac{RT}{2nF}\ln\left(\frac{D_{red}}{D_{ox}}\right)\right] + \frac{RT}{nF}\ln\left(\frac{\tau^{1/2} - t^{1/2}}{t^{1/2}}\right) \tag{3.71}$$

Here the second term on the right-hand side becomes zero when $t^{1/2} = \tau/4$, while the first term agrees with the polarographic $E_{\frac{1}{2}}$ value (cf., eqn. 3.37a); hence for the chronopotentiogram we preferably write

$$E = E_{\tau/4} + \frac{RT}{nF}\ln\left(\frac{\tau^{1/2} - t^{1/2}}{t^{1/2}}\right) \tag{3.72*}$$

This equation applies to all reversible electrode reactions with soluble ox and red, so it includes cathodic chronopotentiometry of ions of amalgamating metals such as Cd^{2+}, Cu^{2+}, Pb^{2+} and Zn^{2+} at stationary Hg electrodes. For a redox couple such as Fe^{3+}/Fe^{2+} the diffusion coefficients D_{red} and D_{ox} will not differ much, so that $E_{\tau/4}$ is approximately equal to E^0 (770 V).

A complicated situation arises when ox or red is insoluble, for instance in the reductive chronopotentiometry of an ion of a non-amalgamating metal; then the metal precipitates on the stationary Hg electrode surface without further diffusion, so that

$$E = E^0 + \frac{RT}{nF}\ln\,[\mathrm{Me}^{n+}]_{(x=0)}$$

Considering the situation when $[\mathrm{Me}^{n+}]_{(x=0)} = C_{ox}(0, t)$, we can substitute eqns. 3.68 and 3.68a, respectively, which yields

$$E = E^0 + \frac{RT}{nF}\ln\,[(\tau^{1/2} - t^{1/2})(2i/\pi^{1/2}nFAD_{ox}^{1/2})]$$

or

$$E = \left[E^0 + \frac{RT}{nF}\ln\left(\frac{2i}{\pi^{1/2}nFAD_{ox}^{1/2}}\right)\right] + \frac{RT}{nF}\ln\,(\tau^{1/2} - t^{1/2})$$

In a simplified form we obtain

$$E = E' + \frac{RT}{nF}\ln\,(\tau^{1/2} - t^{1/2}) \tag{3.73}$$

where E' is only approximately constant because the data in its logarithmic term possess fixed experimental values, and E^0, although an apparently standard potential, is not well-defined (Hg is only incompletely covered by metal precipitate).

Another situation occurs in the case of a totally irreversible electrode reaction, notwithstanding the solubility of ox and red; for an irreversible cathodic wave it means that the current i is determined only by the first term on the right-hand side in eqn. 3.8, so that

*For the equations concerning E as a function of t and τ in chronopotentiometry, see Delahay and Berzins[70].

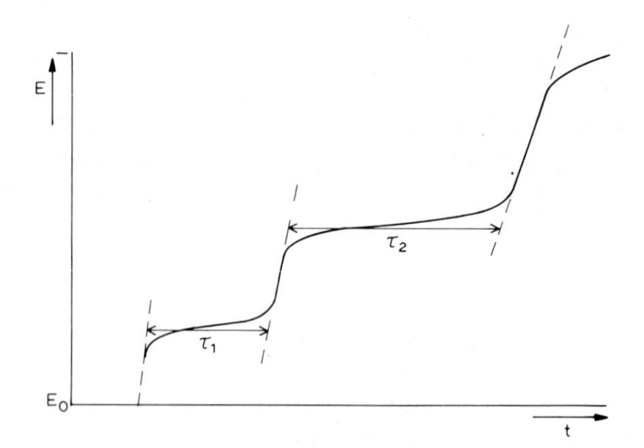

Fig. 3.58. Chronopotentiogram (at Pt) of a two-component mixture.

$$i = i_C = nFAk_{s,h}\,[\mathrm{ox}]\mathrm{e}^{-\alpha nF(E-E_0)/RT}$$

where [ox] represents the concentration at time t at the electrode surface, i.e. $c_{ox}(0, t)$. In chronopotentiometry and according to the Sand equation, the relationship between $c_{ox}(0, \mathrm{t})$ and bulk concentration C_{ox} is given by

$$\frac{c_{ox}(0,\mathrm{t})}{C_{ox}} = \frac{\tau^{1/2} - t^{1/2}}{\tau^{1/2}}$$

and its substitution into the previous equation then yields

$$E = E^0 + \frac{RT}{\alpha nF} \ln \left(\frac{nFAk_{s,h}}{i}\right)\left(\frac{\tau^{1/2} - t^{1/2}}{\tau^{1/2}}\right) C_{ox}$$

or

$$E = \left[E^0 + \frac{RT}{\alpha nF} \ln \left(\frac{nFC_{ox}Ak_{s,h}}{i}\right)\right] + \frac{RT}{nF} \ln \frac{\tau^{1/2} - t^{1/2}}{\tau^{1/2}} \qquad (3.74)^*$$

Eqns. 3.71–3.74 show the possibility of checking the mechanism of the respective electrode reactions on the basis of their logarithmic term containing $t^{1/2}$ and $\tau^{1/2}$; in all the instances the product $i\tau^{1/2}$ is constant as a function of the constant i chosen; however, when a chemical reaction necessarily precedes the electron transfer reaction, i.e. in a CE system, $i\tau^{1/2}$ decreases with increasing i.

In the chronopotentiogram of mixtures (see Fig. 3.58), the reactive components will yield different inflection points if their standard potentials show a sufficient difference (at least 0.1 V). To take a simple example, let us consider a reversible electrode process for both ox_1 and ox_2 in a solution of the supporting electrolyte. Then eqn. 3.72 is simply valid for the first reacting ox_1 with τ_1 up to the first inflection point; however, beyond this point the last traces of ox_1

*Note that the term with E^0 still changes with the analyte concentration, the transfer coefficient α and the specific heterogeneous rate constant $k_{s,h}$ (cf., eqn. 3.7).

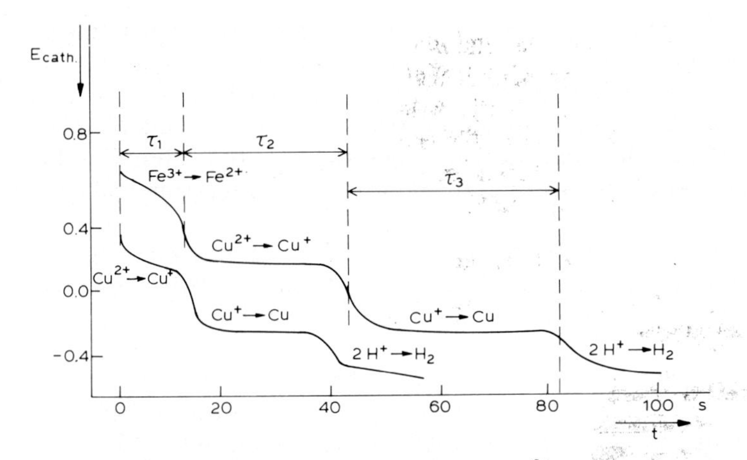

Fig. 3.59. Two-step chronopotentiogram (at Pt) of $Cu^{2+} \rightarrow Cu^{+} \rightarrow Cu$, with and without a preceding $Fe^{3+} \rightarrow Fe^{2+}$ step.

will go on to react together with ox_2 with τ_2 up to the second inflection point. Thus Berzins and Delahay[71] obtained the following Sand relationship:

$$(\tau_1 + \tau_2)^{1/2} - \tau_1^{1/2} = \frac{\pi^{1/2} nFC_2AD_2^{1/2}}{2i}$$

Moreover, $\tau_1^{1/2} = \pi^{1/2}nFC_1AD_1^{1/2}/2i$, so that for the special situation where $[ox_1] = [ox_2]$, both possessing the same values of n and D: $(\tau_1 + \tau_2)^{1/2} - \tau_1^{1/2} = \tau_1^{1/2}$; hence $(\tau_1 + \tau_2)^{1/2} = 2\tau_1^{1/2}$ or $\tau_1 + \tau_2 = 4\tau_1$, so that $\tau_2 = 3\tau_1$ {analogously for another equal $[ox_3]$ one may expect $\tau_3 = 3(\tau_1 + \tau_2)$ and hence $\tau_3 = 12\tau_1$}.

These results are plausible since according to Sand a two-fold concentration of a component yields a four-fold transition time. Now, these features show, in contrast to the net separation and pure additivity of polarographic waves and their diffusion-limited currents as concentration functions, that in chronopotentiometry the transition times of components in mixtures are considerably increased by the preceding transition times of any other more reactive component, which complicates considerably the concentration evaluation of chronopotentiograms.

Such a complication was strikingly illustrated by Lingane[72], as shown in Fig. 3.59* for 100 ml of a 5 mM $CuCl_2$ solution in 1 M HCl and for the same solution containing an additional 5 mM $FeCl_3$; he concluded that using chronopotentiometry the analysis of solutions of concentration less than 10^{-3} M would be unreliable[73]. The results in Fig. 3.59 look even more ideal than those in Figs. 3.57 and 3.58, where the tangents to the curves through the inflection points are not parallel. Bos and Van Dalen[74] and others made extensive studies of the evaluation of chronopotentiograms in order to establish τ

*Ordinarily we prefer the reversal or negative potential scale on the ordinate, especially in cathodic reduction where i is positive according to convention (cf., Figs. 3.57 and 3.58).

more accurately so as to improve the precision of analysis. Several methods of evaluation were considered, among which that of Delahay and Mattax[75] in Figs. 3.57 and 3.58 appeared to be the best provided that a fixed initial potential E_0 (200 mV positive vs. $E_{\tau/4}$) was chosen. Nevertheless, the following disturbances remain, notwithstanding that the usual conditions for the validity of the equations in chronopotentiometry are fulfilled (absence of migration and convection), viz.:

(1) some minor disturbances (which can be avoided by careful working):

(a) with solid electrodes: roughness of the electrode surface (with small τ), incorrect electrode orientation with respect to diffusion (with large τ) or oxide layer on electrode;

(b) with mercury electrodes: electrode surface vibration [more cumbersome at Hg pool than at mercury thin-film electrode (MTFE)], no linear diffusion near walls of electrode, AE not parallel to WE, or RE too close to WE;

(2) two main disturbances: (a) occurrence of contaminants and (b) capacitive currents at mercury.

(a) In theory the electrode reaction of one substance is assumed to be completed before that of the second substance begins; however, Bos stated that this assumption holds true only if the standard reduction potentials differ sufficiently, depending on factors such as the concentration ratio of the components and the number of electrons involved; he also found that the distinction between chronopotentiometric waves obtained by reduction of ion mixtures of amalgamating metals is more difficult at an MFE than at an Hg pool[76], whereas the distinction between these waves obtained by oxidation of these metals is easier from an MFE than from an Hg pool[77].

From the foregoing remarks it will be clear that contaminants which may be more reactive than the analyte can seriously disturb the establishment of the transition time τ required analytically; oxygen is a most inconvenient contaminant because its last traces are difficult to remove and can vary during serial measurements. Some investigators tried to prevent an increase in τ by using prior reduction or oxidation; further, an increase in the current proportional to $t^{1/2}$, so-called programmed current chronopotentiometry, renders the waves additive[78], but requires more sophisticated apparatus.

A more usual procedure for overcoming the disturbances from contaminants is current reversal chronopotentiometry; here the current is reversed at the initial transition time τ_f of the forward reaction and the next transition time τ_b of the backward reaction is measured; as a rule the reversal wave will not be influenced by the contaminant because it will react either before the forward or after the backward reaction of the analyte (see Fig. 3.60a); the entire procedure can be even repeated as cyclic chronopotentiometry (see Fig. 3.60b), which may provide a further check on the reliability. The reversal technique can be applied to initial reduction followed by re-oxidation and also to initial oxidation followed by re-reduction[79].

For a more detailed explanation, let us consider the example of initial reduction where, after current reversal, the following phenomena may occur:

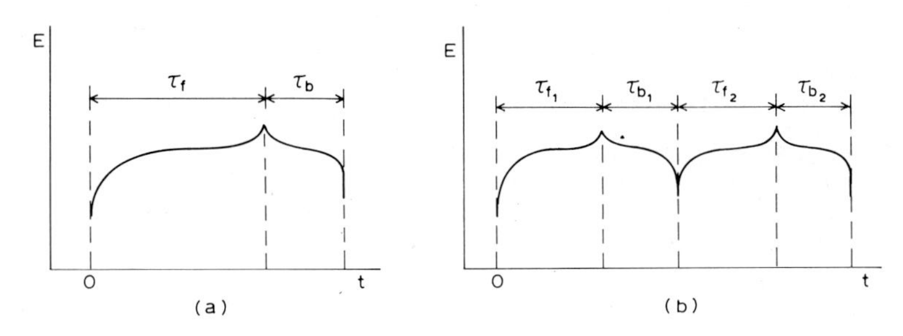

Fig. 3.60. Current reversal chronopotentiometry (at Pt), (a) one cycle, (b) two cycles.

(a_1) metals reversibly precipitated on the electrode will re-dissolve, yielding a re-oxidation wave as long as the reduction wave;

(a_2) amalgamating metals reversibly reduced at an Hg pool will re-dissolve, yielding a re-oxidation wave with a length one third of that of the reduction wave (cf. the equation derived for semi-infinite linear diffusion conditions[80]; there may be deviating results at the MTFE);

(a_3) after an irreversible reduction the re-oxidation wave will show a much less negative potential, or there will be no wave at all.

In the reversal technique in general the currents and transition times are of the order of 0.1–1 mA cm^{-2} and 10–20 s, respectively.

(b) In chronopotentiometry the capacitive or charging current at mercury electrodes causes a problem that is difficult to overcome as the electrode potential is changing all the time. The effect is a wave distortion as depicted in Fig. 3.61, which clearly shows the lengthening of the transition time owing to the charging current. Shults et al.[81] used a blank cell with a base solution in which the variation of the double-layer capacity with the potential was taken into account; this was used as a positive feed-back to the measuring cell. However, in the first steep part of the chronopotentiogram the electrolysis potential will be reached in the shortest time possible with the greatest possible total current, which may destroy the apparatus within the first few moments.

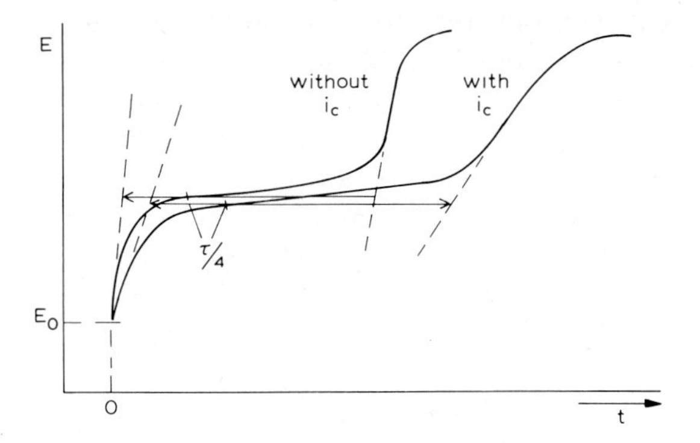

Fig. 3.61. Chronopotentiographic wave at Hg, distorted by charging current.

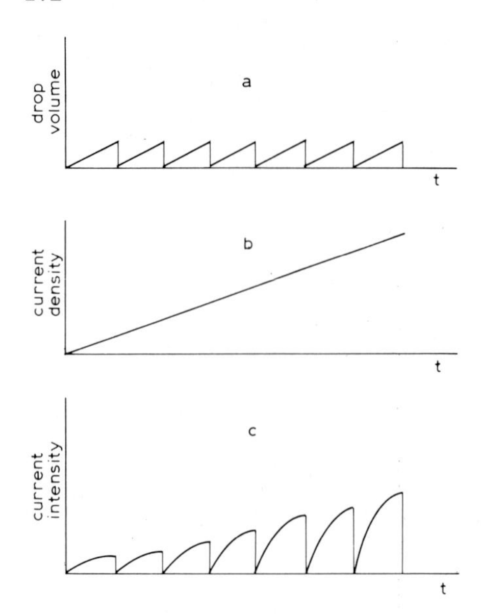

Fig. 3.62. Controlled current density synchronized with Hg drop fall control.

By means of special precautions Bos[74] prevented this difficulty in his additional capacity-current device, which also allows short transition times (down to 0.1 s) at low concentrations. Thus, together with the current reversal technique, the analysis of 10^{-5} M solutions appeared possible.

Chronopotentiometry at a dme appeared to be impossible until Kies[82a] recently developed polarography with controlled current density, i.e., with a current density sweep. He explained the method as follows. The high current density during the first stage of the drop life results in the initiation of a secondary electrolysis process at a more negative electrode potential followed by a reverse reaction with rapid (reversible) systems because of the increase in the electrode potential.

The use of a circuit with controlled current density synchronized with the drop fall control (Fig. 3.62) overcomes this difficulty. The current density sweep is slow and the Hg drop expands at the normal rate. When the apparatus has just been started, the transition time is too long to be contained within the drop life, so that only a small part of the chronopotentiogram is covered by the recorder. The increasing current density involves a reduction in the transition time and a moment comes at which the drop time and transition time coincide. From this time on, the electrode potential undergoes a considerable shift in the negative direction during the last period of the drop life. This indicates that a second process (reduction of a second component or decomposition of the supporting electrolyte) starts during this period. By analogy with the term "limiting current" in ordinary polarography, "limiting time" may be used for the moment when a considerable change in the course of the envelope is observed.

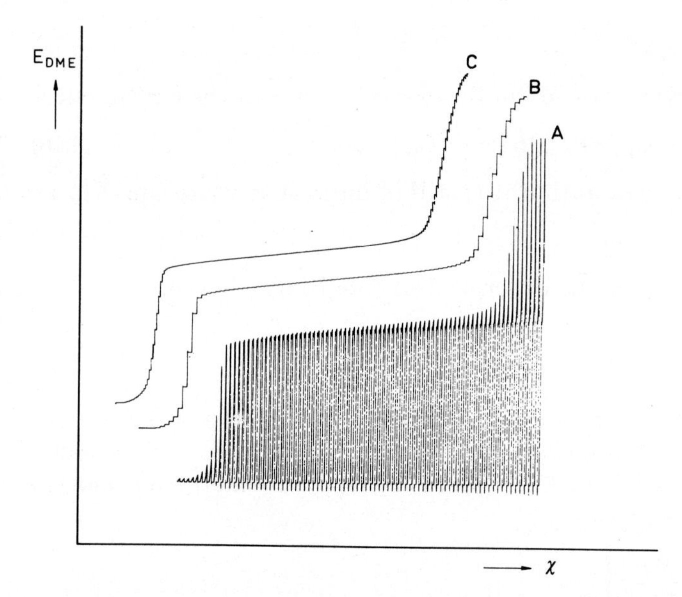

Fig. 3.63. Polarography with controlled current density (Kies method).

Instead of the original figure of Kies, we prefer the set-up shown in Fig. 3.63 (Fig. 3 in the paper on instrumentation), as it allows a more direct comparison with Fig. 3.57. For curve A the current density $j = i/A$ was 1/3 of the maximum value at a recorder-paper velocity $X = 10\,\mathrm{s\,cm^{-1}}$ with the memory switched off, for curve B j was 1/2.236 of the maximum value at a recorder-paper velocity $X = 10\,\mathrm{s\,cm^{-1}}$ with the memory switched on, and for curve C j was 1/5.590 of the maximum value at a recorder-paper velocity $X = 25\,\mathrm{s\,cm^{-1}}$ with the memory switched on.

An advantage of the method is that the envelope of curve A and also curves B and C are free of charging current interference; however, χ does not represent a direct scale of the transition time τ, nor is χ automatically protected against interference by a contaminant, if any is present.

In order to establish the relationship between χ and τ, Kies developed a theory[82b] based on a current source which, in view of the constant current density during drop growth, should obey the equation $i = at^{2/3}$, so for the symbol χ in connection with the current density j converted into $i = b\chi t^{2/3}$. This means that by inserting the drop surface

$$q = 4\pi\left(\frac{3mt}{4\pi d_{\mathrm{Hg}}}\right)^{2/3}$$

(cf., Section 3.3.1.1.2) we obtain

$$j = i/q = 6^{-2/3}\pi^{-1/3}d_{\mathrm{Hg}}^{2/3}bm^{-2/3}\chi$$

For the Hg mass flow-rate $m = k'h_{\mathrm{eff}} = h_{\mathrm{eff}}/\kappa$, where h_{eff} is the effective Hg column height (corrected for capillary back-pressure, see p. 118) and is known as the capillary constant, we then find

$$j = 6^{-2/3}\pi^{-1/3}d_{\mathrm{Hg}}^{2/3}\kappa^{2/3}b\chi h_{\mathrm{eff}}^{-2/3} \tag{3.75}$$

Further elaboration by Kies yielded (for the example of ox) for the limiting time

$$\chi_1 = 2^{-1/3}3^{2/3}\pi^{5/6}\kappa^{-2/3}h_{\mathrm{eff}}^{2/3}d_{\mathrm{Hg}}^{-2/3}nFD_{\mathrm{ox}}^{1/2}b^{-1}\tau^{-1/2}C_{\mathrm{ox}} \tag{3.76}$$

(In the opinion of the present author it is still of interest to write eqn. 3.67 as

$$nFD_{\mathrm{ox}}^{1/2}\tau^{-1/2}C_{\mathrm{ox}} = 2i/\pi^{1/2}A$$

so that eqn. 3.76 becomes, for the experimental conditions chosen,

$$\chi_1 = k_{\mathrm{dme}}nFD_{\mathrm{ox}}^{1/2}\tau^{-1/2}C_{\mathrm{ox}} = k_{\mathrm{dme}} \cdot \frac{2i_{\mathrm{max}}}{\pi^{1/2}A_{\mathrm{max}}} = \frac{2k_{\mathrm{dme}}}{\pi^{1/2}} \cdot j$$

which agrees with the linear relationship between χ and j in eqn. 3.75.)

Analogously to eqn. 3.72 for stationary electrodes and a reversible redox couple of soluble ox and red, Kies derived for chronopotentiometry at a dme via insertion in the Nernst equation

$$E = \mathrm{E}_0 + \frac{RT}{nF}\ln\left[\frac{C_{\mathrm{ox}}(0,\ t)}{C_{\mathrm{red}}(0,\ t)}\right]$$

the E–χ relationship

$$E = E_{\frac{1}{2}} + \frac{RT}{nF}\ln\left(\frac{\chi_1 - \chi}{\chi}\right) \tag{3.77}$$

For their current density technique at the dme, Kies and Van Dam[82c] reported a standard deviation for a single solution ($8 \cdot 10^{-5}$ M $TlNO_3$), based on 11 recordings, of 0.5% and rectilinear calibration graphs in the range 20–1000 μM.

Regarding chronopotentiometry in general, the early optimistic expectations of accuracy and precision (10^{-6} M)[83] were not realized[84]; however, thanks to the aforementioned studies by Bos[69,74;76,77,79] and Kies[82] essential improvements have been obtained; these results, together with the interesting aspects of rapidity (complete curve often obtained within 1 min) and information on the kinetics of the electrode processes (both reductive and oxidative), have urged us to pay considerable attention to chronopotentiometry.

5. *Stripping voltammetry*

The technique of stripping voltammetry (SV) has gained considerable analytical importance owing to its sensitivity and selectivity; it also became known under several other names such as linear-potential sweep stripping chronoamperometry[85], stripping analysis[86], anodic amalgam voltammetry and inverse voltammetry[87] and (occasionally) re-dissolution voltammetry[88]. It is ordinarily used as an anodic stripping technique, so that we shall discuss it mainly as this variant, referring to the extensive publication by Barendrecht[89], to own work and to studies by others on SV in general.

Anodic SV consists in a concentration (pre-electrolysis or cathodic deposition) step, in which the analyte metal ion is reduced with stirring and at a

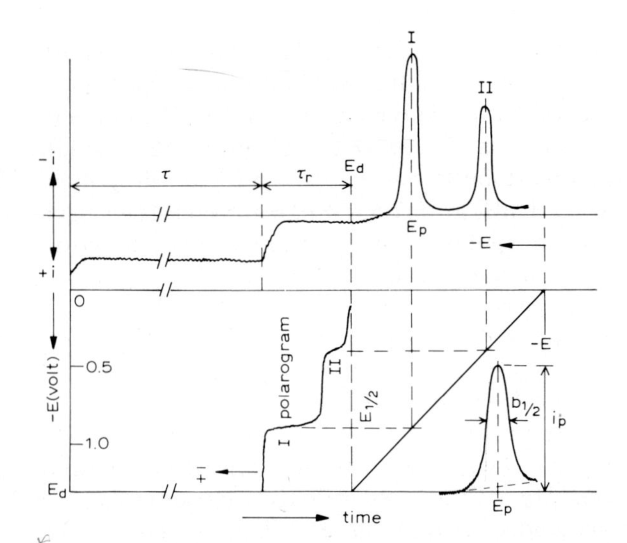

Fig. 3.64. Idealized picture (by Barendrecht) of anodic stripping voltammetry.

controlled potential whilst being deposited as the metal on a solid microelectrode (plating) or amalgamated at the surface of a mercury drop or at an MTFE, a subsequent rest period, in which the stirring or rotation is stopped, and finally the stripping step, in which the metal, usually without stirring and by means of a potential sweep, is anodically oxidized and re-dissolved (the pause is not necessary at stripping rates lower than $10\,\mathrm{mV\,s^{-1}}$).

The principle of the entire procedure (usually with thermostating at 25° C) and its effects has been instructively illustrated by Barendrecht as shown in Fig. 3.64[90]. This represents a stripping voltammogram in relation to a corresponding idealized polarogram [$E_d = -1.3\,\mathrm{V}$ vs. SCE, $\tau = 180\,\mathrm{s}$, $\tau_r = 60\,\mathrm{s}$, $\tau_{(\mathrm{stripping})} = 130\,\mathrm{s}$ ($v = 10 \cdot 10^{-3}\,\mathrm{V\,s^{-1}}$) and the peak characteristics are i_p, E_p and $b_{\frac{1}{2}}$].

Generally the peak height, i_p, is linearly dependent on the analyte concentration; however, the impression given by the idealized Fig. 3.64 that E_p agrees with $E_{\frac{1}{2}}$ is less correct, but the mid-value between E_p and the half-peak potential at the high side $E_{\frac{1}{2}p}$ appears to be a better, although not strict, approximation of $E_{\frac{1}{2}}$ (see later).

The great attraction of SV lies in the effect of pre-concentration of the analyte at the electrode with, as a consequence for the stripping current, a very high ratio of faraday current to charging and impurity currents; it is this high ratio which has made SV the most sensitive voltammetric analysis method to date.

We shall now consider the electrochemical apparatus used (electrodes and electrolytic cell) and the three steps in its procedure more closely, especially for anodic stripping voltammetry (ASV) as most important application.

The electrodes usually consist of mercury or deposited mercury or occasionally of inert solid material; further, they are mainly of a stationary type (in the stripping step as the crucial analytical measurement, but not in the concentration step, where often the solution is stirred or the electrode is rotated). Considering the mercury, only exceptionally has a sessile mercury drop electrode (SMDE)[91,*] or a slowly growing DME (drop time 18 min and phase-selective recording of stripping curve)[92] been applied. Most popular are the hanging mercury drop electrode (HMDE) and the mercury film or thin-film electrode (MFE or MTFE).

Barendrecht used an HMDE with screw micrometer[88,89] according to Von Sturm; after being cleaned the inner wall of the capillary must be siliconized in order to prevent analyte solution from intruding between the Hg and the glass wall; before its measurement the drop must be renewed, but is fairly reproducible. This type of electrode, i.e., with an uninterrupted connection between the Hg drop and the Hg filling, accordingly characterized as HDME(Hg), was strongly recommended by Kemula et al.[93], because no intermetallic compounds can be formed with contacting materials such as Pt or Au, nor can hydrogen be evolved by lowering of the hydrogen overpotential at these materials. However, there are some disadvantages, viz. (1) the low ratio of surface area to volume and diffusion of the analyte metal into the capillary filling at long plating times and (2) limitation of the stirring rate of the solution in order to avoid dislodging of the drops. This situation led many investigators to prefer an HMDE (Pt) with an Hg drop suspended from a previously mercury-wetted platinum contact[94], while Underkofler and Shain[95] used an HMDE (Pt), in which the end of the Pt wire (diameter 0.2–0.4 mm) protruding from a glass bar (see Fig. 3.65) was ground flat with the glass tip polished as smooth as possible, and etched with aqua regia to about 0.1 mm under the glass surface. An Hg drop from a DME is caught on a small glass spoon and transferred to the electrode.

No intermetallic compound with Pt can be formed if prior to each determination a new drop is attached; further, the electrode can be rotated[96] during both concentration and stripping[97,**], yielding an increase in selectivity and sensitivity. However, the aforementioned HMDEs still retain the disadvantage of a low ratio of surface area to volume; this ratio is increased considerably in a mercury film electrode (MFE), and more especially in its thin-film version (MFTE) with a film thickness down to 100 Å or less. An additional advantage of the MFEs is their excellent sensitivity and high resolution owing to the extremely fast diffusion from within the Hg film to the surface[98]. As substrate for the Hg film materials such as Pt, Ni or Ag were unsuccessful; better

*Not to be confused with the PARC SMDE, referred to as static (see Fig. 3.28), which virtually represents a temporary HMDE.

**If during measurement in the stripping step the electrode is rotated, in principle we have to deal with a hydrodynamic electrode (see Section 3.3.2.2), but in practice it is useful to treat their application in SV in this section.

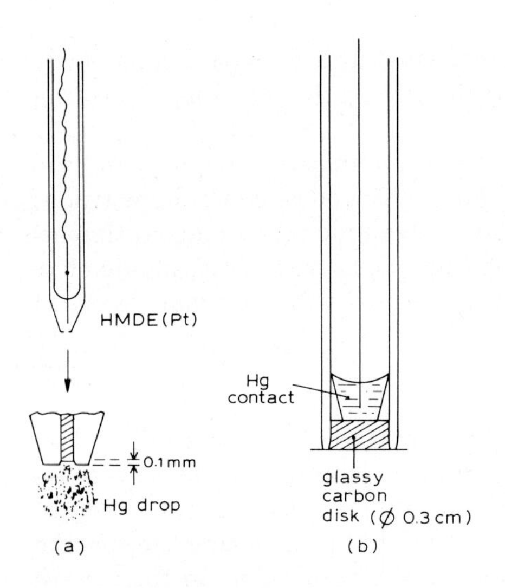

Fig. 3.65. Stationary electrodes for anodic SV, (a) HMDE (Pt), (b) GCE.

results were obtained with Hg deposited on carbon paste, wax-impregnated graphite, pyrolytic graphite or glassy carbon. Most successful to date has been the MFE of Florence[99] (see Fig. 3.65b); he sealed a polished glassy carbon disc into a glass tube with epoxy cement and prepared the Hg film in situ, i.e., mercury(II) nitrate was added (10^{-5}–10^{-4} M) to the analyte solution, so that during concentration both Hg and analyte metal were simultaneously electrodeposited, also at a high velocity of rotation (several thousand r.p.m.) if desirable.

Many workers have tried to omit the mercury film by depositing the analyte directly on inert metals such as Pt, Au, Rh, Ir or Ag or on carbon materials such as glassy carbon or wax-impregnated graphite; however, in general this was not successful (lack of selectivity for mixtures as a consequence of interdiffusion) and therefore it is rarely applied except for those nobler analyte metals that cannot be measured at mercury, such as Au, Ag and Hg itself. Nevertheless, metals such as Ni and Cr, which do not amalgamate, can be determined on an HMDE[100].

The electrolytic cell should preferably consist, as in polarography, of a three-electrode system of working, auxiliary and reference electrodes in order to minimize the fall in IR drop within the solution. In addition, a potentiostat (galvanostat) mode within the electrical circuit is desirable; in addition to the necessity for a stirrer and a nitrogen flush, the latter is of special importance in the concentration step. In the case of insufficient purity of the supporting electrolyte its contaminants can be previously removed by cathodic (i.e. at about -1.5 V) pre-electrolysis in a separate cell at a mercury pool vs. a saturated calomel electrode (SCE) or a Pt gauze anode with stirring, as recommended by Barendrecht[88,89].

In the concentration step the metal is deposited at a potential sufficiently negative to maintain its limiting current while stirring the solution or rotating the working electrode.

Levich[124] has given the relationships between the limiting current i_l and the bulk concentration C of the metal ion for plate electrodes, conical electrodes and rotated disc electrodes (RDEs) under hydrodynamic conditions; anticipating his well known equations treated in Section 3.3.2.2 on hydrodynamic electrodes, we may assume the relationships concerned using the more general equation

$$i_l = km(nFCA) \tag{3.78}$$

where k is a constant depending of the geometrical type of the electrode and on the nature of the hydrodynamics chosen (stirring or rotation) and m is the mass transfer coefficient, determined by the motion intensity. Hence, according to eqn. 3.78, in all situations i_l depends linearly on C, but considering the specific conditions of stirring or rotation i_l depends via the value of m on the square root of the flow velocity u or the angular velocity ω, respectively (at the hmde i_l is of the order of 0.5 μA for a metal ion at a concentration of 10^{-5} M[89]).

Now, if we assume throughout the entire concentration step a current yield of 100% and no perceptible alteration of the bulk concentration, in view of its function as an indicator in the stripping step the electrode area cannot be taken as large, Faraday's law can be applied to calculate the final amalgam concentration according to

$$C_a = \frac{i_l \tau}{nFV} \tag{3.79}$$

where τ is the cathodic reduction time and V the volume of the mercury drop or film. So, by substituting in eqn. 3.78 we obtain the following expressions for C_a: for HMDE (and its drop radius r_0):

$$C_a = \frac{3k'mC\tau}{r_0} \tag{3.78a}$$

for MTFE (and its layer thickness L):

$$C_a = \frac{k''mC\tau}{L} \tag{3.78b}$$

Of course, it is not essential that the bulk concentration does not alter during reduction*, provided that a completely reproducible SV is carried out in the usual calibration step.

In the rest period the electrode should become free of vibration (with an hmde) and any undesirable liquid convection at its surface should be stopped;

*If one reduces (as recommended) at a constant potential, i_l will diminish with time as a function of decreasing C; hence eqn. 3.78 becomes a differential equation the integration of which yields C_a, and eqns. 3.78a and b change accordingly.

the main purpose of the rest period remains that its duration τ_r suffices for complete homogenization of the metal concentration C_a in the amalgam to be achieved; this takes place fairly quickly, e.g. within 2 s for an MFE[101]. The cell circuit can usually remain uninterrupted, because the current becomes instantly zero; would electro-deposition still go on, then change of medium is recommended.

In the stripping step anodic oxidation and re-dissolution usually take place without stirring and by means of a linear potential sweep from a negative to a positive value, as illustrated in Fig. 3.64, so we have to deal with oxidative linear sweep voltammetry (LSV) as depicted by the anodic wave in Fig. 3.32. In this specific case the surface area of the dme may be considered to remain virtually constant during the rapid potential sweep, which represents a situation comparable to that of the hmde in the stripping step, albeit with a slower sweep. Considering chemical analysis it is now of interest to know the quantitative relationship between the peak current, i_p, of the wave and the original bulk concentration of the analyte; moreover, and especially for selective measurements in mixtures, some knowledge of the position of the peak potential, E_p, versus the polarographic $E_{\frac{1}{2}}$ is required.

From the complexity of the subject and from many studies carried out in this connection, it is almost impossible to draw clear conclusions; so we must confine ourselves to some general and fairly superficial statements about the peak characteristics (see Fig. 3.64).

(1) i_p increases with increasing $1/r_0$ (HMDE) or $1/L$ (MFE), as for i_l in the concentration step, and also with v (potential sweep in V s^{-1}); however, $b_{\frac{1}{2}}$ increases with increasing r_0 or L and with increasing v, thus causing a selectivity decrease by overlap of successive oxidation waves, which difficulty can be effectively overcome by a so-called interrupted potential sweep (cf., interrupted LSV depicted in Fig. 3.35).

(2) E_p may shift to more negative values with increasing r_0 or L and with increasing v, which is accompanied by the aforementioned increase in $b_{\frac{1}{2}}$.

For the HMDE and for a solution that contains only ox of a reversible redox couple, Reinmuth[102], on the basis of Fick's second law for spherical diffusion and its initial and boundary conditions, derived the quantitative relationship (at 25° C)

$$i_p = D^{1/2}CA\left[(2.69 \cdot 10^5)n^{3/2}v^{1/2} - \frac{(0.725 \cdot 10^5)nD^{1/2}}{r_0}\right] \qquad (3.80)^*$$

The second term on the right-hand side accounts merely for the spherical contribution and may amount to 20% of i_p; however, it becomes negligible at high sweep rates and large r_0, thus leaving the first term as the planar contribution, which apparently agrees with the well known Randles–Ševčik equation[103a] of LSV:

*The derivation of eqns. 3.80–3.84 by solving the differential equations using Laplace transformation and numerical methods has been clearly and comprehensively presented by Bard and Faulkner[103].

$$i_p = kn^{3/2}D^{1/2}CAv^{1/2} \tag{3.81}$$

where k is a constant at constant t. In fact, this equation is not strictly representative for SV and although a few authors simply used it for this purpose, Barendrecht[88,89] argued its deficiency in SV as follows:

(1) as far as the sphericity term of eqn. 3.80 is concerned, i_p is proportional to the drop radius r_0 and not to its area A as in the Randles–Ševčik equation[103a];

(2) instead of representing a constant bulk concentration of the solution as in LSV, C in SV (i.e. its stripping step) is the amalgam concentration C_a, which during stripping diverges from the drop to the solution and thus diminishes at high v according to a quasi-infinite diffusion, and at low v and for a small drop even becoming zero;

(3) the above arguments (1) and (2) have a greater influence in SV, because a high sweep rate and/or a large drop radius r_0 hamper selective measurements on analyte mixtures.

It can be mentioned that in the derivation of eqn. 3.80 (at 25° C) and for the planar term only (cf., eqn. 3.81), the following relationships for 25° C were obtained:

$$E_p - E_{\frac{1}{2}} = -1.109 \cdot \frac{RT}{nF} = -\frac{1.109 \cdot 25.69}{n} \text{mV} \quad \text{or} \quad -\frac{28.5}{n} \text{mV} \tag{3.82}$$

and

$$E_{p/2} - E_{\frac{1}{2}} = -1.09 \cdot \frac{RT}{nF} = \frac{1.09 \cdot 25.69}{n} \text{mV} \quad \text{or} \quad \frac{28.0}{n} \text{mV} \tag{3.83}$$

where $E_{\frac{1}{2}}$ represents the polarographic half-wave potential (cf., eqn. 3.37a):

$$E_{\frac{1}{2}} = E^0 + \frac{RT}{nF} \ln \left(\frac{D_{red}}{D_{ox}}\right)^{1/2}$$

hence

$$E_{p/2} - E_p = 2.2 \cdot \frac{RT}{nF} = \frac{56.5}{n} \text{mV} \tag{3.84}$$

This means that $E_{\frac{1}{2}}$ lies about midway on the SV wave between E_p and $E_{p/2}$ at the high, i.e. positive, side, as we have already mentioned in connection with Fig. 3.64. However, in practice and on the basis of the above arguments (1)–(3), the position of $E_{\frac{1}{2}}$ will deviate from this more or less theoretical approximation where the sphericity term of eqn. 3.80 has been neglected. Further, with incompletely reversible redox couples, e.g. for quasi-reversible systems (cf., pp. 125–126), there will of course be more deviations, but at any rate the linear relationship between i_p and C appears to remain.

For the MFE and a solution again that contains only ox of a reversible redox couple, De Vries and Van Dalen[101,104] studied anodic SV at a plane MTFE with film thicknesses L of 4–100 μm, assuming linear diffusion; on the basis of their mathematical model, Barendrecht[88,89] made additional numerical calculations, from which he drew the following conclusions:

(1) i_p at $L < 25\,\mu m$ is nearly proportional to v (pure thin-layer diffusion), whereas at $L > 25\,\mu m$ i_p becomes proportional to $v^{1/2}$ (semi-infinite diffusion; cf., eqn. 3.81);

(2) i_p at small v values ($< 1/3\,V\,min^{-1}$) is hardly influenced by L;

(3) $b_{1/2}$, either at small L and/or at low v (e.g., $2\,mV\,s^{-1}$), can be very narrow (about 40 mV), which favours peak resolution in the case of analyte mixtures;

(4) E_p depends on both v and L to such an extent that its position can be even 40 mV before (more negative than) the polarographic half-wave potential $E_{\frac{1}{2}}$.

The above conclusions clearly indicate that the use of very thin mercury films gives maximal sensitivity as it enables fast sweep rates to be combined with excellent peak resolution.

De Vries and Van Dalen[104b] gave the following more explicit quantitative relationships between i_p, E_p and $E_{\frac{1}{2}}$:

$$i_p = -k_1'' n^2 C_a A L v \qquad (3.85)^*$$

or

$$i_p = -k_2'' m n^2 C A v \tau \qquad (3.86)^*$$

obtained by substituting eqn. 3.78b in eqn. 3.85, and further

$$E_p - E_{\frac{1}{2}} = \frac{2.3\,RT}{nF} \log \left(\frac{\delta n F L v}{D_{ox} RT} \right) \qquad (3.87)$$

where δ is the diffusion layer thickness.

Eqn. 3.87 confirms the effect of v and L on E_p, as already mentioned in conclusion (4), and moreover an additional effect of δ.

Eqn. 3.86 shows the linear relationship between i_p and the original analyte concentration C, and also the influence of the mass transfer coefficient, i.e. its type of motion (stirring or rotation). However, basically eqn. 3.85 indicates that i_p is directly proportional to $ALC_a = VC_a$, i.e. the mass of metal dissolved in the ultra-thin mercury; this relationship was also confirmed by Roe and Toni[105] in their approximate equation

$$|i_p| = \frac{n^2 F^2 C_a A L |v|}{2.7\,RT} \qquad (3.88)$$

In fact it means that i_p is determined by the mass of metal deposited, rather than its concentration in the thin film[106].

Interferences in SV. As a consequence of the pre-concentration, the influence of charging and impurity currents will be slight. Moreover, as the surface area of stationary electrodes is constant, their capacity C can only

*The negative sign of i_p (in anodic SV) is in agreement with the convention of considering an anodic current as negative (cf., Fig. 3.64).

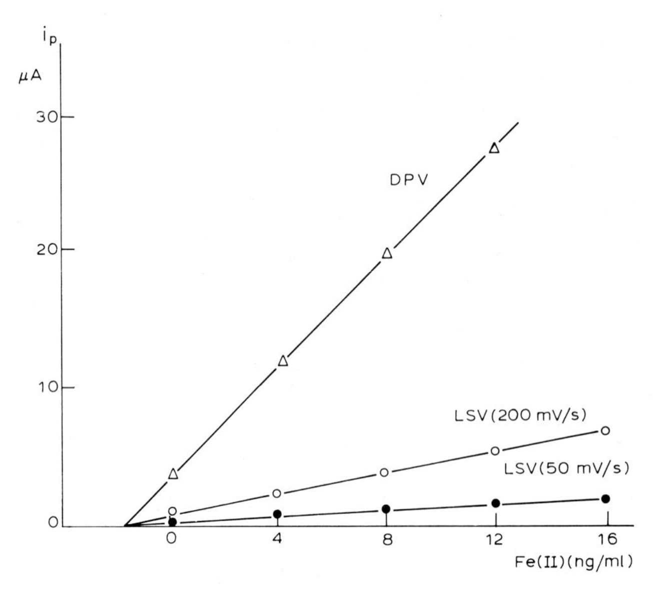

Fig. 3.66. Effect of various stripping modes in anodic SV of Pb(II).

change with the potential E by alteration of the double layer at mercury around its electrocapillary zero point E_m (see Figs. 3.18 and 3.20). However, if in the determination of extremely low analyte concentrations (10^{-10} M or less) the effect of i_C on i_p cannot be neglected, one can use other stripping modes instead of ordinary linear sweep potentiometry. For instance, the aforementioned interrupted potential sweep voltammetry completely avoids i_C interference; more often one applies the more common voltammetric variants such as derivative linear sweep, stair-case, pulse, sinusoidal AC or square-wave voltammetry. By means of the latter stripping modes one minimizes not only the charging current but also impurity current interferences. In Fig. 3.66[107] as an example the extreme sensitivity of the differential pulse vs. the linear sweep stripping technique at an MFE (mercury on wax-impregnated graphite) is shown for the determination of lead; by this method the selective simultaneous determination of 4 ng ml^{-1} of Pb(II) and 4 ng ml^{-1} of Cd(II) could easily be achieved.

A completely different technique is chemical stripping, which represents non-faradaic chronopotentiometry, that is, without any current during stripping. As an explanation we describe the most usual procedure given originally by Bruckenstein and co-workers[108] and more recently extensively developed by Jagner and co-workers[109]. On a Pt microelectrode (WE), the amalgamating metals are electro-deposited from their ionic solution to which Hg^{2+} has been previously added and the current is stopped by opening the cell circuit; from that moment on the metals are successively stripped from the Hg film by oxidation by means of either the Hg^{2+} in the solution or dissolved oxygen, during which the potential change between the working and reference

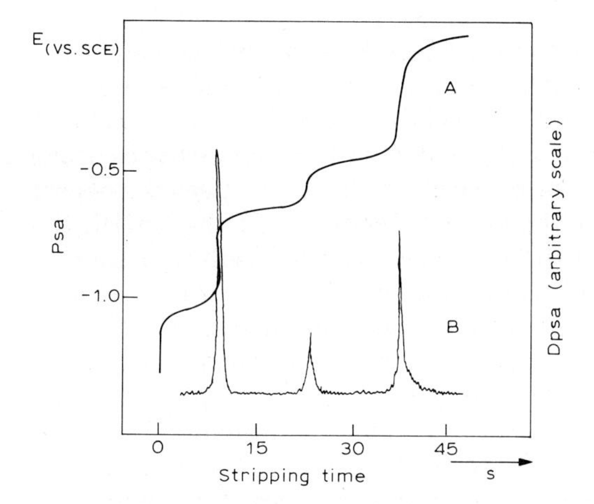

Fig. 3.67. *E–t* curves of potentiometric stripping analysis (Psa) and its derivative (Dpsa) for 0.1 μM Zn, Pb and Cd in 0.1 *M* NaCl (chemical stripping).

electrodes is recorded. The *E–t* curves for potentiometric stripping analysis (PSA) and its derivative (DPSA) for 0.1 μM Zn, Pb and Cd in 0.1 *M* NaCl, depicted in Fig. 3.67 (A and B), clearly illustrate the selectivity of the chemical stripping according to the half-wave potentials and in the amounts corresponding to the respective transition times.

In this specific case the MFE was rotated, an inert gas flow of nitrogen held at about 100 ml min^{-1} and the working electrode potential decreased in steps of 0.1 V from -0.5 to -1.0 V with plating for 1 min at each potential; next the potentiostat was disconnected, the other conditions being maintained, and the potentiometric stripping curve was recorded. In 1979 Jagner described the PSA of Hg itself by electro-depositing it on a glassy carbon electrode plated with a thin layer of copper and next stripping by oxidation with $KMnO_4$ previously added to the sample. Originally Jagner and co-workers used an HMDE for the analysis of amalgamating metals with some success, but later an MFE proved to be even more suitable; further, they applied MFE rotation or mechanical or magnetic stirring of the solution. It is on the basis of all this experience that Radiometer (Copenhagen)[110] developed their ISS 820 Ion Scanning System (more or less automatic, see Part C), consisting of a three-electrode circuit with a PTFE-covered glassy carbon electrode (GCE) for the determination of trace amounts of Cu, Pb, Cd, Zn, Bi, In, Tl and Hg (detection limit $< 0.2 \mu g\,l^{-1}$) in environmental and chemical analysis.

Irrespective of the specific choice of the stripping step, some interferences may still remain[88,89], as follows.

(a) Formation of intermetallic compounds, either between an analyte metal and the metal of the electrode or on the electrode, or between analyte metals themselves, e.g., Pt with Sb, Sn and Zn; Au with Cd, Mn, Sn and Zn; Ag with Cd and Zn; Hg with Co, Fe, Mn and Ni; Cu with Cd, Mn, Ni and Zn; Fe with Mn; Ni with Mn, Sb, Sn and Zn. However, Pb and Tl give no problems. The interference is caused by the often occurring effect, known since about 1900[111], that the standard potential E^0 of the intermetallic compound can be shifted to a more positive value than those of its respective metals; the intermetallic compound yields its own i_p at the cost and possibly with the disturbance of the i_ps of the separate metals involved. For research on this subject see the papers by Kemula and co-workers[85,93,112] and Ficker and Meites[113]. At the HMDE the interference by the intermetallics can be avoided by using solution concentrations of 10^{-5}–10^{-6} M and at the MFE by using lower concentrations; these conditions can be best tried out by previous standard experiments.

(b) The occurrence of maxima in anodic stripping has been reported by Hickling et al.[114]; the phenomenon seems analogous to that of the polarographic maxima (see p. 132 and ref. 115) and of the maximum in cathodic LSV at an HMDE (for $3 \cdot 10^{-3}$ M Cu^{2+} in acetate buffer at 3.7 mV s^{-1}), which maximum was suppressed by 0.005% phenol red[116]. It is clear that traces of surface-active agents or organic substances such as gelatin may have such a beneficial effect in the stripping step of SV in some instances, but in other instances can cause disturbances (cf., tensammetry, p. 152); therefore this should always be verified because the maxima concerned may be difficult to distinguish from the desired i_ps.

(c) Interaction with anions of the solution may hamper the evaluation of the stripping voltammogram in the presence of halides, as the mercury of the electrode may be oxidized to mercury(I) halide (Hg_2Cl_2, Hg_2F_2), yielding a broad peak in the neighbourhood of 0 V vs. SCE; therefore, the connection with the SCE should be a nitrate bridge.

(d) Exchange reaction between a metal from the amalgam with the ion of another less noble metal during the concentration step, e.g. when Zn^{2+} and Tl^+ are simultaneously present in the analyte solution:

$$Zn(Hg) + 2Tl^+ \rightarrow 2Tl(Hg) + Zn^{2+}$$

As a result, the Tl(Hg) concentration at the end of the electro-deposition time τ has become relatively greater than that corresponding to the original concentration ratio Tl/Zn of the solution.

Thus far we have considered primarily anodic SV, as it represents the most important application of SV; however, cathodic SV is also of increasing interest, especially for the analysis of halides and sulphide, although the more irregular course of the different steps in this technique makes its sensitivity and quantitative reliability less favourable. In treating this subject we shall confine ourselves more or less again to the experience of Barendrecht[88,89], but without omitting to mention the more recent publications on chloride and bromide[117], sulphide[118] and iron(II) ions[119]. In cathodic SV the electrode type and the electrolytic cell set-up are the same as in the anodic mode, but the electrodics in the concentration and stripping steps are different, and the rest period is omitted.

In the concentration step the anodic oxidation must yield between the electrode material (generally Hg) and the analyte anion an insoluble compound or complex that adheres strongly to the electrode surface; with mercury one may favour the adherence by anodically polarizing for a pre-set time (5 min). As no diffusion of the insoluble layer or formation of intermetallic compounds need be feared, an MFE with its larger surface to volume ratio and its rigid foundation to the insoluble film (no tearing) is more attractive to the HMDE; in iodide analysis an Ag electrode is suitable[120]. The stirring rate should not be high in order to avoid irregularities due to intermediate disruption of the deposited film; τ is 5–15 min and must not be longer as determinations below 10^{-6} *M* are unsatisfactory, while the anodic deposition potential, E_d, for this concentration should not be lower than + 0.4 V vs. SCE. On the other hand, a concentration above 10^{-4} *M* may cause passivation of the Hg anode[121], unless very low τ values are applied. To prevent analyte contamination by chloride a nitrate bridge to the SCE should be used, as well as an acidified 0.1 *M* KNO_3 supporting electrolyte; for both purposes the nitrate used must be previously freed from traces of chloride by electrolysing its concentrated solution at an Hg anode and a 1 *N* mercury(I) sulphate auxiliary electrode (at + 0.4 V vs. SCE); however, in the long term the Hg is oxidized so that a certain chloride blank always remains, resulting in the relatively high detection limit of $5 \cdot 10^{-6}$ *M* chloride.

In the stripping step the cathodic redissolution sets the anion free from the deposited film of insoluble compound or complex. For chloride a linear sweep of 2–5 mV s^{-1} is applied from + 0.4 to + 0.1 V vs. SCE without stirring; the E_p of 10^{-5} *M* Cl^- then lies at about + 0.3 V vs. SCE. As the redissolution is based on the direct removal of the anion from the crystal lattice the current peak is unsymmetrical, viz., in the initial part the curve rises exponentially, next becomes horizontally rounded off and subsequently, when a monolayer thickness of the deposit has been reached, falls steeply proportionally to the decreasing fraction of the electrode surface covered. Because the solubility products of Hg_2Br_2 and Hg_2I_2 are lower than that of Hg_2Cl_2, the E_p values of Br^- and I^- are more negative than for Cl^-; therefore, when simultaneously present, these halides yield badly reproducible anodic oxidation currents superimposed on the cathodic chloride peak between + 0.4 and + 0.1 V vs. SCE. However, on the basis of the above-described mechanism of redissolution and even with chloride only one can hardly expect a strictly linear relationship between i_p and the analyte concentration *C*; for this reason a chronocoulometric evaluation, i.e., the integration of current as a function of time for the entire peak, is the best approach; nevertheless, relative standard deviations for chloride (10^{-5} *M*) below 10% have not yet been obtained.

In the stripping voltammetry, in general it is anodic SV which, owing to its extreme sensitivity and selectivity together with its cheapness, has gained so much analytical importance that for instance the Kernforschungsanlage Jülich (F.R.G.) recently (1983) replaced their atomic absorption spectrometer with an SV system for the simultaneous determination of Cu, Cd, Pb, etc.

References pp. 224–228

Experimental details on the anodic stripping analysis of metals such as Bi, Pb, Cd, Zn, Mn, Sb and Sn in Cu at the HMDE, and Ag in Cd at a glassy carbon electrode (GCE), have been given by Verbeek and co-workers[122].

3.3.2.2. *Voltammetry at hydrodynamic electrodes*

In all previous sections on voltammetry the analytical measurement proper is carried out in principle without rotation or strirring, so mass transfer takes place merely by diffusion and only occasionally slight rotation or stirring is applied simply to sustain and thus to improve and/or accelerate the diffusion process by some convection; here one finds oneself somewhere on a borderline (see the note about stripping with rotation on p. 176). However, in so-called hydrodynamic voltammetry, i.e., voltammetry at hydrodynamic electrodes, the working electrode is rotated at a fairly high speed or the solution is vigorously stirred, so that mass transfer takes place by forced convection rather than by diffusion. To obtain reproducible measurement results (i–E curves usually of sigmoidal shape) these electrodes, among which we can distinguish (1) rotating disc, ring and wire* electrodes, (2) plate, conical and tubular electrodes and (3) vibrating dme and streaming mercury electrodes, require careful and precise construction of the electrode itself and of the measuring cell. They are also used for detection in the continuous analysis of flowing streams (see Part C). We shall treat hydrodynamic voltammetry more specifically for rotating electrodes on the one hand and other hydrodynamic electrodes on the other.

3.3.2.2.1. *Voltammetry at rotating electrodes*

Here we have to deal with three types (see Fig. 3.68), viz.: (a) the rotating disc electrode (RDE), and (b) the rotating ring electrode (RRE) and the rotating ring-disc electrode (RRDE). The construction of the latter types suits all purposes, i.e., if the disc or the ring is not included in the electric circuit, it yields an RRE or an RDE, respectively, and if not an RRDE, where either the disc forms the cathode and the ring the anode, or the reverse.

Fig. 3.68 shows the longitudinal section and the bottom view of the electrode and also a section of the streamlines of flows in the solution near the bottom and symmetrical to the axis of rotation of the electrode; the electric circuit is provided for instance by brush contacts at the top of the brass shaft of the disc and/or at the brass collar of the ring; it is essential for the proper functioning of the electrode that its rotation is kept rigidly centralized by means of bearings in order to exclude swinging as far as possible.

In order to asses the analytical aspects of the rotating electrodes we must consider the convective–diffusion processes at their bottom surface, and in view of this complex matter we shall confine ourselves to the following conditions:

(1) as a model of electrode process we take the completely reversible equilibrium reaction:

---

*For a rotating wire electrode see under amperometric titrations.

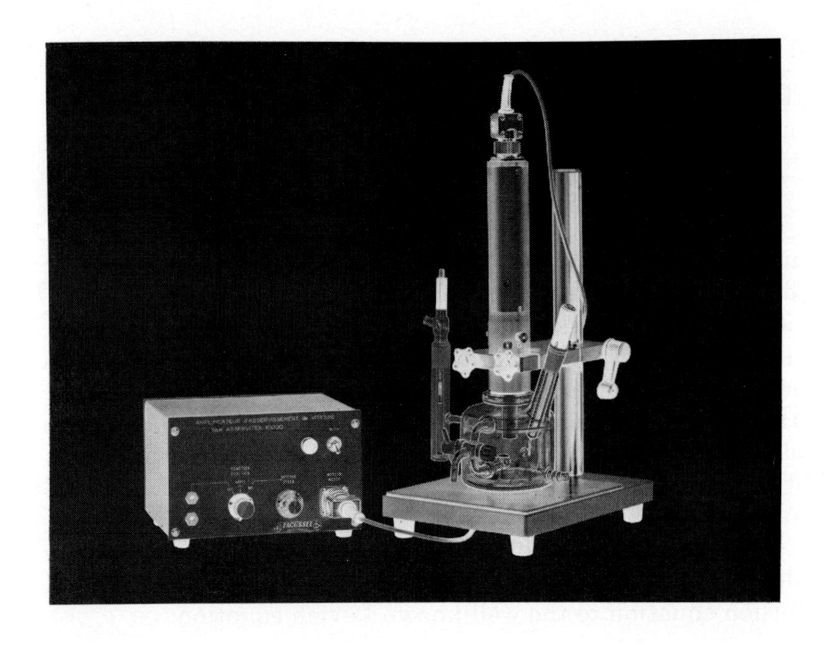

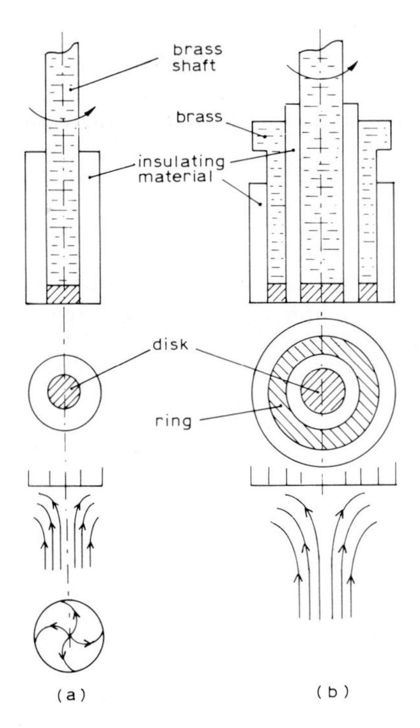

Fig. 3.68. Rotating electrodes for voltammetry, (a) RDE, (b) RRE and RRDE.

ox + $ne \rightleftharpoons$ red

(2) The rotation rate should ensure forced convection on the one hand, but laminar flow on the other, so that it remains well below the conditions of the critical Reynolds number, above which turbulent flow sets in:

$$Re_{cr} = v_{ch} l/\nu$$

where v_{ch} is the characteristic velocity (cm s^{-1}) chosen, l the characteristic length (cm) and the kinematic viscosity $\nu = \eta_s/d_s$;

(3) a steady state has been attained, i.e., $dv/dt = 0$;

(4) for the relationship between i and the bulk concentration C we refer to elaborate mathematical derivations elsewhere[123]; first we are interested in the resulting equations for i and C at sufficiently negative potentials, i.e. for the limiting cathodic current $i_{l,c}$; next we shall consider the influence of less negative potentials; the equations concerned will be given for the RDE, RRE and RRDE in succession.

At the RDE the velocity profile obtained by Karman and Cochran (see ref. 124) and depicted in Fig. 3.68a leads via solution of its differential convection–diffusion equation to the well known Levich equation:

$$i_{l,c} = 0.620\, nFAD_{ox}^{2/3} \omega^{1/2} \nu^{-1/6} C_{ox} \tag{3.89}$$

where $i_{l,c}/\omega^{1/2} C_{ox}$ is often called the Levich constant. Comparison of eqn. 3.89 with the more general eqn. 3.78, $i = km(nFCA)$, indicates that $k = 0.620$, the constant appropriate for the RDE, and $m = D^{2/3}\omega^{1/2}\nu^{-1/6}$, the mass transfer coefficient for the RDE.

Now we can write the foregoing Levich eqn. 3.89 as a steady-state diffusion layer equation (cf., eqn. 3.4):

$$i_{l,c} = km_{ox}(nFCA) = nFAD_{ox}^{2/3} C_{ox}/\delta$$

where δ is the thickness of such a hypothetical stagnant layer.

However, at a less negative potential the limiting current is not attained, so that there is a dynamic equilibrium between c_{ox} and c_{red} at the electrode; therefore, according to eqn. 3.3 we obtain for ox

$$i_c = nFAD_{ox}^{2/3}(C_{ox} - c_{ox})/\delta$$

and for red

$$i_a = -nFAD_{red}^{2/3}(C_{red} - c_{red})/\delta$$

while $i_{l,a}$ would have been

$$i_{l,a} = -nFAD_{red}^{2/3} C_{red}/\delta$$

The negative sign of i_a (according to convention) illustrates the counteraction of i_a versus i_c. If we follow the same arguments as given in Section 3.2.2 (p. 106), but now applying the above equations for $i_{l,c}$, i_c, i_a and $i_{l,a}$, we obtain for rapid electron transfer the i–E relationship (cf., eqn. 3.14):

$$E = E^0 + \frac{RT}{nF} \ln \left(\frac{D_{red}}{D_{ox}}\right)^{2/3} + \frac{RT}{nF} \ln \left(\frac{i_{l,c} - i}{i - i_{l,a}}\right) \quad (3.90)$$

For a solution with only ox or only red, then $i_{l,a}$ or $i_{l,c}$, respectively, disappears from this equation and for the cathodic or anodic half-wave potentials we obtain the same relationship (cf., eqns. 3.15 and 3.16):

$$E_{\frac{1}{2}} = E^0 + \frac{RT}{nF} \ln \left(\frac{D_{red}}{D_{ox}}\right)^{2/3} \quad (3.90a)$$

Hence the picture of the cathodic and anodic waves obtainable for a completely reversible redox couple by means of the RDE corresponds fully with that in Fig. 3.9; the value of i, i.e., the height of the sigmoidal waves, is linearly proportional to $\omega^{1/2}$ and to C (see eqn. 3.89 and the Levich constant). If for a well chosen combination of C and E a plot of i against $\omega^{1/2}$ deviates from a straight line passing through the origin, then in the kinetics of the electrode reaction we have to deal only with a rapid electron transfer (cf., Fig. 3.10) or even with a slow electron transfer (cf., Fig. 3.11), in which latter instance the transfer coefficient α plays an appreciable role (cf., eqns. 3.17 and 3.18).

At the RRE the derivation[123] of the Levich equation requires reconsideration of the convection–diffusion equation, which results in

$$i_{R,l,c} = 0.620 nF\pi(r_3^3 - r_2^3)^{2/3} D_{ox}^{2/3} \omega^{1/2} \nu^{-1/6} C_{ox} \quad (3.91)$$

where r_3 and r_2 are the outer and inner ring radii, respectively. Therefore, if the radius of the disc of an RRDE is r_1, the ratio of i_R and i_D at any potential and with all other experimental conditions remaining the same would be

$$i_R/i_D = \frac{(r_3^3 - r_2^3)^{2/3}}{r_1^2} = \left(\frac{r_3^3}{r_1^3} - \frac{r_2^3}{r_1^3}\right)^{2/3} = \beta^{2/3} \quad (3.92)$$

At the RRDE, however, so-called collection experiments, the actual purpose of its integrated construction, are often carried out, i.e., in a solution of, for instance, only ox of a reversible redox couple ox + $ne \rightleftharpoons$ red, a cathodic current at the disc i_D produces red, which reaches the ring where it is completely reoxidized to ox, because the ring is maintained at a sufficiently positive potential E_R; an RE, generally an SCE, is used together with a bipotentiostat for precise adjustment of the potentials E_D and E_R required.

Our analytical interest now is to know i_R as a direct measure of disc-generated red collected at the ring. Again the solution[123] of the differential convection–diffusion equation needs a complex mathematical treatment, resulting in an involved equation for the so-called collection efficiency

$$N = -i_R/i_D \quad (3.93)$$

which is dependent on the radii of the disc (r_1) and the ring (r_2, r_3), but independent of all other relevant parameters such as ω and C; in practice, N values for different ratios r_3/r_2 and r_2/r_1 can be read directly from a published table[125]. The advantage of N is that it can be applied directly to the basic Levich equation (cf., eqn. 3.89) so that

$$i_R = 0.620NnFA_R D^{2/3}\omega^{1/2}\nu^{-1/6}C \quad (3.94)$$

where $A_R = \pi(r_3^2 - r_2^2)$, the surface area of the ring. Of course, N can be measured experimentally in a calibration sample; for commercial RRDEs the N value is often given, e.g., 0.25 ± 0.05 for the Tacussel[126] types EAD 4000 and EAD 10000. Thus far in collection experiments we have treated primarily oxidation at the ring of disc-generated red; in principle one can equally well carry out the reverse process, i.e., reduction at the ring of disc-generated ox. However, the situation becomes different if, for instance, the reduction at the ring of ox of an analyte solution is preceded by a partial decrease of ox by reduction at the disc, in which event one talks of shielding experiments. To explain these, let us reconsider eqn. 3.92, which for $i_{R,l}/i_{D,l} = \beta^{2/3}$ in fact indicates the value of $i^0_{R,l}$, i.e., its relationship with $i_D = 0$; hence, as soon as i_D acquires a finite value, the flux of ox to the partly shielded ring has decreased, so that the limiting ring current becomes

$$i_{R,l} = i^0_{R,l} - Ni_D = i^0_{R,l}(1 - N\beta^{-2/3}) \quad (3.94)$$

where $(1 - N\beta^{-2/3})$ represents the so-called shielding factor.

The aforementioned experiments at rotating electrodes concerned merely steady-state conditions; so-called transients[123] at these electrodes, e.g., with potential or current steps, as well as with hydrodynamic modulation, i.e., variation of ω with time, are, as a consequence of their non-steady-state conditions, less important in analysis and therefore will not be treated here.

With regard to the electrode materials and the working reliability, it suffices to refer to the Tacussel RRDEs[126], which can be delivered with PTFE as an insulating material and to the standard model with pure platinum as the disc ($r_1 = 2.00 \pm 0.01$ mm) and the ring ($r_2 = 2.20 \pm 0.01$ mm, $r_3 = 2.40 \pm 0.01$ mm), all having been set in the PTFE stem ($r_4 = 7.00 \pm 0.10$ mm); different materials can be ordered, viz., for both the disc and ring Pt–Rh alloy, Au, Ag, Cu, Fe or Ni, and for the disc high-quality glassy carbon; further, within limits of machining possibilities, the customer can furnish other materials for the disc and, in certain instances, also for the ring. The speed adjustment ranges are 40–4000 and 100–10000 rpm with a regulation accuracy of ± 10 and ± 25 rpm for the EAD 4000 and EAD 10000, respectively; the maximum deviation between the geometric axis and the rotation axis at the active end does not exceed 0.03 mm.

Interesting fundamental studies and analytical applications of RRDEs have been published by Bruckenstein and co-workers; the attraction of collecting experiments with the RRDE lies more especially in the fact that metals with different oxidation states are becoming more accessible to analysis, e.g., Cu(II) and Cu(I)[127], U(VI) and U(V), Fe(III) and Fe(II). Shielding experiments were carried out for Bi(III) and Bi(0)[128]. Special use of stripping voltammetry with collection at a glassy carbon RRDE for the determination of tin in the presence of lead was proposed by Kiekens et al.[129]; after cathodic electrode-deposition

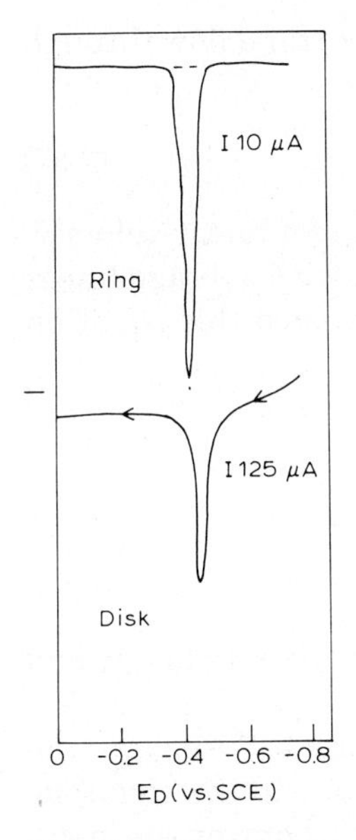

Fig. 3.69. SV with collection at GC-RRDE for Sn determination in presence of Pb.

(E_{dep} = -1 V) of both metals on the disc from an Sn(II) and Pb(II) solution the deposit is anodically stripped by a linearly scanning potential (up to 0.6 V), which enables a fraction of the oxidized species reaching the ring to be selectively detected at a fixed potential (E_R = +1 V) (see Fig. 3.69); $\tau_{dep.}$ = 5 min, v = 0.1 V s^{-1}; ω = 188 rad s^{-1}; when working in a 1 M HCl solution of 10 μM Sn(II) and 2 μM Pb(II) it was shown that the metal deposit became reoxidized to the divalent ions with an anodic peak i_R at about -0.45 V, but that only the Sn(II) was collected at the ring, yielding a sharp peak i_R in the analysis of Sn (calibration vs. dotted baseline).

3.3.2.2.2. *Voltammetry at other hydrodynamic electrodes*

The particular features of this technique are (a) plate, conical and tubular electrodes in contact with the flowing solutions and (b) vibrating dme and streaming mercury electrodes.

(a) For solutions with flow velocity U and kinematic viscosity ν Levich[124] derived the following limiting current equations (for flow along the electrode surface): for a plate electrode

$$i_l = 0.68nFCbl^{1/2}D^{2/3}U^{1/2}\nu^{-1/6} \tag{3.95}$$

where b is the width normal to the flow direction and l is the length of the plate in the flow direction, and for a conical electrode

$$i_l = 0.77nFCAl^{-1/2}D^{2/3}U^{1/2}\nu^{-1/6} \tag{3.96}$$

where A is the lateral area of the cone and l is its slant height.

Blaedel et al.[130] derived, according to the Levich procedure for a flow through a tubular electrode, the equation

$$i_l = 2.01nFC\pi r^{2/3} l^{2/3} D^{2/3} V_0^{1/3} \tag{3.97}$$

where r is the internal radius, l its length and V_0 the maximum linear velocity. In view of its use as a voltammetric detector and for a tube with average linear velocity $\bar{v} = \frac{1}{2}V_0$, Hanekamp and Van Nieuwkerk[131] rearranged this equation to

$$i_l = 2.01 \cdot 2^{1/3}\pi nFCD(\nu D^{-1})^{1/3} l^{1/3} r^{2/3} (\bar{v} l \nu^{-1})^{1/3}$$

or

$$i_l = 7.96nFCD(Sc)^{1/3} r^{2/3} l^{1/3} (Re_x)^{1/3} \tag{3.98}$$

where $Sc = \nu D^{-1}$ and $Re_x = \bar{v} l \nu^{-1}$ represent the dimensionless Schmidt and Reynolds numbers, respectively.

There are other variants of the above stationary electrodes[131], e.g. a two-plate type or constructions with a flow stream perpendicular to the electrode surface; all these electrodes are of great importance for the continuous monitoring (see Part C) of industrial processes and in sensing streams from analytical techniques such as in high-performance liquid chromatography (HPLC) with an electrochemical detector (ElCD); these detectors require careful design not only of the microelectrodes and microcell, but also of their mutual positioning (cf. the metrohm EA 1096 detector cell, with glassy carbon or carbon paste WE and an Ag–AgCl RE). Blaedel and co-workers originally developed the tubular Pt electrode (TPE) for the non-faradaic potentiometry of ox/red ratios in solutions passing through a rubber tube externally pulsed in order to avoid laminar flow[132]; later they[133] and others[134] applied the tubular electrode as a voltammetric detector. Further, a tubular mercury-covered graphite electrode for performing anodic stripping voltammetry in a flowing system was applied by Napp et al.[135].

(b) In fact, with the non-stationary vibration dme and a streaming Hg electrode we come back to the domain of rapid DC polarography (see Section 3.3.1.2.1) for, by leaving the long drop times in conventional polarography, we can increase the potential scan rate considerably. Thus with very short controlled drop times one even arrives at a transition between a dme and a streaming electrode; then scan rates of potential up to several hundred mV s^{-1} are possible, resulting in an extremely rapid analytical technique.

Cover and Connery[136] called their dme with a drop time of only 5 ms a vibrating dropping mercury electrode (VDME) and mentioned its advantages over the dme (cf., p. 133) as the suppression of maxima in the polarographic wave, the minimization or elimination of catalytic kinetic waves[23] and a diminution of adsorption effects[137]. An additional advantage of the short drop times in particular is the minimization of the anomalous dme response due to insoluble reaction products at the electrode surface, e.g. Hg_2I_2[138].

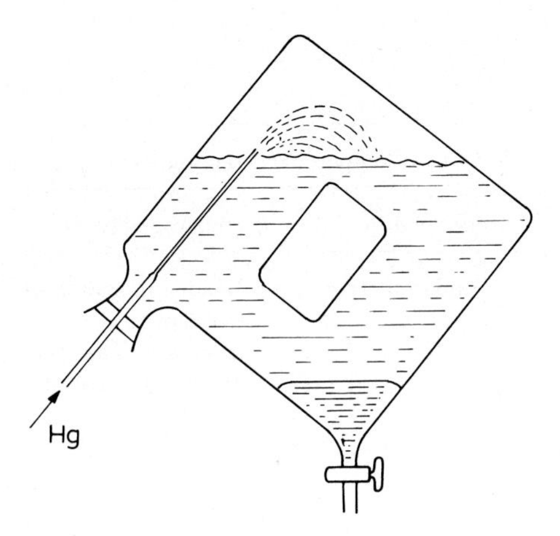

Fig. 3.70. Streaming mercury electrode.

One can imagine that to go from the VDME to a streaming mercury electrode would be another obvious step; many experiments in this connection have been carried out, e.g., the use of a jet-type Hg electrode as depicted in Fig. 3.70. However, the results obtained so far are not sufficiently reliable, so that the streaming Hg electrode has not found general acceptance.

Finally, hydrodynamic electrodes remain of much interest especially in continuous analysis; the equations concerned, although complex, show that the i_l values are linearly proportional to the analyte concentration.

3.3.2.3. *Voltammetric titration*

In voltammetric titration the reaction is pursued by means of voltammetry; interest is sometimes taken in the complete titration curve, but mostly in its part around the equivalence point in order to establish the titration end-point.

In a wider sense the subject of voltammetric titration would include the polarographic mode as a type of amperometric titration; however, we have already treated this in Section 3.3.1.3, because we prefer to use the term "voltammetric titration" in the strict sense, i.e., for faradaic non-polarographic titration techniques, a survey of which is given in Table 3.3.

As titrations require rapid and complete chemical reactions, it is mandatory to work under hydrodynamic conditions; hence, with stationary electrodes, which are often used in voltammetric titrations, the solution is stirred by a mechanical or magnetic stirrer; with rotating electrodes an additional stirring may sometimes be superfluous.

For a better understanding of voltammetric titrations it is useful to consider first the change in the voltammetric curve at a Pt redox electrode when titrating ox_1 with red_2, thus converting them gradually into red_1 and ox_2. As a classical example we take the following completely reversible redox couples:

TABLE 3.3
VOLTAMMETRIC TITRATION TECHNIQUES

I. Controlled-current techniques		II. Controlled-potential techniques	
Potentiometry	Biamperometry	Amperometry	Biamperometry
1. Constant-current potentiometric titration	2. Differential electrolytic potentiometric titration with the same electrodes of (a) equal surface area or (b) unequal surface area 5. Potentiometric dead-stop end-point titration (potentiometric reversed dead-stop end-point titration)	3. Amperometric titration	4. Dead-stop end-point titration (reversed dead-stop end-point titration)

(1) Ce(IV) ⇌ Ce(III) and (2) Fe(III) ⇌ Fe(II)* with standard reduction potentials, E^0, of 1.443 and 0.770 V, respectively. In Fig. 2.22 the curve B′B′ represents the zero-current potentiometry of the titration, Ce(IV) + Fe(II) → Ce(III) + Fe(III), under consideration. In Fig. 3.71 the voltammetric curves for this titration are given as a function of the titration parameter λ; in contrast with the presentation in Fig. 3.9, where E on the abscissa is plotted in the negative direction in view of following treatment of the polarographic cathodic wave, E is now plotted in the positive direction; however, according to convention $i_{cathodic}$ and i_{anodic} remain positive and negative, respectively.

Returning to the changes during titration in Fig. 3.71, it can be seen that the original cathodic wave of Ce(IV) becoming more and more anodic shift upwards until finally (i.e., at $\lambda = 1$) it has been converted into the anodic wave of Ce(III); at the same time the cathodic wave of Fe(III) comes into being and shifts downwards (versus the Ce wave) more and more, first reaching the steepness of the purely cathodic wave of Fe(III) (at $\lambda = 1$) and next becoming also partly anodic via the increasing excess of the Fe(II) titrant (beyond $\lambda = 1$); the latter increase also causes a continuing upwards shift of the entire voltammetric wave.

Fig. 3.71 offers an instructive possibility for considering the voltammetric titration techniques mentioned in Table 3.3, and we shall do so in the sequence given there.

*In this argumentation complete reversibility of the Fe(III)/Fe(II) couple has been assumed, although not really true under all circumstances.

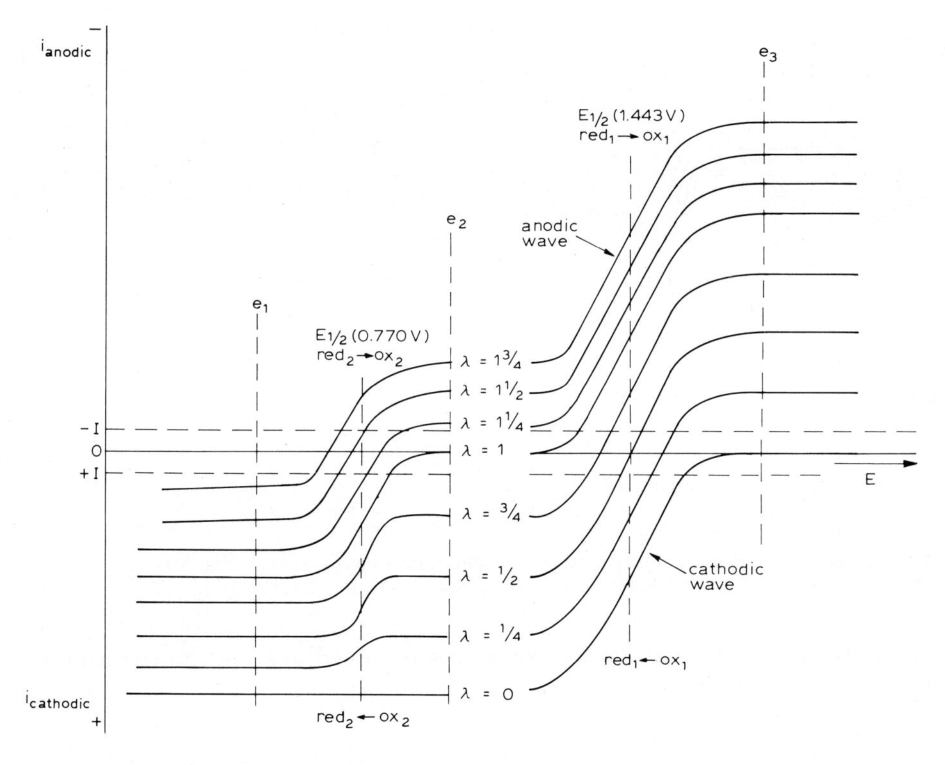

Fig. 3.71. Voltammetric curves at Pt electrode during titration of Ce(IV) with Fe(II).

3.3.2.3.1. *Constant-current potentiometric titration*

Let us introduce into the titrant one Pt indicator electrode vs. an SCE and maintain in the electric circuit a low constant current $+I$, as indicated by the broken horizontal line in Fig. 3.71. For this line we shall consider the successive points of its intersection with the voltammetric curves during titration and observe the following phenomena as expressed in the corresponding electrode potentials. Immediately from the beginning of the titration E remains high (nearly 1.44 V), but falls sharply just before the equivalence point (E = 1.107 V), and soon approaches a low E value (below 0.77 V) (see Fig. 3.72, cathodic curve $+I$).

With a low constant current $-I$ (see Fig. 3.71) one obtains the same type of curve but its position is slightly higher and the potential falls just beyond the equivalence point (see Fig. 3.72, anodic curve $-I$). In order to minimize the aforementioned deviations from the equivalence point, I should be taken as low as possible. Now, it will be clear that the zero current line (abscissa) in Fig. 3.71 yields the well known non-faradaic potentiometric titration curve (B′B′ in Fig. 2.22) with the correct equivalence point at 1.107 V; this means that, when two electroactive redox systems are involved, there is no real need for constant-current potentiometry, whereas this technique becomes of major advantage

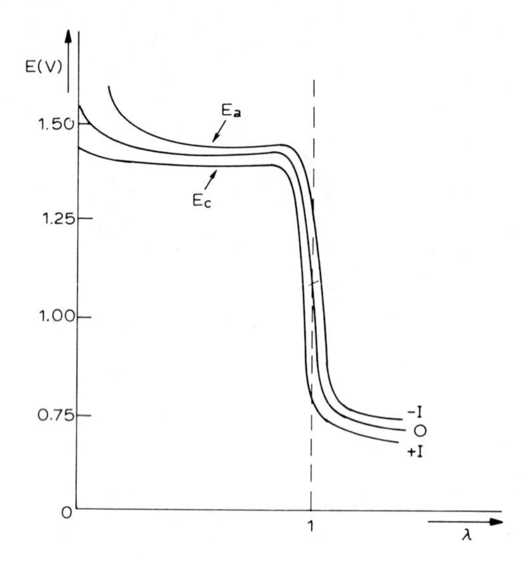

Fig. 3.72. Constant current potentiometric titration curves obtained from Fig. 3.71.

when only one such electroactive system is involved, e.g., in the titration of iodine with thiosulphate.

3.3.2.3.2. *Differential electrolytic potentiometric titration*

Here we have to deal with bipotentiometry, i.e., the use of two indicator electrodes, one cathodic and one anodic.

Such a system for the above titration of Ce(IV) with Fe(II) would imply the application of two Pt electrodes, which according to Fig. 3.72 would yield a $+I$ curve vs. a $-I$ curve; if the corresponding potentials were to be measured separately versus one SCE and if for each λ the potential values of both curves are averaged, the resulting mean curve represents virtually the zero-current potentiometric curve; hence such a procedure is again superfluous compared with simple non-faradaic potentiometry. However, bipotentiometry becomes more useful if under constant current conditions we measure the potential differences across the two Pt electrodes, i.e., differential electrolytic potentiometry. In reversible systems the electrodes ($A \approx 0.5\,cm^2$) should be of equal area; for irreversible systems it may be advantageous to make the inactive electrode slightly smaller than the active electrode[139]. In Fig. 3.73 the result of the differential method ($\Delta E = E_a - E_c$) is depicted for the titration of Ce(IV) with Fe(II) in Fig. 3.72; the rapid establishment of potentials for both reversible systems allows an equal electrode area [method (a)]. If one of the systems is comparatively slow, then the lag in the establishment of its potential can be largely prevented by a smaller, i.e., unequal area [method (b)] of the inactive electrode, which acquires stronger polarization through its higher current density.

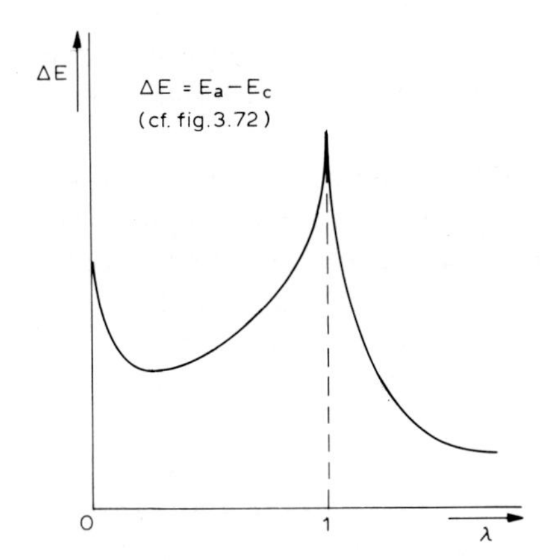

Fig. 3.73. Differential titration curve obtained from Fig. 3.72.

3.3.2.3.3. *Amperometric titration*

Again for the titration of Ce(IV) with Fe(II) we shall now consider constant-potential amperometry at one Pt indicator electrode and do so on the basis of the voltammetric curves in Fig. 3.71. One can make a choice from three potentials e_1, e_2 and e_3, where the curves are virtually horizontal. Fig. 3.74 shows the current changes concerned during titration; at e_1 there is no deflection at all as it concerns Fe(III) and Fe(II) only; at e_2 and e_3 there is a deflection at $\lambda = 1$ but only to an extent determined by the ratio of the i_l values of the Ce and Fe redox couples. The establishment of the deflection point is easiest at e_2 as it simply agrees with the intersection with the zero-current abscissa as being the equivalence point; in fact, no deflection is needed in order to determine this intersection point, but if there is a deflection, the amperometric method is not useful compared with the non-faradaic potentiometric titration unless the concentration of analyte is too low.

Of much greater importance, however, is the amperometric method for precipitation and complex-formation titrations. Here the advantages are as follows:

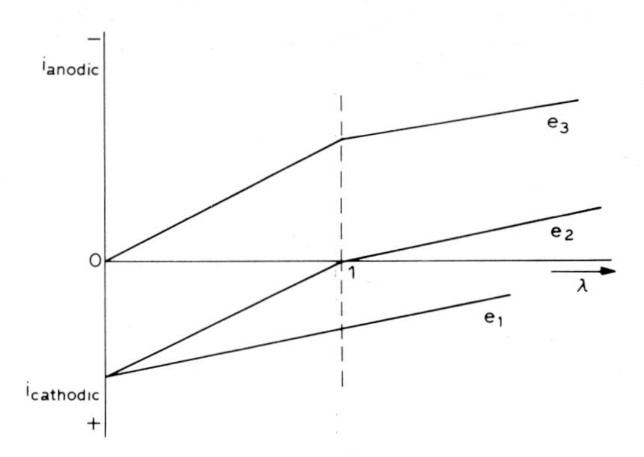

Fig. 3.74. Amperometric titration curves obtained from Fig. 3.71.

(1) rapid titration is possible on the basis of graphical evaluation where only a few current measurements (in the horizontal plateaux of the voltammogram) well before and after the end-point suffice;

(2) even where the potentiometric method may fail because of too low a solubility of the precipitate or moderate stability of the complex, the amperometric method still works with its measurements outside the end-point region, so that the excesses of analyte and titrant, by their mass action, can still suppress solubility or promote complexation, respectively;

(3) dilutions of even down to 10^{-4} M are still acceptable where potentiometry becomes inaccurate;

(4) strong electrolytes, if not taking part in the titration reaction, do not interfere and as the supporting electrolyte eliminate the migration current;

(5) a thermostat is not required if the temperature remains constant during the titration;

(6) as the horizontal plateaux of the voltammogram, between which the current measurements take place, are generally and preferably long, no severe requirements with regard to constancy of the potential have to be met.

For a full understanding of the above we shall consider the choice of electrodes and potentials.

Most successful is a rotating Pt wire microelectrode as illustrated in Fig. 3.75; as a consequence of the rotation, which should be of a constant speed, the steady state is quickly attained and the diffusion layer thickness appreciably reduced, thus raising the limiting current (proportional to the rotation speed to the 1/3 power above 200 rpm[140] and 15–20-fold in comparison with a dme) and as a result considerably improving the sensitivity of the amperometric titration.

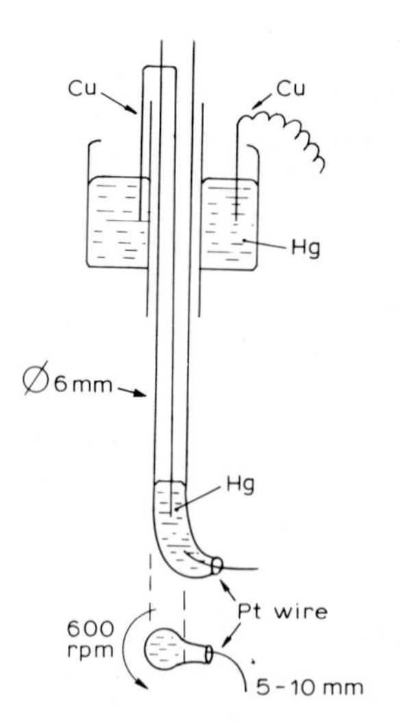

Fig. 3.75. Rotating Pt wire microelectrode.

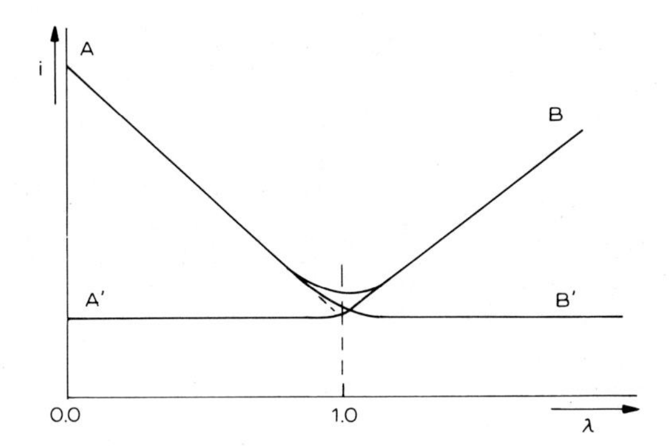

Fig. 3.76. Typical amperometric titration curves at a rotating Pt electrode (Fig. 3.75), A → B as precipitation titration at 1.0 V of Pb^{2+} with $Cr_2O_7^{2-}$, A′ → B as same titration but at 0.0 V, A → B′ as precipitation titration at 1.0 V of Pb^{2+} with SO_4^{2-} or a complex-formation titration with EDTA.

Only when discharge of hydrogen at the Pt electrode interferes, the dme with its large hydrogen overpotential must be used (see polarographic titration, Fig. 3.55). The following titration examples concern the use of a rotating Pt wire electrode; in order to obtain straight lines for accurate graphical establishment of the end-point from intersections, a volume addition correction is again recommended.

The typical curves illustrated in Fig. 3.76 can represent by A → B the precipitation titration at 1.0 V of Pb^{2+} with $Cr_2O_7^{2-}$ → $PbCr_2O_7$↓, by A′ → B the same titration but at 0.0 V, where Pb^{2+} is not cathodically reduced, and by A → B′ the precipitation titration at 1.0 V of Pb^{2+} with SO_4^{2-} → $PbSO_4$↓, or the complex-formation titration with EDTA. The last titration can be carried out for other metal ions, of course, provided that the potentials and other experimental conditions required for their i_l are applied. All the aforementioned titrations with the typical curves given can be carried out in the reverse direction. An additional advantage of amperometric titration is the admissibility of an approximate constancy of the applied potential, so that instead of an external emf source often an RE with an adapted potential can act as an internal emf source in short-circuit with the rotating Pt electrode. The effect of this principle in practice has been most strikingly demonstrated by Laitinen et al.[140] in the successive titration of iodide, bromide and chloride with silver nitrate (see Fig. 3.77). Here, a saturated mercury(I) iodide and a saturated calomel RE are applied with a KNO_3 bridge between the REs and the analyte solution. In the titration first AgI is precipitated in an ammoniacal solution (pH 9) with the former RE ($Hg/Hg_2I_2/KI$) in the measuring circuit, which yields a potential (−0.23 V vs. SCE) just negative enough to reach the i_l of $Ag(NH_3)_2^+$ → Ag without reducing O_2 to H_2O_2; one continues the titration up to 3–4 points in the rising current curve beyond the end-point.

Next AgBr is precipitated after addition of 0.8 *N* nitric acid and the titration is continued with the SCE, whose potential now suffices to reach the i_l of Ag^+ → Ag, again up to 3–4 points beyond the end-point; in this solution AgCl

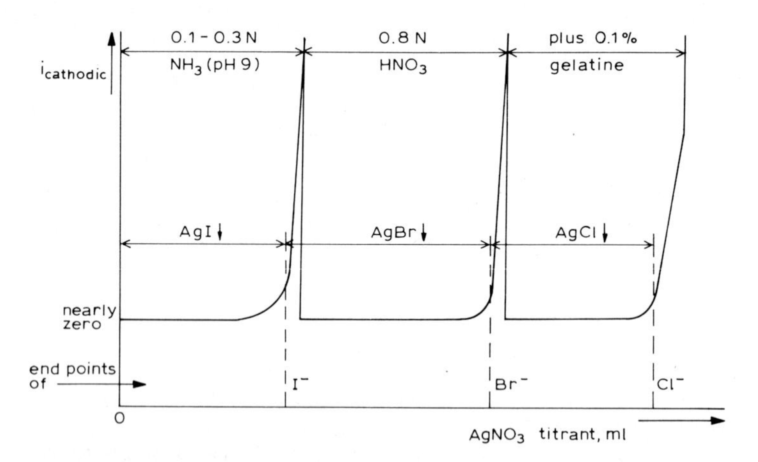

Fig. 3.77. Successive titration of I^-, Br^- and Cl^- with $AgNO_3$ according to Laitinen, using an internal emf source in short-circuit between a rotating Pt electrode and two reference electrodes as well as intermediate solution alterations.

particles, if any, would still cause a cathodic current of $Ag^+ \rightarrow Ag$ and dissolve, notwithstanding a large chloride excess. Finally, AgCl is precipitated after the addition of 0.1% gelatin and the titration continued with the same SCE until sufficiently beyond the end-point.

3.3.2.3.4. *Dead-stop end-point titration*

Here we deal with biamperometry for detection in the titration, i.e., a technique involving the measurement of a current across two indicator electrodes with a constant potential difference, ΔE, so that one could also talk of differential amperometric titration. To understand the possibilities of the method, let us consider the titration of Fe(II) with Ce(IV) at two Pt electrodes of potential difference $\Delta E = 100\,\text{mV}$ and examine the shift of the voltammetric curve with the titration parameter λ (see Fig. 3.78)*. As one of the two electrodes must be the cathode and the other the anode, the resulting current passing across them will be found around the abscissa as a function of ΔE shifting along this axis; Fig. 3.78 therefore shows that during the titration the current starts at zero, becomes maximum at $\lambda = \frac{1}{2}$, goes back to zero at $\lambda = 1$ and increases again beyond this end-point. This phenomenon is depicted qualitatively in Fig. 3.79 and can also be simply understood on the basis of the argument that a current can only pass if at the same time the cathode meets an oxidant and the anode a reductant, which condition is optimal at $\lambda = \frac{1}{2}$; but the cathode receives no oxidant at $\lambda = 0$ and the anode no reductant at $\lambda = 1$, so that under both conditions the system is polarized.

*In contrast with Fig. 3.71, it was assumed in Fig. 3.78 for simplicity that i_l for $Ce^{4+} \rightleftharpoons Ce^{3+}$ is lower than that for $Fe^{3+} \rightleftharpoons Fe^{2+}$.

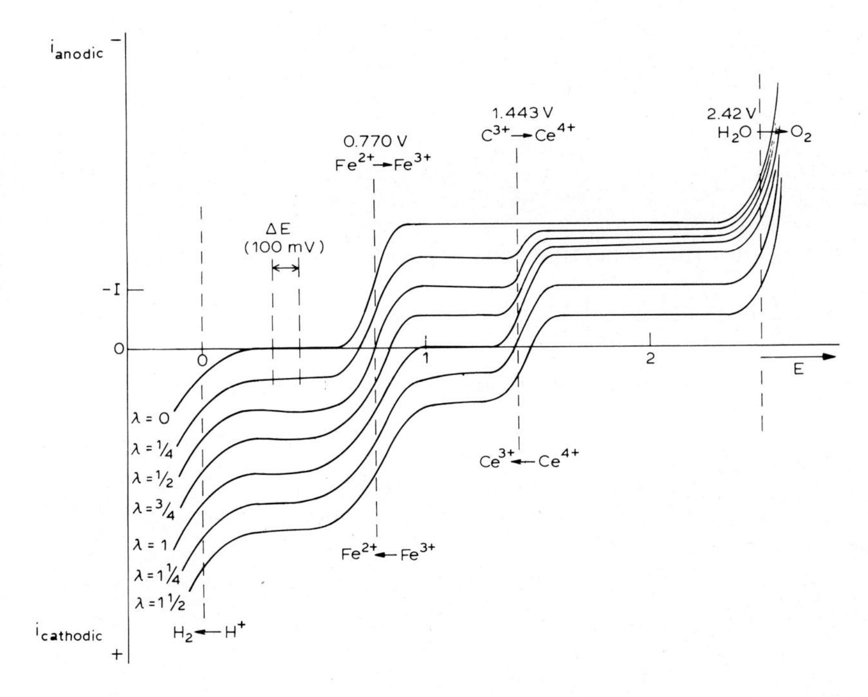

Fig. 3.78. Voltammetric curves at Pt electrode during titration of Fe(II) with Ce(IV).

It has been proved that in a redox titration between completely reversible couples the bulbous curve in Fig. 3.79 represents half of a hyperbola; here we shall follow the reasoning of Kies[141].

For the cathode potential we derived eqn. 3.14, but we now simplify it by assuming $D_{red} = D_{ox}$:

$$E = E_{\frac{1}{2},c} + \frac{RT}{nF} \ln \left(\frac{i_{c,d} - i}{i - i_{a,d}} \right)$$

and could in its derivation define $i_{c,d}$ as

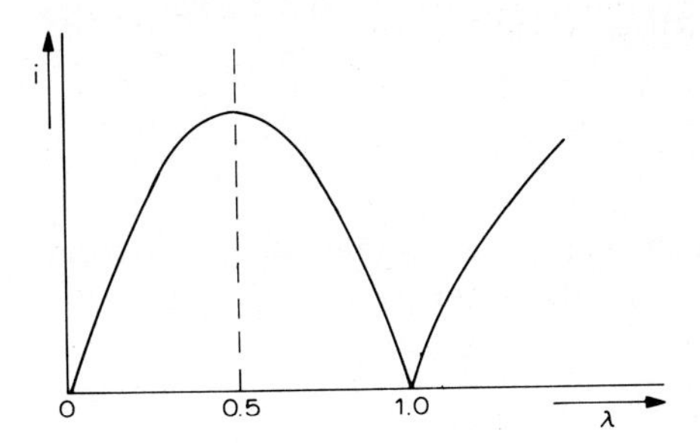

Fig. 3.79. Dead-stop end-point titration, i.e. measuring the current across two Pt-IE's with constant potential difference ΔE (differential amperometric titration), curves being obtained from Fig. 3.78.

$$i_{c,d} = \frac{nFD_{ox}A}{\delta} \cdot C_{ox} = k_{ox}C_{ox}$$

and $i_{a,d}$ as

$$i_{a,d} = -\frac{nFD_{red}A}{\delta} \cdot C_{red} = -k_{red}C_{red}$$

where k_{ox} and k_{red} are the respective proportionality constants. Hence

$$E_c = E_{\frac{1}{2},c} + \frac{RT}{nF} \ln \left(\frac{k_{ox}C_{ox} - i}{k_{red}C_{red} + i} \right) \tag{3.99}$$

For the anode potential we can write eqn. 3.16 in its original more extended form:

$$E_a = E_{\frac{1}{2},a} + \frac{RT}{nF} \ln \left(\frac{i - i_{c,d}}{i_{a,d} - i} \right)$$

where now

$$i_{c,d} = -k_{ox}C_{ox}$$

and

$$i_{a,d} = k_{red}C_{red}$$

and so

$$E_a = E_{\frac{1}{2},a} + \frac{RT}{nF} \ln \left(\frac{k_{ox}C_{ox} + i}{k_{red}C_{red} - i} \right) \tag{3.100}$$

As for the reversible couple $E_{\frac{1}{2},a} = E_{\frac{1}{2},c}$ (D_{ox} and D_{red} are assumed to be equal), the potential difference between both indicator electrodes will be

$$\Delta E = E_a - E_c = \frac{RT}{nF} \ln \left[\frac{(k_{ox}C_{ox} + i)(k_{red}C_{red} + i)}{(k_{ox}C_{ox} - i)(k_{red}C_{red} - i)} \right] \tag{3.101}$$

The previous assumption that $D_{ox} = D_{red}$ also makes $k_{ox} = k_{red} = k$; further, if the startng analyte concentration [ox] was C_a, and if a fraction x has been titrated, then $C_{red} = xC_a$ and $C_{ox} = (1 - x)C_a$, so that

$$\Delta E = \frac{RT}{nF} \ln \left\{ \frac{[k(1 - x)C_a + i][kxC_a + i]}{[k(1 - x)C_a - i][kxC_a - i]} \right\} \tag{3.102}$$

Next Kies introduced $\Delta' = e^{\Delta E / \frac{RT}{nF}}$, which via eqn. 3.102 yields the following expression for i as a function of the titration parameter x:

$$i = \frac{kC_a}{2} \cdot \frac{\Delta' + 1}{\Delta' - 1} \left[1 - \sqrt{\left\{ 1 - 4\left(\frac{\Delta' - 1}{\Delta' + 1} \right)^2 x(1 - x) \right\}} \right] \tag{3.103}$$

Further mathematical treatment shows that eqn. 3.103 represents the lower part of a hyperbola whose symmetry axis is the vertical at $x = \frac{1}{2}$; it clearly shows that $i = 0$ at $x = 0$ and $x = 1$, while the calculation of the top of the curve, i.e., its maximum at $x = \frac{1}{2}$, yields

$$i_m = \frac{kC_a}{2} \cdot \frac{\sqrt{\Delta'} - 1}{\sqrt{\Delta'} + 1} \tag{3.104}$$

Kies verified the above equations for instance on the titration of 1 mmol hexacyanoferrate(III) in 5.5 N NaOH with arsenite:

$$2Fe(CN)_6^{3-} + AsO_2^- + 4NaOH \rightarrow 2Fe(CN)_6^{4-} + AsO_4^{3-} + 2H_2O + 4Na^+$$

After the usual corrections for analyte impurity, potential drop in the solution and volume increase during titration, the experimental results were in perfect agreement with the theoretical hyperbolic curve.

For reverse titration, i.e., of arsenite with hexacyanoferrate(III), the situation will be the same as in Fig. 3.79; the part of the curve beyond the end-point is a hyperbola again, for which Kies gave the following equation:

$$i = (y + 1)\frac{kC_{red}}{2} \cdot \frac{\Delta' + 1}{\Delta' - 1}\left[1 - \sqrt{\left\{1 - 4\left(\frac{\Delta' - 1}{\Delta' + 1}\right)^2 \cdot \frac{y}{(y + 1)^2}\right\}}\right] \tag{3.105}$$

where the excess of hexacyanoferrate(III) is given by $y = C_{ox}/C_{red}$, while for high y values the curve approaches a horizontal asympote; eqn. 3.105 is obtained from eqn. 3.104 by replacing C_a by $(C_{ox} + C_{red}) = (y + 1)C_{red}$ and the C_{ox} fraction x by $y(y + 1)$. However, when Foulk and Bawden[142] titrated iodine in KI solution with thiosulphate at two Pt electrodes, they obtained Fig. 3.80 and introduced the term "dead-stop end-point method", as beyond the end-point ($\lambda = 1.0$) the current fails to appear; the explanation is that, while the redox couple I_3^-/I^- is reversible at the cathode, the tetrathionate appears to be electrochemically irreducible.

The reverse titration of thiosulphate with iodine is depicted in Fig. 3.81 and is often called the "reversed dead-stop end-point method".

Instrumentally it is common to apply a "magic eye" as a very sensitive and easily perceptible indication of the dead-stop transition of the titration end-point.

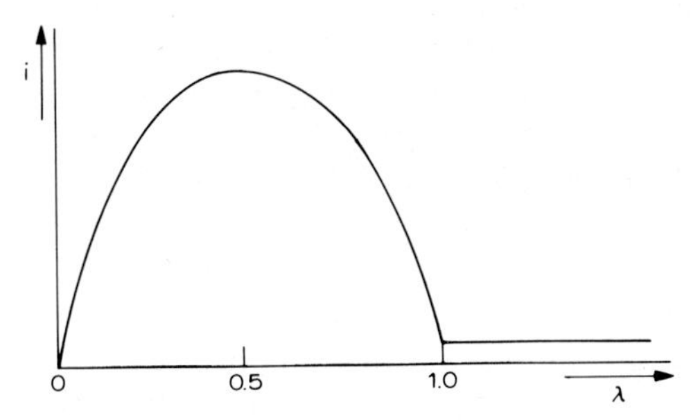

Fig. 3.80. Dead-stop end-point titration of I_3^- with thiosulphate.

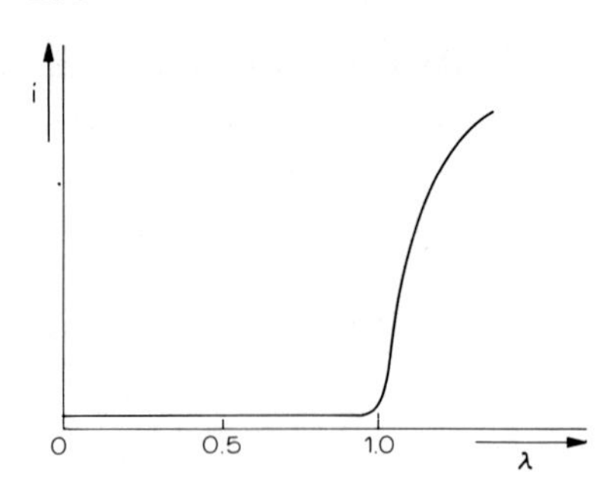

Fig. 3.81. Reversed dead-stop end-point titration of thiosulphate with I_3^-.

When we compare Figs. 3.80 and 3.81 with Fig. 3.79, it does not seem entirely logical to call the titration in Fig. 3.79 a dead-stop method, although this has been done; the term differential amperometric titration might be more useful.

One of the most important applications of the dead-stop end-point method is the Karl Fischer titration of water; the titrant usually consists of I_2 amd SO_2 with pyridine in methanol, which reacts with H_2O as follows:

$$C_5H_5NI_2 + C_5H_5NSO_2 + C_5H_5N + H_2O \rightarrow 2C_5H_5NHI + C_5H_5N\begin{matrix} SO_2 \\ | \\ O \end{matrix}$$

and

$$C_5H_5N\begin{matrix} SO_2 \\ | \\ O \end{matrix} + CH_3OH \rightarrow C_5H_5N\begin{matrix} H \\ SO_4CH_3 \end{matrix}$$

Essentially it concerns the reaction

$$H_2O + SO_2 + I_2 \rightarrow 2HI + SO_3$$

During the KF titration of water as the analyte, e.g., in an organic solvent, with a controlled potential difference (ΔE) across two Pt wire electrodes, one obtains a curve as depicted in Fig. 3.81, so in effect it represents a reversed dead-stop end-point method; it also means that the titration of KF reagent with a methanolic solution with a known water content represents the normal dead-stop end-point method as depicted in Fig.3.80. The importance of the KF method moreover lies in the fact that many organic functional groups react with stoichiometric liberation or consumption of water, which can subsequently measured by aquametry[143] with KF reagent. Sometimes such a water liberation reaction may occur although it is undesirable, e.g., aldehydes and ketones yield water through acetal or ketal formation with the methanol of the KF reagent, which thus interferes with the determination of free water in these compounds. One can avoid this either by determination of water at low temperature ($< 5°$ C) or, more attractive, by displacing the methanol of the KF reagent with methyl Cellosolve ($CH_3OC_2H_4OH$)[144].

In contrast with former opinions about the reaction mechanism in KF titration, more recent investigations by Verhoef and co-workers[145] have shown that neither SO_2 nor a pyridine–SO_2 complex is oxidized by iodine in the presence of water, but the monosulphite ion:

$$\left.\begin{array}{l} 2CH_3OH \rightleftharpoons CH_3OH_2^+ + CH_3O^- \\ CH_3O^- + SO_2 \rightleftharpoons CH_3OSO_2^- \end{array}\right\} \text{equilibria in solution}$$

$$\left.\begin{array}{l} CH_3OSO_2^- + I_2 \rightarrow CH_3OSO_2I_2^- \\ CH_3OSO_2I_2^- + H_2O \rightarrow CH_3OSO_3^- + 2H^+ + 2I^- \end{array}\right\} \text{titration reactions}$$

Verhoef and co-workers suggested omitting the foul smelling pyridine completely and proposed a modified reagent, consisting of a methanolic solution of sulphur dioxide (0.5 *M*) and sodium acetate (1 *M*) as the solvent for the analyte, and a solution of iodine (0.1 *M*) in methanol as the titrant; the titration proceeds much faster and the end-point can be detected preferably bipotentiometrically (constant current of 2 μA), but also biamperometrically (ΔE about 100 mV) and even visually as only a little of the yellow sulphur dioxide–iodide complex SO_2I^- is formed (for the coulometric method see Section 3.5).

Kies[141] made an extensive study of oxidimetric titrations of iodide with iodate according to Andrews in the presence of hydrochloric acid. The shape (in 1 *N* HCl) of the two hyperbolic sections followed by a flat part of the curve is explained by the occurrence of two (nearly) reversible systems, I_2Cl^-/I^- and ICl_2^-/I_2Cl^-, according to the titration reactions

$$5I^- + IO_3^- + 6H^+ + 3\,Cl^- \rightarrow 3I_2Cl^- + 3H_2O$$
$$2I_2Cl^- + IO_3^- + 6H^+ + 8\,Cl^- \rightarrow 5ICl_2^- + 3H_2O$$
$$\text{overall}\quad 2I^- + IO_3^- + 6H^+ + 6\,Cl^- \rightarrow 3ICl_2^- + 3H_2O$$

Fig. 3.82 illustrates most impressively the sensitivity of the method.

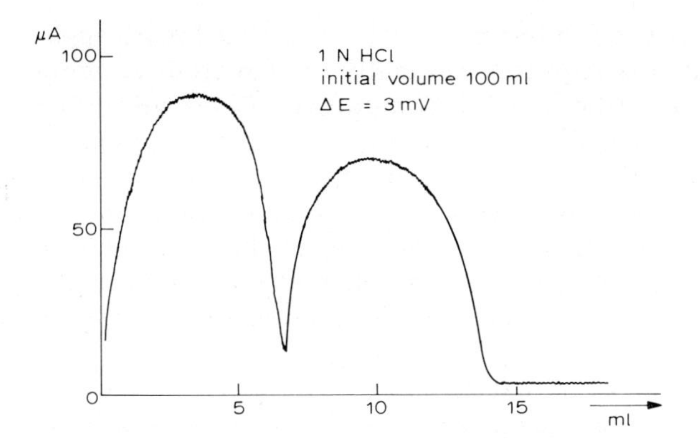

Fig. 3.82. Dead-stop end-point titration of (5 ml 0.0983 *M*) I^- with (0.01662 *M*) IO_3^- in 1 *N* HCl by Kies (titration reactions according to Andrews).

3.3.2.3.5. *Potentiometric dead-stop end-point titration*
Instead of following a dead-stop end-point titration as is usual in amperometry, it is often attractive to do so potentiometrically, i.e., at constant current, for two reasons:

(1) it is electronically more simple, also if a "magic eye" indication is used;

(2) the approach to the dead-stop end-point is manifested by a sharp increase in the potential difference (ΔE) across the two indicator electrodes, and this increase can be plotted against the titration parameter (Fig. 3.83).

In fact, this has already been illustrated in Fig. 3.73 for the differential electrolytic potentiometric titration of Ce(IV) with Fe(II), both being reversible systems. This technique can be usefully applied, for instance, to the aforementioned KF titration of water and its reverse titration (cf., Verhoef and co-workers' preference for bipotentiometric detection); in these instances the potentiometric dead-stop end-point titration and the reversed potentiometric dead-stop end-point titration, respectively, yield curves as depicted in Fig. 3.83.

3.3.2.3.6. *Polarovoltry*
This technique with two indicator electrodes, proposed in 1956 by Dubois and Walisch[146], is an intermediate between bipotentiometry and biamperometry, because neither the current nor the potential across the electrodes of the same metal are kept strictly constant. The authors' opinion, e.g., in the titration of iodide with bromate in hydrochloric acid, that the ΔE value does not matter much, has been contradicted by Kies (ref. 141a, p. 18). As a method of alkalimetry or acidimetry it cannot be preferred, like any other dead-stop technique to the usual glass electrode methods[147]. Nevertheless, the fact that the apparatus permits a choice of adjustment towards constant current or constant potential can be useful, but then the method approaches either bipotentiometry or biamperometry.

3.3.3. Practice of voltammetry in general

From the foregoing treatment of voltammetry it is clear that for its application sophisticated apparatus is required, especially the electronics; moreover, the construction of the electrodes and their mutual positions within the measuring vessel, of adapted size and without or with stirring, requires great care and experience. For this reason it is often advisable to purchase commercial apparatus, which has reached a high level of sophistication and reliability. Here the most desirable property is smooth recording of the voltammetric curve, which necessitates the kind of automation inherently required in the voltammetric method; this is different from advanced automation, which is treated in Part C.

The question arises of which is the best apparatus available. The answer depends on the experimenter's requirements as to the type of analyte and the accuracies desired; so the specifications provided by the manufacturers should be considered.

(a)

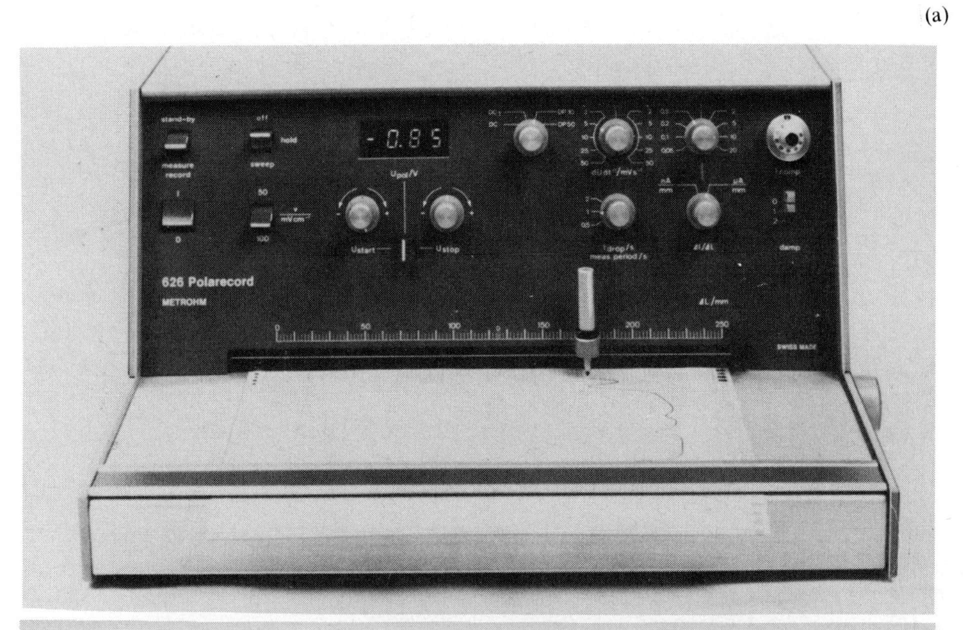

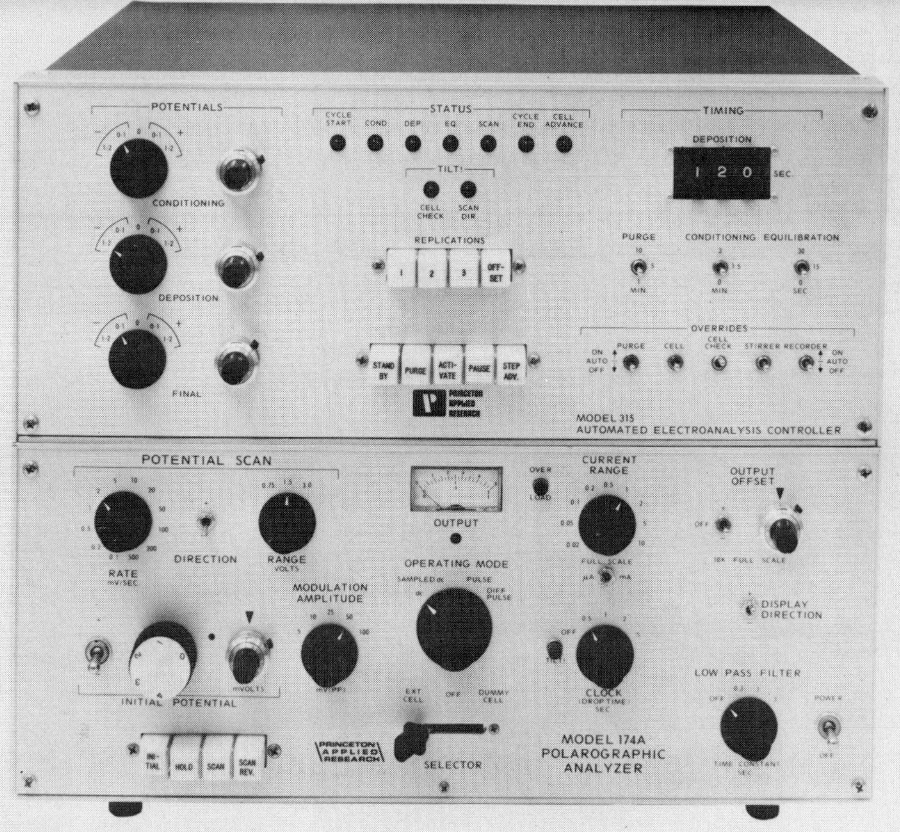

Fig. 3.83.

(Continued on p. 208)

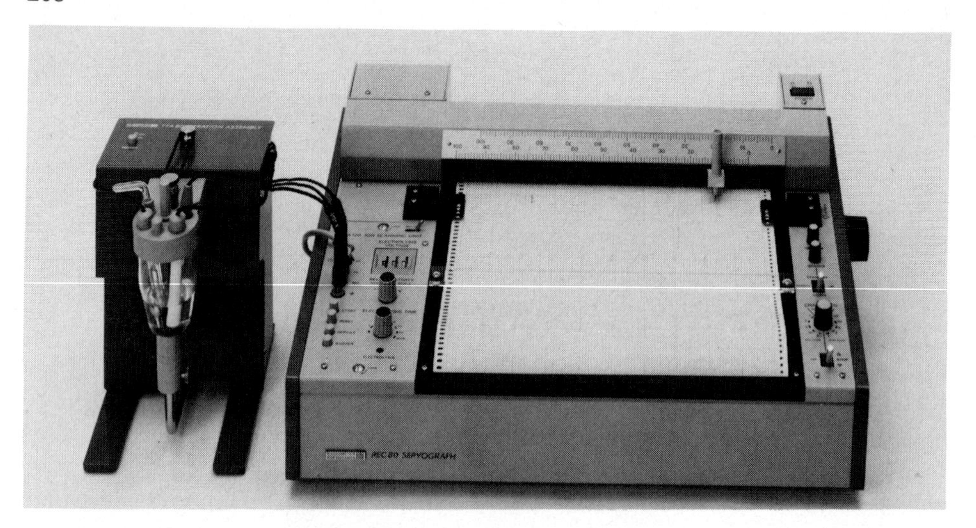

(b)

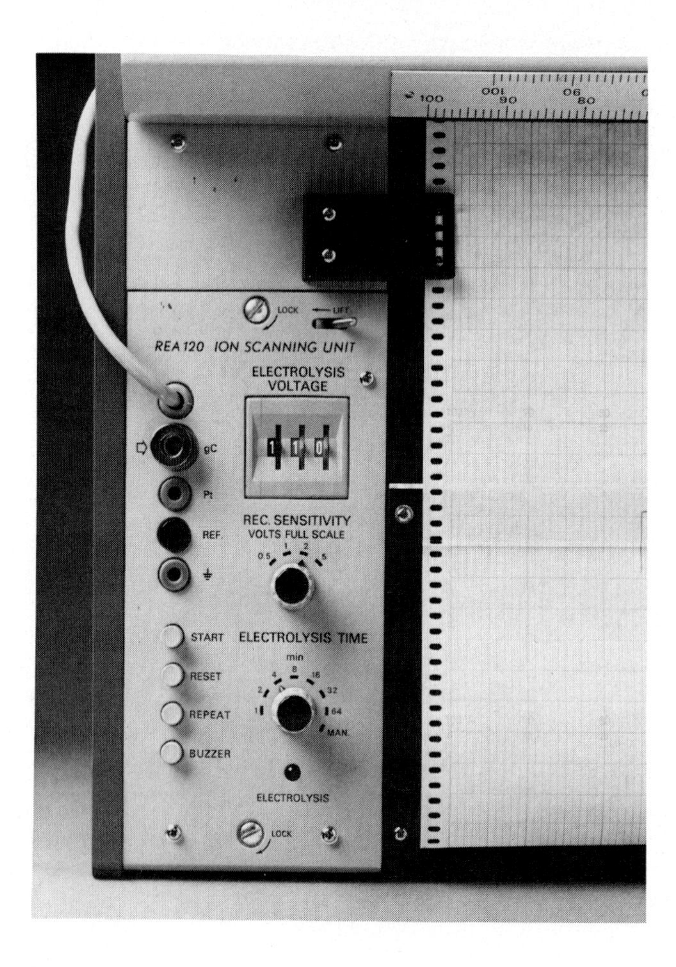
LOCK
LIFT
REA 120 ION SCANNING UNIT
ELECTROLYSIS
VOLTAGE
gC
Pt
REF.
REC. SENSITIVITY
VOLTS FULL SCALE
0.5
1
2
5
START
RESET
REPEAT
BUZZER
ELECTROLYSIS TIME
min
1
2
4
8
16
32
64
MAN.
ELECTROLYSIS
LOCK
100
90
80

Fig. 3.83.

(c)

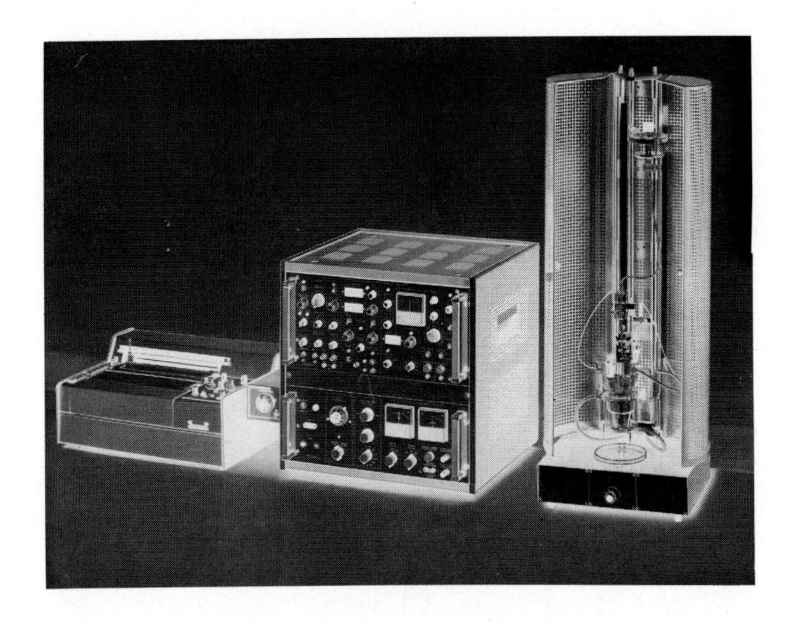

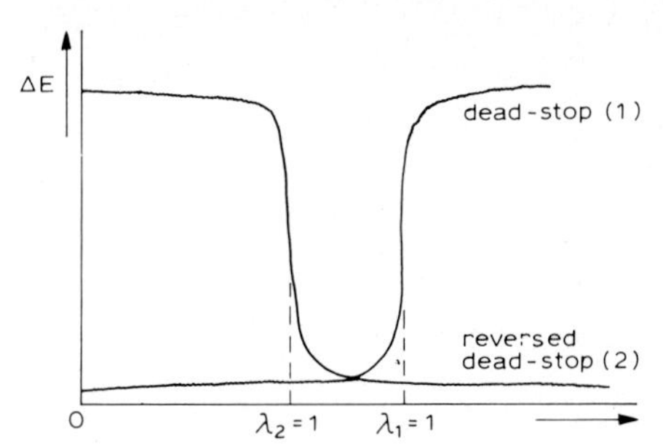

Fig. 3.83. Potentiometric dead-stop (1) and reversed dead-stop (2) end-point titrations. (a) Metrohm Polarecord 626; PARC Model 174A Polarographic Analyzer. (b) Radiometer ISS 820 Ion Scanning System. (c) Tacussel PRG4 Polarograph.

It is an advantage of electroanalysis and its apparatus that the financial investment is low in comparison, for instance, with the more instrumental spectrometric methods; real disadvantages are the need to have the analyte in solution and to be familiar with the various techniques and their electrochemistry; it is to be regretted that the knowledge of chemistry and the skill needed often deter workers from applying electroanalysis when this offers possibilies competitive with more instrumental methods (cf., stripping voltammetry versus atomic absorption spectrometry).

The manufacturers of electroanalytical instrumentation cannot always immediately satisfy the experimenter's requirements, but often this can be remedied by instrumental development in mutual contact. An illustrative example is the construction of the carbon-fibre microelectrodes (MFC) by

TABLE 3.4

COMMERCIAL VOLTAMMETRIC (VA) APPARATUS

Measuring cells are supplied by all firms in various forms according to the measuring techniques, including thermostating, stirring and inert gas introduction; GCE – glassy carbon electrode, CPE – carbon paste electrode.
Key: +, supplied; –, not supplied; no entry, not known. Key to numbering:
1, (A) means HMDE size automatically reproduced, if required.
2, SMDE = static (quasi-stationary) MDE, although often not clearly defined; for PARC this is their DME temporarily stationary, the current being sampled before the drop fall (Model 303 SMDE, see Fig. 3.28, etc.)
3, Dead-stop mostly amperometric, sometimes also potentiometric.
4, RDE with screw-on electrode tips of glassy carbon (GC) or precious metals (Pt, Au, Ag, etc.); RDE control unit to be added.
5, NPP and DPP as well as NPV and DPV, respectively, are usually provided.
6, In the built-in pulse technique for DC, the stripping step of SV can also use the pulse technique.
7, Phase sensitive, but requires 174/50 Interphase and lock-in amplifier.
8, Only for Model 384B, which in addition to square-wave VA also provides cyclic staircase VA.
9, iR compensation (in addition to i_C compensation).
10, Inputs and outputs for function generator and cathodic-ray oscilloscope (or rapid x y recorder) in single-sweep and cyclic VA.
11, Phase-selective AC polarography of 1st and 2nd harmonics.
12, Additional 612 VA scanner provides single-sweep triangular-wave and cyclic triangular-wave VA.
13, Redeveloped multimode electrode (MME), i.e. DME/SMDE + HMDE combined in one design.
14, Metals concentrated on GCE together with Hg as their amalgams and next chemically stripped off while recording the potential time curve.
15, Phase-selective AC VA possible, as well as tensammetry and cyclic VA.
16, Potential sweep can be single or triangular, allowing cyclic VA.
17, Pulses can be direct or reverse on linear potential, or they can increase linearly at a constant potential; biopulse is a separate pulse VA system providing DPV and DNPV (according to Brumleve et al.[63a]) at MFC (carbon-fibre microelectrode) and two working electrodes in vivo.
18, iR compensation provided.
19, The BIPAD potentiostat provides the separate potentials for the ring and disc in addition to the Asservitex 10.000 control unit for RRDE.
For abbreviations, see Section 1.2.2.

Manufacturer	Model	DME (A)	HMDE[1]	SMDE[2]	M(T)FE	GCE	CPE	Dead stop[3]	RDE[4]	RRDE	Three electrode	Advanced DC			LSV	pulse[5]	AC superimposed			SV[6]
												rapid	Tast	derivative			sinus	square	Kal	
Metrohm	Polarecord 506	+	+[1]	+	+	+	+		+		[9]	+	+		+[10]	+	+[11]		+	+
	Polarecord 626	+	+[1]	+	+	+	+		+		+		+		+[12]	+			+	
	646 VA processor (see also Part C)	+	+	+[13]	+	+	+		+				+			+		+		
	KF titrator							+												
Mettler	Memotitrator DL40 (KF titration)							+												
PARC	310/364 (formerly 171)	+	+	+					–	–	–	–	+	–	+	+	–	–	–	+
	174A (formerly 170)	+	+	+								–	+		+	+	+[7]	–	–	+
	384 (see also Part C)	+	+	+								–	+	+	+	+	–	+[8]	–	+
Radiometer	ISS 820					+[14]														+[14]
Tacussel	PRG 3	+							+		+		+				+[15]			
	PRG 4	+											+		+[16]	+[17]				
	PRG 5	+	+		+	+	+		+		[18]		+							+
	BIPAD potentiostat									+[19]										

Tacussel and their application by Gonon et al.[148] to differential pulse voltammetry (DPV) and differential normal pulse voltammetry (DNPV) in vivo, also called the biopulse technique; the microelectrodes are implanted in the living animal brain and variations in the concentrations of some molecules can be followed via the Tacussel PRG 5 and BIPAD instruments (see also the selection of commercial models in Table 3.4).

3.4. ELECTROGRAVIMETRY

Electrogravimetry is one of the oldest electroanalytical methods and generally consists in the selective cathodic deposition of the analyte metal on an electrode (usually platinum), followed by weighing. Although preferably high, the current efficiency does not need to be 100%, provided that the electrodeposition is complete, i.e., exhaustive electrolysis of the metal of interest; this contrasts with coulometry, which in addition to exhaustive electrolysis requires 100% current efficiency.

As the first experimental condition in electrogravimetry one makes a choice from two possibilities, viz.:

(a) slow electrolysis with the use of a stationary electrode without stirring; in this instance only at low current densities is satisfactory adherence of the metal deposit on the electrode obtained;

(b) rapid electrolysis, where either an independent mechanical stirrer is used or, more usually, a rotating anode (platinum gauze cylinder according to Sand) is applied, which with an intermediate space of 3–5 mm is surrounded by a non-rotating platinum gauze cylinder as the cathode; in this instance a considerably higher current density is admissible, which reduces the electrolysis time quite appreciably.

One must realize that once complete metal deposition has been attained, the emf across the electrodes cannot be switched off before the cathode has been taken out of the solution and rinsed with water, otherwise the metal deposit may start to redissolve in the solution as a consequence of internal electrolysis by the counter emf. After disconnection the electrode is rinsed with acetone and dried at 100–110° C for 3–4 min. The analytical result is usually obtained from difference in weight of the dry cathode before and after electrolysis. In a few instances a copper- or silver-plated Pt cathode or even an Ag cathode is used, e.g., Zn and Bi are difficult to remove entirely from Pt, as they leave black stains and on heating form an alloy with the noble metal; for this and other reasons (see below) the experimenter should consult the prescriptions in handbooks[149].

As a second experimental condition, the metal deposit should be pure and adhere well to the electrode, for which the following two factors are of most importance:

(1) the choice of solution in relation to the chemistry of the metal ion, the stirring, the temperature and, if necessary, the addition of a depolarizer to prevent gas evolution;

(2) the electrolysis must be controlled, i.e. with a constant current, constant potential or controlled potential.

In many instances the presence of metals other than the one of interest require the determination to be preceded by electroseparation.

3.4.1. Constant-current electrolysis

This technique is applied to mixtures of metal ions in an acidic solution for the purpose of electroseparation; only the metal ions with a standard reduction potential above that of hydrogen are reduced to the free metal with deposition on the cathode, and the end of the reduction appears from the continued evolution of hydrogen as long as the solution remains acidic. Considering the choice of the cathode material and the nature of its surface, it must be realized that the method is disturbed if a hydrogen overpotential occurs; in that event no hydrogen is evolved and as a consequence metal ions with a standard reduction potential below that of hydrogen will still be reduced; a classic example is the electrodeposition of Zn at an Hg electrode in an acidic solution.

3.4.2. Constant-potential electrolysis

This technique allows the selective electro-deposition of a metal from a solution in the presence of ions of a less noble metal, provided that there is a sufficient difference between their standard reduction potentials; the latter condition suggests remaining on the safe side (less negative) with the cathodic potential, so that the analysis may lose much in velocity; on the other hand, the simplicity of procedure and apparatus is an advantage.

3.4.3. Controlled-potential electrolysis

Controlled-potential electrolysis (CPE) represents an improvement over the previous constant-potential method; this is attained by the application of an emf across the electrodes that yields a cathodic potential as negative as is acceptable in view of current density limitations and without taking the risk that the less noble metal is deposited; hence the technique requires non-faradaic control of the cathodic potential versus the solution.

In order to explain the principle of the CPE and its practical effects we shall consider the electrolysis of a sulphuric acid solution of 2 g-ions l^{-1} of Ag^{+} and 1 g-ion l^{-1} of Cu^{2+} at two Pt electrodes and 25° C; soon after the electrolysis has been started the cathode becomes covered first with Ag and then with Cu; at the anode we have to deal with oxygen generation, yielding a counter potential vs. solution (cf. p. 42) of

$$E_{O_2 \rightarrow OH^-} = E^0_{O_2 \rightarrow OH^-} - 0.059 \log[OH^-] = 0.397 - 0.059 \log 10^{-13} = 1.16 \, V$$

In addition, the emf to be applied must overcome the ohmic resistance of the solution and the cathode counter potential, which are respectively,

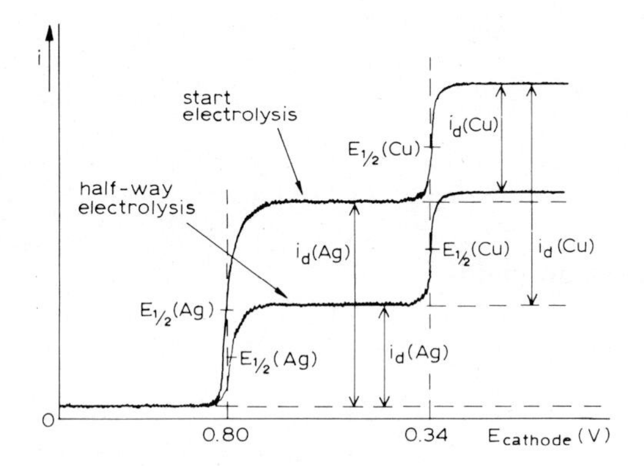

Fig. 3.84. Controlled potential electrolysis on behalf of electrogravimetric determination of Ag in presence of Cu.

$$E_{\text{appl.(Ag)}} = 1.16 - (0.800 + 0.059 \log[\text{Ag}^+]) + i_{(\text{Ag})}R$$

and

$$E_{\text{appl.(Cu)}} = 1.16 - \left(0.340 + \frac{0.059}{2} \log[\text{Cu}^{2+}]\right) + i_{(\text{Cu})}R$$

The values 0.397, 0.800 and 0.340 V are the respective standard reduction potentials of $O_2 \rightarrow OH^-$, $Ag^+ \rightarrow Ag$ and $Cu^{2+} \rightarrow Cu$ at 25° C. In general, the ohmic resistance of the solution (with sufficient supporting electrolyte) can be neglected. Further, the zero-current Nernst potentials belonging to the initial concentrations are $E_{Ag^+ \rightarrow Ag} = 0.818$ V and $E_{Cu^{2+} \rightarrow Cu} = 0.340$ V. Hence the selective deposition of Ag requires a cathode potential between these two values, i.e., an emf applied across the electrodes of between 0.34 and 0.82 V.

The attraction of controlled-potential electrolysis now is that it can start with a cathode potential well below 0.818 V, because this increases considerably the current throughput and, as a consequence, the rapidity of depostion; however, there are two limiting conditions: (1) the current density should not be too high and (2) in particular the final potential must remain well above 0.340 V in order to prevent Cu deposition.

Considering the first condition, the cathode surface area should be above a certain size depending on the initial analyte concentration in order to collect the limiting current i_d; during the electrolysis the current will decrease rapidly as shown in Fig. 3.84, and this occurs, as we know from coulometry (see later), exponentially with time.

Considering the second condition, we realize that the deposition of the metal on a solid electrode is in a way comparable to the polarography of ions of non-amalgamating metals (cf., eqn. 3.38 and Fig. 3.16). Hence $E_{\frac{1}{2}}$ alters with the analyte bulk concentration, which can be seen from

$$i = nFDA(C - c)/\delta \tag{3.3}$$

and

$$i_d = nFDAC/\delta \qquad (3.4)$$

so that

$$i_d - i = nFDAc/\delta$$

which substituted in the Nernst equation yields

$$E = E^0 + \frac{RT}{nF} \ln \left(\frac{i_d - i}{nFDA/\delta} \right)$$

and gives for $E_{\frac{1}{2}}$, when $i = i_d/2$:

$$E_{\frac{1}{2}} = E^0 + \frac{RT}{nF} \ln \left(\frac{i_d/2}{mFDA/\delta} \right) = E^0 + \frac{RT}{nF} \ln \left[\frac{(nFDA/\delta)C/2}{nFDA/\delta} \right]$$

or

$$E_{\frac{1}{2}} = E^0 + \frac{RT}{nF} \ln C/2 \qquad (3.106)$$

Hence the alteration appears small, e.g., the transition from $2\,M$ (where $E_{\frac{1}{2}} = E^0$) to $2 \cdot 10^{-3}\,M$ produces a shift for $E_{\frac{1}{2}}$ (to the right in Fig. 3.84) of ca. -180 mV; for a $2 \cdot 10^{-7.5}\,M$ solution $E_{\frac{1}{2}(Ag)}$, with a shift of ca. -450 mV, approaches $E_{\frac{1}{2}(Cu)}$.

Eqn. 3.106 must be considered as an approximate relationship for at least two reasons: first, the assumption of a rapid complete coverage of the Pt electrode surface by Ag right from the start of the electrolysis is certainly incorrect (cf., Bard and Faulkner[150]); second, at the end of the electrolysis the remaining Cu^{2+} solution is virtually in contact with a silver electrode instead of a copper electrode, for which $E^0_{Cu^{2+} \to Cu} = 0.340$ V is valid. Practice has shown that by means of CPE, selective electro-deposition and thus electrogravimetry of silver in addition to copper is possible down to $10^{-8}\,M\,Ag^+$, as the above calculation indicates.

3.4.4. Electrolysis at a mercury pool

In many instances electrogravimetry must be preceded by a separation between metals; suitably this can be an electroseparation by means of constant-current electrolysis as previously described, but more attractively an electroseparation by means of controlled-potential electrolysis at a mercury pool or sometimes at an amalgamated Pt or brass gauze electrode. In this way one can either concentrate the metal of interest on the Hg or remove other metals from the solution; alternatively, it can be a rougher separation, i.e., the concentration of a group of metals such as Fe, Ni, Co, Cu, Zn and Cd on the Hg whilst other metals such as alkali and alkaline earth metals, Be, Al, Ti and Zr remain in solution[151]. In all these procedures specific separation effects can be

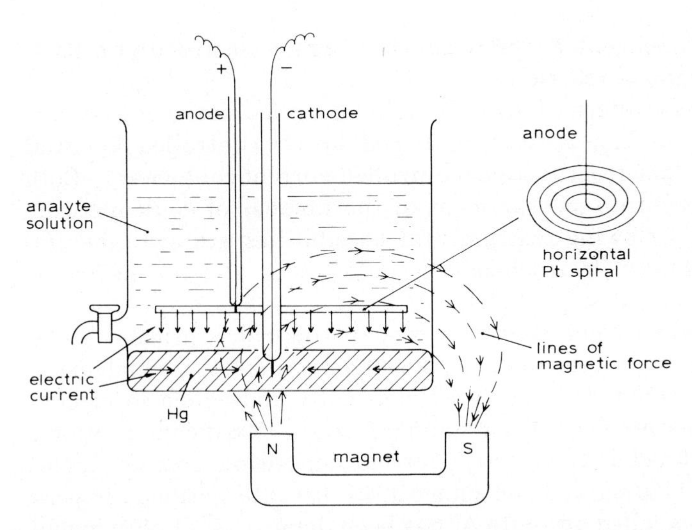

Fig. 3.85. Electroseparation between metal groups at a magnetically stirred Hg pool.

obtained by variation of pH and/or complexation. Various types of apparatus are used, also commercially available; most interesting is the method with a magnetically stirred mercury pool according to Center et al.[151a], depicted in Fig. 3.85. Here we obtain the following mutual effects between electric current and magnetic field:

(1) the electro-deposited ferromagnetic metals are drawn below the Hg surface, thus remaining clean, and cannot redissolve;

(2) application of the corkscrew rule (or Fleming's left-hand rule) reveals directions of the Lorentz forces such that looking from above into the vessel we see the electrolyte turning clockwise and the mercury anticlockwise; here the central introduction of the cathodic current is of course advantageous.

The aforementioned effects increase the speed of electrolysis and make any mechanical stirring superflous.

3.5. COULOMETRY

In electroanalysis, coulometry is an important method in which the analyte is specifically and completely converted via a direct or indirect electrolysis, and the amount of electricity (in coulombs) consumed thereby is measured. According to this definition there are two alternatives: (1) the analyte participates in the electrode reaction (primary or direct electrolysis), or (2) the analyte reacts with the reagent, generated (secondary or indirect electrolysis) either internally or externally.

The condition of specific and complete conversion of the analyte means for alternative 1 an exclusive and complete electrolytic reaction of the analyte at the working electrode with 100% current efficiency (exhaustive electrolysis), and for alternative 2 preferential and detectable complete conversion of the analyte by means of the reagent, generated at the working electrode with 100% current efficiency (complete reaction).

The way in which these alternatives with their particular measuring characteristics are carried out can be best described by (1) controlled-potential coulometry and (2) coulometric titration (controlled-current coulometry). Both methods require an accurate measurement of the number of coulombs consumed, for which the following instrumental possibilities are available: (a) chemical coulometers, (b) electrochemical coulometers and (c) electronic coulometers.

(a) Among the chemical coulometers we may distinguish a primary and a secondary type. The primary type includes the silver coulometer, which determines the number of coulombs electrogravimetrically and which in electrochemical practice represents the primary standard, i.e., in 1908 the international ampere ($C\,s^{-1}$) was defined as the electric current depositing from an $AgNO_3$ solution per second 1.1180 mg of silver (since 1940 this international ampere, A_{int} = 0.99985 A, the so-called absolute A, has been defined as 0.1 electromagnetic unit of electric current). The copper coulometer works on the same principle. In the gas coulometer the amount of gas electrodically generated is measured volumetrically, i.e., for the hydrogen–oxygen coulometer the total volume of gas liberated from aqueous 0.5 *M* K_2SO_4 solution[152], for the nitrogen–hydrogen coulometer the total amount of gas evolved at the anode (N_2) and cathode (H_2) from aqueous 0.1 *M* hydrazinium sulphate solution (the net reaction is $N_2H_5^+ \rightarrow N_2 + 2H_2 + H^+$)[153]. The secondary type includes the coulometric coulometer, whose operation is based on a kind of stripping analysis, viz., the instrument in series with the coulometric analysis of the sample follows the consumption of electricity, Q, by means of internal electro-deposition on the working electrode; next in a separate circuit and by means of a precisely constant current in the opposite direction the deposit is stripped off, the completion of which is indicated by a sharp potential jump; the product of current and time (chronoamperometry) represents Q of the coulometric analysis. For this type of coulometer a few different systems are in use, e.g., an Ag spiral (WE) reacts anodically to AgCl with a weakly acidic 0.3 *M* KCl solution vs. Ag (AE) and is next cathodically stripped off by potentiometric control vs. an Ag–AgCl RE[154], or analogously a Cu^{2+} solution reacts cathodically to Cu, which deposit is next anodically stripped off[155].

In contrast with the other chemical coulometers, which are fairly accurate especially with high Q values, but cumbersome, the coulometric coulometer is rapid and sensitive, especially with low Q values.

(b) The electrochemical coulometers virtually represents current–time integrators; in coulometric analysis two types of these integrators have been used;

(1) a small, low-inertia electric motor[156] provided with a counter and switched across both ends of a sensitive low resistance, which is placed within the electrolysis circuit;

(2) a ball-and-disk integrator[157], connected to a recorder which registers the current measured by the ammeter within the electrolysis circuit; this integrator

has the advantages not only of being more readily commercially available than the former, although much cheaper, but also of yielding the current–time curve in controlled-potential coulometry; in view of this, however, preference is often given today to the more flexible and informative method of the virtually electronic system below.

(c) Electronic coulometers can work in two different ways, i.e., involving either potential coulometry or coulometric titration:

(1) For controlled-potential coulometry the voltage drop over a standard resistor is measured as a function of time by means of a voltage-to-frequency converter; the output signal consists of a time-variant and integrally increasing number of counts (e.g., 10 counts mV^{-1}), which by means of an operational amplifier–capacitor yields the current–time curve and integral[158].

(2) For coulometric titration the potential change can be followed by direct measurement, while the current–time integral indicates the number of coulombs consumed at any point of the potentiometric curve, thus including its titration end-point[159].

The electronic coulometers allow extremely accurate determinations even of small current or voltage effects; there may be some, although low, noise interference, but with today's computerization this and other background signals (e.g., residual current) can be easily eliminated.

Now returning to the coulometric analysis proper we can say that any determination that can be carried out by voltammetry is also possible by coulometry; whether it should be done by means of the controlled-potential or the titration (constant-current) method much depends on the electrochemical properties of the analyte itself and on additional circumstances; both methods, because they are based on bulk electrolysis, require continuous stirring.

3.5.1. Controlled-potential coulometry

According to the example of Fe(III) $\rightleftharpoons$ Fe(II) and its cathodic wave $ox_2 \rightarrow red_2$ in Fig. 3.71 ($E_{\frac{1}{2}} \approx E^0 = 0.770$ V), controlled-potential coulometry of Fe(III) to Fe(II) can be carried out at +0.20 V vs. SCE (RE)[160] at a Pt electrode (WE) in 1 M H_2SO_4 (AE = Pt); the potential chosen is sufficiently below $E_{\frac{1}{2}}$ to maintain the limiting current i_l. When i_l of the initial Fe(III) concentration, C_0, in a volume V is i_{l_0}, the velocity of decrease of i_l will be

$$\mathrm{d}i_l/\mathrm{d}t = -ki_l$$

or

$$\mathrm{d}i_l/i_l = -k\mathrm{d}t$$

so that i_l decreases exponentially with time:

$$i_{l_t} = i_{l_0}\mathrm{e}^{-kt} \qquad (3.107)$$

This also means that the consumption of coulombs at time t is

$$Q_t = Q_0(1 - \mathrm{e}^{-kt}) \qquad (3.108)$$

where $Q_0 = nFN_0 = nFVC_0$, which represents the consumption for completion of the electrolysis ($t \rightarrow \infty$). Eqn. 3.108 can be written as

$$Q_t = Q_0(1 - e^{-t/\tau})$$

where τ is the time constant of the decrease; in general practice it is assumed that $Q_t \approx Q_0$ after $t = 7\tau$, because continuation beyond that stage would require too much time.

Many important applications of controlled-potential coulometry are known[160], as the WE not only Pt, but also sometimes Ag on Pt or often Hg have been used; the electrode conversion is mostly reductive (for metal ions, organic nitro compounds, etc.), sometimes oxidative ($2I^- \rightarrow I_2$, nitrite → nitrate, ascorbic acid → dehydroascorbic acid); even selective sequential reduction is applied, e.g., between Ag and Cu (Pt–WE at +0.13 and −0.12 V vs. SCE, respectively) and in more difficult separations between Cu and Bi, Cd and Zn, Ni and Co (Hg–WE at −0.16 and −0.40 V, −0.85 and −1.45 V and −0.95 and −1.20 V vs. SCE, respectively[161], as well as between different valences of one metal [Sb(V) → SB(III) and Sb(III) → Sb(0), respectively[162]]. Such a sequential reduction coulometry usually requires a long time (e.g., for Ni and Co 2–3 h for each metal have been reported), but with today's computerized data treatment the time of completion of the electrolysis can be predicted from only part of the smoothed exponential curve, although in this instance two separate analytical procedures have to be carried out, i.e., one for the easier reducible ion selectively and one for both ions simultaneously, so that a subtractive calculation can be applied. It remains an advantage of controlled-potential coulometry that the method is direct and accurate as well as sensitive, and that the analyte solution is not diluted during the analysis; moreover, by its purely electric operation and resulting electric signal it is directly adapted to remote process control and connected automation (see Part C).

3.5.2. Coulometric titration (controlled-current coulometry)

In the controlled (constant) potential method the procedure starts and continues to work with the limiting current i_l, but as the ion concentration and hence its i_l decreases exponentially with time, the course of the electrolysis slows down quickly and its completion lags behind; therefore, one often prefers the application of a constant current. Suppose that we want to oxidize Fe(II); we consider Fig. 3.78 and apply across a Pt electrode (WE) and an auxiliary electrode (AE) an anodic current, $-I$, of nearly the half-wave current; this means that the anodic potential (vs. an RE) starts at nearly the half-wave potential, $E_{\frac{1}{2}}$, of Fe(II) → Fe(III) (= 0.770 V), but increases with time, while the anodic wave height diminishes linearly and halfway to completion the electrolysis falls below $-I$; after that moment the potential will suddenly increase until it attains the decomposition potential (nearly 2.4 V) of $H_2O \rightarrow O_2$. The way to prevent this from happening is to add previously a small amount of a so-called redox buffer, i.e., a reversible oxidant such as Ce(IV) with a standard reduction potential higher than that of Fe(II); hence, as long as any Fe(II) remains in the solution, that Fe(II) will be oxidized according to

Fe(II) + Ce(IV) → Fe(III) + Ce(III)

The Ce(III) thus obtained will next be oxidized at its decomposition potential (nearly 1.4 V); this overall process of direct and indirect oxidation continues until Fe(II) has been completely converted into Fe(III); at this stage no Ce(III) remains, but only the amount of Ce(IV) previously added; in order to stop the anodic oxidation in time the potential of the Pt–WE (vs. SCE–RE) should automatically interrupt the electrolysis beyond say 1.5 V, i.e., well below the decomposition potential of H_2O.

The result of the entire procedure, being a 100% conversion of Fe(II) to Fe(III), thus represents a so-called coulometric titration with internal generation; of course, it seems possible to titrate Fe(II) with Ce(IV) generated externally from Ce(III), but in this way one would unnecessarily remove the solution of the 100% conversion problem; hence the above titration with internal generation in the presence of a redox buffer as an intermediary oxidant represents an extremely reliable method, unless occasional circumstances are prohibitive; for the remainder internal generation offers the advantage of no dilution of the analyte solution.

A special advantage of the internal generation may be that the reagent is generated if necessary only where and when it is required directly, which is especially useful if the reagent generated is less stable (e.g., Br_2, Ti^{3+}, Cr^{2+}); at any rate, conservation and standardization are superfluous. However, situations occur where external generation is compulsory, for instance, when conditions for optimal generation are not compatible with those for rapid reaction of the analyte, or when some accompanying substances in the sample cannot be prevented from interfering with the electrode generation reaction. When either the catholyte or the anolyte from the generation cell would interfere with the analyte reaction, a double-arm generator[163] cell can be used, into the top of which the generator electrolyte is fed continuously and is next divided at a T-joint, with Pt electrodes on either side, over both arms of the cell; either the anolyte or the catholyte is delivered to the titration cell, the other going to waste. In, for instance, the well known titration of an azo dye in a hot solution with Ti(III), the latter can be externally generated from a Ti(IV) solution at room temperature in a single-arm generator cell[164] with a Pt foil anode and an Hg pool cathode (with its favourable hydrogen overpotential).

It is certainly clear that a coulometric titration, like any other type of titration, needs an end-point detection system; in principle any detection method that chemically fits in can be used, be it electrometric, colorimetric, photoabsorptionmetric, etc.; for instance, in a few cases the colour change of the reagent generated (e.g., I_2) may be observed visually, or after the addition of a redox, metal or pH indicator the titration end-point can be detected photoabsorptiometrically by means of a light source and photocell combination. Concerning the aforementioned coulometric titration of Fe(II), it is

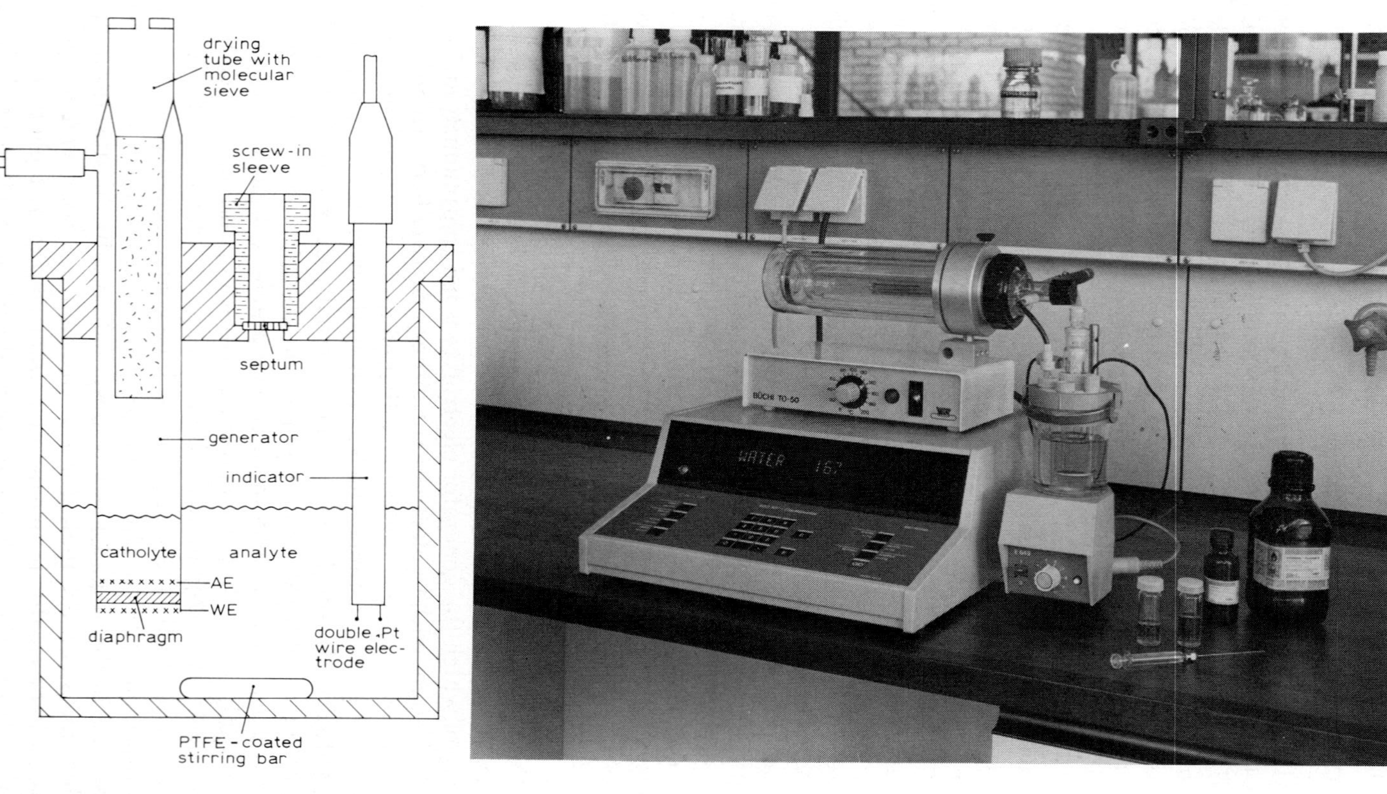

Fig. 3.86. Coulometric Karl Fischer water titration (Courtesy of Metrohm).

obvious that one should use detection based on zero-current potentiometry of the Pt–WE vs. an SCE–RE; in the procedure concerned we chose for convenience an automatic stop at a pre-set end-point of about 1.5 V; if this had been omitted a steep potential jump at nearly 1.5 V would have occurred and the electrolysis should have been stopped by manual operation; from the potentiometric curve thus obtained and its inflection point the time of complete titration and so the corresponding number of coulombs consumed must be established afterwards; in this mode the method can be characterized as a potentiometric coulometric titration.

It must be realized that the constant current ($-I$) chosen virtually determines a constant titration velocity during the entire operation; hence a high current shortens the titration time, which is acceptable at the start, but may endanger the establishment of equilibrium of the electrode potentials near the titration end-point; in an automated potentiometric titration the latter is usually avoided by making the titration velocity inversely proportional to the first derivative, dE/dt. Now, as automation of coulometric titrations is an obvious step, preferably with computerization (see Part C), such a procedure can be achieved either by such an inversely proportional adjustment of the current value or by a corresponding proportional adjustment of an interruption frequency of the constant current once chosen. In this mode the method can be characterized as a potentiometric controlled-current coulometric titration.

Another important point in coulometric titration is the necessity in many instances for internal generation to keep the electrolyte around the auxiliary electrode separated from the analyte solution around the working electrode by means of a diaphragm or a special type of membrane; we describe below two important examples of this situation.

Coulometric Karl Fischer titration of water. Fig. 3.86 shows the measuring cell of the Metrohm 652 KF coulometer[165]. The catholyte with the Pt gauze cathode (AE) is situated above the diaphragm, separated from the analyte solution) with the Pt gauze anode (WE) below it; it is claimed that all types of KF reagents can be used, which means that the cell solution consists of pyridine, SO_2 and methanol or methyl Cellosolve, I_2 being instantaneously generated during the analysis only. The method closely resembles that of the automatic KF coulometric titration originally published by Barendrecht[166], who used the biamperometric dead-stop end-point method for detection and automatic regulation.

The Metrohm apparatus applies a comparable, i.e., a bivoltammetric end-point indication measuring the AC voltage U_{ind} appearing on the double platinum wire electrode owing to the impressed current. The titration is regulated by means of a micoprocessor and consists of pulse coulometry with automatic control optimized for time and accuracy, i.e., constant-current pulses are applied in such a way that a long-lasting pulse is given at the beginning for the major part of the titration followed by several short pulses near the end-point

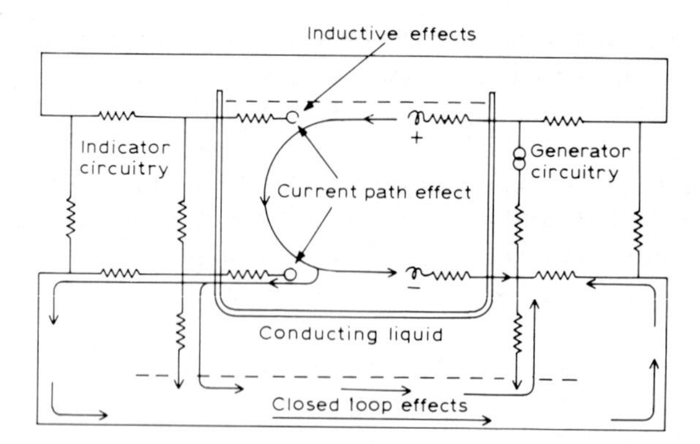

Fig. 3.87. Disturbance of electrometric indicator function by generating current.

decreasing with time; with its maximum titration speed of 2 mg min^{-1} of water the system yields a short overall titration time without the risk of overshoot.

Coulometric acid–base titration. The method below, described by Dahmen and Bos[167], was applied to non-aqueous media (see part B), where precautions should be even more strictly adhered to than in aqueous media. The instrumental set-up, as depicted in Figs. 3.87–3.89, was originally developed by their former colleague Dubbeling together with Van den Enk-van Twillert).

If one considers Fig. 3.87, the following observations can be made: (a) an inductive effect on higher resistance leads must be eliminated by appropriate shielding; (b) a current path effect, i.e., a voltage drop across the indicator electrodes, caused by the generating current, should be avoided by positioning the indicator electrodes close together and far from the generator electrode and its counter electrode, as shown in Fig. 3.88 (cf., also Fig. 3.86); (c) a closed loop effect should be eliminated by galvanic separation of the indicating and generating circuits.

It is most important, of course, also to solve the diaphragm problem of mutually separating the generator and counter electrode compartments, either

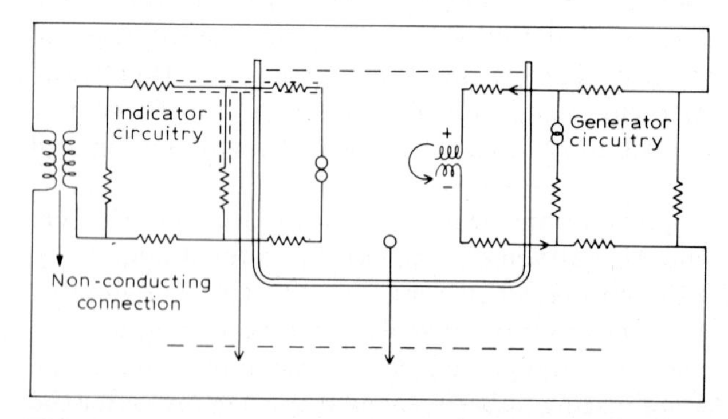

Fig. 3.88. Connection diagram for non-disturbed indicator performance.

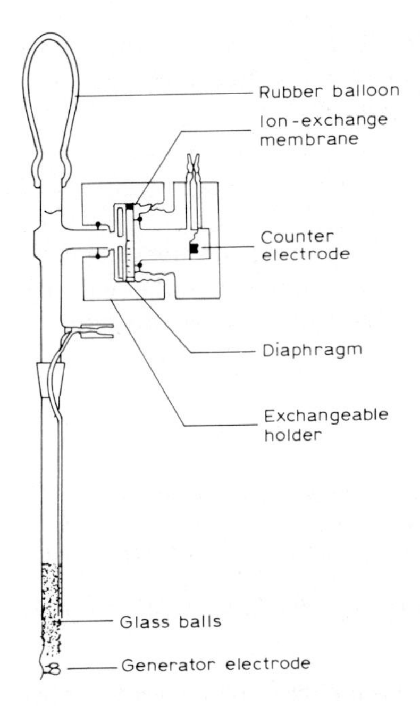

Fig. 3.89. Coulometry electrode set.

(1) by use of porous materials, such as sintered-glass discs of adequate porosity, when using solvents of high viscosity such as *m*-cresol or dimethyl sulphoxide, or (2) by ion-exchange membranes when using solvents of low viscosity such as water or acetonitrile; Du Pont supply special membranes for organic media. With regard to the ion-exchange membranes, a particular coulometry electrode set as shown in Fig. 3.89 was used; a pH generator of platinum wire is wound around the tip of a glass stem, which, by pressing the rubber balloon and allowing it to expand after each coulometric titration, remains filled with fresh end-point liquid; this contacts the membrane and its counter-electrode compartment behind. One can also use a combination of two membranes, each with its own counter-electrode compartment, opposite one another, one serving the acid titration and the other the base titration. In the experience of the authors a sodium perchlorate solution appeared the most useful electrolyte around the counter electrode. Considering the ion-exchange membranes (see pp. 47–48) the following procedure is recommended: in acid titration, take an anion-exchange membrane in the perchlorate form and use the generator electrode as the cathode; in base titration, take a cation-exchange membrane in the Na^+ form and use the generator electrode as the anode. Considering the counter electrode compartment, H^+ ions are stopped by the anion membrane where the ClO_4^- ions provide the charge transfer, OH^- ions being stopped by the cation membrane where the Na^+ ions provide the change transfer.

For end-point detection, any method usual in acid–base titration can be used; with electrometric indication the precautions for protection against the generation system should be taken into account. The authors used the above double generation system in a permanent assemby for titrant standardization where the coulomb functions as the primary standard.

3.5.3. Conclusion

Coulometry provides a most powerful means of electroanalysis, not only with regard to its general application to various types of analytes, but also with its possibilities for continuous and automatic analysis and in remote control applications (see Part C).

REFERENCES

1 W. J. Moore, Physical Chemistry, Longmans, London, 5th ed., 1972, pp. 160–161.
2 J. J. Lingane, Electroanalytical Chemistry, Interscience, New York, 2nd ed., 1958, p. 226.
3 A. J. Bard and L. R. Faulkner, Electrochemical Methods, Fundamentals and Applications, Wiley, New York, 1980, Ch. 3.3, Fig. 3.3.2.
4 J. A. V. Butler, Trans. Faraday Soc., 19 (1924) 729 and 734; T. Erdey Gruz and M. Volmer, Z. Phys. Chem., Abt. A, 150 (1930) 203.
5 Ref. 1, pp. 556–557, Table 12.6.
6 Ref. 1, pp. 563–564; see also H. Gerischer and W. Vielstich, Z. Phys. Chem. (Frankfurt am Main), 3 (1955) 16.
7 J. Tafel, Z. Phys. Chem., 50 (1905) 641.
8 A. C. Testa and W. H. Reinmuth, Anal. Chem., 33 (1961) 132; cf., ref. 3, Ch. 11 and also ref. 9, Ch. 2.2.2.
9 A. M. Bond, Modern Polarographic Methods in Analytical Chemistry, Marcel Dekker, Basle, 1980, p. 226.
10 R. C. Weast (Editor), Handbook of Chemistry and Physics, CRC Press, Cleveland, OH, 61st ed., 1980–81, pp. D155–160.
11 J. Heyrovský and J. Kůta, Principles of Polarography, Academic Press, New York 1966, Ch. VI, pp. 59–61.
12 Ref. 3, pp. 132, 147–148 and 150–151.
13 Ref. 3, pp. 154–155 (for a more detailed discussion).
14 Ref. 9, p. 96.
15 J. Tomeš, Collect. Czech. Chem. Commun., 9 (1957) 12, 81 and 150.
16 Ref. 3, pp. 163–164.
17 Ref. 9, pp. 104–106.
18 Ref. 9, p. 75.
19 P. Delahay, J. Amer. Chem. Soc., 75 (1953) 1430; Advan. Polarogr., 1 (1960) 26.
20 Ref. 9, pp. 99–108 and references cited therein.
21 Ref. 9, pp. 112–117 (for further discussion).
22 R. E. Cover and J. G. Connery, Anal. Chem., 41 (1969) 1797.
23 R. E. Cover, Rev. Anal. Chem., 1 (1972) 141.
24 S. Wolf, Angew. Chem., 72 (1960) 449.
25 A. M. Bond and R. J. O'Halloran, J. Electroanal. Chem., 68 (1976) 257.
26 G. C. Barker, Advan. Polarogr., 1 (1960) 144.
27 D. R. Ferrier, D. H. Chidester and R. R. Schroeder, J. Electroanal. Chem., 45 (1973) 361.
28 W. M. Peterson, "The Static Mercury Drop Electrode", Amer. Lab. Dec. (1979) 69 (available from EG & G PARC as Note T-2).
29 E. P. Parry and R. A. Osteryoung, Anal. Chem., 37 (1965) 1634.
30 Ref. 9, pp. 142–150.

31 G. Semerano and L. Riccoboni, Gazz. Chim. Ital., 72 (1942) 297.
32 Ref. 9, pp. 154–156.
33 J. Heyrovský, Analyst (London), 72 (1947) 229; Anal. Chim. Acta, 2 (1948) 537.
34 H. Vandebroek and F. Verbeek, Anal. Lett., 5 (1972) 317; R. Dewolfs and F. Verbeek, Z. Anal. Chem., 269 (1974) 349.
35 P. Vermeiren, E. Steeman, E. Temmerman and F. Verbeek, Bull. Soc. Chim. Belg., 86 (1977) 499.
36 Ref. 3, pp. 142–143.
37 PARC Anal. Instruments Division, Basics of Voltammetry and Polarography, Application Note-P2, Princeton Applied Research, Princeton, NJ, 1980, p. 7, Fig. 10.
38 Ref. 9, Ch. 7.6 and 7.7, and references cited therein.
39 Ref. 3, Ch. 9.1.3–9.4.1.
40 A. M. Bond, Anal. Chem., 44 (1972) 315.
41 A. M. Bond, Talanta, 21 (1974) 591.
42 A. M. Bond and R. J. O'Halloran, Anal. Chem., 47 (1975) 1906.
43 B. Breyer and H. H. Bauer, Alternating Current Polarography and Tensammetry, Interscience, New York, 1963; see also Z. Galus, Fundamentals of Electrochemical Analysis, Ellis Horwood, Chichester, 1976, p. 503–504, and H. Jehring, J. Electroanal. Chem., 20 (1969) 33 and 21 (1969) 77.
44 M. D. Booth and B. Fleet, Talanta, 17 (1970) 491.
45 M. Bos, Anal. Chim. Acta, 122 (1980) 387; 135 (1982) 249.
46 H. Jehring, J. Electroanal. Chem., 21 (1969) 77; H. Jehring, E. Horn, A. Reklat and W. Stolle, Collect. Czech. Commun., 33 (1968) 1038 and 1670.
47 M. Bos and W. H. M. Bruggink, Anal. Chim. Acta, 152 (1983) 35.
48 G. C. Barker and I. L. Jenkins, Analyst (London), 77 (1952) 685; G. C. Barker and R. L. Faircloth, in I. S. Longmuir (Editor), Advances in Polarography Pergamon Press, Oxford, 1960, p. 313.
49 T. Takahashi and E. Niki, Talanta, 1 (1958) 245.
50 G. C. Barker, Anal. Chim. Acta, 18 (1958) 118.
51 G. C. Barker, G. W. C. Milner and H. I. Shalgosky, Polarography, Proceedings of Congress on Modern Analytical Chemistry in Industry, St. Andrews, 1957, p. 199; G. C. Barker, A. W. Gardner and M. J. Williams, J. Electroanal. Chem., 42 (1973) App. 21.
52 W. H. Reinmuth, Anal. Chem., 36 (1964) 211R.
53 G. C. Barker, Proc. Anal. Div. Chem. Soc., (1975) 171.
54 R. Kalvoda, Techniques of Oscillographic Polarography, Elsevier, Amsterdam, 2nd ed., 1965.
55 V. Majer, Z. Electrochem., 42 (1936) 120.
56 R. Přibil and B. Matyska, Collect. Czech. Chem. Commun., 16 (1951) 139; D. J. Myers and J. Osteryoung, Anal. Chem., 46 (1974) 356.
57 J. W. Dieker, Thesis, Amsterdam, 1980; J. W. Dieker, W. E. van der Linden and G. den Boef, Talanta, 24 (1977) 321 and 597; 26 (1979) 193 and 973.
58 A. M. Bond, R. J. O'Halloran, I. Ruzic and D. E. Smith, Anal. Chem., 48 (1976) 872; 50 (1978) 216.
59 S. P. Perone and T. R. Mueller, Anal. Chem., 37 (1965) 2.
60 G. L. Connor, G. H. Boehme, C. J. Johnson and K. H. Pool, Anal. Chem., 45 (1973) 437.
61 J. W. Dieker, W. E. van der Linden and H. Poppe, Talanta, 25 (1978) 151.
62 K. B. Oldham and E. P. Parry, Anal. Chem., 38 (1966) 867.
63 R. F. Lane and A. T. Hubbard, Anal. Chem., 48 (1976) 1287.
63a T. R. Brumleve, R. A. Osteryoung and J. Osteryoung, Anal. Chem., 54 (1982) 782.
64 A. M. Bond, R. J. O'Halloran, I. Ruzic and D. E. Smith, Anal. Chem., 50 (1978) 216.
65 L. Gierst and A. Juliard, Com. Intern. de Thermodynamique et de Cinétique Electrochimique, Compt. Rend de la IIme Réunion, Milan, 1950, pp. 117 and 279; J. Phys. Chem., 57 (1953) 701.
66 P. Delahay and C. Mamantov, Anal. Chem., 27 (1955) 478.
67 H. J. S. Sand, Philos. Mag., 1 (1901) 35.
68 Ref. 3, pp. 252–253.

69 P. Bos, Chem. Weekbl., 61 (1965) 533; Thesis, Free University, Amsterdam, 1970; P. Bos and E. van Dalen, J. Electroanal. Chem., 45 (1973) 165.
70 P. Delahay and T. Berzins, J. Amer. Chem. Soc., 75 (1953) 2486, and references to C. L. Weber, H. J. S. Sand and Z. Karaoglanov cited therein.
71 T. Berzins and P. Delahay, J. Amer. Chem. Soc., 75 (1953) 4205.
72 Ref, 2, p. 623 (Fig. XXII-3).
73 Ref, 2, p. 633.
74 P. Bos and E. van Dalen, J. Electroanal. Chem. Interfacial Electrochem., 45 (1973) 165.
75 P. Delahay and C. C. Mattax, J. Amer. Chem. Soc., 78 (1954) 874; P. Delahay and T. Berzins, J. Amer. Chem. Soc., 75 (1953) 2486.
76 P. Bos, J. Electroanal. Chem. Interfacial Electrochem., 33 (1971) 379.
77 P. Bos, J. Electroanal. Chem. Interfacial Electrochem., 34 (1972) 475.
78 H. Hurwitz, J. Electroanal. Chem., 2 (1961) 328; ref. 3, p. 256.
79 P. Bos and E. van Dalen, J. Electroanal. Chem. Interfacial Electrochem., 17 (1968) 21.
80 Ref. 3, p. 265.
81 W. D. Shults, F. E. Haga, T. R. Mueller and H. C. Jones, Anal. Chem., 37 (1965) 1415.
82 (a) H. L. Kies, J. Electroanal. Chem., 16 (1968) 279 (short communication); (b) J. Electroanal. Chem. Interfacial Electrochem., 45 (1973) 71; (c) H. L. Kies and H. C. van Dam, J. Electroanal. Chem. Interfacial Electrochem., 48 (1973) 391.
83 P. Delahay, New Instrumental Methods in Electrochemistry, Interscience, New York, 1954, ch. 8, p. 214, and ref. 66.
84 Ref. 2, p. 633; W. H. Reinmuth, Anal. Chem., 38 (1966) 270R.
85 W. Kemula and Z. Kublik, Advan. Anal. Chem. Instrum., 2 (1963) 123; P. Delahay, G. Charlot and H. A. Laitinen, Anal. Chem., 32 (1960) 103A.
86 I. Shain, Treatise Anal. Chem., 4 (1964) 2533.
87 R. Neeb, Z. Anal. Chem., 171 (1959) 321, 330; Angew. Chem., 74 (1962) 203.
88 E. Barendrecht, Chem. Weekbl., 60 (1964) 345; 61 (1965) 537.
89 E. Barendrecht, "Stripping Voltammetry", in A. J. Bard (Editor), Electroanalytical Chemistry, Vol. II, Marcel Dekker, New York, 1967, p. 53.
90 Ref. 89, p. 57, Fig. 2.
91 J. G. Nikelly and W. D. Cooke, Anal. Chem., 29 (1957) 933; R. Neeb, Z. Anal. Chem., 180 (1961) 161.
92 N. Velghe and A. Claeys, J. Electroanal. Chem., 35 (1972) 229; cf., ref. 9, p. 439.
93 W. Kemula and Z. Kublik, Anal. Chim. Acta, 18 (1958) 104; Nature (London), 182 (1958) 1128; J. Electroanal. Chem., 1 (1959) 91, 205; Nature (London), 189 (1961) 57.
94 R. D. DeMars, Anal. Chem., 34 (1962) 259; J. W. Olver and J. W. Ross, Anal. Chem., 34 (1962) 791.
95 W. L. Underkofler and I. Shain, Anal. Chem., 33 (1961) 1966.
96 E. Barendrecht, Nature (London), 181 (1958) 764.
97 J. J. Engelsman and A. M. J. M. Claassens, Nature (London), 191 (1961) 240; M. van Swaay and R. S. Deelder, Nature (London), 191 (1961) 241.
98 W. T. de Vries, J. Electroanal. Chem., 9 (1965) 448.
99 T. M. Florence, J. Electroanal. Chem., 27 (1970) 273; W. D. Ellis, J. Chem. Educ., 50 (1973) A131.
100 M. van Swaay and R. S. Deelder, Nature (London), 191 (1961) 241.
101 W. T. de Vries and E. van Dalen, J. Electroanal. Chem., 8 (1974) 366.
102 W. H. Reinmuth, Anal. Chem., 33 (1961) 185.
103 Ref. 3, pp. 416, 215–220 and 222–224.
103a Ref. 9, p. 451.
104 (a) W. T. de Vries, J. Electroanal. Chem., 9 (1965) 448; (b) W. T. Vries and E. van Dalen, J. Electroanal. Chem., 14 (1967) 315.
105 D. K. Roe and G. E. Toni, Anal. Chem., 37 (1965) 1503; ref. 3, p. 418.
106 R. A. Osteryoung and J. H. Christie, Anal. Chem., 46 (1974) 351.
107 T. R. Copeland, J. H. Christie, R. A. Osteryoung and R. K. Skogerboe, Anal. Chem., 45 (1973) 2171.

108 S. Bruckenstein and T. Nagai, Anal. Chem., 33 (1961) 1201; S. Bruckenstein and J. W. Bixler, Anal. Chem., 37 (1965) 786.
109 D. Jagner and A. Granelli, Anal. Chim. Acta, 83 (1976) 19 (HMDE); D. Jagner and K. Arén, Anal. Chim. Acta, 100 (1978) 375 (MTFE); D. Jagner, Anal. Chim. Acta, 105 (1979) 33 (Hg determination); D. Jagner, L. G. Danielson and K. Arén, Anal. Chim. Acta, 106 (1979) 15 (Pb in urine); D. Jagner and K. Arén, Anal. Chim. Acta, 107 (1979) 29 (sea water).
110 Radiometer, Copenhagen, A. M. Graabaek and O. J. Jansen, Ind. Res. Dev., Sept. 1979.
111 H. C. Bijl, Z. Phys. Chem., 41 (1902) 641; W. Reinders, Z. Phys. Chem., 54 (1906) 609.
112 W. Kemula, in I. S. Longmuir (Editor), Advances in Polarography, Vol. I, Pergamon Press, Oxford, 1960, p. 105.
113 H. K. Ficker and L. Meites, Anal. Chim. Acta, 26 (1962) 172.
114 A. Hickling, J. Maxwell and J. V. Shennan, Anal. Chim. Acta, 14 (1956) 287.
115 H. H. Bauer, Electroanal. Chem., 8 (1975) 170.
116 J. W. Olver and J. W. Ross, Anal. Chem., 34 (1962) 791.
117 G. Colovos, G. S. Wilson and J. L. Mogers, Anal. Chem., 46 (1974) 1051.
118 T. Miwa, Y. Fujü and A. Mizuike, Anal. Chim. Acta, 60 (1972) 475.
119 C. C. Young and H. A. Laitinen, Anal. Chem., 44 (1972) 457.
120 I. Shain and S. P. Perone, Anal. Chem., 33 (1961) 325.
121 Th. Kuwana and R. N. Adams, Anal. Chim. Acta, 20 (1959) 51 and 60.
122 G. van Dijck and F. Verbeek, Z. Anal. Chem., 249 (1970) 89 (HMDE); Anal. Chim. Acta, 54 (1971) 475 (HMDE); E. Temmerman and F. Verbeek, Anal. Chim. Acta, 58 (1972) 263 (GCE); G. van Dijck and F. Verbeek, Anal. Chim. Acta, 66 (1973) 251 (HMDE).
123 Ref. 3, Ch. 8.
124 V. G. Levich, Physicochemical Hydrodynamics, Prentice-Hall, Englewood Cliffs, NJ, 1962.
125 W. J. Albery and M. L. Hitchman, Ring-Disc Electrodes, Clarendon Press, Oxford, 1971, Ch. 3.
126 Tacussel Electronique, Notice No. 6–EDA, Tacussel, Lyon, Jan. 1981.
127 D. T. Napp, D. C. Johnson and S. Bruckenstein, Anal. Chem., 39 (1967) 481, see also W. J. Albery and S. Bruckenstein, Trans. Faraday Soc., 62 (1966) 1920; Anal. Chem., 40 (1968) 482; J. Electrochem. Soc., 117 (1970) 460.
128 S. Bruckenstein and P. R. Gifford, Anal. Chem., 51 (1979) 250.
129 P. Kiekens, H. Verplaetse, L. deCock and E. Temmerman, Analyst (London), 106 (1981) 305.
130 W. J. Blaedel, C. J. Olsen and L. R. Sharma, Anal. Chem., 35 (1963) 2100.
131 H. B. Hanekamp and H. J. van Nieuwkerk, Anal. Chim. Acta, 121 (1980) 13.
132 W. J. Blaedel and G. P. Hicks, Anal. Chem., 34 (1962) 388; W. J. Blaedel and C. Olsen, Anal. Chem., 36 (1964) 343.
133 W. J. Blaedel and D. G. Iverson, Anal. Chem., 49 (1977) 1563; W. J. Blaedel and Z. Yim, Anal. Chem., 50 (1978) 1722.
134 D. N. Armentrout, J. D. McLean and M. W. Long, Anal. Chem., 51 (1979) 1039.
135 D. I. Napp, D. C. Johnson and S. Bruckenstein, Anal. Chem., 39 (1967) 48.
136 R. E. Cover and J. G. Connery, Anal. Chem., 41 (1969) 918.
137 J. G. Connery and R. E. Cover, Anal. Chem., 41 (1969) 1191.
138 D. R. Canterford, A. S. Buchanan and A. M. Bond, Anal. Chem., 45 (1973) 1327.
139 E. Bishop, Mikrochim. Acta, (1956) 619; Analyst (London) 83 (1958) 212; 85 (1960) 422.
140 H. A. Laitinen, W. P. Jennings and T. D. Parks, Ind. Eng. Chem., Anal. Ed., 18 (1946) 355 and 358.
141 (a) H. L. Kies, Thesis, Technological University Delft, 1956; (b) H. L. Kies, Anal. Chim. Acta, 18 (1958) 14.
142 C. W. Foulk and A. T. Bawden, J. Amer. Chem. Soc., 48 (1926) 2045.
143 J. Mitchell, Jr. and D. M. Smith, Aquametry, Interscience, New York, 1948.
144 E. D. Peters and J. L. Jungnickel, Anal. Chem., 27 (1955) 450.
145 J. C. Verhoef, Thesis, Free University, Amsterdam, 1977; J. C. Verhoef and E. Barendrecht, J. Electroanal. Chem., 71 (1976) 305; 75 (1977) 705; J. C. Verhoef, W. P. Cofino and E. Barendrecht, J. Electroanal. Chem., 93 (1978) 75; J. C. Verhoef and E. Barendrecht, Electrochim. Acta, 23 (1978) 433.

146 J. E. Dubois and W. Walisch, C. R. Acad. Sci., 242 (1956) 1161 and 1289; J. E. Dubois, La Polarovoltrie et al Polarommétrie, Mises au Point-de Chimie Analytique Pure et Appliqueé et d'Analyse Bromatologie, Publiées sous la Direction de J. -A. Gautier, 53Série, Masson, Paris, 1957.
147 Ref. 83, p. 263.
148 Tacussel Electronique, Notice No. 2 -MFC, April 1982, Toko WD (France), and references cited therein; Notice No. 2-Biopulse-E, Sept. 1983, Impr. Ferréol (France), and references cited therein; F. Gonon, M. Buda, R. Cespuglio, M. Jouvet and J. F. Pujol, Nature (London), 286 (1980) 902; F. Gonon, C. M. Fombarlet, M. J. Buda and J. F. Pujol, Anal. Chem., 53 (1981) 1386.
149 A. I. Vogel, A Textbook of Quantitative Inorganic Analysis, Including Elementary Instrumental Analysis, Longmans, London, 3rd ed., 1962, Ch. VI.
150 Ref. 3, p. 373 [Ch. 10.2.1(c)].
151 Paper NEN 3105-1, Dutch Normalisation Institute, Delft, May 1964.
151a E. J. Center, R. C. Overbeek and D. L. Chase, Anal. Chem., 23 (1951) 1134.
152 J. J. Lingane, J. Amer. Chem. Soc., 67 (1945) 1916; cf., ref. 149, p. 668; H. H. Willard, L. L. Merritt and J. A. Dean, Instrumental Methods of Analysis, Van Nostrand, New York, 5th ed., p. 708.
153 J. A. Page and J. J. Lingane, Anal. Chim. Acta, 16 (1957) 175; cf., ref. 152.
154 W. Jaenicke and K. Hauffe, Z. Naturforsch., Teil A, 4 (1949) 353; V. B. Ehlers and J. W. Sease, Anal. Chem., 26 (1954) 513.
155 S. W. Smith and J. K. Taylor, J. Res. Natl. Bur. Std., 63C (1959) 65.
156 J. J. Lingane, Anal. Chim. Acta, 18 (1958) 349.
157 H. H. Willard, L. L. Merritt and J. A. Dean, Instrumental Methods of Analysis, Van Nostrand, New York, 5th ed., 1974, pp. 534–535.
158 G. L. Booman, Anal. Chem., 29 (1957) 213; L. L. Merritt, E. L. Martin and R. D. Bedi, Anal. Chem., 30 (1958) 487.
159 M. T. Kelley, H. C. Jones and D. J. Fisher, Anal. Chem., 31 (1959) 488.
160 Ref. 3, p. 384, Table 10.3.2, and references cited therein, e.g., G. A. Rechnitz, Controlled Potential Analysis, Pergamon Press, New York, 1963; J. E. Harrar, Electroanal. Chem., 8 (1975) 1.
161 Ref. 149, pp. 673–674.
162 L. B. Dunlap and W. D. Schults, Anal. Chem., 34 (1962) 499.
163 D. D. DeFord, J. N. Pitts and C. J. Johns, Anal. Chem., 23 (1951) 938; N. Bett, W. Nock and G. Morris, Analyst (London), 79 (1954) 607.
164 J. N. Pitts, D. D. DeFord, Th. W. Martin and E. A. Schmall, Anal. Chem., 26 (1954) 6; W. Fuchs and W. Quadt, Z. Anal. Chem., 147 (1955) 184.
165 652 KF-Coulometer, Metrohm, Herisau, Switzerland.
166 E. Barendrecht, Nature (London), 183 (1959) 1181; Chem. Weekbl., 56 (1960) 37.
167 E. A. M. F. Dahmen and M. Bos, Proc. Anal. Div. Chem. Soc., (1977) 86.

B. ELECTROANALYSIS IN NON-AQUEOUS MEDIA

Chapter 4

Introduction

In the general introduction of this book, we promised to pay special attention to the electrolysis in non-aqueous media in view of its growing importance as a result of the fact that in comparison with aqueous media it extends the electroanalytical possibilities considerably for the following reasons:

(i) many analytes such as organic products inherently require non-aqueous solvents for solubility and stability reasons;

(ii) non-aqueous media often permit special determination reactions and/or appropriate discrimination between analyte components that cannot be achieved in aqueous media;

(iii) additional flexibility of analysis is frequently obtained by the use of solvent mixtures, sometimes even with mixtures of organic solvents and water.

In treating non-aqueous electroanalysis separately, there are two major aspects to be considered: (1) the theory of its chemistry and (2) its practical application.

4.1. THEORY OF ELECTROCHEMISTRY IN NON-AQUEOUS MEDIA

An extensive and up-to-date treatment of this theory is beyond the scope of this chapter and superfluous in view of the large amount of literature available. Moreover, although non-aqueous electrochemistry was first developed in the analytical field, i.e., in acid–base titrations[1], further development was also achieved by physical chemists from a more theoretical point of view, and in both the organic and inorganic chemistry areas, where even solutes in molten salts were investigated. An excellent review of these developments was given by Trémillon[2] in a systematic treatment of the theory with literature references to the various investigators in alphabetical order and with a separate list of the most important books in chronological order.

More of less on the basis of own experience we shall confine ourselves in this chapter to a gradual build-up of the theory, as it may be helpful to the electroanalytical chemist to understand the phenomena and analytical possibilities of non-aqueous media in relation to their inherent general concepts; wherever it is useful, literature references will be cited.

As in electroanalysis both ionic and possible electrode aspects are of major interest, both aspects of solutes in non-aqueous solvents have to be considered; this can best be done by dividing the theory of the solutions concerned into two parts, viz. (1) the exchange of ionic particles (ionotropy), which leads to acid–base systems, and (2) the exchange of electrons only, which leads to redox systems.

4.1.1. Acid–base systems in non-aqueous media

Of the several theories of acid–base reactions, three are most important in common practice:

(1) the Franklin–Arrhenius theory or solvents theory (A);

(2) the Brønsted–Lowry theory or proton theory (B);

(3) the Lewis theory or electon theory (C).

(A) In addition to the more modern Brønsted and Lewis theories, it is important not to forget the classic Arrhenius theory in its modern form, the so-called solvents theory, where it can be applied, i.e., with solvents that undergo self-dissociation; in this form it was originally formulated in 1949 by Jander[3] in Germany and is illustrated by the following reaction equations:

$$H_2O \rightleftharpoons H^+ + OH^- \quad (4.1)$$

$$H^+ + H_2O \rightleftharpoons H_3O^+ \quad (4.2)$$

$$2H_2O \rightleftharpoons H_3O^+ + OH^- \quad (4.3)$$

$$2ROH \rightleftharpoons ROH_2^+ + RO^- \quad (4.4)$$

$$2HF \rightleftharpoons H_2F^+ + F^- \quad (4.5)$$

$$2AcOH \rightleftharpoons AcOH_2^+ + AcO^- \quad (4.6)$$

$$2NH_3 \rightleftharpoons NH_4^+ + NH_2^- \quad (4.7)$$

$$2RNH_2 \rightleftharpoons RNH_3^+ + RNH^- \quad (4.8)$$

Regarding these equations, several important points can be made, as follows:

(1) The sequential reactions 4.1 and 4.2 represent the self-dissociation of water as the exchange of a proton between water molecules, where hydration of the proton according to reaction 4.2 is the driving force for its separation (reaction 4.1); although the proton hydration is not limited to one H_2O (hydration number 1), nor is the occurrence of unhydrated OH^- ion realistic, the overall reaction 4.3 is generally written as the simplest form to show the principle of proton acidity.

(2) The overall reactions of proton exchange in other so-called water-resembling solvents can be written in an analogous way (reactions 4.4–4.8).

(3) For all the amphiprotic solvents in the pure state the cation (or acid) concentration is by definition equal to the anion (or base) concentration; in other words, the pure solvent is neutral; therefore, adding a solute such as HCl makes the solution acidic, whereas a solute such as NaOH makes it basic.

(4) The occurring concentration of acid and base in the pure solvent is a specific property of the solvent and depends on a number of factors such as temperature, viscosity, dielectric constant and solvation of the proton; the dielectric constant has a major influence, as the coulombic force of attraction between two particles of opposite charge is reversely proportional to ε; consequently solutes, once ionized, will become fully dissociated in dilute aqueous solution, whereas in solvents of low ε appreciable ion pair formation (association) will occur in addition to dissociation (Bjerrum postulate). Solvation depends on a possible dipole moment and/or polarizability of the solvent molecule; a higher degree of solvation means a larger distance between the solvoions of opposite charge, thus promoting dissociation.

In order to obtain the ion concentrations in the pure solvent, we can consider the equilibrium constant K'_w of the overall reaction 4.3, as separation step 4.1 is followed almost instantaneously by solvation step 4.2, and so

$$K'_w = \frac{a_{H_3O^+} \cdot a_{OH^-}}{a^2_{H_2O}} = \frac{[H_3O^+][OH^-]}{[H_2O]^2} \cdot \frac{f_{H_3O^+} \cdot f_{OH^-}}{f^2_{H_2O}}$$

As the pure solvent is only slightly ionized, both the activity coefficients and the concentration of the non-ionized solvent molecule may be regarded as unity, and one prefers to use $K_w = [H_3O^+][OH^-]$, the so-called ionic product of water. It was determined for the first time by Kohlrausch and Heydweiller at 18° C from the conductivity, $\kappa = 0.0384 \cdot 10^{-6}$ (cf., Ch. 2), which is given by

$$\kappa = \eta_{H_3O^+} \cdot \Lambda_0^{H_3O^+} + \eta_{OH^-} \cdot \Lambda_0^{OH^-}$$

where $\eta_{H_3O} = \eta_{OH^-}$, expressed in equiv. cm^{-3}, and $\Lambda_0^{H_3O^+}$ and $\Lambda_0^{OH^-}$ represent the equivalent ionic conductivities at 18° C of 315 and 174; hence

$$\eta = \frac{\kappa}{\Lambda_0^{H_3O^+} + \Lambda_0^{OH^-}} = \frac{0.0384}{489} \cdot 10^{-6} = 0.785 \cdot 10^{-10} \text{equiv. cm}^{-3}$$

and $[H_3O^+] = [OH^-] = 0.785 \cdot 10^{-7}$ equiv. l^{-1}, and hence

$$K_{w(18°C)} = (0.785 \cdot 10^{-7})^2 = 0.616 \cdot 10^{-14}$$

so that

$$pK_{w(18°C)} = -\log K_{w(18°C)} = 14.2104$$

Table 4.1 (cf. Handbook of Chemistry and Physics, 61st ed., p. D168) shows how K_w depends on temperature; according to these data, K_w equals precisely 10^{-14} at 24° C and $1.008 \cdot 10^{-14}$ at 25° C, which agrees with the general acceptance of $pK_w = 14$ at 25° C and the electroneutrality of pure water at $\frac{1}{2}pK_w = 7$.

For all amphiprotic solvents, reactions like the examples in the overall reactions 4.4–4.8 apply, often expressed by the equilibrium equation*

* H_2S^+ has sometimes been called (e.g., by Kolthoff) lyonium ion and S^- lyate ion.

TABLE 4.1
IONIC PRODUCT OF WATER (K_w) VS. TEMPERATURE

pK_w	Temperature (°C)	pK_w	Temperature (°C)
14.9435	0	13.8330	30
14.7338	5	13.6801	35
14.5346	10	13.5348	40
14.3463	15	13.3960	45
14.1669	20	13.2617	50
14.0000	24	13.1369	55
13.9965	25	13.0171	60

$2HS \rightleftharpoons H_2S^+ + S^-$, while the electroneutrality of pure solvents lies at $\frac{1}{2}pK_s$. An extensive list of pK_s values (mainly at 25° C) for various amphiprotic solvents and for mixtures of water and organic solvents (together with relevant literature data) has been given by Trémillon[4].

Starting from the foregoing theory of amphiprotic solvents, we shall next consider their interaction with some reactive (inorganic or ionophoric) solutes as illustrated below.

In H_2O:

$$HCl(g) + H_2O \xrightarrow{\text{dissociation}} H_3O^+ + Cl^- \tag{4.9}$$

$$AcOH(l) + H_2O \rightleftharpoons H_3O^+ + AcO^- \tag{4.10}$$

- -

$$NaOH \rightleftharpoons Na^+ + OH^- \tag{4.11}$$

$$NH_3(g) + H_2O \rightleftharpoons NH_4^+ + OH^- \tag{4.12}$$

In AcOH:

$$\underset{\text{ansolvoacid}}{HClO_4} + AcOH \xrightleftharpoons{\text{ionization}} \underset{\text{solvoion pair}}{(AcOH_2)^+(ClO_4)^-} \xrightleftharpoons{\text{dissociation}} \underset{\substack{\text{solvoacid}\\ \text{solvated ions}}}{(AcOH_2)^+} + (ClO_4)^- \tag{4.13}$$

- -

$$\underset{\text{ansolvobase}}{AcONa} \rightleftharpoons AcO^-Na^+ \rightleftharpoons Na^+ + \underset{\text{solvobase}}{AcO^-} \tag{4.14}$$

In liquid NH_3:

$$\underset{\text{ansolvoacid}}{HCl + NH_3} \rightleftharpoons NH_4^+Cl^- \rightleftharpoons \underset{\text{solvoacid}}{NH_4^+} + Cl^- \tag{4.15}$$

- -

$$\underset{\text{ansolvobase}}{NaNH_2} \rightleftharpoons Na^+NH_2^- \rightleftharpoons Na^+ + \underset{\text{solvobase}}{NH_2^-} \tag{4.16}$$

The terms solvoacid and solvobase have been introduced originally by Cady and Elsey[8] in 1928.

The proton exchange is called protolysis; solvation as a general term means hydration in the case of water, the dielectric constant* ($\varepsilon = 78$) of which is so high

* Dielectric constant values are at 25° C unless stated otherwise.

that it leads mainly to dissociation. H_3O^+ is often called the hydronium ion* in analogy with the ammonium ion NH_4^+; however, solvonium ion is the more general term. In solvents such as glacial acetic acid (ε = 6.2) and liquid ammonia at − 33° C (ε = 22.4) the coulombic force between cation and anion attains a value such that we must discriminate between ionization, providing an ion pair (associate), and dissociation into solvated ions. Ion pair formation and diminished dissociation will decrease the acid or base strength of solutes,e.g., hydrochloric and nitric acid, although strong in water, become weak in solvents of low ε. In fact, the phenomenon of self-dissociation of amphiprotic solvents is based on autoprotolysis; here the term autoionization might be used for water also, but for non-aqueous solvents it is generally not equivalent to self-dissociation.

$$\underbrace{H_3O^+(+Cl^-}_{(4.9)} + \underbrace{Na^+) + OH^-}_{(4.11)} \xrightarrow{\text{aq.neutralization}} 2H_2O(\underbrace{+Cl^- + Na^+}) \qquad (4.17)$$

$$\underbrace{AcOH_2^+[+(ClO_4)^-}_{(4.13)} + \underbrace{Na^+] + AcO^-}_{(4.14)} \xrightarrow{\text{solvoneutralization}} 2AcOH[\underbrace{(+ClO_4)^- + Na^+}] \qquad (4.18)$$

$$\underbrace{NH_4^+(+Cl^-}_{(4.15)} + \underbrace{Na^+) + NH_2^-}_{(4.16)} \xrightarrow{\text{solvoneutralization}} 2NH_3(\underbrace{+Cl^- + Na^+}) \qquad (4.19)$$

As in water, neutralization in all amphiprotic solvents represents the backward reaction of self-dissociation down to the equilibrium level of the ionic product in the pure solvent.

A further example of resemblance to water is given by

$$AcONa + \underset{\text{(solvent)}}{H_2O} \overset{\text{hydrolysis}}{\rightleftharpoons} \underset{\text{(weak acid)}}{AcOH} + Na^+ + OH^- \qquad (4.20)$$

$$KNO_3 + \underset{\text{(solvent)}}{AcOH} \overset{\text{solvolysis}}{\rightleftharpoons} \underset{\text{(weak acid)}}{HNO_3} + K^+ + AcO^- \qquad (4.21)$$

As in water (reaction 4.20), a salt of a weak acid in AcOH reacts alkaline owing to an increased base (AcO^-) concentration as a consequence of solvolysis.

Comparison of reactions 4.9, 4.10, 4.12, 4.13 and 4.15 leads to another important conclusion, viz., in an amphiprotic solvent its own solvonium cation represents the strongest acid possible, and its own anion the strongest base. Even when a very strong foreign acid or base is dissolved, excessive proton donation to and proton abstraction from the solvent molecule yield the respective acid or base; this phenomenon is generally known as the levelling effect, which in an amphiprotic solvent takes place on both the acid and the basic

*H_3O^+ is also called the hydroxonium ion because it expresses (analogously with the ammonium ion) the donation of a proton to a lone pair of electrons of the oxygen; it explains also the term alkoxonium ion for ROH_2^+ (reaction 4.4) like the alkylammonium ion for RNH_3^+ (reaction 4.8); $CH_3COOH_2^+$ or $AcOH_2^+$ may be called the acetoxonium ion

sides. This means, therefore, that acid–base titrations (at 25° C) in water will be found within the whole pH range of 14, but in acetic acid within a pH range of only about 6. Some alteration to the pH range can be obtained by adding a second solvent, although often this is done with the purpose of improving the solubility of the analyte; thus, addition to water of 70 wt.% of 1,4-dioxan (ε = 2.2) leads to pK_s = 17.85[4] and of 64 wt.% of ethanol (ε = 24.3) leads to pK_s = 15.3[4]; in the latter instance, where both solvents are amphiprotic, we may define $K_s = ([H_3O^+] + [EtOH_2^+])([HO^-] + [EtO^-])$. However, taking another single solvent, wherever possible, is a more effective means of increasing pK_s (see later). The measurement of pH in non-aqueous media in comparison with water will also be treated later.

In addition to the systems involving proton exchange (protonotropy), Jander[3] included among water-resembling non-aqueous media also systems involving exchange of another ion, which Gutmann and Lindqvist[5] called ionotropy, such systems being cationotropic (Ac_2O) or anionotropic (SO_2), as illustrated by equations 4.22–4.27.

$$Ac_2O \rightleftharpoons Ac^+ + AcO^- \quad (4.22)$$

$$Ac^+ + Ac_2O \rightleftharpoons Ac_3O^+ \quad (4.23)$$

$$2Ac_2O \rightleftharpoons Ac_3O^+ + AcO^- \quad (4.24)$$

- -

$$SO_2 \rightleftharpoons SO^{2+} + O^{2-} \quad (4.25)$$

$$O^{2-} + SO_2 \rightleftharpoons SO_3^{2-} \quad (4.26)$$

$$2SO_2 \rightleftharpoons SO^{2+} + SO_3^{2-} \quad (4.27)$$

Thus, in acetic anhydride (Ac_2O) acetyl chloride (AcCl) is a solvoacid (here acetoacid) and potassium acetate a solvobase (here acetobase); in liquid SO_2 thionyl chloride ($SOCl_2$) is a solvoacid (here oxoacid) and potassium sulphite a solvobase (here oxobase).

In the cationotropic system of anhydride [ε = 22 at 20° C; $pK_s = -\log 3 \cdot 10^{-15} = 14.52$ (ref. 3)] pyridine acts as a strong base, as its lone pair of electrons make it readily accept Ac^+, yielding an excess of the solvobase AcO^- according to the equation

$$C_5H_5N: + Ac_2O \rightleftharpoons C_5H_5N\text{–}Ac^+ + AcO^- \quad (4.28)$$

This also explains the use of pyridine in acetic anhydride as an acetylating reagent[6] in the determination of primary or secondary alcohols and amines according to the following sequential reactions (with ROH as an example):

$$\underset{\text{alcohol}}{R\ddot{O}H} + \underset{\text{acetylpyridinium ion}}{C_5H_5\overset{+}{N}COCH_3} \longrightarrow R\overset{+}{O}(H)OCCH_3 + :NC_5H_5 \longrightarrow \underset{\text{acetyl ester}}{ROOCCH_3} + \underset{\text{pyridinium ion}}{H\text{–}\overset{+}{N}C_5H_5} \quad (4.29)$$

where the role of the Ac^+ ion as the active principle of acetylation is clearly shown* (see also ref. 7).

In the anionotropic system of sulphur dioxide (condensed below $-10°$ C; $\varepsilon \approx 16$) we are concerned with an oxidotropic solvent with self-dissociation; here the O^{2-} ion does not exist in the free state, but occurs as the solvated ion SO_3^{2-}. A comparable situation is found in many systems of halotropy with self-dissociation, e.g., according to the overall reactions 4.30–4.33

$$2\,NOCl \rightleftharpoons NO^+ + Cl(NOCl)^- \text{ (NOCl condensed} < -8°\text{ C)} \quad (4.30)$$

$$2\,SbCl_3 \rightleftharpoons SbCl_2^+ + SbCl_4^- \text{ (}SbCl_3\text{ molten} > 73°\text{ C)} \quad (4.31)$$

$$2\,HgBr_2 \rightleftharpoons HgBr^+ + HgBr_3^- \text{ (}HgBr_2\text{ molten} > 236°\text{ C)} \quad (4.32)$$

$$2\,BrF_3 \rightleftharpoons \underset{\text{solvoacids}}{BrF_2^+} + \underset{\text{solvobases}}{BrF_4^-} \text{ (}BrF_3\text{ liquid 8.8° C)} \quad (4.33)$$

with the definition of solvoacidity by Cady and Elsey[8].

We shall not consider these amphitropic systems any further because they are not of analytical interest.

(B) From the foregoing, it is clear that the Arrhenius or solvents theory cannot work for aprotic solvents; most adequate here is the Brønsted–Lowry or proton theory, in which an acid is defined as a proton donor and a base as a proton acceptor, and under conditions such that the acid by donating its proton is converted into its conjugate base, and the base by accepting a proton is converted into its conjugate acid. This mutual relationship is illustrated by the following equilibrium reaction:

$$\underset{\text{(donor)}}{\text{acid}} \rightleftharpoons \underset{\text{(acceptor)}}{\text{base}} + H^+ \quad (4.34)$$

In the simplest case the acid may be positively charged and its base uncharged, or the acid is uncharged and its base negatively charged; e.g., for water as an amphoteric solvent, we have

$$H_3O^+ \rightleftharpoons H_2O + H^+ \quad (4.35)$$

$$H_2O \rightleftharpoons OH^- + H^+ \quad (4.36)$$

This example shows also that the proton theory, in addition to being valid for aprotic solvents, also works for amphiprotic solvents, and so represents a more general theory. How in an acid–base titration the theory works out can be followed from the titration of a certain amount of HCl gas introduced into pyridine as an aprotic solvent:

* After the acetylation, the analytical finish is not based on non-aqueous titration, but consists in acid titration after complete conversion of the anhydride with water and by comparison with a blank determination.

$$\underset{\text{acid 1}}{HCl} \rightleftharpoons \underset{\text{base 1}}{Cl^-} + H^+ \tag{4.37}$$

$$H^+ + C_5H_5N: \rightleftharpoons C_5H_5NH^+ \tag{4.38}$$

$$\underset{\text{acid 1}}{HCl} + \underset{\text{base 2}}{C_5H_5N:} \longrightarrow \underset{\text{acid 2}}{C_5H_5NH^+} + \underset{\text{base 1}}{Cl^-} \tag{4.39}$$

$$\underset{\text{acid 2}}{C_5H_5NH^+} + \underset{\text{base 3}}{OH^-} \longrightarrow \underset{\text{base 2}}{C_5H_5N:} + \underset{\text{acid 3}}{H_2O} \tag{4.40}$$

The pyridinium ion (acid 2) as the analyte can be titrated with quaternary ammonium hydroxide (base 3); as it concerns the determination of H^+ of the Brønsted acid pyridinium, a potentiometric measurement of the pH titration curve and its inflection point is most obvious. In the aprotic, but protophilic, solvent pyridine no stronger acid can exist (see reactions 4.37 and 4.38) than the pyridinium ion itself; hence there is a levelling effect but in theory only on the acid side.

An analogue of the proton theory is found in systems where the oxygen ion O^{2-} is merely split off (without solvation), as illustrated by the following equilibrium reaction:

$$\underset{\text{(donor)}}{\text{oxobase}} \rightleftharpoons \underset{\text{(acceptor)}}{\text{oxoacid}} + O^{2-} \tag{4.41}$$

This concept of oxoacidity was proposed by Lux[9] and Flood and co-workers[10], but it applies mainly to media of molten salts, where the free O^{2-} ion can exist, e.g.,

$$MgO \rightleftharpoons Mg^{2+} + O^{2-} \tag{4.42}$$

$$UO_3 \rightleftharpoons UO_2^{2+} + O^{2-} \tag{4.43}$$

$$CO_3^{2-} \rightleftharpoons CO_2 + O^{2-} \tag{4.44}$$

For such systems an analogue of pH has been used:

$$pO^{2-} = -\log a_{O^{2-}}$$

which means that pO^{2-} increases with increasing oxoacidity, because it decreases with increasing oxobasicity, just as pH decreases with increasing proton acidity. For practical electroanalysis the definition of Lux and Flood and co-workers is mainly of theoretical interest.

The above solvents theory (A) and proton theory (B) have shown that in theory the neutrality point (of the pure solvent) lies for the amphiprotic solvents at pH $= \frac{1}{2}pK_s$ and for the aprotic protophilic solvents at a pH somewhere between the highest acidity (of the protonated solvent) and an infinitely high pH. However, the true pH of the neutrality point of the solvent can only be obtained from a reliable pH measurement and the problem is whether and how this can be achieved. For water as a solvent, the true pH $= -\log a_{H^+} =$ colog a_{H^+} is fixed by the internationally adopted convention $E^0_{2H^+ \rightarrow H_2(1\,atm)} = 0$

(at any temperature) and by the use of measuring cell (see 2.41), preferably at 25° C (see under hydrogen electrode, p. 33) so that pH = 7.00 ± 0.02. In view of the precision, the pH measurement in practice is standardized on buffer solutions with the pH ranges of interest (see pp. 41–42), while the hydrogen electrode is usually replaced by a glass pH electrode [see Section 2.2.1.3.2.1 and Figs. 2.10(1), 2.11 and 2.13].

In measuring cell 2.41, where a saturated calomel electrode is used as a reference, a diffusion potential E_j on contact between the aqueous analyte and the saturated KCl bridge is negligible owing to the nearly identical mobilities of K^+ and Cl^- ions in water (cf., Table 2.2 and Section 2.2.1, eqn. 2.33); however, in the various non-aqueous solvents the mobilities of K^+ and Cl^- ions can become different from each other because of the differing degrees of solvation of the cation and the anion in such media as may be expected from the Gutmann donor number (DN) and the Gutman–Mayer acceptor number (AN) (see later). Nethertheless, and simply as a first approach to the behaviour of acids and bases in various organic solvents in comparison with that in water, we applied the aforementioned measuring cell (glass pH electrode and saturated calomel electrode)[11], but with the following precautionary measures:

(1) the potentiometric results were expressed in millivolts (0 mV corresponded to pH = 7.80 in aqueous solution);

(2) Bu_3MeNI was added ($1.5 \cdot 10^{-3}$ mol^{-1}) to solvents of low conductivity;

(3) the concentration of the acid or base was about 0.01 *N*;

(4) use was made of a 0.1 *N* titrant, mostly quaternary ammonium hydroxide or *p*-toluenesulphonic acid in a solvent such that it avoids as far as possible changes in the medium (for more specific data, see ref. 11).

For the solvents used ε was < 40, so that owing to ion associate formation the ion dissociation even in dilute solution remains limited; therefore, the Henderson equations 2.102 (for acid) and 2.103 (for base) are valid, yielding symmetrical pH curves (see Fig. 2.18), where at half-neutralization pH = pK_a, irrespective of concentration; as our measuring cell could not determine a ture pH value, we simply registered the half-neutralization potentials (h.n.p.). In each specific solvent the strongest possible acid and base provide the limiting h.n.p.s of the potential range within which the h.n.p.s of the various weaker acids and bases are found. The results of these measurements are shown in Fig. 4.1a, b and c; the vertical lines at the background in a and b, and also the vertical blocks in c, represent the titration potential ranges of the specific solvents; the points on the lines are the h.n.p.s of individual acids or bases within those ranges, and the connecting cross-lines illustrate the change of such an h.n.p. over the different solvents.

From Fig. 4.1 various important conclusions can be drawn, as follows.

(1) As would be expected, the larger titration potential ranges offer much more scope for mutually distinguishing between individual acids or bases; in amphiprotic solvents, as a consequence of self-dissociation, the potential ranges are rather limited, whereas in the aprotic protophilic solvents and "aprotic" inert solvents these ranges are considerably more extensive.

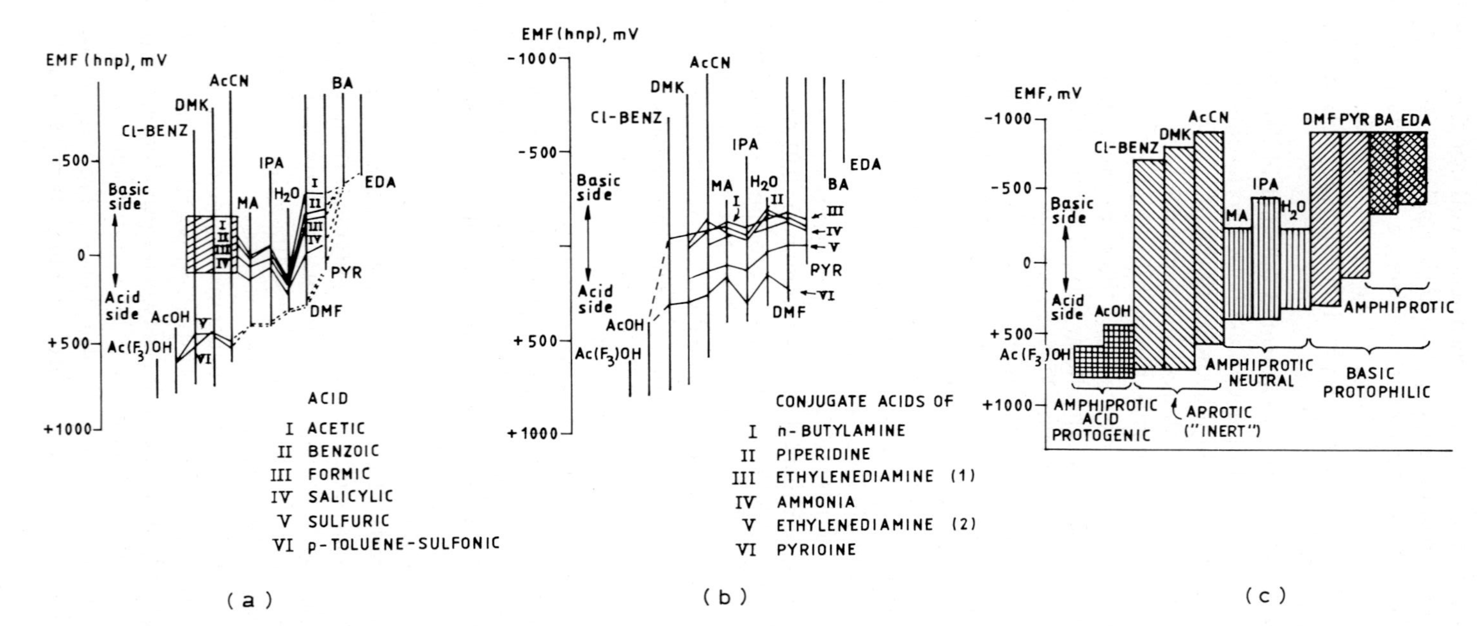

Fig. 4.1. Potential ranges of solvents. (a) h.n.p.s of acids. I, Acetic acid; II, benzoic acid; III, formic acid; IV, salicylic acid; V, sulphuric acid; VI, *p*-toluenesulphonic acid. (b) h.n.p.s of conjugate acids of I, *n*-butylamine; II, piperidine; III, ethylenediamine (1); IV, ammonia; V, ethylenediamine (2); VI, pyridine.

(2) In Fig. 4.1a one sees the occurrence of an increasing levelling effect on the acid side when changing from protogenic solvents via neutral to protophilic solvents (dotted lines below).

(3) In Fig. 4.1b one sees the occurrence of the levelling effect on the basic side only in protogenic solvents (dotted lines above).

(4) The above three conclusions lead to the following recommendations in titration analysis:

(a_1) for total acidity in an organic material, use a basic solvent in order to increase the acidity of all the acids present to the same level as the conjugate acid of the solvent;

(a_2) for differentiation between acids simultaneously present, use a less basic or even an inert solvent or a mixture of both;

(b_1) for total basicity in an organic material, use an acid solvent in order to increase the basicity of all bases present to the same level as the conjugate base of the solvent;

(b_2) for differentation between bases simultaneously present, use a less acidic or even inert solvent or a mixture of both.

Reconsidering the overall picture (apart from the levelling effects) in Fig. 4.1a and b, it appears that if the h.n.p. measurements in non-aqueous media could achieve a better correspondence with the true pH, they would probably yield a characteristic pK_a value for each acid and base (or its conjugate acid), i.e., this pK_a value would represent a so-called intrinsic acidity; its reality seemed more probable for the positively charged conjugate acids of bases (Fig. 4.1b) than for the uncharged acids (Fig. 4.1a), but in the latter instance the dissociation into two ions of opposite charge is severely affected by the appreciably varying coulombic forces caused by the widely differing dielectric constants of the solvents.

In order to arrive at values of the virtually intrinsic acidity, i.e., an acidity expression independent of the solvent used (Trémillon[12] called it the absolute acidity), Schwarzenbach[13] used the normal acidity potential as an expression for the potential of a standard Pt hydrogen electrode (1 atm H_2), immersed in a solution of the acid and its conjugate base in equal activities; analogously to eqn. 2.39 for a redox system and assuming $n = 1$ for the transfer of one proton, he wrote for the acidity potential

$$E_{ac} = E^0_{ac} + \frac{RT}{F} \cdot \ln\left(\frac{a_{acid}}{a_{base}}\right) \tag{4.45}$$

where E^0_{ac} represents the normal acidity potential*

As most of the solvents in Fig. 4.1 are amphiprotic and it is interesting to know the correct vertical position of their potential ranges compared with that of water, we shall apply eqn. 4.45 to a solvent HS with self-dissociation:

* There may be preference for the term "standard acidity potential", analogous to "standard redox potential", as we are concerned with equal instead of normal concentrations of the conjugates.

$$HS + HS \rightleftharpoons H_2S^+ + S^-$$

or, according to Brønsted:

$$\text{acid 1} + \text{base 2} \rightleftharpoons \text{acid 2} + \text{base 1}$$

(cf. eqn. 4.39).

In case of equilibrium $E_{ac_1} = E_{ac_2}$, so that the resulting standard acidity potential is

$$E^0_{HS} = E^0_{ac_1} - E^0_{ac_2} = \frac{RT}{F} \cdot \ln \left(\frac{a_{acid_2} \cdot a_{base_1}}{a_{base_2} \cdot a_{acid_1}} \right)$$

which can be simply written as

$$E^0_{HS} = \frac{RT}{F} \cdot \ln \frac{a_{HS^+} \cdot a_{S^-}}{a^2_{HS}} = \frac{RT}{F} \cdot \ln K_{\text{self-diss.}}$$

E^0_{HS} is negative, which means that the pure solvent is only slightly ionized, so that

$$E^0_{HS} = \frac{RT}{F} \cdot \ln [H_2S^+][S^-] = \frac{RT}{F} \cdot \ln K_S \qquad (4.46)$$

where K_S is the ionic product of the solvent.

In order to make a comparison with water as a solvent, the problem of how to determine the true E^0_{HS} in non-aqueous media has still to be solved; as the inherent difficulties mainly concerned differing degrees of solvation and varying junction potentials, Pleskov[14] proposed measuring the potential at the 1 atm H_2/Pt electrode versus a Rb (amalgam) electrode placed directly in the solvent with a constant Rb^+ ion concentration added. He assumed for his so-called standard ion method that ions having large diameters, such as Rb^+ or Cs^+, possess the same solvation energy in any solvent, so that there respective electrodes yield potentials that depend solely on their ion concentration. The assumption of Pleskov has appeared to be justified, but sometimes as a consequence of attack of the alkaline metal by acidic media the proper function of the Rb electrode could be disturbed; therefore, searching for a redox system of a monovalent spherical cation with large diameter and without structural change during electron exchange, Strehlow and co-workers[15] found a mixture of ferrocene [dicyclopentadienyliron(II)]–ferricinium$^+$ and similar pairs such as cobaltocene–cobaltocinium$^+$ to be much more suitable, in that they exhibit considerable chemical stability and, although slightly but still sufficiently soluble in water, they are much more soluble in organic solvents and represent with their solution at a Pt or Hg electrode a very convenient half-cell*.

In order to show the effect of the standard ion method we give in Table 4.2 (cf. also Kucharský and Šafařík[17]) some results reported by Pleskov[14], simply as

* For an excellent comparison of the Pleskov and Strehlow electrodes, see ref. 16.

TABLE 4.2

COMPARISON OF POTENTIALS AND pH VALUES OF WATER AND HS-TYPE SOLVENTS

Parameter	NH_3	N_2H_4	H_2O	CH_3CN	HCOOH
$E^0_{HS} - E^0_{H_2O}$	−1.00	−0.92	0	0.24	0.52
$\frac{F(E^0_{HS} - E^0_{H_2O})}{2.3RT}$	−17 (at −50° C)	−16	0	4	9

an example of how it works; it concerns a comparison between water and other solvents of the HS type.

A negative value of $E^0_{HS} - E^0_{H_2O}$ means, according to Fig. 4.1, a vertical shift of the acidic side of the titration potential range to a more negative potential (upwards), in agreement with the basic (protophilic) character of the solvent; thus a positive value of $E^0_{HS} - E^0_{H_2O}$ means a corresponding shift to a more positive potential (downwards), in agreement with the acidic (protogenic) character of the solvent. Where with its nearly complete dissociation a 1 N acid solution in water is assumed to possess a pH $= -\log a_{acid}$ value of zero, the standard acidity potential difference $E^0_{HS} - E^0_{H_2O}$ for a non-aqueous solvents vs. water indicates a pH shift on the acidic side of

$$\mathrm{pH} = \frac{F(E^0_{HS} - E^0_{H_2O})}{2.3026RT} = \frac{E^0_{HS} - E^0_{H_2O}}{0.05916} \tag{4.47}$$

(at 25° C; cf., eqn. 2.45). Eqn. 4.47 yields values of −15.55, 4.06 and 8.79 for hydrazine, acetonitrile and formic acid, respectively, at 25° C; a calculation for NH_3 would yield −16.90 at 25° C but −22.58 at −50° C, so that the significance of −17 in Table 4.2 remains obscure. Further, the measurements by Pleskov[14] concerned the shift of pH = 0 (i.e., at the acidic side) of amphiprotic solvents (type HS) only; if in addition pK_s has been determined (at 25° C), the complete vertical position of the pH scale (similarly to the scale 0–14 units for water) up to the basic side of all these solvents will be known. One certainly realizes that the procedure for pH = 0 on the acidic side remains the same for an aprotic purely protophilic solvent (type S) such as pyridine, for which Mukherjee[18] measured $E^0_S - E^0_{H_2O}$ at 25° C by means of a 1 atm hydrogen electrode and the Strehlow standard ferrocene–ferrocinium$^+$ electrode and obtained a value of −0.228 V, i.e., a positive pH shift of 3.85. Theoretically the pH range of pyridine is unlimited on the basic side, but in practice it is limited somewhere in that area as a consequence of the protogenic effect of contaminating traces of water.

Fig. 4.2 shows the data of Pleskow and Mukherjee together with pK_s values from other sources (cf., ref. 19) as vertical pH titration ranges, the overall picture of which appears reasonably comparable to that of the potential ranges in Fig. 4.1.

From a physico-chemical point of view, it is also interesting to determine by means of the Pleskov or Strehlow measuring system the pH titration curve of

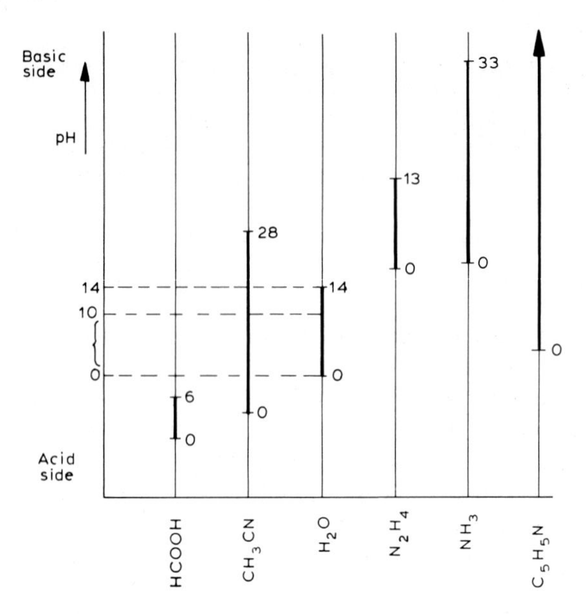

Fig. 4.2. Intrinsic pH ranges of solvents.

acids and bases of moderate acidity such that the latter can manifest itself well within the pH range of the solvent; under these conditions the half-neutralization pH (h.n.pH) corresponds, according to the Henderson equation 2.102 (and E^0_{ac} in eqn. 4.45 at $a_{acid} = a_{base}$), with the true pK_a of the intrinsic acidity. Nevertheless, from an electroanalytical point of view, it remains satisfactory to determine a voltage titration curve by means of a measuring system such as that used for Fig. 4.1; so it may be useful to return now to the conclusions drawn (see p. 241) from Fig. 4.1 and more especially to conclusion (4) (a_2) and (b_2), where the possibility of differentiation between acids or bases has been mentioned.

4.1.1.1. *Differentiation between acids*

Owing to hydrogen-bridge formation between the COOH and the OH groups*:

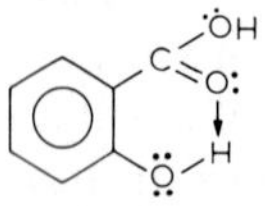

salicylic acid is much stronger than benzoic acid (Fig. 4.4); the OH group is so weakly acidic that salicylic acid and phenol added can be separately determined (Fig. 4.3).

In maleic acid, hydrogen-bridge formation extra strengthens one COOH

*In fact, it is an intramolecular hydrogen bonding, usually called chelation, and thus yielding a chelate ring.

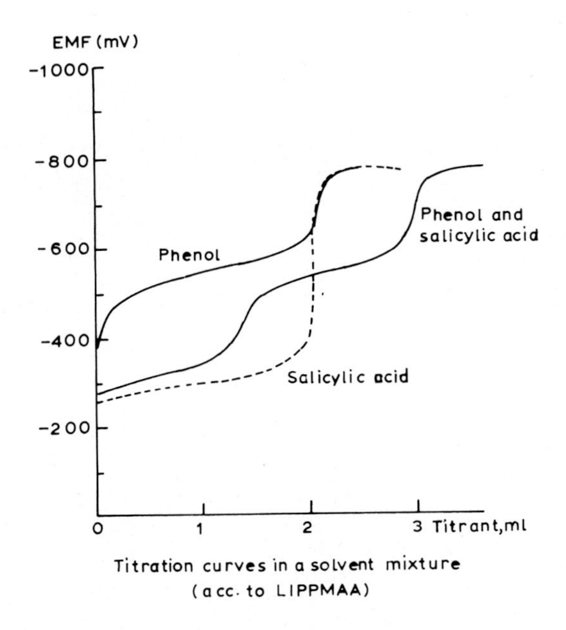

Fig. 4.3. Potentiometric titration of phenol and salicylic acid. Solvent: pyridine–dimethylfuran–diethylamine (4 : 7 : 9, v/v); titrant $(C_4H_9)_3CH_3NOH$, 0.1 N in IPA; glass-calomel electrode (aq.).

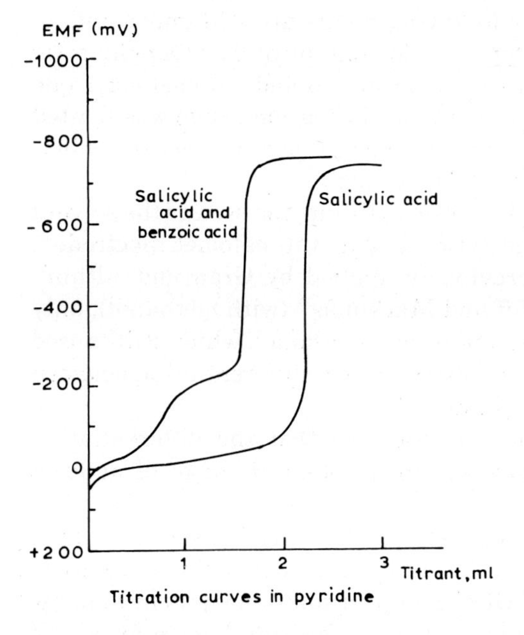

Fig. 4.4. Potentiometric titration of salicylic and benzoic acid. Titrant: $(C_4H_9)_4NOH$, 0.1 N in methanol–benzene; glass-calomel electrode (pyridine).

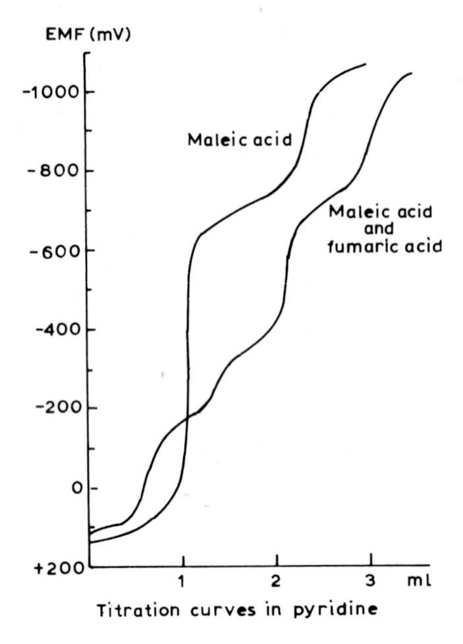

Fig. 4.5. Potentiometric titration of maleic and fumaric acid. Titrant: $(C_4H_9)_3CH_3NOH$, 0.1 *N* in pyridine glass-calomel electrode (pyridine).

group and weakens the other owing to the *cis*-structure, whereas the difference in fumaric acid owing to the *trans*-structure is only weak (Fig. 4.5).

Comparing Fig. 4.3, 4.4 and 4.5, the following points are still important.

(a) In Fig. 4.3, a solvent with an appreciable content of basic amphiprotic diethylamine, a titrant with amphiprotic isopropanol and calomel electrode with KCl in water were used. Hence the potential titration range was limited on both the acidic and basic sides but, owing to their great difference in acidity, salicylic acid and phenol could still be differentiated.

(b) In Fig. 4.4, the aprotic more weakly basic pyridine was used as the solvent and as the filling solution saturated with KCl in the calomel electrode[20]. Similar non-aqueous fillings were previously applied by Gran and Althin[21] (with ethylenediamine) and by Cundiff and Markunas[22] (with methanol); only the titrant in Fig. 4.4 contained some amphiprotic methanol, which still caused a limitation on the basic side without hindrance of the differentiation between salicylic and benzoic acid at the beginning.

(c) In Fig. 4.5 exclusively pyridine was used, so that the differentiation between four acid steps was obtained within a potential range of at least 1200 mV.

In Fig. 4.6, the titration of *p*-hydroxybenzoic acid in pyridine shows that the COOH and OH groups can be clearly determined. However, in acetonitrile there is half-way the titration of the COOH group an additional potential jump; this can be explained by a phenomenon which was already known for acetic acid[23], viz., in the inert solvent acetonitrile intermolecular hydrogen-bridge

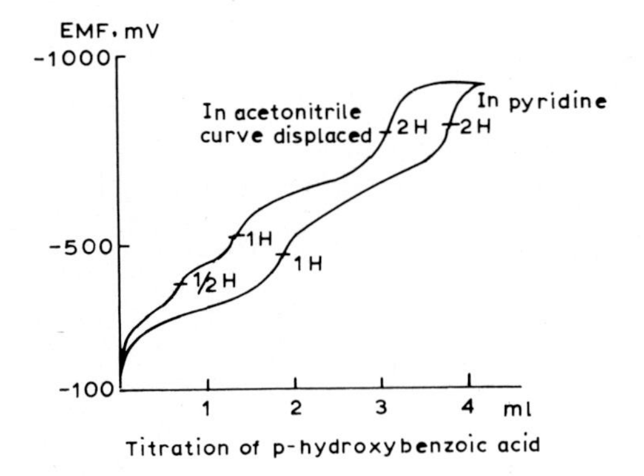

Fig. 4.6. Potentiometric titration of *p*-hydroxybenzoic acid. Titrant: $(C_4H_9)_3CH_3NOH$, 0.1 *N* in pyridine; glass-calomel electrode (pyridine).

formation yields a dimeric acid, which in titration manifests itself as a divalent acid:

$$CH_3C\begin{matrix}\diagup O\cdots H-O \diagdown \\ \diagdown O-H\cdots O \diagup\end{matrix}CCH_3 \longrightarrow H^+ + \left\{ \begin{matrix} CH_3C(=O)O{-}H\cdots O{=}C(O^-)CH_3 \\ \updownarrow \\ CH_3C(=O)O{-}H\cdots O^-{-}C(=O)CH_3 \end{matrix} \right\} \longrightarrow H^+ + 2\,CH_3COO^-$$

Such dimerization is impossible in a protophilic solvent such as pyridine.

Van der Heijde[24] mentioned the frequent occurrence of similar irregularities, especially for titrations of acids in inert or weakly basic solvents, and we shall return to this later.

4.1.1.2. *Differentiation between bases*

For the purpose of potentiometric differentiation between bases we have found an indifferent solvent such as nitrobenzene to be most attractive[25]; optimum results were obtained both in pure nitrobenzene and in mixtures containing equal volume fractions of this solvent and dioxan or a hydrocarbon such as xylene.

Fig. 4.7 shows the titration with perchloric acid of a mixture of piperidine, ethylenediamine and *p*-toluidine. Fig. 4.8 illustrates the effect of different chain lengths on the titration (with perchloric acid) of diamines in nitrobenzene containing 2.5% (v/v) of methanol; once the first amino group of EDA has been protonated, the resulting proton bridge with the lone pair of electrons of the second amino group lowers the basicity of the latter considerably; the effect decreases on the introduction of more intermediate CH_2 groups until complete disappearance when six are present.

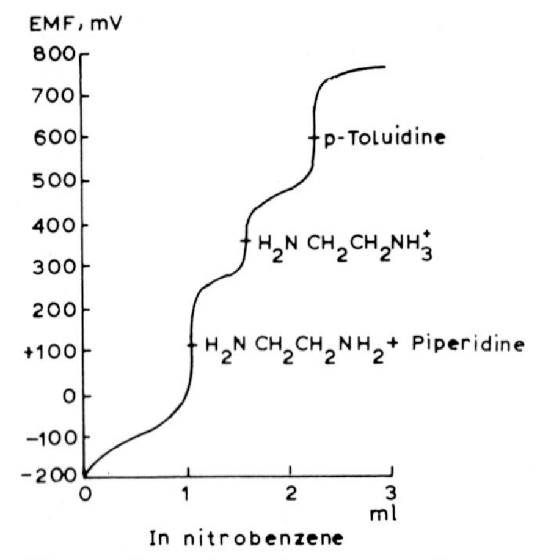

Fig. 4.7. Potentiometric titration of organic bases. Titrant: $HClO_4$, 0.1 *N* in dioxan; glass-calomel electrode (2-propanol).

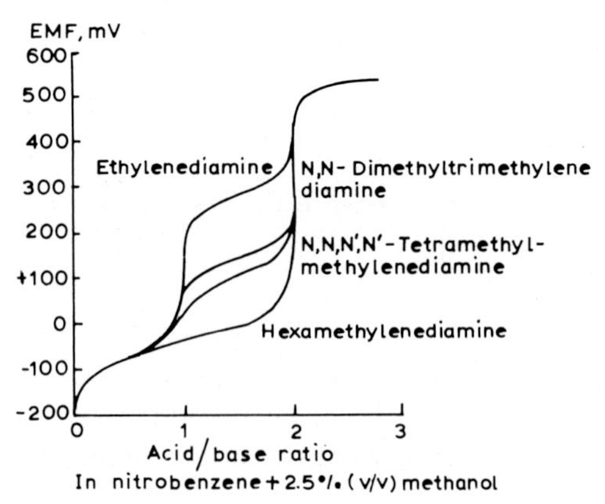

Fig. 4.8. Potentiometric titration of α,ω-diamines. Titrant: $HClO_4$, 0.1 *N* in dioxan; glass-calomel electrode (2-propanol).

H
NH
CH_2
|
CH_2 H^+
NH
H

(C) The Brønsted–Lowry or proton theory interprets the acid–base reaction as a mere proton exchange between the acid (proton donor) and the base (proton acceptor); however, the Lewis theory or electron theory interprets the reaction as a donation and acceptance of a lone pair of electrons*, where the

---

* Mullikan[26] was the first to draw attention to the analogy between π-complexes and Lewis complexes, both being "charge-transfer complexes"; e.g., an aromatic such as phenanthrene (π-base) donates its π-system to an acceptor such as tetracyanoethene (π-acid), yielding a coloured π-complex; such a complex permits the colorimetric determination of aromatics[27] and provides a colour indicator in the titration of cyclopentadiene[28].

base is donor and the acid is acceptor; the following equations illustrate its application.

$$\underset{\text{Brønsted base}}{C_5H_5N} + H^+ \longrightarrow \underset{\text{Brønsted acid}}{C_5H_5N\text{–}H^+} \tag{4.48}$$

$$\underset{\text{base 1}}{C_5H_5N:} + \underset{\text{acid 1}}{B(OC_2H_5)_3} \longrightarrow \underset{\text{complex 1}}{C_5H_5N\text{–}B(OC_2H_5)_3} \tag{4.49}$$

$$\text{in pyridine}\left\{\underset{\text{complex 1}}{C_5H_5N\text{–}B(OC_2H_5)_3} + \underset{\text{base 2}}{R_3N:} \rightleftharpoons \underset{\text{complex 2}}{R_3N\text{–}B(OC_2H_5)_3} + \underset{\text{base 1}}{C_5H_5N:}\right. \tag{4.50}$$

Lewis

$$\text{in } B(OC_2H_5)_3\left\{\underset{\text{complex 1}}{C_5H_5N\text{–}B(OC_2H_5)_3} + \underset{\text{acid 2}}{B(C_2H_5)_3}\right.$$

$$\rightleftharpoons \underset{\text{complex 3}}{C_5H_5N\text{–}B(C_2H_5)_3} + \underset{\text{acid 1}}{B(OC_2H_5)_3} \tag{4.51}$$

Comparison of Brønsted reaction 4.48 with Lewis reaction 4.49 shows that the Lewis theory is more generally applicable, but its interpretation is different in terms of the definition of acids and complexes. In fact, the Lewis theory is valid for all acid–base reactions (cf., eqns. 4.39 and 4.40).

From eqns. 4.50 and 4.51 it can be seen that complex 1 is converted into complex 2 by a stronger base 2 or into complex 3 by a stronger acid 2; in other words, complexes 2 and 3 are much more stable than complex 1. Whereas reactions 4.48 and 4.49 are addition reactions, reactions 4.50 and 4.51 are exchange reactions; often Lewis titrations must be carried out in completely inert solvents such as alkanes or benzene because of instability of the titrants and titrands in other media. Examples of potentiometric Lewis titration curves are given in Fig. 4.9 for CS_2 and CO_2[20], where one of their resonance structures can react as a Lewis acid with OH^- as a Lewis base:

$$\text{O}{=}C^+\text{–}O^- + {}^-{:}\ddot{O}\text{–H} \longrightarrow \underset{\text{bicarbonate}}{\text{O}{=}C(\text{–}O^-)\text{–}\ddot{O}\text{–H}}$$

and CS_2 yields bithiocarbonate analogously.

Probably the solubility of CO_2 and CS_2 can be explained by the formation of a Lewis complex with pyridine, so that the titration equation must be written as an exchange reaction with OH^-.

Often a suitable potentiometric indication for Lewis titrations is not available, whereas a conductometric indication can still be applied; a well known example is the Bonitz titration[29] of triethylaluminium (Et_3Al) with an azine, such as isoquinoline, for determination of active alkylaluminium in the precatalysts of the Ziegler synthesis of polyethene or polypropene; beyond the titration parameter λ of the 1 : 1 complex, the conductivity suddenly decreases,

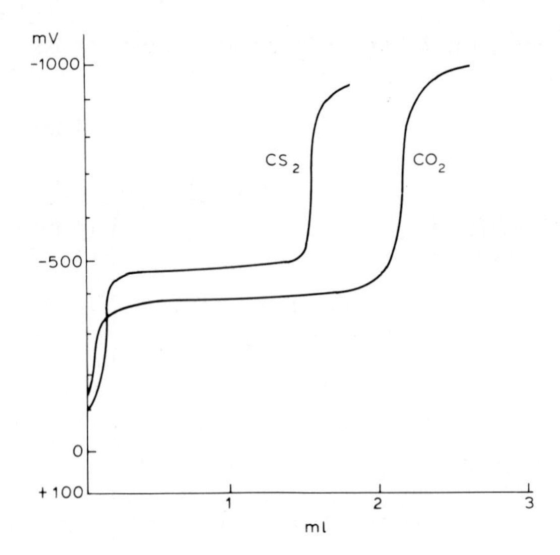

Fig. 4.9. Potentiometric Lewis titration of CS_2 and CO_2. Solvent: pyridine; titrant: $(C_4H_9)_4NOH$ in pyridine (CO_2); $(C_4H_9)_3CH_3NOH$ in pyridine (CS_2); added $(C_4H_9)_3CH_3NI$; glass-calomel electrodes.

and as diethylmonoethoxyaluminium ($Et_2(EtO)Al$] does not yield a complex the method provides a reliable check on the activity decrease due to oxidation of Et_3Al.

The aforementioned application of conductometry in Lewis titrations was an incentive, in addition to our potentiometric studies, to investigate also conductometric titration in non-aqueous media more thoroughly. Figs. 4.10 and 4.11 show two selected examples of the study.

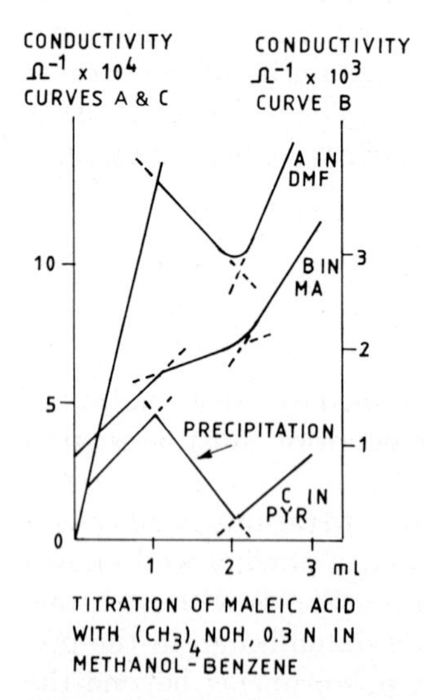

Fig. 4.10. Conductometric titration of divalent acid.

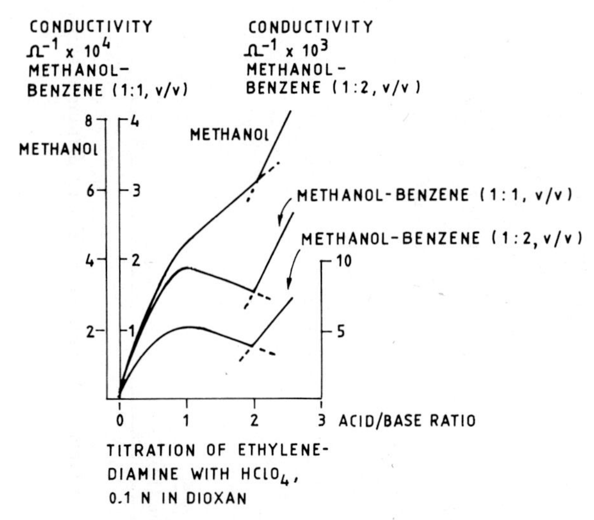

Fig. 4.11. Conductometric titration of divalent base.

These figures indicate that, depending on the choice of the titration medium, divalent acids[30] or bases[31] yield chair- and even N-type curves. The picture of N-type curves was reported in 1955 by Higuchi and Rehm[32] for sulphuric acid in acetic acid when titrating with an alkali metal acetate; they suggested that the degree of ionization of the neutral secondary salt is lower than that of the acidic primary salt. This explanation seems correct, if more strictly defined by replacing the term "ionization" by "dissociation" in the sense of Bjerrum, i.e., the dissociation of ion pairs in acetic acid[33].

Whereas in many instances potentiometric non-aqueous titrations of acids can show anomalies[24] depending on the type of solvents and/or electrodes (owing to preferential adsorption of ions, ion pairs or complexes on the highly polar surface of the indicator electrode, or even adherence of precipitates on the latter), conductometric non-aqueous titrations, in contrast, although often accompanied by precipitate formation[30], are not hindered by such phenomena; sometimes, just as in aqueous titrations, the conductometric end-point can even be based on precipitate formation[34].

4.1.2. Choice of solvents for acid–base reactions

From our previous treatment of the Arrhenius, Brønsted and Lewis acid–base theories, the importance of the choice between the divergent solvent types clearly appeared; if we now confine ourselves to solvents to which the proton theory in general is applicable, this leads to a classification of eight classes as already proposed by Brønsted[35,36] (Table 4.3).

With regard to this classification, Brønsted made the following very important remark: "Solvents of class 4, on account of their comparatively slight acidic and basic character, are the most nearly similar to class 8 in their

TABLE 4.3

CLASSIFICATION OF SOLVENTS BY BRØNSTED AND THEIR CHARACTERIZATION

Class No.	ε*	Relative acidity*	Relative basicity*	Examples** (ε, approx.)	Characterization	
					Amphiprotic or purely protophilic	Aprotic
1	+	+	+	H_2O (78), CH_3OH (33)	Neutral (dissociating)	
2	+	+	–	HCN (115), HF (84 at 0° C)	Acid (protogenic)	
3	+	–	+	$C_2H_5CONHCH_3$ (176) $(CH_3)_2NCON(CH_3)_2$ (23)	Basic (protophilic)	
4	+	–	–	CH_3COCH_3 (21), CH_3CN (36) DMF (37), DMSO (46), $C_6H_5NO_2$ (35)	Rather inert (dipolar)	
5	–	+	+	*t*-BuOH (11), cyclohexanol (15), *t*-PeOH (6)	Neutral (associating)	
6	–	+	–	CH_3COOH (6), C_2H_5COOH (3)	Weakly acid	
7	–	–	+	*n*-$BuNH_2$ (5), THF (7), dioxan (2), C_5H_5N (12), TMG (11), EtOEt (4)	Basic (protophilic)	
8	–	–	–	C_6H_6(2.3), cyclohexane (2.0), CCl_4 (2.2) 1,2-DCE (10), 2,2,4-trimethylpentane (1.9)		inert

* + = high, – = low, which for ε mean > 20 and < 20, respectively.

** DMF = dimethylformamide, DMSO = dimethyl sulphoxide, THF = tetrahydrofuran, TMG = tetramethylguanidine, DCE = 1,2-dichloroethane.

influence on solute behaviour, and useful parallels can be drawn between the two classes". As to application in practice, we can say, "Solvents such as acetonitrile and dimethyl sulphoxide, notwithstanding their high dielectric constants, are fairly inert, so that their pH titration range is nearly as large as that of a real inert solvent such as benzene".

Parker[37] defined class 4 as solvents "which cannot donate suitable labile hydrogen atoms to form strong hydrogen bonds with an appropriate species" and proposed the designation dipolar aprotic solvents; he extended their range down to $\varepsilon > 15$ and quoted as examples acetone, acetonitrile, benzonitrile, dimethylformamide, dimethyl sulphoxide, nitrobenzene, nitromethane (41.8) and sulfolane (tetramethylene sulphone) (44), where ε varies from 21 to 46.5, and the dipole moment μ from 2.7 to 4.7 debye.

The above statements on the "inertness" of solvents in classes 4 and 8 towards solutes in contrast with the "activity" of the other classes of solvents in Table 4.3 are in agreement with the results depicted in Fig. 4.1a, b and c together with the conclusive recommendations (4) (a_1), (a_2), (b_1) and (b_2) with regard to titration analysis. Davies[38], therefore, made a distinction between differentiating solvents and levelling solvents. However, there is much more to be said about solute behaviour in solvents when we only consider phenomena such as solvation, dissociation vs. association, different types of associates (ion pairs, triple ions, etc.), intramolecular hydrogen bonding (distinction between two acidic and/or basic groups within one molecule), intermolecular hydrogen bonding (mid-way titration inflection points in dimers of acids or bases), etc. An excellent review of studies by many investigators in this field before 1968 was given by Davis[36]. We shall link up with much of that information and complete our argumentation on the basis of later work and partly the work of our own research group.

In the first place, it is generally recognized that dissociation of electrolytes into cations and anions in water and other good ionizing media is predominant on the basis of mainly two phenomena, viz., (1) ion–solvent interaction and (2) a high dielectric constant of the solvent. Considering (1), which represents ion solvation, it is of importance that the mutual approach of the ion and solvent molecules is favoured by a small size of both and by their polarity and polarizability; considering (2), a value of $\varepsilon > 40$ (solvent classes 1, 2 and partly 3) is required according to Fuoss[39] for strong electrolyte behaviour in the sense of the classical dissociation theory. In fact, water is the medium most favourable to dissociation; however, it would be an oversimplification to say that in water association never occurs; for instance, Davies[38] found on the basis of pK values representing $-\log K_{\text{diss.}}$ (which also means $\log K_{\text{ass.}}$) that most uni-univalent electrolytes showed no evidence of ion pairing (negative pK) whereas, e.g., for many electrolytes with multivalent anions there was appreciable ion pairing (positive pK); in the latter instance, moreover, addition of methanol ($\varepsilon = 33$) to the aqueous solution did increase the degree of ion pairing further, as expected.

In the second place, it must still be recognized that the solvating power of a specific solvent remains of great importance also for the less ionizing media. Burger[40] argued that the complexity of the solvating power is a combination of specific and non-specific interactions; hence various empirical solvent scales were proposed, among which the most suitable are those which distinguish between donor and acceptor solvation, such as the measurement of the Gutmann donor number (DN)[41] and the Gutmann–Mayer acceptor number (AN)[42].

The DN is determined by calorimetric measurement of the enthalpy of the reaction with the reference acceptor, antimony pentachloride, in dichloroethane; weak donor solvents are hydrocarbons and halogenated hydrocarbons (DN $\approx$ 0), nitromethane (2.7), nitrobenzene (4.4), acetic anhydride (10.5), acetone (17.0) and water (18.0); stronger than water are diethyl ether (DN 19.2), THF (20.0), DMF (26.6) and DMSO (29.8); and strongest are pyridine (DN 33.1) and hexamethylphosphotriamide (38.8). The AN is determined from the ^{31}P NMR chemical shift of the complex with the reference donor, triethylphosphine oxide; on the AN scale the complex of the acceptor, antimony pentachloride, with this donor is taken as 100, while in all determinations the acceptor solvent under investigation functions as the medium. Most solvents have both donor and acceptor properties, particularly amphiprotic solvents such as water and alcohols (oxygen being the donor and the hydroxy proton the acceptor by forming hydrogen bonds), but one of these properties can predominate.

4.1.3. Associates (donor–acceptor complexes or salts, self-aggregates and hydrogen-bonded conjugates)

According to the previous discussion, association becomes predominant in Brønsted classes 4–8 ($\varepsilon \lesssim 40$); however, the associates obtained are of a quite different nature depending on the medium and the species becoming associated.

4.1.3.1. *Donor–acceptor complexes or salts*

As we have seen from reaction 4.49 donor–acceptor complexes (Lewis- or π-type) occur in a fairly inert medium (such as cyclohexane) via charge transfer between a base (electron donor) and an acid (electron acceptor by its electron deficiency). In a few instances, e.g., in the Bonitz titration[29] of the precatalyst diethylalaminium chloride with isoquinoline, the complex constists of an ion-pair ionizate.

$$\text{C}_9\text{H}_7\text{N:} + (C_2H_5)_2\text{Al–Cl} \longrightarrow \text{C}_9\text{H}_7\text{N}\rightarrow\text{Al}(C_2H_5)_2\text{–Cl} \rightleftharpoons \text{C}_9\text{H}_7\text{N–Al}^+(C_2H_5)_2\,\text{Cl}^- \quad (4.52)$$

In many other situations of donor–acceptor solutes in aprotic solvents, such as quaternary alkylammonium salts (R_4NX), a UV absorption shift to higher wavelength has proved the occurrence of simple cation–anion ion pairs ($R_4N^+X^-$)* in contrast with the situation of incompletely substituted salts (e.g., R_3NHX), where hydrogen bonding is possible[43]. Sometimes the simultaneous presence of two types of ion pair can be identified in one solution, e.g., $Co(NH_3)_5SO_4^+X^-$ ("inner" spheric ion pair) in addition to $Co(NH_3)_5(H_2O)SO_4^+X^-$ ("outer" spheric ion pair), where the exchange of a molecule (water) in the first coordination sphere is slow[44].

4.1.3.2. *Self-aggregates*

Rossotti and co-workers[45] introduced the term "oligomers" for dimers and even multimers or *n*-mers in non-aqueous solutions; real examples of such self-aggregates (non-hydrogen bonded) have been reported[46] for quaternary ammonium thiocyanates, iodides and perchlorates with *n*-mers with *n* even up to 20.

Further, in the case of virtually non-existent ion–solvent interactions (low degree of solvation), so that solute–solute interactions become more important, Kraus and co-workers[47] confirmed that in dilute solutions ion pairs and some simple ions occurred, in more concentrated solutions triple ions of type $M^+X^-M^+$ or $X^-M^+X^-$ and in highly concentrated solutions even quadrupoles; the expression triple ions was reserved by Fuoss and Kraus[48] for non-hydrogen-bonded ion aggregates formed by electrostatic attraction.

4.1.3.3. *Hydrogen-bonded conjugates*

In 1963 Kolthoff and Chantooni[49] introduced the term "conjugate ions", among which they distinguished: (a) hydrogen bonding of cations to neutral proton acceptors and (b) hydrogen bonding of anions to neutral proton donors; at the same time they proposed the more specific designations "homoconjugate" and "heteroconjugate" as shown by the following scheme:

homoconjugates	heteroconjugates
(a) $B–H^+–B$	(a) $B–H^+ \cdots B'$
(b) $A^-–H–A$	(b) $A^- \cdots H–A'$

All conjugates are hydrogen bonded; in the homoconjugates the hydrogen may be considered to be equally divided between the donors B (a) or the acceptors A^- (b), in the hetroconjugates as depicted in the scheme BH^+ is a weak proton donor (a) and A^- a weak proton acceptor (b). In fact, in the preceding conjugates we have to deal with a proton bond in contrast with a hydride bond as exists in some less common compounds or a complex ion such

*Bacelon et al.[43a] have confirmed via far-IR studies that R_4N halides exist in CCl_4 in large aggregates, which are dissociated, however, by phenol into ion pairs hydrogen bonded to phenol and multi-solvated anions.

as $Cl_3AlBH_4^-$, obtainable by reaction of aluminium chloride with sodium borohydride[50]:

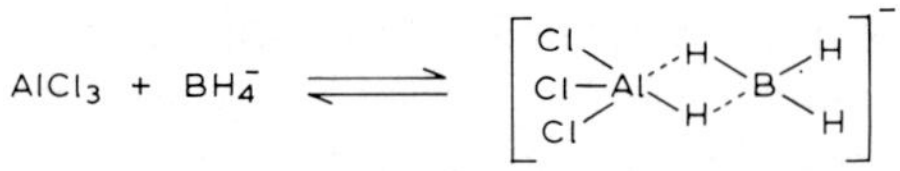

$$AlCl_3 + BH_4^- \rightleftharpoons \left[Cl_3Al \langle {}^{H}_{H} \rangle B \langle {}^{H}_{H} \right]^-$$

Returning to the hydrogen-bonded conjugates according to Kolthoff and Chantooni[49], we can mention the following examples of homoconjugate ions.

(a) By mid-way titration of nitrogen bases with perchloric acid in acetonitrile some homoconjugate cation (a) will be obtained, e.g., $R_3NH^+NR_3$ from a tertiary amine; however, in general such cations of the ammonium type are of low stability.

(b) The homoconjugate anions (b) of aromatic acids in acetonitrile are considerably more stable; the stronger the acids ArCOOH, the more stable are their homoconjugate anions ArCOOH---HOCOAr in acetonitrile; moreover, their stabilities vary inversely with those of the corresponding acid dimers $(ArCOOH)_2$ in benzene (see later).

The occurrence of homoconjugate anions was reported first by Maryott[51] in 1947 for picric, trichloroacetic and camphorsulphonic acids on the basis of a conductance maximum at mid-point of the titration in benzene and dioxan, and subsequently by Van der Heijde[52] for acetic acid in acetone, for phenol in pyridine and for *p*-hydroxybenzoic acid in acetonitrile (see Fig. 4.6), this for the first time from a potentiometric curve (with mid-point inflection) obtained by titration with Bu_3CH_3NOH in pyridine as a titrant. Bruss and Harlow[53] also found a conductance maximum during titrations of some unhindered phenols in pyridine, and Bryant and Wardrop[54] recorded potentiometric curves of titrations of carboxylic acids in solvents such as acetone and toluene, and extracted with toluene the conjugate anion $C_6H_5COO^-$–$HOCOC_6H_5$ after mid-way titration of C_6H_5COOH.

In addition to the aforementioned types of intermolecular homoconjugates, there are also intramolecular homoconjugates, examples of which we have already mentioned with regard to the titrations of (a) ethylenediamine with acid (p. 248) and (b) of salicylic acid with base (p. 245). This phenomenon of intramolecular hydrogen bonding is usually called chelation, and so the conjugates concerned are preferably designated chelates.

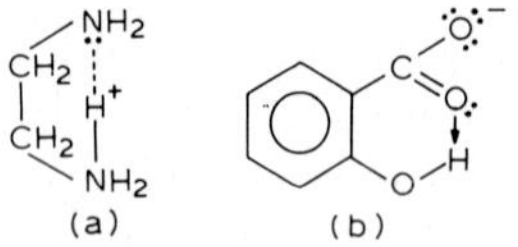

NH_2 CH_2 H^+ CH_2 NH_2 (a) (b)

With regard to heteroconjugate ions, let us consider a medium in classes 4–8 (Table 4.3) again. Now, if we dissolve the protonated form BH^+ of a moderately strong base B and we add another moderately strong base B′, then a heteroconjugate ion will occur, which depending on the base strength ratio may be represented as type (a) by B–H$^+$-----B′ or B---H$^+$–B′; accordingly, if we dissolve the anion A^- of a moderately strong acid HA and we add another moderately strong acid HA′, then the resulting heteroconjugate may be represented as type (b) by $A^- \cdots H{-}A$ or $A{-}H \cdots A'$. It should be noted that in instances where B′ and HA consist of the solvent or part of the solvent, respectively, the resulting heteroconjugates are solvated ions of BH^+ and A^-, respectively.

In our previous treatment of the Kolthoff conjugates we concentrated more particularly on conjugate ion formation; as the latter requires media of a relatively low dielectric constant (classes 4–8), one must also expect ion pairing; hence the question arises of whether one has to deal with a conjugate ion pair, i.e., with hydrogen bonding, or with an ion pair with merely electrostatic attraction. In this connection, Davis[55] stated that in a structure $X{-}H \cdots Y$ a hydrogen bond can be commonly recognized by the following changes in the X–H stretching band in the IR absorption spectrum: a shift to lower wavenumbers, an increase in intensity, a considerable broadening and often a splitting into two or more component bands; such changes of the N–H stretching band (3100–3500 cm^{-1}) for amine salts $N{-}H^+X^-$ have been reported by Sandorfy and co-workers[56]. Shifts to lower frequencies were also found in the UV spectrum, e.g., for the ion pair $R_4N^+X^-$ on going to the hydrogen-bonded ion pair $R_3NH^+X^-$ (ref. 57).

However, as most investigations of analytical interest concerned the titration of amines with organic, mostly carboxylic, acids or vice versa, instead of changes of the N–H stretching band, which can be hindered by the 3385 cm^{-1} band of the OH group in undissociated carboxylic acid, preference was given to carboxylate bands in the range 1200–2000 cm^{-1}, as was done by Bruckenstein and Saito[58]; their study was most revealing because they showed (see below) that during the titration the changes within the solution are so complex that they could be only detected by a sophisticated combination of physico-chemical methods.

4.1.4. Exploratory acid–base titrations in solvents with very low dielectric constant (ε)

In order to increase our knowledge of the complex electrochemistry in non-aqueous media, we shall consider the phenomena that occur during titrations in solvents with very low ε; in this connection, it is useful to treat these titrations separately for protogenic and the aprotic solvents, with the latter being subdivided into protophilic and inert solvents.

4.1.4.1. *Acid–base titrations in protogenic solvents with low ε*

Kolthoff and Bruckenstein[59] were the first to have placed titrations in these solvents (class 6) on a sound theoretical basis; they did so for glacial acetic acid (HOAc, ε = 6.13). We shall briefly describe the method* and the main conclusions of the authors.

* HOAc in the equations of the original papers will be replaced in our description by HS as a designation of protogenic solvents in general (see later its use for *m*-cresol).

The interaction of acid HX with solvent HS is a two-step reaction:

$$HX + HS \underset{}{\overset{K_i^{HX}}{\rightleftharpoons}} H_2S^+X^- \underset{}{\overset{K_d^{HX}}{\rightleftharpoons}} H_2S^+ + X^- \qquad (4.53)$$

which yields the equations

$$K_i^{HX} = \frac{[H_2S^+X^-]}{[HX]} \qquad (4.54)$$

for the ionization constant of HX and

$$K_d^{HX} = \frac{[H_2S^+][X^-]}{[H_2S^+X^-]} \qquad (4.55)$$

for the dissociation constant of HX.

The approximation of taking concentrations instead of activities was justified by the authors on the basis of the small degree of dissociation owing to the low ε, and by the assumption that activity coefficients of neutral molecules and ion pairs are not significantly different from unity in the low range of concentrations studied. From eqns. 4.54 and 4.55 one obtains for the overall dissociation constant of HX

$$K_{HX} = \frac{[H_2S^+][X^-]}{[HX] + [H_2S^+X^-]} \qquad (4.56)$$

which yields

$$K_{HX} = \frac{K_i^{HX} K_d^{HX}}{1 + K_i^{HX}} \qquad (4.57)$$

Analogously for the two-step reaction of base B with solvent HS:

$$B + HS \underset{}{\overset{K_i^{B}}{\rightleftharpoons}} BH^+S^- \underset{}{\overset{K_d^{B}}{\rightleftharpoons}} BH^+ + X^- \qquad (4.58)$$

one obtains the equations for the ionization and dissociation constants of B:

$$K_i^B = \frac{[BH^+S^-]}{[B]} \qquad (4.59)$$

and

$$K_d^B = \frac{[BH^+][S^-]}{[BH^+S^-]} \qquad (4.60)$$

respectively, and so for the overall dissociation constant of B we have

$$K_B = \frac{[BH^+][S^-]}{[B] + [BH^+S^-]} \qquad (4.61)$$

which yields

$$K_B = \frac{K_i^B K_d^B}{1 + K_i^B} \qquad (4.62)$$

Owing to the low ε of the solvent HS, the degree of dissociation of the solutes is so small that not only does K_d become different depending on the various ion pairs, but explicitly K_i gives an exact expression of acid or base strength vs. the solvent; arbitrarily, an acid or base may be called strong when K_i is equal to or greater than unity (e.g. in acetic acid, perchloric acid is strong).

Kolthoff and Bruckenstein[59], considering the determination of K_i and K_d of acids and bases, applied a spectrophotometric measurement of the interaction of the acid or base of interest with a well chosen colour indicator base I; referring for details to the original papers, we must confine ourselves to a concise explanation for the cases of an acid HX and a base B.

Measurement for an acid HX

All acid forms of an indicator base, IH^+, IH^+X^-, $X^-IH^+X^-$, etc., appeared to have the same absorption spectra and molar absorptivities; therefore for a given amount of indicator $(C_I)_t$ added to an acid solution of known concentration [HX], one obtains the following relationships from absorption spectrometry:

$$\Sigma[IH^+] = [IH^+] + [IH^+X^-] \tag{4.63}$$

from $(C_I)_t$ added:

$$[I] = (C_I)_t - \Sigma[IH^+] \tag{4.64}$$

and from the total concentration of acid $(C_I)_t$ added:

$$C_{HX} = (C_{HX})_t - \Sigma[IH^+] \tag{4.65}$$

so that

$$C_{HX} = [HX] + [H^+X^-] + [H^+] \tag{4.66}$$

Further, we may write

$$K_i^{IHX} = \frac{[IH^+X^-]}{[I][HX]} \tag{4.67}$$

and

$$K_d^{IHX} = \frac{[IH^+][X^-]}{[IH^+X^-]} \tag{4.68}$$

With a weak acid such as hydrochloric and *p*-toluenesulphonic acids, $[H^+X^-]$ and $[H^+]$ are negligibly small compared with [HX], so that eqn. 4.66 becomes $C_{HX} = [HX]$; also as in acid solution $[S^-]$ from autoprotolysis of the solvent H_2S can be neglected, the electroneutrality rule reduces to

$$[H^+] + [IH^+] = [X^-] \tag{4.69}$$

Considering substitution of the above relationships into the experimentally determined ratio $\Sigma[IH^+]/[I]$, Kolthoff and Bruckenstein derived the equation

$$\frac{\Sigma[\mathrm{IH}^+]}{[\mathrm{I}]\sqrt{C_{\mathrm{HX}}}} = K_\mathrm{i}^{\mathrm{IHX}}\sqrt{C_{\mathrm{HX}}} + \frac{K_\mathrm{i}^{\mathrm{IHX}}K_\mathrm{d}^{\mathrm{IHX}}}{\sqrt{K_{\mathrm{HX}} + K_\mathrm{i}^{\mathrm{IHX}}K_\mathrm{d}^{\mathrm{IHX}}[\mathrm{I}]}} \qquad (4.70)$$

For each constant value of [I] eqn. 4.70 can be represented by a straight line $y = mx + b$, where $x = \sqrt{C_{\mathrm{HX}}}$, $m = K_\mathrm{i}^{\mathrm{IHX}}$ is the slope of the line and $b = K_\mathrm{i}^{\mathrm{IHX}}K_\mathrm{d}^{\mathrm{IHX}}/\sqrt{K_{\mathrm{HX}} + K_\mathrm{i}^{\mathrm{IHX}}K_\mathrm{d}^{\mathrm{IHX}}[\mathrm{I}]}$ is the intercept; b can be transformed into

$$[\mathrm{I}]b^2 = K_\mathrm{i}^{\mathrm{IHX}}K_\mathrm{d}^{\mathrm{IHX}} - \frac{K_{\mathrm{HX}}}{K_\mathrm{i}^{\mathrm{IHX}}K_\mathrm{d}^{\mathrm{IHX}}}\cdot b^2 \qquad (4.71)$$

representing another straight line $y' = m'x' + b'$, where $x' = b^2$, $m' = -K_{\mathrm{HX}}/K_\mathrm{i}^{\mathrm{IHX}}K_\mathrm{d}^{\mathrm{IHX}}$ is the slope and $b' = K_\mathrm{i}^{\mathrm{IHX}}K_\mathrm{d}^{\mathrm{IHX}}$ is the intercept; from its slope and intercept, $K_\mathrm{d}^{\mathrm{IHX}}$ and K_{HX} can now be obtained as $K_\mathrm{i}^{\mathrm{IHX}}$ is already known from eqn. 4.70.

With a strong acid such as perchloric acid, all terms in eqn. 4.66 have to be maintained. The authors[59] experimentally found the value of $\Sigma[\mathrm{IH}^+]/[\mathrm{I}]C_{\mathrm{HX}}$ to be constant at different concentrations of perchloric acid; by way of derivation with respect to the full eqn. 4.66 this could be explained only when $K_\mathrm{d}^{\mathrm{IHClO_4}} = K_{\mathrm{HClO_4}}$, which means that $HClO_4$ in glacial acetic acid is strong. Hydrobromic acid is fairly strong, which explains the use of HBr in HOAc for the determination of α-epoxy compounds[60] and of a tertiary alcohol[61]; however, although approaching $HClO_4$ in strength, it is definitely a weaker acid[59].

Measurement for a base B

In this instance I should be a relatively strong base in glacial acetic acid, e.g., *p,p'*-bis(dimethylamino)azobenzene, and its K_i^I and K_d^I must be previously determined; hence, according to reaction 4.58, eqns. 4.59 and 4.60 yield

$$K_\mathrm{i}^\mathrm{I} = \frac{[\mathrm{IH}^+\mathrm{S}^-]}{[\mathrm{I}]} \qquad (4.72)$$

and

$$K_\mathrm{d}^\mathrm{I} = \frac{[\mathrm{IH}^+][\mathrm{S}^-]}{[\mathrm{IH}^+\mathrm{S}^-]} \qquad (4.73)$$

Usually there is a small amount of water in the solvent where it behaves as a base also, so that according to eqn. 4.61 we may write for its overall dissociation constant

$$K_{\mathrm{H_2O}} = \frac{[\mathrm{H_3O}^+][\mathrm{S}^-]}{[\mathrm{H_2O}] + [\mathrm{H_3O}^+\mathrm{S}^-]} \qquad (4.74)$$

The electroneutrality rule yields

$$[\mathrm{H_3O}^+] + [\mathrm{IH}^+] = [\mathrm{S}^-] \qquad (4.75)$$

or

$$\frac{[\mathrm{S}^-]}{[\mathrm{IH}^+]} = \frac{[\mathrm{H_3O}^+]}{[\mathrm{IH}^+]} + 1 \qquad (4.76)$$

and in eqn. 4.74 the sum

$$[H_2O] + [H_3O^+S^-] = C_{H_2O}$$

if we neglect $[H_3O^+]$.

By substituting eqns. 4.72, 4.73 and 4.74 into eqn. 4.76, we obtain

$$K_i^I K_d^I \cdot \frac{[I]}{[IH^+]^2} = \frac{K_{H_2O} C_{H_2O}}{K_i^I K_d^I [I]} + 1$$

or

$$\frac{[I]}{[IH^+]^2} = \left[\frac{K_{H_2O}}{(K_i^I K_d^I)^2}\right]\left[\frac{C_{H_2O}}{[I]}\right] + \frac{1}{K_i^I K_d^I} \tag{4.77}$$

which again represents a straight line, $y = mx + b$, where $x = C_{H_2O}/[I]$, $m = K_{H_2O}/(K_i^I K_d^I)^2$ and $b = 1/K_i^I K_d^I$. Here C_{H_2O} is found by Karl Fischer titration of the solvent, [I] is the difference between $(C_I)_t$ and the spectrophotometrically determined $\Sigma[IH^+]$, while $[IH^+] = \Sigma[IH^+] - K_i^I[I]$.

Next, by trial and error an estimate value of K_i^I is sought such that the $[IH^+]$ values obtained yield the best fit with the straight line for eqn. 3.77 (e.g., by means of the least-squares method); finally, K_d^I and K_{H_2O} are calculated from the intercept and the slope of this line.

Once the above system of I in solvent HS has been studied, one can determine the properties of an additional base B (e.g. pyridine) in this medium. Here the electroneutrality rule yields

$$[IH^+] + [BH^+] = [S^-]$$

because $[H_3O^+]$ from traces of water and ion triplets, if any, can now be neglected. Analogously to the transformation of eqn. 4.76 into eqn. 4.77, by inserting K_B (eqn. 4.61), K_i^I (eqn. 4.72) and K_d^I (eqn. 4.73) we obtain

$$\frac{[I]}{[IH^+]^2} = \left[\frac{K_B}{(K_i^I K_d^I)^2}\right]\left[\frac{C_B}{[I]}\right] + \frac{1}{K_i^I K_d^I} \tag{4.78}$$

which represents another straight line, $y = mx + b$, where K_B is calculated from the slope $x = K_B/(K_i^I K_d^I)^2$. However, the values of K_i^B and K_d^B in solvent HS must be determined in a separate way, i.e. to the solution of a known amount of B in HS is added an excess of $HClO_4$, by means of which B is entirely converted into the ionized form BH^+S^- and $BH^+ClO_4^-$ without perceptible dissociation. B is usually colourless, and especially so is BH^+, so that $\Sigma[BH^+]$ must and can be measured by UV absorptiometry at one suitable wavelength of maximum absorptivity; via the molar absorptivity thus obtained, the ionized form of B in solvent HS without $HClO_4$ added can now be measured; hence one can calculate K_i^B (cf., eqn. 4.59) and then K_d^B from eqn. 4.62 by inserting the values of K_i^B and K_B (see the previous experiment with eqn. 4.78).

In this way Kolthoff and Bruckenstein[59] determined spectrophotometrically at 25° C for acid–base equilibria in glacial acetic acid the following ionization and dissociation constants of the bases:

p,p'-bis(dimethylamino)azobenzene (I = DMAAB) ($K_{H_2O} = 8.4 \cdot 10^{-11}$):

$$(\lambda_{IH^+} = 517\,\text{nm}) \begin{cases} K_i^I = 0.100 \\ K_d^I = 5.0 \cdot 10^{-6} \\ K_I = 4.6 \cdot 10^{-7} \end{cases}$$

pyridine (B = Py):

$$(\lambda_{PyH^+} = 255.5\,\text{nm}) \begin{cases} K_i^{Py} = 5.37 \\ K_d^{Py} = 9.4 \cdot 10^{-7} \\ K_{Py} = 7.9 \cdot 10^{-7} \end{cases}$$

Final conclusions from the aforementioned investigations by Kolthoff and Bruckenstein[59]

The authors studied, as they call it, "acid–base equilibria in glacial acetic acid"; however, as they worked at various ratios of indicator–base concentration to HX or B concentration, we are in fact concerned with titration data. In this connection one should realize also that in solvents with low ε the apparent strength of a Brønsted acid varies with the reference base used, and vice versa. Nevertheless, in HOAc the ionization constant predominates to such an extent that overall the picture of ionization vs. dissociation remains similar irrespective of the choice of reference; see the data for I and B (Py) already given, and also those for HX, which the authors obtained at 25°C with I = *p*-naphtholbenzein (PNB) and $K_i^{PNB} \leqslant 0.0042$, i.e., for hydrochloric acid $K_i^{IHCl} = 1.3 \cdot 10^2$, $K_d^{IHCl} = 3.9 \cdot 10^{-6}$ and $K_{HCl} = 2.8 \cdot 10^{-9}$ and for *p*-toluenesulphonic acid $K_i^{IHTs} = 3.7 \cdot 10^2$, $K_d^{IHTs} = 4.0 \cdot 10^{-6}$, $K_{HTs} = 7.3 \cdot 10^{-9}$.

In the meantime, it should not be forgotten that all the equations involving the proton have been written in a conventional, i.e., the most concise, way as the degree of solvation of the proton in a solvent such as HOAc is unknown (cf., reaction 4.53).

Another more important point is the observation by the authors on the basis of their spectrophotometric measurements that in HOAc higher ionic aggregates containing the indicator I had been formed such as ion-triplet, quadrupole, etc.; e.g., with increasing HX they found homoconjugates such as $X^-IH^+X^-$, $IH^+X^-H^+X^-$, etc., with H_2O in the solvent hetroconjugates such as $IH^+Ac^-H_3O^+$, $IH^+Ac^-H_3O^+Ac^-$, etc., in addition to the homoconjugate $Ac^-IH^+Ac^-$, and with a base B the quadrupole $IH^+Ac^-BH^+Ac^-$. These phenomena were expected by the authors, in agreement with former cryoscopic investigations by Oddo and Anelli[62] and Turner and co-workers[63].

In our laboratory, Bos repeated the spectrophotometric method of Kolthoff and Bruckenstein[59] on a few more indicator bases I, with the following results:

for DMAAB: $K_i^I = 0.0812$; $K_d^I = 7.04 \cdot 10^{-7}$

(Kolthoff and Bruckenstein[59] found 0.100 and $5 \cdot 10^{-6}$, respectively);

for tropaeoline OO: $K_i^I = 0.0059$; $K_d^I = 2.37 \cdot 10^{-7}$

for methyl red: $K_i^I = 0.118$; $K_d^I = 1.79 \cdot 10^{-7}$

for methyl orange: $K_i^I = 0.228$; $K_d^I = 1.38 \cdot 10^{-7}$.

However, for *m*-cresol purple, thymol blue and *o*-nitroaniline, many estimated K_i values yielded straight lines with eqn. 4.77, so that with these indicators the spectrophotometric method failed. This result led Bos to check some of the above results by means of the potentiometric method of Tanaka and Nakagawa[64] applied to titrations in glacial acetic acid also; he obtained the following data:

Indicator (base)	Spectrophotometric method: $pK_I = pK_B$	Titration (with $HClO_4$): pK_B
DMAAB	7.3	7.2
Methyl orange	7.6	8.2

In the titration method Bos used the value $K_d = 4 \cdot 10^{-6}$ obtained by Kolthoff and Bruckenstein[59] for the perchlorate formed. It is doubtful whether the difference between the two methods for methyl orange is of any significance, especially as calculation (eqn. 4.62) from the data of Kolthoff and Bruckenstein[59] for DMAAB yields $pK_B = 6.3$.

Subsequently, Bos and Dahmen used in *m*-cresol[65] ($\varepsilon = 12.29$ at 25° C) a potentiometric titration method combined with conductometry. Essential precautions were the preparation of water-free *m*-cresol* (< 0.01% of water), the use of a genuine Brønsted base B, e.g., tetramethylguanidine (TMG), and the application of a glass electrode combined with an Ag–AgCl reference electrode filled with a saturated solution of Me_4NCl in *m*-cresol. The ion product of the self-dissociation of *m*-cresol, K_s, was determined from the part beyond the equivalence point of the potentiometric titration curve of HBr with TMG; comparison with titration curves calculated with various K_s values showed the best fit for $K_s = 2 \cdot 10^{-19}$.

Under the aforementioned circumstances, the two-step reaction 4.53 and the associated eqns. 4.54–4.62 are equally valid on the understanding that HS represents Hcres, etc.; further, it must be realized that during titration various amounts of HX and B are simultaneously present. Therefore, from previous measurement of the conductivities (κ) of dilution series of the separate acids, bases and salts in *m*-cresol, the overall constants K_{HX}, K_B and $K_{BH^+X^-}$ were calculated by the Fuoss and Kraus method[66,67] (with the use of $\varepsilon = 12.5$ and viscosity $= 0.208$ P for *m*-cresol). For $C_6H_5SO_3H$ and HCl it was necessary to calculate the equivalent conductivity at zero concentration from the equation

* Although *m*-cresol shows predominant acidity, and so is mainly protogenic, its basicity is still appreciable (cf., $K_s = 2 \cdot 10^{-19}$) and therefore it belongs to Brønsted class 5; further, it shows excellent solubility for a wide range of organic compounds.

$$(\Lambda_0)_{HX} = (\Lambda_0)_{2,4\text{-}(NO_2)_2C_6H_3SO_3H} - (\Lambda_0)_{TMG\text{-}2,4\text{-}(NO_2)_2C_6H_3SO_3H} + (\Lambda_0)_{TMG\text{-}HX}$$

During the conductometric measurements it was also observed that for various acids the equivalent conductivity became constant at higher concentrations, which can be explained[67] by the formation of their homoconjugate ions* $HX_2^- = X^-H^+X^-$; for a number of acids we estimated by means of the method of French and Roe[68] the formation constants concerned:

$$K_{HX_2^-} = \frac{[X^-H^+X^-]}{[HX][X^-]} \tag{4.79}$$

It is most probable that HX_2^- occurring at higher concentrations is present as the non-hydrogen-bonded ion pair formed by electrostatic attraction bewtween the homoconjugate anion HX_2^- and the protonated base cation BH^+ (see also ref. 58, p. 709, Conclusions). From the curves of the potentiometric titrations of the various acids with TMG, determined with a calibrated glass electrode, K_{HX} and $K_{HX_2^-}$ were found in the following way: the known value of K_S ($2 \cdot 10^{-19}$), K_B (eqn. 4.61), $K_d^{BH^+X^-}$ (cf., eqn. 4.68) and estimated values of K_{HX} (eqn 4.56) and $K_{HX_2^-}$ (eqn. 4.79) were applied in the relevant equilibria in order to calculate the titration curves; as $K_d^{BH^+X^-}$ was not known from conductometric measurements, a plausible value of 10^{-4} was used. The estimated values of K_{HX} and $K_{HX_2^-}$ were changed by one pK unit at a time until the best fit between the calculated and experimental titration curves occurred. The same procedure was followed with the curves of the potentiometric titrations of the various bases with 2,4-dinitrobenzenesulphonic acid with respect to an estimated value of K_B.

Generally, the results of the measurements indicated that, where dissociation constants were determined by conductometry and also potentiometric titration, they were in agreement with each other; further, K_{HX} is low, e.g., about 10^{-4}–10^{-6} mol l^{-1} for aromatic sulphonic acids and 10^{-13}–10^{-16} mol l^{-1} for carboxylic acids, $K_{HX_2^-}$ is high, e.g., 10^2–10^4, and K_B is low again, e.g., 10^{-5}–10^{-6} for aliphatic amines and 10^{-10} for aromatic amines.

4.1.4.2 *Acid–base titrations in protophilic solvents with low ε*

One of the most interesting purely protophilic solvents with low ε (class 7) is pyridine ($\varepsilon = 12.3$ at 25° C). For instance, Tsuji and Elving[69] showed by polarographic measurement of the relative strengths of Brønsted acids in pyridine with a background electrolyte consisting of a large univalent ion such as $(C_2H_5)_4NClO_4$ that for acids with an aqueous pK_a value above ca. 3 there is a linear correlation between $E_{\frac{1}{2}}$ of the one-electron attack and pK_a^{aq}, whilst for acids with p$K_a < 3$ a levelling effect occurs; the authors concluded on the basis of diffusion control and concentration proportionality of the polarographic waves that the acids were mainly present as the associated species of the

*In the original paper these ions were designated triple ions; however, the author prefers to use this designation according to Fuoss and Kraus[48] only for $M^+X^-M^+$ or $X^-M^+X^-$, formed merely by electrostatic attraction (see p. 255).

non-ionic pyridine–acid adduct (Py–HX) and its undissociated ion pair (PyH^+X^-). More detailed information about this in pyridine was obtained by Bos and Dahmen[70] by means of potentiometric titrations combined with spectrophotometry and differential vapour pressure (DVP) measurement of colligative properties. For an acid HX titrated with a nitrogen base B, in view of the various equilibria, we had to deal with the constants K_i^{HX}, K_d^{HX}, K_{HX}, $K_d^{BH^+}$ and K_d^{salt}; here, as the solvent Py is weakly-basic but non-protogenic, the acid–solvent interaction, instead of being reaction 4.53 as HS, now becomes

$$HX + Py \underset{}{\overset{K_i^{HX}}{\rightleftharpoons}} PyH^+X^- \overset{K_d^{HX}}{\rightleftharpoons} PyH^+ + X^- \qquad (4.80)$$

so the corresponding equilibrium constants (cf., eqns. 4.54, 4.55 and 4.56) are

$$K_i^{HX} = \frac{[PyH^+X^-]}{[HX]} \qquad (4.81)$$

$$K_d^{HX} = \frac{[PyH^+][X^-]}{[PyH^+X^-]} \qquad (4.82)$$

and

$$K_{HX} = \frac{[PyH^+][X^-]}{[HX] + [PyH^+X^-]} \qquad (4.83)$$

During the titration with the nitrogen base B (which as a titrant must be a stronger base than the solvent Py), there are the competitive acid–base reaction

$$HX + B \rightleftharpoons BH^+X^- \overset{K_d^{salt}}{\rightleftharpoons} BH^+ + X^- \qquad (4.84)$$

and the dissociation reaction

$$BH^+ \overset{K_d^{BH^+}}{\rightleftharpoons} B + H^+ \qquad (4.85)$$

with constants

$$K_d^{BH^+} = \frac{[B][H^+]}{[BH^+]} \qquad (4.86)$$

and

$$K_d^{salt} = \frac{[BH^+][X^-]}{[BH^+X^-]} \qquad (4.87)$$

As a first starting point, K_{HX} was determined in three different ways:

(1) By spectrophotyometry according to Kolthoff and Buckenstein[59] via K_i and K_d; with polynitrophenols and sulphonphthaleins the method did not yield a single value for K_i, but here K_{HX} was obtained from the simple dissociation equation

$$\{\Sigma[\text{base}]\}^2/[HIn] = K_{HX}$$

where [HIn] is the concentration of the indicator in the acid colour and Σ[base] is the spectrophotometrically determined concentration of the indicator in the basic colour. For sulphonphthaleins this means that the first step of the acid was neutralized with TMG. Further, molar absorptivities were determined in indicator solutions after addition of excess of $(C_2H_5)_4NOH$.

(2) By DVP measurement at 37° C in a Mechrolab Model 301 A osmometer, calibrated for pyridine with benzil. Here

$$K_{HX} = (m_d - m_s)^2/(2m_s - m_d)$$

where m_d is the molarity measured by DVP and m_s is the stoichiometric HX concentration.

(3) By potentiometry at a calibrated glass electrode.

As a second starting point, $K_d^{salt} = K_d^{BH^+X^-}$ was determined by DVP measurement at 37° C again according to

$$K_d^{salt} = (m_d - m_s)^2(2m_s - m_d)$$

for a salt solution (see above).

Finally, as a third starting point, the potentiometric curve of the titration of an acid HX with a nitrogen base B was determined; on the basis of K_{HX}, K_d^{salt} and $K_d^{BH^+}$ values one might have calculated[71] the titration curve with the aid of a computer. By repeatedly inserting estimated values for the still unknown $K_d^{BH^+}$ we obtained one value with the best fit between the experimental and the calculation curves. In this way we obtained the $pK_{HX} = pK_a$ values in pyridine for a number of acids such as perchloric, picric and hydrochloric acid and dinitrophenols from titration with TMG; moreover, for titrations of perchloric*, picric and hydrochloric* acid each with the different bases TMG, *n*-butylamine, triethylamine and morpholine, the $pK_d^{BH^+} = pK_a$ values of the latter in pyridine could be determined; by averaging the results with the three acids we obtained $pK_a = 9.6$ for TMG-H^+, 5.5 for *n*-butylamine-H^+ 3.8 for triethylamine-H^+ and 3.5 for morpholine-H^+; this illustrates once more the attraction of TMG as a titrant, being an appreciably stronger Brønsted base than the other amines. The values of $K_d^{salt} = K_d^{BH^+X^-}$ appeared to lie between 10^{-3} and 10^{-6}, and mostly at 10^{-4} to 10^{-5}, for all acid–base combinations, which means that the ion-pair formation BH^+X^- is predominant.

For the different values of pK_{HX} and $pK_d^{BH^+}$ see the summary Table 4.5 later of pK_a data in various solvents of low ε in comparison with $pK_a(H_2O)$. The mutual agreement of pK_{HX} values obtained by spectrophotometry, DVP, potentiometry and titration was reasonably good; the typical form of the curves for titration of the dinitrophenols with TMG can be explained by homoconjugation and more especially by its influence on the potentiometric measurements, calculated on the basis of simple dissociation; hence the major discrepancies in the spectrophotometric and potentiometric p*K* values. In order to

* In the calculations of these titrations the value $pK_{HClO_4} = 3.2$ and $pK_{HCl} = 5.66$, taken from the literature[72], were used.

TABLE 4.4
EQUILIBRIUM CONSTANTS* OF DINITROPHENOL AT 25°C IN PYRIDINE

Acid HX	Overall dissociation constant pK_{HX}				K_{HX}	$K_{HX_2^-}$
	Spectro-photometry	DVP at 37°C	Potentio-metry	Titration with TMG		
2,4-Dinitrophenol	–	4.0	4.4	4.3	$5 \cdot 10^{-5}$	10
2,5-Dinitrophenol	5.0	4.6	5.3	6.5	$2.8 \cdot 10^{-7}$	100
2,6-Dinitrophenol	4.2	3.7	4.8	4.7	$1.1 \cdot 10^{-5}$	10^{-1}

* K_{HX} = dissociation constant; $K_{HX_2^-}$ = homoconjugate formation constant.

account for homoconjugation, we had to consider in our set of equilibria, in addition to K_{HX}, K_d^{salt} and $K_d^{BH^+}$, also $K_{HX_2^-}$ (eqn. 4.79); here we varied both K_{HX} and $K_{HX_2^-}$ until the best fit between the experimental and calculated titration curves occurred. Table 4.4 illustrates the results for dinitrophenols.

In the case of an acid HX titrated with a quarternary ammonium base R_4NOH there are a few complications as a consequence of some additional equilibrium constants and of a certain degree of instability of the titrant tetramethylammonium hydroxide (TMAOH)* in pyridine; for this reason the titrant solution was standardized against benzoic acid and used only on the day of preparation. Now, in addition to K_{HX} and $K_{HX_2^-}$, we had to deal with

$$K_{TMAOH} = \frac{[TMA^+][OH^-]}{[TMAOH]} \tag{4.88}$$

$$K_{TMAX} = \frac{[TMA^+][X^-]}{[TMAX]} \tag{4.89}$$

and K_{H_2O} in pyridine, because the water arises from the acid–base reaction itself.

At first we determined, by means of the DVP method, K_{TMAX} of 2,4-dinitrophenolate, 2,5-dinitrophenol picrate, acetate and benzoate, which lay between 10^{-3} and 10^{-5}. Next, separate potentiometric titrations of 2,5-dinitrophenol and picric acid were carried out; on the basis of the previously known (see above) $pK_{HX} = 6.5$ and $pK_{HX_2^-} = 100$ for 2,5-dinitrophenol and $pK_{HX} = 3.0$ for picric acid, we calculated titration curves for estimated values of K_{H_2O} and obtained, for the best fit between the experimental and calculated curves, $K_{H_2O} = 10^{-21}$ for both 2,5-dinitrophenol and picric acid. In both instances changing K_{TMAOH} for 1 to 10^{-5} did not alter the calculated titration curve. Finally, for potentiometric titrations of other acids with TMAOH while using K_{TMAX} values from DVP results, in addition to $K_{H_2O} = 10^{-21}$, we obtained the best fit between the experimental and calculated curves again when $pK_{benzoic\ acid} = 1$ (see Fig. 4.12)

* According to Harlow[73], it is nevertheless one of the least unstable quaternary ammonium hydroxides (see under practice of non-aqueous titration, p. 281).

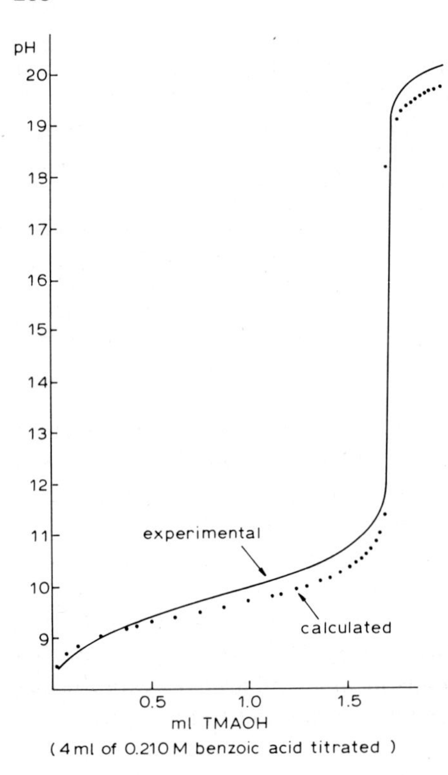

Fig. 4.12. Potentiometric titration curves of benzoic acid in pyridine.

and $pK_{\text{acetic acid}} = 12$, respectively; here we could describe the curve of acetic acid best when taking homoconjugation with a value of $K_{HX_2^-} = 10^2$ into account.

4.1.4.3. *Acid–base titrations in inert solvents with low ε*

Here we shall confine ourselves to the solvents benzene and 1,2-dichloroethane (class 8). Considering benzene, many investigators have demonstrated since the 1930s the feasibility of titrations in this solvent using both potentiometric and spectrophotometric methods, paying much attention to acid–base indicator reactions under the influence of primary, secondary and tertiary amines*. Association of carboxylic acids in benzene was studied at a later stage, mainly on the basis of colligative properties, IR spectroscopy and solvent extraction*.

Of great interest is the study of Gur'yanova and Beskina[74] on dielectrometric and cryoscopic titrations of benzoic acids in benzene with amines; they concluded that benzoic acid reacts as a dimer $(HX)_2$ yielding with primary and

* Cf., literature cited by Bruckenstein and Saito[58].

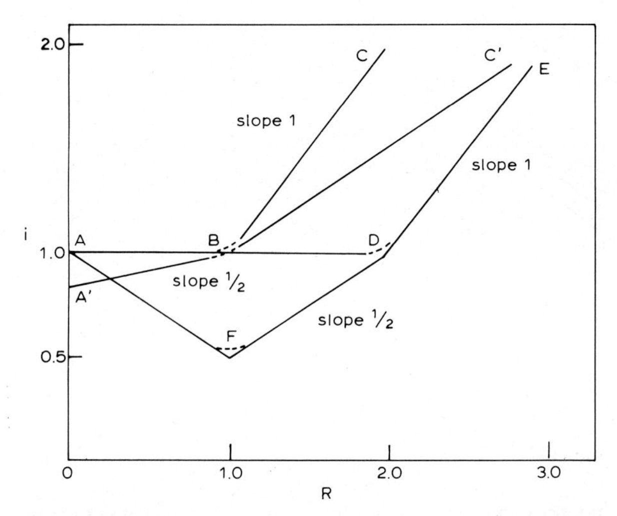

Fig. 4.13. Hypothetical titration curves of Van 't Hoff i value.

secondary amines species $B(HX)_2$ and $(BHX)_2$ and with tertiary amine species $B(HX)_2$ and BHX.

However, the most revealing investigation was made by Bruckenstein and Saito[58] in 1964; they pursued acid–base reactions in the reverse direction, i.e., during titration of bases with acids, applying the DVP technique to the colligative properties and IR spectroscopy of the carboxylate bands within the range 1200–2000 cm^{-1} to judge between hydrogen bonding or mere ion-pair formation (cf., p. 257). For DVP measurement, ΔE versus concentration curves were constructed in the range 0.001 M of naphthalene in benzene; if a titration is started with a total base concentration $(C_B)_t$, equivalent to $a\,M$ naphthalene, then one obtains the Van 't Hoff i value from DVP according to

$$i = \frac{\text{temperature rise for mixtures of B and HX, where } (C_B)_t = a\,M}{\text{temperature rise for } a\,M \text{ naphthalene solution}}$$

By plotting i versus the ratio $R = (C_{HX})_t/(C_B)_t$ during the titration, they determined simultaneously the extent of acid–base interaction, the stoichiometry of that interaction and the degree of association of the acid–base adduct. Fig. 4.13 shows hypothetical titration curves; line ABC corresponds to the interaction between B and HX as monomers without further reaction between BHX and HX, and the subsequent occurrence of the latter reaction to a small extent is indicated by the line ABC′ and to the full extent by line ABDE, when no more HX can react with BHX · HX; line AFDE arises when formation of BHX · HX starts right away; in the case of previous partial dimerization of B, the various lines will begin at A′ instead of A.

Fig. 4.14 represents typical experimental curves obtained for the titration of BA (benzylamine), DDA (dodecylamine), DBA (dibenzylamine), DPG

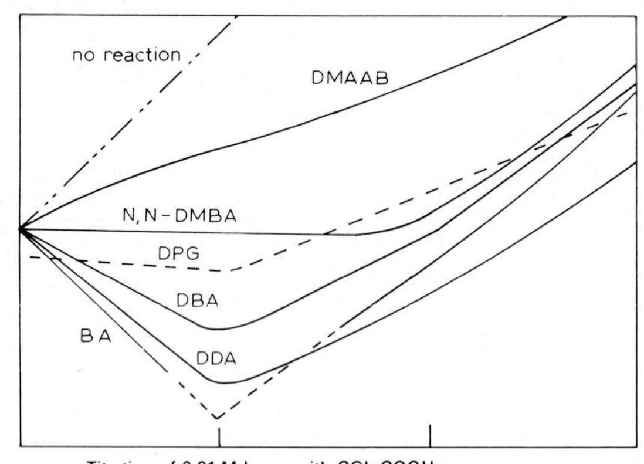

Fig. 4.14. Base titration curves; cf., Fig. 4.13.

(1,3-diphenylguanidine), N,N-DMBA (N,N-dimethylbenzylamine) and DMAAB (dimethylaminoazobenzene) with trichloroacetic acid*.

Summarizing their results with DVP and IR spectroscopy, Bruckenstein and Saito[58] drew the following main conclusions:

(a) Mere dimerization in benzene appeared almost negligible for the bases investigated, but, although small ($K_{dim} \approx 10^{-3}$), increased in the order of trichloroacetic < dichloroacetic < salicylic < *m*-nitrobenzoic ≈ benzoic ≈ phenylacetic ≈ *m*-chlorobenzoic acid.

(b) The principal adducts formed from acid–base reaction were ion pairs and other uncharged ionic aggregates.

(c) Basically the ion pairs consist of three types: (1) BH^+X^-, the ionic aggregate** of protonated base cation BH^+ with carbonxylate anion X^-; (2) $BH^+HX_2^-$, the one with homoconjugate anion HX_2^-, and (3) $BH^+X(HX)_2^-$, the one with complex homoconjugate anion $X(HX)_2^-$.

(d) These ion pairs tend to form oligomers; the stronger the acid or base, the greater is the extent of oligomer formation, e.g., at a 0.01 *M* base concentration

$$B + HX \rightleftharpoons BH^+X^- \rightleftharpoons \frac{1}{m}(BH^+X^-)_m$$

where when B = benzylamine *m* for the oligomer was ca. 6 with trichloroacetic acid, ca. 3 with *m*-chlorobenzoic and 2.3 with salicylic acid.

(e) The occurrence ratio of the three basic types (1), (2) and (3) depends on the choice of bases and acids; in mixtures of bases strong interaction can occur between the ion pairs BH^+X^- and $B'H^+X^-$, suggesting that ionic aggregates

* Comparable graphs were obtained by Bruckenstein and Vanderborgh[75] by means of cryoscopic detection also.

** The authors claim for the adducts in benzene a merely electrostatic attraction between the BH^+ cation and the anions, instead of hydrogen bonding which Barrow and Yerger proposed for adducts in carbon tetrachloride and chloroform.

containing different cations are more stable than aggregates containing several identical cations; surprising effects for titration of a DBA–DLA (dilaurylamine) mixture with benzoic acid were formed, e.g., formation of a compound DBA · DLA · 3HX.

The solvent 1,2-dichloroethane (ε = 10.60 at 20° C) was used by Bos and Dahmen[77] in a study of the titrations of mono- and dibasic acids with TMG (tetramethylguanidine) and a few other bases.

For monobasic acids we are concerned (as with benzene) with the following constants in a simple form. i.e., without interaction of the inert solvent:

$$K_{HX} = \frac{[H^+][X^-]}{[HX]} \tag{4.90}$$

$$K_d^{BH^+} = \frac{[B][H^+]}{[BH^+]} \tag{4.86}$$

$$K_d^{salt} = \frac{[BH^+][X^-]}{[BH^+X^-]} \tag{4.87}$$

and at half-neutralization sometimes

$$K_{B(HX)_2} = \frac{[BH^+ \cdot HX_2^-]}{[BH^+X^-][HX]} \tag{4.91}$$

which has been written as a homoconjugation constant. With carboxylic acids the compound $B(HX)_2$ formed at half-neutralization more probably represents a complex than a homoconjugate (cf., study by Barrow and Yerger[76] in carbon tetrachloride and chloroform); moreover, carboxylic acids have shown dimerization with

$$K_{dim} = \frac{[(HX)_2]}{[HX]^2} \tag{4.92}$$

For dibasic acids, e.g., sulphonphthaleins, there are the following six equilibrium constants:

$$K_{H_2X} = \frac{[H^+][HX^-]}{[H_2X]} \tag{4.93}$$

$$K_{HX^-} = \frac{[H^+][X^{2-}]}{[HX^-]} \tag{4.94}$$

$$K_d^{BH^+} = \frac{[B][H^+]}{[BH^+]} \tag{4.86}$$

$$K_{BH_2X} = \frac{[BH_2X]}{[BH^+][HX^-]} \tag{4.95}$$

$$K_{BHX^-} = \frac{[BHX^-]}{[BH^+][X^{2-}]} \tag{4.96}$$

and

References pp. 295–299

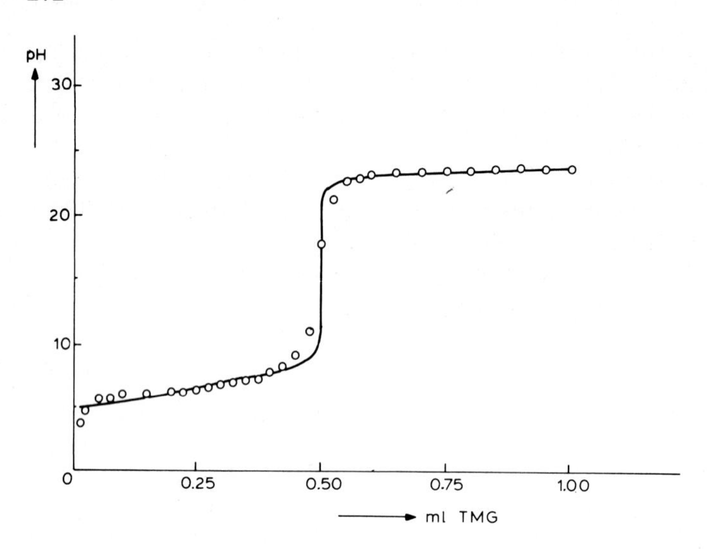

Fig. 4.15. Experimental (———) and theoretical (O) pH titration curves of HI in 1,2-dichloroethane.

$$K_{B_2H_2X} = \frac{[B_2H_2X]}{[BH^+][BHX^-]} \tag{4.97}$$

In the investigation we determined by means of DVP measurement (at 37° C) K_{dim}, by means of conductometry (at 20° C) K_d^{salt} according to Fuoss[66] ($\varepsilon = 10.60$ and viscosity $\eta = 0.00787$ P) and K_{HX} according Walden's[78] rule from

$\Lambda_0[\Lambda_0\eta(\text{dichloroethane}) = \Lambda_0\eta(m\text{-cresol})]$,

and subsequently by potentiometric titration (at 25° C) $K_d^{BH^+}$ and $K_{B(HX)_2}$ while choosing values estimated and adjusted until the best fit between the experimental curve and the titration curve calculated with the computer program EQUIL of Bos and Meershoek[79] was obtained.

As an example, Fig. 4.15 shows the titration of 3.500 ml of 0.01644 *M* hydriodic acid with 0.1168 *M* TMG in 1,2-DCE, the points (O) calculated concerning $K_{HX} = 1.4 \cdot 10^{-8}$, $K_f^{BH^+} = 1/K_d^{BH^+} = 1.25 \cdot 10^{22}$, $K_d^{salt} = K_d^{BHX} = 9 \cdot 10^{-6}$ and $K_{B(HX)_2} = 10^3$; with the use of the equation $E_{mV} = E_{0_{mV}} - 59\text{pH}$ for its behaviour, the glass electrode was calibrated by calculating E_0 from the recorded E values and the computed pH values of the titration curve. The slight inflection at half-neutralization indicates the occurrence of $B(HX)_2$, apparently as the homoconjugate $BH^+ \cdot HX_2^-$.

With the dibasic sulphonphthaleins the situation becomes much more complex, as there are at least six equilibrium constants involved. Fig. 4.16 shows the titration of 3.000 ml of 0.01140 *M* bromocresol green with 0.0855 *M* TMG in 1,2-DCE. It appears that, once $K_f^{BH^+} = 1/K_d^{BH^+}$ is known, the other constants can be found, as each of the four regions in the titration curve enables the possibility of determining one dissociation constant, viz., in:

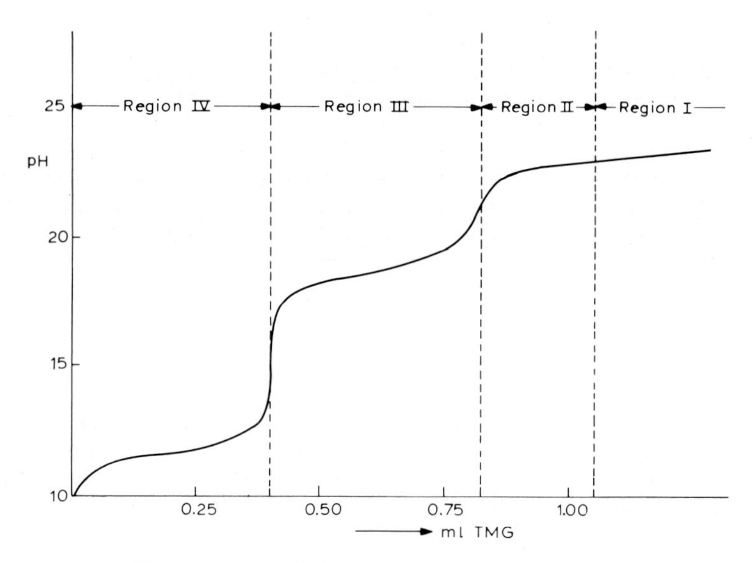

Fig. 4.16. pH titration curve of bromocresol green in 1,2-dichloroethane.

	effective constants
region I, more than 30% beyond the 2nd equivalence point:	$K_f^{BH^+}$, $K_{B_2H_2X}$
region II, just beyond the 2nd equivalence point:	$K_f^{BH^+}$, $K_{B_2H_2X}$, K_{BHX^-}
region III, from the 1st to the 2nd equivalence point:	K_{HX^-}, K_{BH_2X}, K_{BHX^-}, $K_{B_2H_2X}$
region IV, from the start to the 1st equivalence point:	K_{H_2X}, K_{BH_2X}

For region III we assumed $K_{BH_2X} = K_{B_2H_2X}$, so that together with $K_f^{BH^+}$, which is already known, all the six constants could be determined. With bromocresol green titrated with TMG (Fig. 4.16), we obtained the following values: $K_f^{BH^+} = 3 \cdot 10^{21}$ (value used), $K_{H_2X} = 8 \cdot 10^{-14}$, $K_{HX^-} = 6.3 \cdot 10^{-23}$, $K_{BH_2X} = 10^{-5}$, $K_{B_2H_2X} = 10^{-5}$ and $K_{BHX^-} = 10^{-7}$.

In the titration of benzoic acid with TMG, the inflection point at half-neutralization agrees, as mentioned previously for carboxylic acids in general with a complex $B(HX)_2$.

4.1.4.4. *Comparison of acid–base equilibria in solvents with low ε and in water*

Thermodynamically it can be stated, if the differences of solvation of the compounds X^- and HX between two solvents are neglected, that the difference in the pK_a values of compound HX in the two solvents is completely determined by the difference in the proton affinities of the two solvents[80]; hence a comparison of the pK_a values of various compounds in the solvents 1,2-DCE, *m*-cresol, acetic acid, pyridine and water is worth considering (see Table 4.5)[80].

TABLE 4.5

COMPARISON OF pK_a VALUES IN 1,2-DCE, *m*-CRESOL, ACETIC ACID, PYRIDINE AND WATER (SOLVENTS OF BRØNSTED CLASSES 8, 5, 6, 7 AND 1, RESPECTIVELY)

Compound	pK_a				
	1,2-DCE (ε = 10.23)	*m*-Cresol (ε = 12.3)	Acetic acid (ε = 6.13)	Pyridine (ε = 12.3)	Water (ε = 78.5)
Perchloric acid			4.87	3	
Hydriodic acid	7.9	4.4		3.2	
Hydrobromic acid	8.7	4.4		4.4	
Hydrochloric acid	10.8	6.4	8.55	6.1	
Benzenesulphonic acid	12.1	6.3			
Picric acid	13.7	12.0		3.0	0.38
Iodoacetic acid	17.6	13.0			3.12
Benzoic acid	20.0	15.0		11.0	4.19
Acetic acid		16.0		12.0	4.75
DMAABH$^+$ *		11.0	8.13		3.3
Methyl orange-H$^+$		11.0	7.60		3.5
Strychnine-H$^+$	16.0	13.0			8.0
Morpholine-H$^+$	15.0	12.1		3.5	9.6
n-Butylamine-H$^+$	17.0	13.9	8.59	5.5	10.6
Triethylamine-H$^+$	17.7	14.2	9.45	3.8	10.8
Tetramethylguanidine-H$^+$	21.5	14.5		9.6	12.3
Bromophenol blue (2**)	23.8			5.3	4.1
Bromocresol green (2)	22.2			5.5	4.9
Bromocresol purple (2)	24.7			7.5	6.4

* *p,p'*-Bis(dimethylamino)azobenzene-H$^+$.
** Dissociation of the second proton.

For a discussion of the data in Table 4.5, it is useful to distinguish the acids into (1) positively charged or cationic-type acids, (2) uncharged or neutral-type acids and (3) negatively charged or anionic-type acids.

(1) For the cationic-type acids (BH^+) in the solvents of classes 5–8 Table 4.5 shows that in general the pK_a value decreases with increasing basicity of the solvents; thus according to this basicity increase the pK_a decrease shows the following trend

$$pK_{a(DCE)} \xrightarrow{-3} pK_{a(m\text{-cresol})} \xrightarrow{-4} pK_{a(\text{acetic acid})} \xrightarrow{-4} pK_{a(\text{pyridine})}$$

As BH^+ dissociates into H^+ and the uncharged base B, the dielectric constant can exert only a minor effect on the mutual coulombic attraction, so that even in water (ε = 78.5) the pK_a values of aliphatic amines do not differ much from the above picture of the influence of solvent basicity. That $pK_{a(water)}$ lies between $pK_{a(m\text{-cresol})}$ and $pK_{a(\text{acetic acid})}$ instead of between the latter and $pK_{a(\text{pyridine})}$ may be ascribed to effects of solvation; however, the $pK_{a(water)}$ values of the aromatic amines are low owing to effects of mesomerism.

(2) For the neutral-type acids (HX), the dissociation into two ions of opposite

charge and the appreciable differences in ionic radii imply a large influence of the dielectric constant and especially on the pK_a in water; moreover, pH levelling occurs in *m*-cresol, acetic acid and pyridine, as shown for the hydrohalic acids.

(3) For the anionic-type acids (HX^-) there is again an effect of the differences in ionic radii

Finally, the computer program EQUIL of Bos and Meershoek[79] provides a rapid means of judging the accuracy of titrations in solvents with low ε depending on estimated values of the equilibrium constants of the species possibly occurring in the solution[80].

4.1.5. Redox systems in non-aqueous media

Apart from the investigations by Strehlow and co-workers[15] and by Trémillon[81], systematic studies of redox systems in non-aqueous media are scarce; the reason for this may be the shortage of reversible redox systems, especially in aprotic solvents, and even when available they mainly concern organic compounds, mostly with complicated electrochemical behaviour. On the other hand, and by analogy with acid–base equilibria, one must also expect for redox systems a considerable increase in possibilities by the choice of solvents being less levelling and more differentiating than water. Using this analogy, comparison with the Brønsted acid–base reaction (see p. 238)

$$\underset{\text{(donor)}}{\text{acid}} \rightleftharpoons \underset{\text{(acceptor)}}{\text{base}} + H^+ \tag{4.34}$$

we can consider the redox reaction

$$\underset{\text{(donor)}}{\text{red}} \rightleftharpoons \underset{\text{(acceptor)}}{\text{ox}} + ne^- \tag{4.98}$$

In solutions neither H^+ nor e^- can exist in a free state; they will be donated only if they are accepted within the solution, e.g., by another acceptor, which may be the solvent and thus cause solvation; here the mere solvation of electrons is an exceptional case, but may occur, e.g., in liquid ammonia, where according to Kraus[82] the strongly reducing alkali metals dissolve while dissociating into cations M^+ and solvated electrons e^-, which, however, are soon converted into NH_2^- and H_2 gas. Further, from the analogy with acid–base reactions and the definition of

$$\text{pH} = -\frac{F}{2.3RT} \cdot E_{H^+/H_2}$$

it has been proposed to use

$$\text{pe}^- = -\frac{nF}{2.3RT} E_{\text{redox}}$$

(and even pO^{2-}; cf., p. 238); however, until now this proposal has found no real acceptance.

In the meantime, it is of great practical importance to establish how the redox potential ranges can vary from one solvent to another without the occurrence of solvent decomposition. Therefore, we shall first consider the redox potential range in water. Here at the surface of the hydrogen platinized Pt electrode (cf., p. 31) the following reversible reduction can take place:

$$2H_2O + 2e^- \rightleftharpoons H_2 + 2OH^-$$

Hence

$$E_{red(ox)} = E^0_{red(ox)} + \frac{RT}{2F} \cdot \ln\left(\frac{pH_2[OH^-]^2}{[H_2O]^2}\right)$$

As for pure water $[H_2O]$ may be regarded as unity, and in fact we are concerned with a hydrogen electrode, we may write

$$E_{H^+/H_2} = E^0_{H^+/H_2} + \frac{RT}{2F} \cdot \ln\left(\frac{[pH_2]K_W^2}{[H^+]^2}\right)$$

For pH_2 we take 1 atm, so that by definition $E^0_{H^+/H_2} = 0$, which at 25° C yields

$$E_{H^+/H_2} = 0.059\,\text{pH} - 0.059 K_W$$

Therefore, as the pH varies from 0 to 14, the red(ox) potential may vary from $-0.059 \cdot 14 = -0.83$ V to 0 V without risk of water reduction. Further, let us assume for the moment the existence of an inert electrode at the surface of which the following reversible oxidation can take place:

$$2H_2O - 4e^- \rightleftharpoons O_2 + 4H^+$$

Hence

$$E_{(red)ox} = E^0_{(red)ox} + \frac{RT}{4F} \cdot \ln\left(\frac{[pO_2][H^+]^4}{[H_2O]^2}\right)$$

As $[H_2O]$ again equals unity and by analogy with the above we are now concerned with an oxygen electrode, we can write

$$E_{O_2/OH^-} = E^0_{O_2/OH^-} + \frac{RT}{4F} \cdot \ln([pO_2][H^+]^4)$$

For pO_2 we take 1 atm again, but for the non-existent oxygen electrode $E^0_{O_2/OH^-} = 1.23$ V could be determined only indirectly (cf., pp. 42–43); as a result

$$E_{O_2/OH^-} = 1.23\text{ V} - 0.059\,\text{pH}$$

so that the (red)ox potential may vary from 1.23 V to 1.23 − 0.83 = 0.40 V.

Fig. 4.17 shows reduction and oxidation potentials depicted as linear functions of pH, together with their potential ranges, thus yielding a parallelogram; in fact, it shows that pure water starts to be reduced to hydrogen gas or oxidized to oxygen gas as soon as we apply a potential of < -0.83 V at an ideal hydrogen electrode or a potential > 1.23 V at an ideal oxygen electrode; here

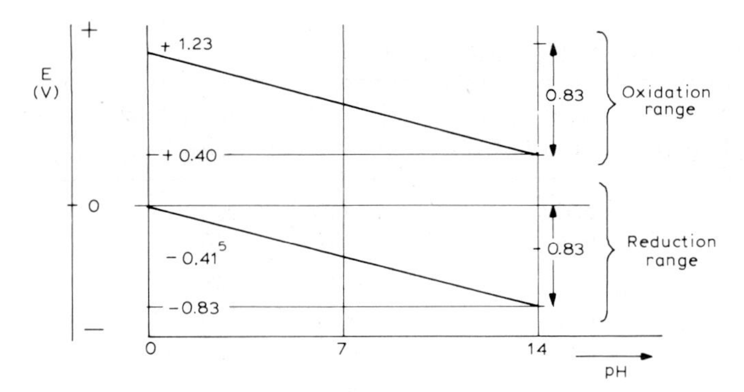

Fig. 4.17. Redox potential ranges between pH 0 and 14 in water.

the designation "ideal" means that for the reduction a platinized Pt electrode must be used, whilst for the oxidation such an electrode does not exist at all. Hence with electrodes used in normal practice we are concerned with the well known phenomenon of an overpotential in excess of the equilibrium potential in order to achieve a definite breakthrough of electrolytic current (cf., pp. 109–111); for reduction the overpotential η_{H_2} is negative and for oxidation η_{O_2} is positive.

Table 4.6 indicates that depending of the type of electrode, the "practical" E_{redox} range is appreciably more extended than the "ideal" range, which according to Fig 4.17 amounts to

$$(E^0_{O_2/OH^-} - 0.059\ pK_W) + 2 \cdot 0.059\ pK_W = 0.40 + 2 \cdot 0.83 \tag{4.99}$$
$$= 2.06\ V$$

As expected the overpotentials remain much lower at rough or spongy than at smooth metal surfaces; the highest values of $-\eta_{H_2}$ (0.78 V) and η_{O_2} (0.52 V) in Table 4.6 yield a redox potential range of at least (0.78 + 0.83) + (1.23 + 0.52) = 3.36 V. Another important effect, already published by Tafel (Ch. 3, ref. 7), is that an increase in the electrolytic current density often results in a further rise of the overpotentials (and therefore of the redox potential range) (cf., eqn. 3.38 and Fig. 3.16 for non-amalgamating metals in polarography).

Considering non-aqueous solvents, one can expect a widening of the redox potential ranges comparable to that of the pH ranges (see Fig. 4.2). For instance, for a solvent (HS) with self-dissociation, $2HS \rightleftharpoons H_2S^+ + S^-$, the primary effect can be seen directly from eqn. 4.99 by substituting pK_S for pK_W; a secondary effect comes from another value of E^0_{anodic} and of $E^0_{H^+/H_2}$, which latter will now be different from the standard hydrogen potential (0 V) in water; finally, a tertiary effect is caused by a different degree of solvation, unless all redox potentials considered were measured versus the ferrocene–ferrocinium$^+$ electrode. As a consequence, the use of a differentiating solvent such as acetonitrile, where pK_S has about twice the values of pK_W, is most favourable, and the absence of a levelling effect as in pyridine on the basic side is even more

TABLE 4.6

OVERPOTENTIALS AND REDOX POTENTIAL RANGES (V)

Electrode	$-\eta_{H_2}$ (in 2 N H_2SO_4)	η_{O_2} (in 1 N KOH)	$\eta_{O_2} - \eta_{H_2}$	E_{redox} range ($2.06 - \eta_{H_2} + \eta_{O_2}$)
Ni (rough)	0.138	0.05	0.188	2.25
Pt (platinized)	0.000	0.24	0.24	2.30
Fe (rough)	0.087*	–	} > 0.33	2.39
(smooth)	-	0.24		
Ni (smooth)	0.29	0.11	0.40	2.46
Pt (smooth)	0.08	0.44	0.52	2.58
Au (rough)	0.016	–	} > 0.54	2.60
(smooth)	–	0.52		
Cu (rough)	0.135	–	–	–
(smooth)	0.415	0.25	0.665	2.73
Ag (rough)	0.097	–	–	–
(smooth)	0.495	0.40	0.895	2.96
Pb (smooth)	0.78	0.30	1.08	3.14

* In 1 N NaOH.

attractive; traces of water, however, may interfere. The value of E^0_{anodic} depends on the mechanism of the electrolytic oxidation of the specific solvent and often its determination requires considerable investigation (e.g., for *m*-cresol[83]; see later). $E^0_{H^+/H_2}$ has been measured by Pleskov[14] and Strehlow[15] versus the ferrocene–ferrocinium$^+$ electrode in many solvents; some of the results of Pleskov, already given in Table 4.2, indicate that for the E_{redox} ranges, instead of eqn. 4.99 for water, the following more general equation can be used for all solvents of type HS (at 25° C):

$$(E^0_{anodic} - 0.059\,pK_S) - E^0_{H^+/H_2} + 2 \cdot 0.059\,pK_S$$
$$= (E^0_{anodic} - E^0_{H^+/H_2}) + 0.059\,pK_S \qquad (4.100)$$

This means, in agreement with Fig. 4.17, that a negative $E^0_{H^+/H_2}$ value yields an additional increase in the E_{redox} range. Finally, solvation in itself contributes to extending the E_{redox} range, as a more solvated ion is less accessible to electrolysis; however, it is doubtful whether separate consideration of this effect makes any sense provided that E^0_{anodic} and $E^0_{H^+/H_2}$ have been determined absolutely, i.e., versus the ferrocene–ferrocinium$^+$ electrode.

On the basis of the above, and more especially owing to the widening of the redox potential ranges, it can be said that the possibilities of redox reactions are considerably extended by the use of non-aqueous media, analogously to the situation with acid–base reactions owing to the widening of the pH ranges. Even in inert solvent, where such ranges seem unlimited, or in other sophisticated media, reactions not only of analytical but also of industrial interest (such as the manufacture of alkali metals or aluminium) can be carried out that cannot be realized in water (cf., Introduction, p. 3).

4.2. PRACTICE OF ELECTROANALYSIS IN NON-AQUEOUS MEDIA

The application of electroanalysis in non-aqueous media to a certain analytical problem requires a well considered selection of the solvent together with a suitable electroanalytical method, which can be carried out on the basis of the solvent classes mentioned in Table 4.3 and of the related theories. The steps to be taken include the preparation of the solvent and the apparatus for the electroanalytical method proper, together with other chemicals, especially when the method includes titration. Much detailed information on the purification of the solvents and on the preparation of titrants and primary standards can be found in the references cited in Section 4.1 and in various commercial brochures[1,84,85] and books[17,86–89]; we shall therefore confine ourselves to some remarks on points of major importance.

Purity of solvents. To avoid possible interfering reactions with analyte components and in order to achieve clear acid–base, redox or other intended reactions, impurities in the solvent must be eliminated as far as possible. In addition to many other authors, Burger[40] provided an attractive and comprehensive treatment of the subject; the most important aspect remains the determination and removal of moisture ($< 0.01\%$), as water, owing to its strongly levelling and solvating effects, can have a detrimental influence on titration results, especially in aprotic solvents. Burger mentioned several possibilities for determining moisture content; nevertheless, the Karl Fischer method still remains the most attractive and reliable, and the reader is again referred to its discussion on pp. 204–205.

Mixtures of solvents. Figs. 4.3, 4.4, 4.8, 4.10 and 4.11 have already illustrated that the use of mixed solvents may sometimes have a beneficial effect on the titration result; some further practical examples are worth mentioning.

The so-called G–H solvents were introduced by Palit et al.[90]; they usually consist of 1 : 1 mixtures of ethylene or propylene glycol and isopropyl or *n*-butyl alcohol (with use of the same mixtures in the titrants); the glycol acts as a solvent for polar groups and the alcohol for the non-polar hydrocarbon groups, which explains the solvating power of their mixtures for both salts and soaps, whereas the alcohol chosen is beneficial for wider pH ranges (see Figs. 4.1 and 2).

Glacial acetic acid, pure or mixed with other solvents, is one of the most attractive solvents for the titration of amines. Commercial acetic acid containing not more than 1% of water (Karl Fischer titration check) can be used in normal practice; for the highest accuracy, however, the water content must be lowered to about 0.01% by addition of acetic anhydride and standing for 24 h; not more than the stoichiometric amount of acetic anhydride should be used in order to avoid possible reactions with active hydrogen-containing analyte components such as primary or secondary amines or alcohols. A similar procedure is followed in the preparation of perchloric acid titrant from the commercial

72% aqueous solution, but taking precautions to slow down the rapid exothermic reaction with the anhydride, which is catalysed by the perchloric acid.

The pK_S value of acetic acid is about the same as that of water; therefore, in order to obtain high accuracy and greater differentiation in potentiometric titrations, Pifer and Wollish[91] preferred to use acetic acid mixed with dioxan in the solvent and also in the perchloric acid titrant; in view of the possible presence of peroxide, the previous purification requires great care[40]. Similarly, Fritz[92] recommended acetic acid mixed with acetonitrile (pK_S = 19.5 at 25° C)[93] in both the solution and the titrant.

Titrants. Titrants can be divided into acid, base, redox and precipitation titrants.

Acid titrants are commonly perchloric acid and sometimes hydrochloric, benzenesulphonic or *p*-toluenesulphonic acid.

Base titrants are commonly alkali metal hydroxides or alkali metal alkoxides (often the methoxide or ethoxide, sometimes the aminoethoxide) and genuine Brønsted amine bases such as aniline, benzidine, morpholine, tris-(hydroxymethyl)aminomethane or guanidines such as 1,3-diphenylguanidine (DGP) and 1,1,3,3-tetramethylguanidine (TMG); sodium acetate, being the strongest base by definition in acetic acid, or potassium hydrogen phthalate in this medium; and finally strong quaternary alkylammonium hydroxides or ethoxides.

Redox titrants (mainly in acetic acid) are bromine, iodine monochloride, chlorine dioxide, iodine (for Karl Fischer reagent based on a methanolic solution of iodine and SO_2 with pyridine, and the alternatives, methyl-Cellosolve instead of methanol, or sodium acetate instead of pyridine (see pp. 204–205), and other oxidants, mostly compounds of metals of high valency such as potassium permanganate, chromic acid, lead(IV) or mercury(II) acetate or cerium(IV) salts; reductants include sodium dithionate, pyrocatechol and oxalic acid, and compounds of metals at low valency such as iron(II) perchlorate, tin(II) chloride, vanadyl acetate, arsenic(IV) or titanium(III) chloride and chromium(II) chloride.

Precipitation titrants include silver nitrate (in methanol or ethanol) and mercury(II) perchlorate (in glacial acetic acid).

When considering the specific choice of a titrant in a non-inert medium, it must fit in with the solvent system; for instance, in a solvent HS with self-dissociation, a metal salt MS (preferably an alkali metal salt) dissolved in HS is the strongest possible and therefore most preferred base titrant; with a protophilic solvent such as pyridine, a genuine Brønsted base that is definitely more protophilic (such as TMG) than the solvent itself is preferable, because from the acid–base reaction no water can occur, which owing to its levelling effect would restrict the titration pH range; the latter can be extended by admixing a more inert, i.e., a strongly differentiating, solvent to the analyte solution and possibly to the titrant itself (cf., Figs. 4.4 and 4.5). However, in the titration of very weak acids (e.g., sulphonamides) we obviously need a quaternary alkylammonium hydroxide as a well known type of strong organic base, especially as its salts show high solubility in organic media. Unfortunately, and apart from the disadvantage of water arising from the acid–base reaction, most of them undergo the so-called Hofmann degradation, with an *E*2 reaction mechanism (cf., Sykes[94]) representing a β-elimination of a proton, induced by the OH^- ion* itself or by any other basic group or base such as pyridine:

$$\overset{\overset{+}{N}R_3}{\underset{H\;\;\;\;OH}{>C-C<}} \longrightarrow H_2O + >C=C< + :NR_3 \quad (4.101)$$

If more than one β-hydrogen in the quaternary ion is present, the Hofmann rule states that in the β-elimination predominance is given to the least substituted ethylene.

In addition to the above olefin formation, another although generally slow *E*1 reaction can take place, yielding a tertiary amine:

$$R_4N^+OH^- \rightarrow ROH + :NR_3 \quad (4.102)$$

which in a way represents an α-elimination of an R group.

If in the quaternary ion a β-hydrogen is not available, as in tetramethyl-, benzyltrimethyl- or phenyltrimethylammonium hydroxide, reaction 4.101 cannot occur, but reaction 4.102 still can; this means that the quaternary ammonium ions without a β-hydrogen are appreciably more stable, so that the type of base concerned is mainly the commercially available one (in alcoholic solution); $(CH_3)_4NOH$ appears especially attractive, although the $(CH_3)_4N$ salts are less soluble in non-aqueous media than, for instance, $(C_4H_9)_4N$ salts.

The formation of a weakly basic tertiary amine in reaction 4.102 does not alter the titrant normality, but in the titration of an acid it may suppress the height of the titration curve on the basic side. In an extensive study of twelve quaternary ammonium titrants in non-aqueous media (mainly isopropyl alcohol), Harlow[73] observed large differences in stability; the presence of water had a profound stabilizing action but at the sacrifice of basic strength; inert and basic solvents increased the rate of decomposition (see Fig. 4.18).

Comparing the symmetrical R_4N bases, Harlow obtained the following stability sequence: $CH_3 \gg C_2H_5 < C_3H_7 \ll C_4H_9$ (the same for amyl, hexyl and heptyl); hence, in addition to $(CH_3)_4NOH$, $(C_4H_9)_4NOH$ is also commercially available in mainly alcoholic media. However, in the past and in fundamental studies, many investigators had to prepare R_4N basic titrants themselves and to use them in the fresh state (see p. 267), especially with a protophilic solvent such as pyridine[70], or to keep them cool[11].

There are three general procedures for preparing R_4NOH titrants: (1) the silver oxide method[11,95] (2) the ion-exchange method[11,96] and (3) the potassium

*In an *E*2 mechanism one normally expects second-order kinetics; however, Harlow[73] observed that the decomposition approached first-order kinetics, in agreement with the assumption of an ion pair, $R_4N^+OH^-$, as an intermediate.

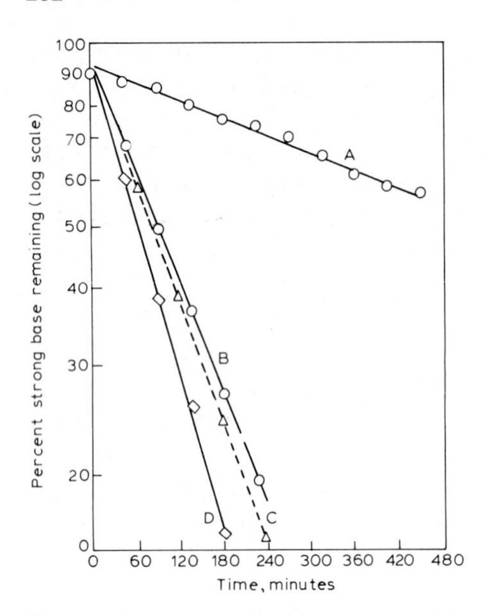

Fig. 4.18. First-order reaction plots of decomposition of tetraethylammonium titrant in (A) isopropyl alcohol, (B) isopropyl alcohol–toluene (1 : 1), (C) benzene and (D) pyridine.

hydroxide method[97]. Here we have confined ourselves to some original specific literature references and at the same time refer to the valuable discussion by Kucharský and Šafařík[17].

In 1961, Heumann[98] proposed the use of a quaternary alkylammonium alkoxide titrant, viz., 0.3 N $(C_4H_9)_4NOEt$ in ethanol–benzene (1 : 10); although there might be some influence of the R_4N ion on the intrinsic basic strength of the ion pair in such a medium, R_4NOR was expected to be a very strong base, as in alkali metal alkoxides the anion is a stronger base than even the OH^- ion in the hydroxides; for the ethoxide it may be an additional advantage that the acid–base reaction yields ethanol ($pK_s = 19$) with a levelling effect less than that of water. With regard to the instability of the R_4N ion in the alkoxide titrant, the situation will probably be similar to that in the hydroxide titrant.

Finally, mention can be made of the more exceptional titrants for inert media, such as sodium triphenylmethane, lithium aluminium hydride and lithium aluminium amide.

Sodium triphenylmethane: treatment of triphenylmethyl chloride with an excess of sodium in benzene leads, via the triphenylmethyl radical (Gomberg, 1900)*, to triphenylmethyl sodium:

---

* The trityl radical (gold-coloured) is readily oxidized to peroxide (white); the comparable 2,4,6-tri-(*tert*-butyl)phenoxy radical (blue) in, e.g., cyclohexane was applied by Paris et al.[99] to so-called free radical titration (either potentiometric or photometric) of oxygen or antioxidant (the latter by hydrogen abstraction).

$$(C_6H_5)_3CCl \xrightarrow{+Na^{\cdot}} (NaCl) + (C_6H_5)_3C^{\cdot} \xrightarrow{+Na^{\cdot}} (C_6H_5)_3\overset{-}{C}\overset{+}{Na}$$

trityl chloride — radical — trityl sodium

Trityl sodium is an extremely strong base, being soluble in ethers and aromatic hydrocarbons to give deep blood-red solutions so that in titration of an acid it can serve as its own visual indicator, although potentiometric detection is also possible; a disadvantage is its high sensitivity to oxygen and moisture.

Lithium aluminium hydride: Higuchi and co-workers[100] introduced it as a titrant, usually in tetrahydrofuran (previously liberated from peroxide), for the titration of alcohols and phenols according to the overall reaction

$$LiAlH_4 + 4\ HOR \rightarrow 4H_2\uparrow + LiAl(OR)_4 \qquad (4.103)$$

As the reaction can be considered as a stepwise attack of the active hydrogen by the four respective H^- ions[101], it is often recommended to add an excess of $LiAlH_4$ and back-titrate with a standard solution of butanol in benzene. Higuchi and co-workers applied either a colour indicator (N-methyl-*p*-amino-azobenzene) or potentiometric detection with a Pt electrode vs. an Ag–AgCl electrode; we found that a dead-stop technique (at 12 V) at two Pt electrodes could also be used[101]. The titrant must be carefully protected against oxygen, moisture and carbon dioxide, and we kept the tetrahydrofuran inhibited with 200 ppm of Ionol (2,6-dimethyl-4-*tert*-butylphenol)[101]. It must be realized that the titration of the hydroxy group with $LiAlH_4$ is restricted, as many other organic functional groups readily react with this reagent (for a review, see ref. 101); moreover, there are many other and more specific analytical methods available for the hydroxy group.

Lithium aluminium amide: this amide was introduced by Higuchi et al.[102] because it yields true acid–base reactions:

$$LiAl(NR_2)_4 + 4\ HOR \rightarrow 4\ HNR_2 + LiAl(OR)_4 \qquad (4.104)$$

and also because this type of reagent is less sensitive to oxygen than the hydride. They used piperidide, dibutylamide and pyrrolidide in tetrahydrofuran, with visual end-point detection.

Primary standards. The compounds mostly used are potassium hydrogen phthalate ($C_8H_5O_4K$), tris(hydroxymethyl)aminomethane (Tris) ($C_4H_{11}O_3N$), 1,3-diphenylguanidine ($C_{13}H_{13}N_3$), anhydrous sodium carbonate (Na_2CO_3), strychnine ($C_{21}H_{22}O_2N_2$), benzoic acid ($C_7H_6O_2$) and *p*-toluenesulphonic acid ($C_7H_8O_3S$).

4.2.1. Non-faradaic methods of electroanalysis in non-aqueous media

In general these methods are restricted to titrations, either conductometric or potentiometric.

TABLE 4.7

EQUIVALENT CONDUCTIVITIES AT INFINITE DILUTION AT 25° C[103]

Solvent	Λ_0 (H^+)	Λ_0 (other ions) (mean)
Water	350	≈ 50
Methanol	146	≈ 50
DMF, pyridine	≈ 40	≈ 40

Conductometric titrations. Van Meurs and Dahmen[25,30,31] showed that these titrations are theoretically of great value in understanding the ionics in non-aqueous solutions (see pp. 250–251); in practice they are of limited application compared with the more selective potentiometric titrations, as a consequence of the low mobilities and the mutually less different equivalent conductivities of the ions in the media concerned. The latter statement is illustrated by Table 4.7[103], giving the equivalent conductivities at infinite dilution at 25° C of the H^+ ion and of the other ions (see also Table 2.2 for aqueous solutions). However, in practice conductometric titrations can still be useful, e.g., (i) when a Lewis acid–base titration does not foresee a well defined potential jump at an indicator electrode, or (ii) when precipitations on the indicator electrode hamper its potentiometric functioning.

(i) The conductometric titration (in an inert solvent such as cyclohexane) of Ziegler precatalysts such as $(C_2H_5)_3Al$ or $(C_2H_5)_2AlCl$ according to Bonitz[29] (see p. 249) is a good working method, especially with $(C_2H_5)_2AlCl$:

$$(C_2H_5)_2AlCl + \text{N(quinoline)} \longrightarrow (C_2H_5)_2Al(Cl){-}\text{N(quinoline)} \rightleftharpoons \left[(C_2H_5)_2Al{-}\text{N(quinoline)}\right]^+ Cl^-$$

where some ionization of the Lewis complex occurs. Other compounds of electron-deficient metals such as boron halides and tin(II) chloride can be titrated as genuine Lewis acids in a similar manner.

(ii) Jander and Kraffczyk[104] determined compounds such as acetone, ethyl acetate, acetonitrile, acetamide and nitrobenzene by means of a conductometric titration with $LiAlH_4$ in diethyl ether, notwithstanding the formation of precipitates.

In a few instances where precipitation prevents conductometry at electrodes in direct contact with the analyte solution, use has been made of high-frequency titration, e.g., with the metal plates outside a measuring capacity cell (see pp. 19–21 and 25); examples are the titration of organic bases with perchloric acid in glacial acetic acid[105] and of strong or weak acids with sodium methoxide in DMF[106].

In the above titrations of Lewis acids or with the use of the highly reactive titrant $LiAlH_4$, one needs inert solvents of Brønsted class 8, having a low dielectric constant (e.g., at 20° C $\varepsilon = 2.0$ for cyclohexane or 4.3 for diethyl ether), which is unfavourable; the choice of dipolar aprotic solvents (class 4), whenever possible, remains recommended in view of their still appreciable ε values (favouring ionic dissociation) and limited levelling properties (permitting greater analyte differentiation). If a high-frequency titration is necessary the solvent, analyte and titrant and their concentrations must satisfy a suitable compromise between conductance and overall capacity (see eqns. 2.22 and 2.23), as for instance has been discussed by Cruse[107].

Potentiometric titrations. Zero-current potentiometric titrations represent by far the most important and extended application of electroanalysis in non-aqueous media, as a consequence of selectivity and high flexibility in combination with the stoichiometry of the reactions. This statement is based not only on the theory of electrochemistry in these media, as already considered in Section 4.1, but also on the methodology of their application, especially in the field of functional group determinations[108–110]. Here in many instances the result of a possible time-consuming quantitative reaction is established by a subsequent finishing titration instead of a direct titration. Using a few examples, partly from our own experience, we shall illustrate that a knowledge of the mechanism of the reactions concerned[111] is most beneficial to the suitability of the analysis.

Addition as a preceding reaction. In comparison with the use of I_2, the more reliable results of the two-step electrophilic addition of ICl in glacial acetic acid with some CCl_4 (Wijs iodine number for edible oils and fats) according to

$$\mathrm{{>}C{=}C{<}} + \overset{\delta+}{\mathrm{I}}-\overset{\delta-}{\mathrm{Cl}} \longrightarrow \mathrm{{>}C\overset{I^+}{—}C{<}}\ (\mathrm{Cl^-}) \longrightarrow \mathrm{{>}C(I)—C(Cl){<}}$$

isolated unsaturation — iodonium ion — *trans*-addition

can be explained by the polarity of ICl, where the electronegativity of I (2.66 kcal mol^{-1}) is less than that of Cl (3.16 kcal mol^{-1})[112]; in this instance the analysis is finished with the usual iodometric titration vs. the blank. However, in case of α,β-unsaturates with a negative group X (esters nitriles or amides), RCH=CHX, with the double bond polarized, a nucleophilic addition must be chosen, e.g., in glacial acetic acid (also as a catalyst) with morpholine at 98 ± 2° C (in capsulated pressure bottles):

$$\mathrm{O(CH_2CH_2)_2N{-}H} + \overset{\delta+}{\mathrm{RCH}}{=}\mathrm{CH}{-}\overset{\delta-}{\mathrm{X}} \longrightarrow \mathrm{O(CH_2CH_2)_2\overset{+}{N}(H){-}CH(R){-}\overset{-}{C}HX} \longrightarrow \mathrm{O(CH_2CH_2)_2N{-}CH(R){-}CH_2X}$$

tetrahydro-1,4-isoxazine (secondary amine) — proton jump via hydrogen bridge — tertiary amine

After cooling and dilution with methyl-Cellosolve (2-methoxyethanol), acetic anhydride is slowly added (in order to convert the secondary amine to amide) and the solution is cooled to room temperature. Finally, the resulting tertiary amine, representing the original unsaturate, is titrated with 0.5 N perchloric acid in methyl-Cellosolve, either visually (with thymol blue + xylene cyanol

FF)[110,113] or potentiometrically. In a micro-method, Terentev et al.[114] used piperidine instead of morpholine.

Another example of interest with regard to the reaction mechanism is the analysis of epoxy groups. Durbetaki[60] titrated α-epoxy compounds with HBr (cf., p. 260) in glacial acetic acid with crystal violet as indicator, but the method was slow for glycidyl esters, $CH_2{-}CHCH_2OOCR$ (with CH_2 and CH bridged by O). As it concerns a two-step electrophilic addition:

$$\rangle C\overset{O}{-}CH_2 \xrightarrow[\text{(1st step)}]{+H^+ \text{ rapid protonation}} \rangle C\overset{\overset{H}{O^+}}{-}CH_2 \xrightarrow[\text{(2nd step)}]{+Br^- \text{ slow anion addition}} \rangle C(OH){-}CH_2Br$$

where the second step is rate determining, Dijkstra and Dahmen[115] considerably improved the titration by accelerating the second step with an excess of bromide, e.g., cetyltrimethylammonium bromide, and by using $HClO_4$ (as a more accessible and even stronger acid than HBr) for the first step. The method can be carried out either visually with crystal violet as indicator or potentiometrically (glass/calomel*) as a direct titration for α-epoxides such as epichlorohydrin, 1,2-epoxypropane and glycidyl ethers or esters, as an absorptive determination of ethylene oxide in gas streams and as an indirect titration for some compounds with non-terminal epoxy groups. The excellent results of the method were confirmed later by others[116].

Substitution as a preceding reaction. In addition to the well known determination of primary and secondary alcohols via esterification with acetic anhydride in pyridine at about 98° C, esterification is possible at room temperature in ethyl acetate with perchloric acid[117] or 2,4-dinitrobenzenesulphonic acid[118] as a catalyst. However, as tertiary alcohols preferably split off their hydroxy group, they can be adequately determined by OH-substitution with HBr in glacial acetic acid according to

$$R_3COH + HBr \rightarrow H_2O + R_3CBr \qquad (4.105)$$

and back-titration of the excess of HBr with sodium acetate in glacial acetic acid using a colour indicator or potentiometry (glass/calomel)[119].

In fact, reaction 4.105 also represents an example of a condensation reaction. A prior redox reaction in non-aqueous medium also often occurs, e.g., in the highly sensitive analysis of peroxides with HI in acetic acid, both under absolutely water-free conditions, where iodine is quantitatively liberated and is subsequently titrated. For much work on non-aqueous redox titrations by Tomíček's school published mainly in the Czech literature, see ref. 17.

On the assumption that the above examples have sufficiently elucidated the methodology of potentiometric titrations in non-aqueous media, some remarks

*With an R_4NCl bridge, avoiding contact between K^+ and ClO_4^- of the analyte solution.

must still be made about the choice and maintenance of the indicator and reference electrodes.

Indicator electrodes. In principle, all types of electrodes can be used in non-aqueous media provided that under the measurement conditions (1) a reversible electrode reaction at the electrode is possible and (2) the electrode surface will not be physically and/or chemically attacked. Many electrodes of the normal type (see pp. 7 and 28–47) fulfil these requirements, e.g., as redox electrodes we can mention the classic hydrogen electrode, H_2 (1 atm)/Pt (platinized), as in the Jackson cell (Fig. 2.8) and in the work of Pleskov (Table 4.2 and Fig. 4.2), and its Pd version coated either with Pd black (Stock et al.[120]) or with Pt black (Tomíček and Heyrovský[121]), which electrodes, however, are more suitable for fundamental than practical analytical work; the quinhydrone electrode (p. 43) and the tetrachloroquinhydrone electrode (often called the chloranil electrode*) originally proposed by Hall and Conant[122] for acid–base titrations in glacial acetic acid, and used for SO_3 titration with water in oleum by Šponar[123] and Van der Heijde[124]; under certain limited conditions metal electrodes other than platinum can also be used for acid–base titrations, e.g., gold, antimony or bismuth; also carbon as a non-metal (e.g., graphite or glassy carbon) has been applied; especially gold (cf., in Table 4.6 its low hydrogen overpotential, $\eta_{H_2} = -0.016$ V, at a rough surface) appeared attractive for the titration of amines in alcohols, acetic acid, acetic anhydride or acetonitrile; however, for redox titrations, usually in glacial acetic acid and sometimes in acetonitrile, platinum provides the most preferred inert indicator electrode[125].

Electrodes of the first kind have only limited application to titration in non-aqueous media; a well-known example is the use of a silver electrode in the determination of sulphides and/or mercaptans in petroleum products by titration in methanol–benzene (1 : 1) with methanolic silver nitrate as titrant. As an indicator electrode of the second kind the antimony pH electrode (or antimony/antimony trioxide electrode) may be mentioned; its standard potential value depends on proton solvation in the titration medium chosen; cf., the equilibrium reaction on p. 46).

Among the membrane electrodes, the glass pH electrode is by far the most important pH electrode in non-aqueous media; in fact, there seem no real limitations to its use, if proper handling is adopted.

On p. 57 it was explained that the working of the glass electrode is based on the occurrence of "swell" films, a kind of hydrated layer, on both sides of the "dry" glass continuum; inside the glass bulb the film remains undisturbed owing to its permanent contact with the aqueous filling solution, but the film outside can dry out by repeated or continuous contact with strongly hygroscopic or non-aqueous solutions; in the latter instance a loss of the correct Nernst potential slope soon occurs which can better be prevented by regular

*In fact, it is based on an equimolar mixture of tetrachloroquinone (chloranil) and tetrachlorohydroquinone (hydrochloranil) at a white Pt electrode.

intermittent rehydration. Here an advisable procedure may be as follows: when not in use, keep the electrode in water or an aqueous buffer of pH 7; when in use, remove the electrode intermittently from the analyte solution, rinse with a low-boiling cleaning solvent (miscible with organic solvents and with water, e.g., acetone or acetonitrile), keep in aqueous buffer of pH 7 for a few hours (up to 24 h, if necessary), remove before use and rinse again with cleaning solvent, blow dry, insert into analyte solution and equilibrate for 5 min.

When the titration solvent (or a component of it) is miscible with water, the solvent itself can be used for cleaning[70], thus excluding dry blowing; in specific cases consult the relevant literature[65,77]. The rehydration procedure thus maintains the external gel film of the glass bulb, which furnishes a rapid proton equilibrium exchange with the analyte solution and accordingly indicates its proton activity by the electrode potential. However, a possible alkali error in the titration of very weak acids with a strong base[126] must be excluded a priori (see below).

At present other types of membrane electrodes are not yet of real importance with regard to non-aqueous media; however, with developments in the field of crowns ethers (see p. 70), extensively reviewed by Kolthoff[127], this may change in the near future.

Reference electrodes. There are two types of reference electrodes (see the scheme in Section 1.3.1): (a) those constructed as a reference type and (b) those used as a reference type; both types fulfil the requirement of a constant reference potential by either being non-polarizable or becoming non-polarized during the measurememt.

Type a, in the shorthand notation of electrochemical cells (pp. 27–28), is described by double bars (‖), which represents a salt bridge or a boundary with negligible junction potential; behind it comes the proper electrode, which may be a redox electrode, e.g., a quinhydrone type, an electrode of the first kind, e.g., Ag–0.10 *M* $AgNO_3$ in acetonitrile[128] and Ag–1 *M* $AgNO_3$ in pyridine[128a], or usually one of the second kind, e.g., Ag–AgCl in a saturated solution of $(CH_3)_4NCl$ in *m*-cresol[65], pyridine[70] or 1,2-dichloroethane[77]; Hg–Hg(II) acetate in a saturated solution of sodium acetate* in methanol or glacial acetic acid; Hg–saturated Hg(I) sulphate solution in concentrated sulphuric acid[124]; Cu–saturated Cu(II) acetate solution in glacial acetic acid[129]; or Hg–saturated Hg_2Cl_2 solution in aqueous (mostly saturated) KCl solution, representing the usual calomel electrode. In these examples of reference electrodes there was an only a single junction (porous disc or sleeve joint) in contact with the non-aqueous analyte solution; however, with the calomel electrode leakage of water may adversely affect the net titration result but, more seriously, leakage of potassium, as was shown by Harlow[126], can disturb the correct potential indication by the glass electrode; this "alkali error", well known also with

*This electrode filling is unsuitable with acetic acid or another related acid as the titration solvent, as in this medium possible leakage of sodium acetate would act as the inherent base.

aqueous media, varies with the type and manufacture of the electrode (see p. 58); therefore, one should avoid this type of error by using a double junction, viz., a salt bridge, between the inner KCl and the outer analyte solution, filled with a solution of a tetraalkylammonium chloride such as $(C_4H_9)_4NCl$[126], which is also soluble in organic solvents; even the bridge solution can be non-aqueous, provided that insufficient electric contact as a consequence of KCl clogging within the intermediate boundary does not occur. Such organic bridge solutions have often been used for other electrodes of the second kind, e.g., 0.1 *M* $(C_2H_5)_4NClO_4$ in *m*-cresol[130] and 0.2 *M* $(C_2H_5)_4NClO_4$ in DMSO[131]; 0.1 *M* $(C_2H_5)_4NClO_4$ in DMSO, DMF and acetonitrile[132]; and 5% methyl-cellulose gel in N-methylpyrrolidone (NMP)[133] saturated with $KClO_4$* (in combination with the internal system Hg–mercury oxides in NMP[134]). Although within the various solvents the cation and anion mobilities of the salt may differ, the diffusion potential, if any, and hence the reference electrode potential remain fairly constant during a titration.

Type b reference electrodes occur, although not often in non-aqueous media, under the following conditions: if in an analyte solution a constant activity of an ion is present or has been obtained by previous addition, and if this activity does not alter during the titration desired, then an indicator electrode of that ion in direct contact with the solution will act as a reference electrode. Examples can be found among all the kinds of electrodes of the normal type, e.g., a redox electrode according to Strehlow (see p. 242) such as Pt or Hg at Fc–Fc^+ (or a similar redox couple) in a titrand solution with DMF or NMP[135]; an electrode of the first kind according Pleskov (see p. 242), i.e., Rb at Rb^+ or Cs at Cs^+ in a titrand solution with an organic solvent; an electrode of the second kind such as Ag–AgCl (often Pt electroplated with silver from a 0.05 *M* argentocyanide solution[136]) in a titrand solution such as acetonitrile or any other organic solvent, preferably with a constant amount of chloride such as a tetraalkylammonium chloride (the objection of Glenn[137] to the dipping of the Ag–AgCl electrode directly into the titrand solution, because of potential reversal vs. the indicator electrode near the titration end-point is not realistic as it can be overcome by the intercalation of a constant potential from a potentiostat or a standard galvanic cell). Finally an example of the membrane type is a glass electrode in a titrand solution in which a constant pH can be maintained.

4.2.2. Faradaic methods of electroanalysis in non-aqueous media

In the faradaic methods the passage of a certain intensity of electric current is essential; for non-aqueous media this means that a minimum level of ionic dissociation together with the addition of salts readily soluble in organic solvents (preferably quaternary ammonium salts) must provide the required

* When applied versus a glass indicator electrode the potassium salt can better be replaced by a tetraalkylammonium salt like $(C_2H_5)_4NClO_4$.

conductance. At the same time, a differentiating medium is often desired, so that generally dipolar solvents (class 4, Table 4.3) are mostly applied to the faradaic methods in non-aqueous media, not only for voltammetry but also and more especially for coulometry; a few examples will illustrate this.

Voltammetry. Here in a non-aqueous medium the three-electrode measuring system is of great importance for recording the current as a function of the true potential of the indicator electrode (IE), and in polarography the dme (see Fig. 3.24). Moreover, a potential drop still remaining between the IE and the reference electrode (RE) must be minimized by keeping the distance between them as small as possible, because the conductance of a non-aqueous solution, even with a supporting electrolyte, is generally low. The RE can be one of the above reference electrodes normally used for zero-current potentiometry. The counter or auxiliary electrode (AE) serves for passage of the current; therefore, under the experimental conditions chosen, the AE should be non-polarizable, e.g., the SCE, or only slightly polarizable, owing to its large electrode surface, such as the mercury pool; it is not essential that during the experiment the AE potential remains strictly constant, provided that the analyte solution is not contaminated by contact with the product generated at the AE or with solution from the AE bridge. In fact, there is plenty choice of suitable reference electrodes[138] and various types are commercially available[139].

From the influence of the non-aqueous medium, in comparison with water on the electrodics of the IE one can expect higher charging currents at the dme (by the lower dielectric constant of the solution) and often, which is more important, more complex electrode processes, especially with organic analytes. In this connection we can refer to Section 3.3.1.1.2, where we considered the aspects of irreversibility vs. reversibility (pp. 124–127) in relation with EC and CE mechanisms, catalysed systems, etc. Whatever the electrode process may be, reductive or oxidative, the analyte should be voltammetrically active or be made so by conversion. For instance, many reducible organic functional groups, containing π- and n-electrons, may be sufficiently polarographically active by itself but, in order to become so, mostly need another activating group, affecting the electron distribution and thus shifting the half-wave potential to a less negative value[140]. A well known example is the conversion of ketones with 2,4-dinitrophenylhydrazine into their hydrazones:

$$O_2N\text{-}C_6H_3(NO_2)\text{-}NHNH_2 + O{=}C\langle \xrightarrow[\text{oxalic acid}]{\text{aqueous}} O_2N\text{-}C_6H_3(NO_2)\text{-}NHN{=}C\langle \downarrow + H_2O$$

The hydrazones precipitate or can be extracted with an inert solvent such as benzene or toluene; after addition of a lower alcohol such as methanol together with supporting electrolyte, the hydrazones can be determined polarographically; the dinitrophenylketimine structure predominates more or less in the values of $E_{\frac{1}{2}}$ and i_d per mole.

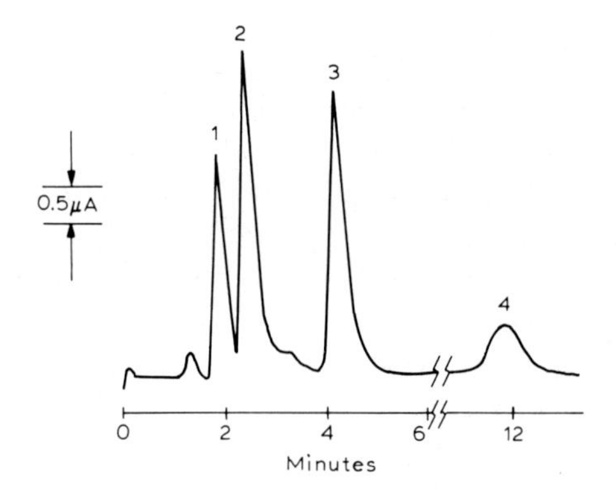

Fig. 4.19. Polarographic detection of a mixture of volatile N-nitrosamines. Sample mixture containing (1) 37 ng of N-nitrosodimethylamine, (2) 34 ng of N-nitrosodiethylamine, (3) 42 ng of N-nitrosodipropylamine and (4) 27 ng of N-nitrosodibutylamine (courtesy of PARC).

Another example concerns the specific conversion of secondary aliphatic amines with HNO_2 into nitrosamines and in situ polarographic determination according to English[141,142]; model compounds of the nitrosamines can be obtained as their hydrochloride salts[142] or as such[143]; PAR[143] have shown that the individual nitrosamines can be separated by liquid chromatography with methanol–water–acetic acid (50 : 47 : 3.0) pH 2.55 as the mobile phase and with their Model 310 polarographic detector using the sampled DC or normal pulse technique (see Fig. 4.19). Lund[144] showed that under these acidic conditions nitrosamine yields its hydrazine:

$$\begin{matrix} R_1 \\ R_2 \end{matrix}\!\!>\!N{-}N{=}O + 4H^+ + 4e^- \rightarrow \begin{matrix} R_1 \\ R_2 \end{matrix}\!\!>\!N{-}NH_2 + H_2O$$

Some carcinogenic nitrosamines have also been determined by differential pulse polarography[145].

Before mentioning some more literature data on non-aqueous voltammetry, we suggest on the basis of our previous discussions that the choice of the experimental conditions used in the techniques must be a compromise between a sufficient solubility of the analyte in the solution, an ample redox potential range of the solvent, a suitable type of indicator electrode and adequate conductance of the solution with supporting electrolyte added. In this connection Fig. 4.20 may be a useful guide.

Here the following selected examples are of interest.

(a) The more fundamental studies of Tsuji and Elving[69], with polarographic measurement of relative strengths of Brønsted acids in pyridine [0.1 M $(C_2H_5)_4NClO_4$], and those by Sinicki and Bréant[146], with a voltammetric study of bromine and bromides in DMF (0.1 N $LiClO_4$) at polished Pt (various irreversible equilibria between Br_2, Br^- and tribromide Br_3^-).

(b) Polarographic research on symmetrical triazacarbocyanines in acetonitrile by Hellrung[147] (the reduction behaviour of these cyanine dyes, studied

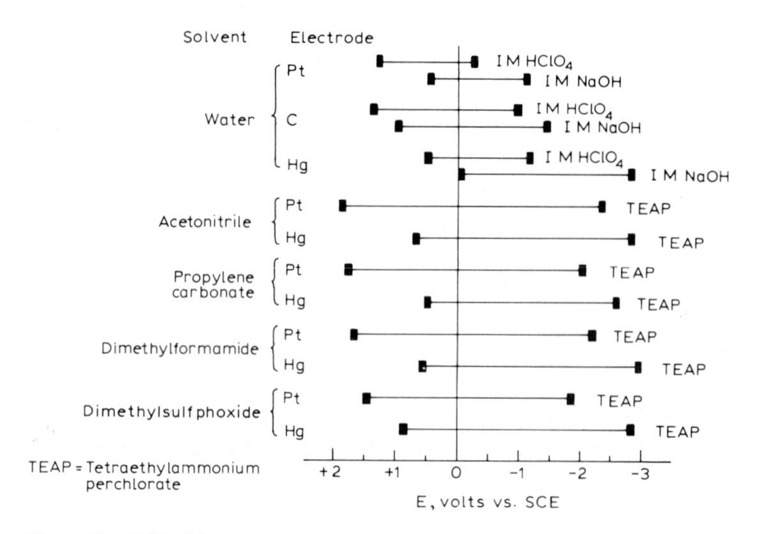

Fig. 4.20. Utilizable ranges of various voltammetric working electrodes (courtesy of Metrohm).

by means of dc and ac polarography, cyclic oscillopolarography and coulometry, revealed an EEC mechanism, the first step being reversible and the second step being reversible to irreversible).

(c) A series of electroanalytical investigations: (i) rapid voltammetric determination of tocopherols and antioxidants in oils and fats, by McBride and Evans[148] [oxidative linear sweep voltammetry (LSV) at a glassy carbon electrode (GCE) of an oxidant such as butylated hydroxyanisole (BHA) in 0.12 *M* H_2SO_4 in ethanol–benzene (1 : 1) and of tocopherols in 0.12 *M* H_2SO_4 in ethanol–benzene (2 : 1) or their naturally occurring member vitamin E in 0.1 *M* H_2SO_4 in ethanol–toluene (2 : 1)]; α-tocopherol and α-tocopheryl acetate can be easily and separately determined by oxidative differential pulse voltammetry (DPV) at a GCE in 0.1 *M* $(C_4H_9)_4NBF_4$ in acetonitrile[149]; (ii) constant-current potentiometric titration of phenols with 0.15 *M* bromine in anhydrous acetic acid in the presence of pyridine and with the use of two Pt-foil electrodes by Huber and Gilbert[150] (the method resembles the differential amperostatic Karl Fischer titration as a purely non-aqueous voltammetric technique; see pp. 204–205); (iii) rotating disc electrode (RDE) determination of nitrogen dioxide concentrated in DMF by Shinozuka and Hayano[151] (reductive voltammetry of the NO_2 absorbed from ppm levels in air); (iv) rotating ring-disc electrode (RRDE) method in non-aqueous media, applied to the study of the behaviour of phenothiazine in acetonitrile by Tacussel and Fombon[152]. Phenothiazine (P) (10^{-3} *M*), dissolved in $HClO_4$ (10^{-1} *M* in water-free acetonitrile) undergoes a two-electron oxidation via a stable free radical (R) to the phenazothionium salt (S):

P $\xrightarrow{-e}$ R $\xrightarrow{-e}$ S

(scanning speed $dE_D/dt = 0.1\,V\,min^{-1}$, rotation speed $W = 3000\,r\,min^{-1}$).

Coulometry. Even in water, controlled potential or potentiostatic coulometry is a difficult and often time-consuming technique, as the analyte must participate in a direct electrode reaction. Therefore, in non-aqueous media there are only a few examples of its application, e.g., the potentiostatic coulometry of nitro and halogen compounds in methanol (99%) with graphical end-point prediction, as described by Ehlers and Sease[153].

Coulometric titration. For this technique, often designated controlled-current or amperostatic coulometry, it is useful to distinguish between redox, complex-formation and precipitation titrations on the one hand and acid–base titrations on the other and to discuss each group separately.

In the first group the titrant is generated either directly from a participating or active electrode, or indirectly from an inert or passive electrode, in which case it is necessary to add previously an auxiliary substance that generates the titrant by either cathodic reduction or anodic oxidation; the end-point detection is usually potentiometric or amperometric. The following selected examples are illustrative of the first group in non-aqueous media:

(a) Potentiometric coulometric titration of secondary amines and mercaptans with mercury(II), as described by Przybylowicz and Rogers[154]. The generator and indicator electrodes consisted of mercury-coated gold, and the various solvents (acetonitrile, DMF, methanol, isopropanol, *tert.*-butanol, acetone and methyl-Cellosolve) contained 5% (v/v) of CS_2 (for prior conversion of secondary amines into dithiocarbamic acids) and anhydrous $NaClO_4$ as a supporting electrolyte; the interference of primary amines, which react incompletely with CS_2, is removed by previous conversion with salicylaldehyde into the Schiff bases. Dithiocarbamic acid reacts quantitatively with the Hg(II) generated and so does the sulphydryl group (as known for mercaptans) in the same way, but without the preceding reaction.

(b) Constant-current coulometric determination of oxygen with viologen radical-cation with biamperometric detection, as described by Van der Leest[155]. All electrodes were made of platinum; the radical-cation was generated by cathodic reduction of the auxiliary substance 1,1′-dimethyl-4,4′-bipyridinium dichloride either in aqueous solution or in an organic solvent such as acetonitrile or DMF; the method can be applied to oxygen in solution and in air (see also the footnote on p. 282 about the free-radical titration by Paris et al.[99]).

(c) Amperostatic coulometric titration of ferrocene and derivatives in acetonitrile with biamperometric detection by Kies and Ligtenberg[156]. All electrodes were made of platinum; the titrant, $Cu(BF_4)_2$, was generated by anodic oxidation of the auxiliary substance $CuBF_4$ with 0.1 *M* $(C_2H_5)_4NBF_4$ as supporting electrolyte; the titrant readily oxidizes the ferrocene compound and the (bi)amperogram shows the normally expected hyperbole (see Fig. 3.80).

(d) Constant-current pulse coulometric Karl Fischer titration with biamperometric detection as used in the Metrohm 652 KF-coulometer (see Fig. 3.86 with description, p. 221).

In the second group, acid–base coulometric titrations in non-aqueous media, the greatest problem is how to achieve 100% current effeciency in generating the titrant from the solvent itself. Serious difficulties do not seem to be encountered with water-resembling solvents with self-dissociation; for instance, in anhydrous acetic acid, with $NaClO_4$ as the supporting electrolyte, anodic oxidation at mercury yields 100% current efficiency according to the reaction[157]

$$2Hg + 2\,CH_3COOH - 2\,e \rightarrow Hg_2(CH_3COO)_2 + 2H^+$$

However, in a similar solvent with autoprotolysis such as *m*-cresol, 100% current efficiency could not be obtained in anodic oxidation with respect to base titrations. According to a further study[158], the following reaction scheme seems most probable:

$$B + HCres \rightleftharpoons BH^+Cres^- \rightleftharpoons BH^+ + Cres^-$$

$$Cres^- \longrightarrow Cres\cdot + e$$

$$2\,Cres\cdot \longrightarrow Cres{-}Cres \quad (H_3C{-}C_6H_3(OH){-}C_6H_3(OH){-}CH_3)$$

The first reaction sequence indicates that near the end-point of the base titration there is virtually no base left to undergo the anodic oxidation with sufficient *m*-cresolate ion. This difficulty can be overcome by previously adding to the titration medium a large amount (e.g., 0.2 *M*) of urea, a base much weaker than that to be titrated but still sufficiently basic to provide the *m*-cresolate ion required; in this way 100% current efficiency could be obtained[159] using Pt gauze with 0.1 *M* $(C_2H_5)_4NClO_4$ as the supporting electrolyte and with potentiometric detection.

Next, in view of the attractive discriminating properties of the dipolar aprotic solvents (class 4) we wanted to establish whether and how coulometric acid–base titrations in dimethyl sulphoxide (DMSO) might be performed. We disregarded the possibility of adding trace amounts of water (e.g., 0.3% in acetonitrile[160]), as its absence is essential in sophisticated titrations in these media. In DMSO with 0.2 *M* $(C_2H_5)_4NClO_4$ (TEAP) as the supporting electrolyte and potentiometric detection, we found that 100% current efficiency could be obtained in the titration of acids by cathodic reduction at a Pt gauze working electrode and in the titration of bases by anodic oxidation at this Pt gauze working electrode with introduction of hydrogen gas as a means of proton transfer which is more effective and less restrictive than the previous addition of *m*-cresol or hydroquinone[161]. It also appeared that the Pt gauze/H_2 working electrode can yield good results in coulometric titrations in acetonitrile and sulfolane[162].

Finally, it must be said that coulometry, more especially the technique in non-aqueous media, requires strict adherence to a number of experimental precautions such as those described on pp. 222–223 and in Figs. 3.87 and 3.88[162].

REFERENCES

1 J. S. Fritz, Acid–Base Titrations in Nonaqueous Solvents, G. F. Smith Chemical Co., Columbus, OH, 1952.
2 B. Trémillon, La Chimie en Solvants Non-Aqueux, Presses Universitaires de France, Paris, 1971.
3 G. Jander, Die Chemie in Wasserähnlichen Lösungsmitteln, Springer, Berlin, 1949.
4 Ref. 2, Table 4, pp. 62–65.
5 V. Gutmann and I. Lindqvist, Z. Phys. Chem., 203 (1954) 250.
6 N. D. Cheronis and T. S. Ma, Organic Functional Group Analysis by Micro and Semimicro Methods, Interscience, New York, 1964, p. 494.
7 E. A. M. F. Dahmen, Chem. Weekbl., 67 (1971) A55.
8 H. P. Cady and H. M. Elsey, J. Chem. Educ., 5 (928) 1425.
9 H. Lux, Z. Electrochem., 45 (1939) 303.
10 H. Flood and T. Førland, Acta Chem. Scand., 1 (1947) 592; H. Flood, T. Førland and B. Roald, Acta Chem. Scand., 1 (1947) 790.
11 H. B. van der Heijde and E. A. M. F. Dahmen, Anal. Chim. Acta, 16 (1957) 378 (cf., XV Congresso International de Quimica Pura e Aplicada, Lisboa, 1957, Actas do Congresso, Vol. I, p. III-27, Potentiometric Titrations in Non-Aqueous Solutions, I: An Empirical Acidity Potential Scale of Twelve Solvents); E. A. M. F. Dahmen, Chim. Anal. (Paris), 10 (1958) 378 and 430 (Exposé presenté au IVe Symposium sur l'Analyse Fonctionelle Organique, Paris, 26–27 Nov., 1956); H. B. van der Heijde, Anal. Chim. Acta, 17 (1957) 512.
12 Ref. 2, pp. 215–231.
13 G. Schwarzenbach, Helv. Chim. Acta, 13 (1930) 870.
14 V. A. Pleskov, Usp. Khim., 16 (1947) 254: J. Phys. Chem. USSR, 22 (1948) 351.
15 H. Strehlow, Z. Electrochem., 56 (1952) 827; J. J. Lagowski (Editor), The Chemistry of Non-Aqueous Solvents, Vol. 1, Academic Press, New York, 1966, Ch. 4; H. M. Koepp, H. Wendt and H. Strehlow, Z. Electrochem., 64 (1960) 483.
16 Ref. 2, pp. 180–187 and 215–219.
17 J. Kucharský and L. Šafařík, Titrations in Non-Aqueous Solvents, Elsevier, Amsterdam, 1965.
18 L. M.. Mukherjee, J. Phys. Chem., 76 (1972) 243.
19 Ref. 2, Table 4, pp. 62–65.
20 E. A. M. F. Dahmen, Chim. Anal. (Paris), 10 (1958) 378 and 430.
21 G. Gran and B. Althin, Acta Chim. Scand., 4 (1950) 967.
22 R. H. Cundiff and P. C. Markunas, Anal. Chem., 28 (1956) 792.
23 G. Allen and E. F. Caldin, Q. Rev. Chem. Soc., 7 (1953) 255.
24 H. B. van der Heijde, Anal. Chim. Acta, 16 (1979) 392.
25 N. van Meurs and E. A. M. F. Dahmen, Anal. Chim. Acta, 21 (1959) 193.
26 R. Mullikan, J. Amer. Chem. Soc., 74 (1952) 811.
27 G. H. Schenk and M. Ozolins, Talanta, 8 (1961) 109.
28 M. Ozolins and G. H. Schenk, Anal. Chem., 23 (1961) 1035.
29 E. Bonitz, Chem. Ber., 88 (1955) 742; cf., R. Dijkstra and E. A. M. F. Dahmen, Z. Anal. Chem., 181 (1961) 399.
30 N. van Meurs and E. A. M. F. Dahmen, Anal. Chim. Acta, 19 (1958) 64; 21 (1959) 10 and 443; J. Electroanal. Chem., 1 (1960) 458.
31 N. van Meurs and E. A. M. F. Dahmen, Anal. Chim. Acta, 21 (1959) 193 (see p. 196).
32 T. Higuchi and C. R. Rehm, Anal. Chem., 27 (1955) 408.
33 S. Bruckenstein and I. M. Kolthoff, J. Amer. Chem. Soc., 78 (1956) 10 and 2974.
34 G. Goldstein, D. L. Manning and H. E. Zittel, Anal. Chem., 34 (1962) 1169.
35 J. N. Brønsted, Chem. Ber., 61 (1928) 2049.
36 M. M. Davis, Acid–Base Behaviour in Aprotic Organic Solvents, NBS Monograph No. 105, National Bureau of Standards, Washington, DC, 1968, p. 9.
37 A. J. Parker, Q. Rev. Chem. Soc., 16 (1962) 263.

38 C. W. Davies, Ion Association, Butterworth, London, 1962.
39 R. M. Fuoss, Trans. Faraday Soc., 30 (1934) 967.
40 K. Burger, Solvation, Ionic and Complex Formation Reactions in Non-Aqueous Solvents (Experimental Methods for their Investigation), Studies in Analytical Chemistry, Vol. 6, Elsevier, Amsterdam, 1983, Ch. 2 and 3 and Ch. 9, pp. 256–257.
41 V. Gutmann and E. Wychera, Inorg. Nucl. Chem. Lett., 2 (1966) 257; V. Gutmann, Coordination Chemistry in Non-Aqueous Solutions, Springer, Vienna, 1968; V. Gutmann, The Donor–Acceptor Approach to Molecular Interactions, Plenum Press, London, 1978.
42 U. Mayer, Pure Appl. Chem., 41 (1975) 291.
43 M. M. Davis, J. Amer. Chem. Soc., 71 (1949) 3544; M. M. Davis and H. B. Hetzer, J. Amer. Chem. Soc., 76 (1954) 4247.
43a P. Bacelon, J. Corset and C. De Loze, Chem. Phys. Lett., 32 (1975) 458.
44 H. Taube and F. A. Posey, J. Amer. Chem. Soc., 75 (1953) 1463; 78 (1956) 15.
45 F. J. C. Rossotti and H. Rossotti, J. Phys. Chem., 65 (1961) 926, 930 and 1376; D. M. W. Anderson, J. L. Duncan and F. J. C. Rossotti, J. Chem. Soc., (1961) 140, 2165 and 4201.
46 Ref. 36, p. 50.
47 C. A. Kraus, J. Chem. Educ., 35 (1958) 324; cf., C. A. Kraus and co-workers, J. Amer. Chem. Soc., 56 (1934) 2017; 59 (1937) 1699; 73 (1951) 4557 and 4732.
48 R. M. Fuoss and C. A. Kraus, J. Amer. Chem. Soc., 55 (1933) 3614; C. A. Kraus, Science, 90 (1939) 281.
49 I. M. Kolthoff and M. K. Chantooni, Jr., J. Amer. Chem. Soc., 85 (1963) 2195.
50 Die Natur des H. C. Brownschen Hydroborierungsreagenz, Nachr. Chem. Tech. 23, No. 6 (1975) 107.
51 A. A. Maryott, J. Res. Natl. Bur. Stand., 38 (1947) 527.
52 H. B. van der Heijde, Anal. Chim. Acta, 16 (1957) 392; see also E. A. M. F. Dahmen, ref. 20, Fig. 11.
53 D. B. Bruss and G. A. Harlow, Anal. Chem., 30 (1958) 1836.
54 P. J. R. Bryant and A. W. H. Wardrop, J. Chem. Soc., (1957) 895.
55 Ref. 36, p. 62.
56 R. C. Lord and R. E. Merrifield, J. Chem. Phys., 21 (1953) 166; B. Chenon and C. Sandorfy, Can. J. Chem., 36 (1958) 1181; C. Brissette and C. Sandorfy, Can. J. Chem., 38 (1960) 34.
57 M. M. Davis, J. Amer. Chem. Soc., 71 (1949) 3544; M. M. Davis and H. B. Hetzer, J. Amer. Chem. Soc., 76 (1954) 4247.
58 S. Bruckenstein and A. Saito, J. Amer. Chem. Soc., 87 (1965) 698.
59 I. M. Kolthoff and S. Bruckenstein, J. Amer. Chem. Soc., 78 (1956) 1; S. Bruckenstein and I. M. Kolthoff, J. Amer. Chem. Soc., 78 (1956) 10 and 2974.
60 A. J. Durbetaki, Anal. Chem., 28 (1956) 2000; 30 (1958) 2024.
61 J. N. Hogsett, cf., method description in F. E. Critchfield, Organic Functional Group Analysis (International Series of Monographs on Analytical Chemistry, Vol. 8), Pergamon Press, Oxford, 1963, p. 95.
62 G. Oddo and G. Anelli, Gazz. Chim. Ital., 411 (1911) 532.
63 W. E. S. Turner and C. C. Busett, J. Chem. Soc., 105 (1914) 1777; W. E. S. Turner and C. T. Pollard, J. Chem. Soc., 105 (1914) 1751.
64 M. Tanaka and G. Nakagawa, Anal. Chim. Acta, 33 (1965) 543.
65 M. Bos and E. A. M. F. Dahmen, Anal. Chim. Acta, 57 (1971) 361; M. Bos, Thesis, Twente University of Technology, Enschede, 1972.
66 R. M. Fuoss, J. Amer. Chem. Soc., 57 (1935) 488.
67 R. M. Fuoss and C. A. Kraus, J. Amer. Chem. Soc., 55 (1933) 476 and 2387.
68 C. M. French and I. G. Roe, Trans. Faraday Soc., 449 (1953) 314.
69 K. Tsui and P. J. Elving, Anal. Chem., 41 (1969) 286 and 1571.
70 M. Bos and E. A. M. F. Dahmen, Anal. Chim. Acta, 53 (1971) 39 and 285; see also ref. 65.
71 D. F. Detar, Computer Programs for Chemistry, Vol. II, W. A. Benjamin, New York, 1969, pp. 65–67.
72 L. M. Mukherjee, J. J. Kelly, W. Barnetsky and J. Sica, J. Phys. Chem., 72 (1968) 3410.

73 G. A. Harlow, Anal. Chem., 34 (1962) 1487.
74 E. N. Gur'yanova and I. G. Beskina, J. Gen. Chem. USSR, Engl. Transl., 33 (1963) 914; see also ref. 36, pp. 48–49.
75 S. Bruckenstein and N. E. Vanderborgh, Anal. Chem., 38 (1966) 687.
76 G. M. Barrow and E. A. Yerger, J. Amer. Chem. Soc., 76 (1954) 5211; E. A. Yerger and G. M. Barrow, J. Amer. Chem. Soc., 77 (1955) 4474, 6204; G. M. Barrow, J. Amer. Chem. Soc., 78 (1956) 5802.
77 M. Bos and E. A. M. F. Dahmen, Anal. Chim. Acta, 63 (1973) 185.
78 P. Walden, Z. Phys. Chem., 55 (1906) 207; Elektrochemie Nicht-Wässriger Lösungen, J. A. Barth, Leipzig, 1924.
79 M. Bos and H. Q. J. Meershoek, Anal. Chim. Acta, 61 (1972) 185.
80 M. Bos and E. A. M. F. Dahmen, Anal. Chim. Acta, 63 (1973) 325.
81 Ref. 2, Ch. IV and V. 3.
82 C. A. Kraus, J. Amer. Chem. Soc., 43 (1921) 749.
83 M. Bos and E. A. M. F. Dahmen, Anal. Chim. Acta, 72 (1947) 169.
84 A. H. Beckett and E. H. Tinley, Titrations in Non-Aqueous Solvents, British Drug Houses, Poole, 2nd ed., 1957; 3rd ed., 1962.
85 Eastman Organic Chemicals Division, Reagents for Non-Aqueous Titrimetry, Eastman-Kodak, Rochester, NY, 1970.
86 A. P. Kreshkov, Practical Textbook on Acid–Base Titrations in Non-Aqueous Media, Khim.-Tekh. Inst., Moscow, 1958; see also Talanta, 17 (1970) 1029.
87 W. Huber, Titration in Nicht-Wässrigen Lösungsmitteln, Akademie Verlag, Leipzig, 1964; Engl. transl.: Titrations in Non-Aqueous Solvents, Academic Press, New York, 1967.
88 I. Gyenes, Titrations in Non-Aqueous Media, Van Nostrand, Princeton, NJ, 1967; Titrations in Non-Aqueous Media, Iliffe, London, 1967; Titrationen in Nicht-Wässrigen Medien, 3rd ed., F. Encke Verlag, Stuttgart, 1970.
89 G. A. Harlow and D. H. Morman, Anal. Chem., 40 (1968) 418R.
90 S. R. Palit, M. N. Das and G. R. Somayajalu, Non-Aqueous Titrations, Indian Association for the Cultivation of Science, Calcutta, 1954; cf., S. R. Palit, Ind. Eng. Chem., Anal. Ed., 18 (1946) 246.
91 C. W. Pifer and E. G. Wollish, Anal. Chem., 24 (1952) 300.
92 J. S. Fritz, Anal. Chem., 22 (1950) 578; 26 (1954) 1701.
93 E. Römberg and K. Cruse, Z. Elektrochem., 63 (1959) 404.
94 P. Sykes, A Guidebook to Mechanism in Organic Chemistry, Longmans, Green, London, 1963.
95 R. H. Cundiff and P. C. Markunas, Anal. Chem., 34 (1962) 584.
96 G. A. Harlow, C. M. Noble and G. E. A. Wyld, Anal. Chem., 28 (1956) 787.
97 G. A. Harlow and G. E. A. Wyld, Anal. Chem., 34 (1962) 172; see also G. A. Harlow, Anal. Chem., 34 (1962) 148.
98 W. R. Heumann, in D. S. Jackson (Editor), Titrimetric Methods, Plenum Press, New York, 1961, pp. 121–128.
99 J. P. Paris, J. D. Gorsuch and D. M. Hercules, Anal. Chem., 36 (1964) 1332.
100 T. Higuchi, C. J. Lintner and R. H. Schleif, Science, 111 (1950) 63; C. J. Lintner, R. H. Schleif and T. Higuchi, Anal. Chem., 22 (1950) 534; C. J. Lintner, D. A. Zuck and T. Higuchi, J. Amer. Pharm. Ass., 39 (1950) 418; T. Higuchi and D. A. Zuck, J. Amer. Chem. Soc., 73 (1951) 2676.
101 E. A. M. F. Dahmen, l'Hydrure de Lithium–Aluminium comme Réactif en Analyse Organique, 10e Série de Mises au Point de Chimie Analytique Pure et Appliquée et d'Analyse Bromatologique (J.-A. Gautier), Masson, Paris, 1962, pp. 23–48.
102 T. Higuchi, J. Concha and R. Kuramoto, Anal. Chem., 24 (1952) 685.
103 N. van Meurs and E. A. M. F. Dahmen, J. Electroanal. Chem., 1 (1960) 458.
104 G. Jander and K. Kraffczyk, Z. Anorg. Allg. Chem., 283 (1956) 217.
105 W. F. Wagner and W. B. Kauffmann, Anal. Chem., 25 (1953) 538.
106 J. A. Dean and C. Cain, Jr., Anal. Chem., 27 (1955) 212.
107 K. Cruse, Z. Anal. Chem., 181 (1961) 180.
108 S. Siggia, Quantitative Organic Analysis via Functional Groups, Wiley, New York, 3rd ed., 1963.

109 N. D. Cheronis and T. S. Ma, Organic Functional Group Analysis by Micro and Semimicro Methods, Interscience, New York, 1964.
110 F. E. Critchfield, Organic Functional Group Analysis (International Series of Monographs on Analytical Chemistry, Vol. 8), Pergamon Press, Oxford, 1963.
111 E. A. M. F. Dahmen, Chem. Weekbl., 67 (1971) A55.
112 W. J. Moore, Physical Chemistry, Longmans, London, 5th ed., 1972, p. 698, Table 15.4.
113 F. E. Critchfield, G. L. Funk and J. B. Johnson, Anal. Chem., 28 (1956) 76.
114 A. P. Terentev, M. M. Buzlanova and S. I. Obtemperomskaya, Zh. Anal. Khim., 14 (1959) 506.
115 R. Dijkstra and E. A. M. F. Dahmen, Anal. Chim. Acta, 31 (1964) 38; E. A. M. F. Dahmen, Chem. Weekbl., 57 (1961) 257.
116 B. Dobinson, W. Hofmann and B. P. Stark, The Determination of Epoxide Groups (Monographs in Organic Functional Group Analysis, Vol. 1), Pergamon Press, New York, 1969, p. 39.
117 J. S. Fritz and G. H. Schenk, Anal. Chem., 31 (1959) 1808.
118 D. J. Pietrzyk and J. Belisle, Anal. Chem., 38 (1966) 1508.
119 Ref. 110, p. 96.
120 J. T. Stock, W. C. Purdy and T. R. Williams, Anal. Chim. Acta, 20 (159) 73.
121 O. Tomíček and A. Heyrovský, Chem. Listy, 44 (1950) 245; Ref. 17. Ch. VIII.
122 N. F. Hall and J. B. Conant, J. Amer. Chem. Soc., 49 (1927) 3047.
123 J. Šponar, Collect. Czech. Chem. Commun., 16 (1951) 526.
124 H. B. van der Heijde, Chem. Weekbl., 51 (1955) 823.
125 Ref. 17, Ch. VIII; W. J. Mergens and G. W. Ewing, Anal. Chim. Acta, 74 (1975) 347.
126 G. A. Harlow, Anal. Chem., 34 (1962) 148.
127 I. M. Kolthoff, Anal. Chem., 51 (1979) 1R.
128 I. M. Kolthoff and M. K. Chantooni, Jr., J. Amer. Chem. Soc., 87 (1965) 4428; J. F. Coetzee and G. R. Padmanabhan, J. Phys. Chem., 66 (1962) 1708; V. A. Pleskov, Zh. Fiz. Khim., 22 (1948) 351.
128a A. Cisak and P. J. Elving, J. Electrochem. Soc., 110 (1963) 160.
129 H. Zeidler, Z. Anal. Chem., 146 (1955) 251.
130 M. Bos and E. A. M. F. Dahmen, Anal. Chim. Acta, 72 (1974) 169.
131 M. Bos, S. T. Ijpma and E. A. M. F. Dahmen, Anal. Chim. Acta, 83 (1976) 39.
132 M. Bos and W. Lengtọn, Anal. Chim. Acta, 76 (1973) 149.
133 M. Bréant and C. Buisson, Electroanal. Chem. Interfacial Electrochem., 24 (1970) 145.
134 M. Bréant, M. Bazouin, C. Buisson, M. Dupin and J. M. Rebattu, Bull. Soc. Chim. Fr., (1968) 5065.
135 M. Bréant, C. Buisson, M. Porteix, J. L. Sue and J. P. Terrat, Electroanal. Chem. Interfacial Electrochem., 24 (1970) 409.
136 I. M. Kolthoff and T. B. Reddy, Inorg. Chem., 1 (1962) 189.
137 R. A. Glenn , Anal. Chem., 25 (1953) 1916.
138 D. J. G. Ives and G. J. Janz (Editors), Reference Electrodes, Academic Press, New York, 1961.
139 R. D. Caton, Jr., J. Chem. Educ., 50 (1973) A571; 51 (1974) A7.
140 P. Zuman, Chem. Eng. News, 46 (March 18, 1968) 94; Substituent Effects in Organic Polarography, Plenum Press, New York, 1967; Topics in Organic Polarography, Plenum Press, New York, 1970.
141 F. L. English, Anal. Chem., 23 (1951) 344.
142 E. A. M. F. Dahmen, D. Vader and J. D. van der Laarse, Z. Anal. Chem., 186 (1962) 161.
143 EG & G Princeton Applied Research (Analytical Instrument Division), Application Note C-3, Operating Parameters for Optimization of the Model 310, 1980.
144 H. Lund, Acta Chem. Scand., 11 (1957) 990.
145 K. Hasebe and J. Osteryong, Anal. Chem., 47 (1975) 2412.
146 C. Sinicki and M. Bréant, Bull. Soc. Chim. Fr., (1967) 3080; see also refs. 133–135.
147 B. Hellrung, Helv. Chim. Acta, 64(1981) 1522.
148 H. D. McBride and D. G. Evans, Anal. Chem., 45 (1973) 446.
149 EG & G Princeton Applied Research (Applied Instruments Group), Application Note F-2, Applications of Voltammetry to the Food Industry, 1982.

150 C. O. Huber and J. M. Gilbert, Anal. Chem., 34 (1962) 247.
151 N. Shinozuka and S. Hayano, Talanta, 28 (1981) 319.
152 J. R. Tacussel and J. J. Fombon, Electrodes Tournantes à Disque et Anneau, Notice No. 6-EAD, Texte No. 702, Tacussel Electronique, Solea, Lyon, 1981.
153 V. B. Ehlers and J. W. Sease, Anal. Chem., 31 (1959), 16 and 22; see also K. Abresch and I. Claassen, Coulometric Analysis, Chapman & Hall, London, 1965.
154 E. P. Przybylowicz and L. B. Rogers, Anal. Chim. Acta, 18 (1958) 596.
155 R. E. van der Leest, Electroanal. Chem. Interfacial Electrochem., 43 (1973) 251.
156 H. L. Kies and H. Ligtenberg, Z. Anal. Chem., 287 (1977) 142.
157 W. B. Mather, Jr. and F. C. Anson, Anal. Chim. Acta, 21 (1959) 468.
158 M. Bos and E. A. M. F. Dahmen, Anal. Chim. Acta, 79 (1974) 169.
159 M. Bos and E. A. M. F. Dahmen, Anal. Chim. Acta, 72 (1974) 345.
160 C. A. Streuli, Anal. Chem., 28 (1956) 130; R. B. Hanselman and C. A. Streuli, Anal. Chem., 28 (1956) 916.
161 M. Bos, S. T. Ypma and E. A. M. F. Dahmen, Anal. Chim. Acta, 83 (1976) 39.
162 E. A. M. F. Dahmen and M. Bos, Proc. Anal. Div. Chem. Soc., (1977) 86.

C. ELECTROANALYSIS IN AUTOMATED CHEMICAL CONTROL

Chapter 5

Introduction

Chemical control provides the most effective means of monitoring hygiene through environmental, clinical, pharmacological and nutrition control, and of the manufacture of utilities through production processes control; at the same time, it stimulates optimization of all regulation processes concerned, be they physiological, physical or chemical. In treating the subject of this chapter, we can distinguish between (1) the nature of chemical control, (2) the character and degree of its automation and (3) the role of electroanalysis.

5.1. NATURE OF CHEMICAL CONTROL

The first question is what kind of information the chemical control can and should provide. Must it be a qualitative and/or quantitative analysis, is it based on a one- or two-dimensional measurement and should the latter consist of an analog and/or digital display? The meaning of all this can be well illustrated by the example of the differential titration (see Fig. 5.1) of equivalent amounts of a strong acid (cf., Fig. 2.17, AA) and a weak acid (cf., Fig. 2.18, BA, $pK_a = 4$).

In practice, the volume of titrant is plotted instead of the titration parameter, λ. At the titration end-points E_1 and E_2, the volumes of titrant consumed indicate the respective amounts of the acids, while their pH values or better the pH heights around the half-neutralization points, $h.n.p_1$ and $h.n.p_2$, are related to the identities of the acids. Therefore, in the two-dimensional figure the abscissa represents the quantitative aspects and the ordinate the qualitative aspects.

It must be realized that the acidity of an acidic solution, expressed by its pH, is a physico-chemical property, which in fact (see calculations on pp. 83–85) represents a resultant of the identity and concentration of the acid; even the overall pH height of the titration curve is still influenced by the concentrations of a strong acid, but for a weak acid that curve height, especially its h.n.pH value, forms a fairly reliable identity indication.

The second question concerns the quality of the chemical control, directed more at the chemical analysis proper and its procedure. Important factors here are sufficient specificity and accuracy together with a short analysis time. In connection with accuracy, we can possible consider the quantization of the analytical information obtainable. For instance, from the above example of titration, if we assume for the pH measurement an accuracy of ± 0.02, an uncertainty remains of 0.04 over a total range of 14.0, which means a gain in information of $n_1 = 14.0/0.04 = 350$ (at least 8 bits); with an accuracy of $\pm 5\%$ as a mean for the titration end-point establishment of both acids, the remaining uncertainty of 1% over a range of $2 \times 100\%$ means a gain in information of $n_2 = 200$ (at least 7 bits), so that the two-dimensional presentation of this titration represents a quantity of information $I = {}^2\log n_1 n_2 = 15$ bits at least.

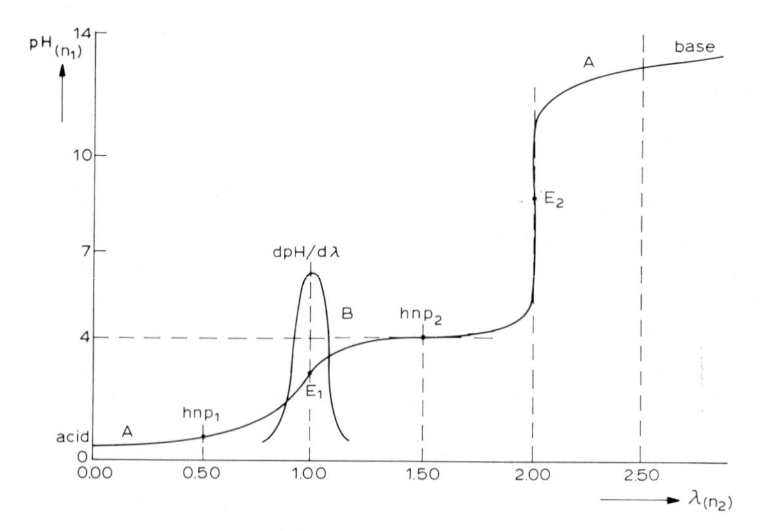

Fig. 5.1. Differential titration of a strong and a weak acid.

In connection with the analysis time, if merely the net duration (T_m) or really the measuring time of the titration is considered, we may talk about an information generation velocity or bit rate (bits s^{-1}) of $I_t = 15/T_m$ baud (Bd = bit s^{-1}).

However, this simplified reasoning does not meet the real situation of the analysis as a whole, where both the preparation of the sample and the renewal of the titration medium cause a certain and often considerable loss of time; therefore, the following set of more or less well defined symbols are used (see Fig. 5.2. a and b for a signal S of first-order response):

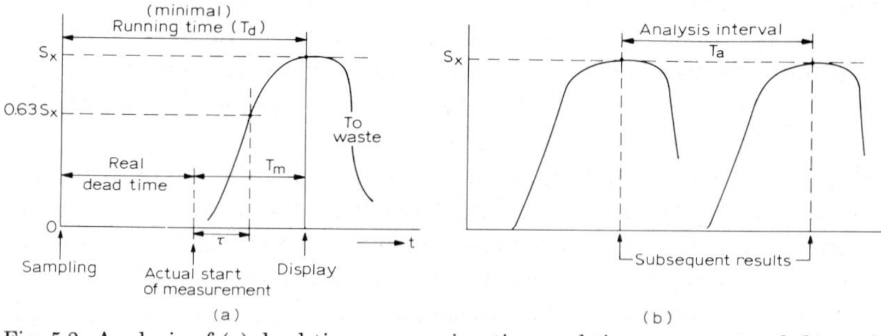

Fig. 5.2. Analysis of (a) dead time, measuring time and time constant and (b) analysis interval.

T_m = real measuring time (from starting at zero signal to the maximum);
T_d = running time, usually called the "dead time" instead of the real dead time;
τ = time constant of first-order response: $y = Sx(1 - e^{-t/\tau})$;
T_a = analysis interval, i.e., the time between successive displayed results;

$1/T_a$ = sampling frequency, i.e., analysis frequency;
σ_x = standard deviation = $\sqrt{\Sigma_{i=1}^{n}(x_i - \bar{x})^2/(n - 1)}$;
D = dynamic range = log(maximal meter deflection/sensitivity).

A two-dimensional picture like that in Fig. 5.1 represents the most common analytical presentation of results; however, multi-dimensional pictures can occur, as in mass spectrometry, when measurements are made for the same analyte but with varying ionization energy and/or magnetic field strength.

5.2. CHARACTER AND DEGREE OF AUTOMATION IN CHEMICAL CONTROL

For a number of practical reasons, including economy, ease of operation and technology, automated chemical control is desirable. In the laboratory the usual practice is often still discrete analysis, which means not being tied to a fixed analysis time, provided that the results are available before a certain date; of course, with regard to laboratory optimization one may convert the analytical apparatus concerned into a robot, which even at night can automatically carry out various selected types of determinations on request. However, manufacturing process control often requires strictly programmed test by means of automated product or stream analysis. Automation in this instance necessitates a balanced decision between discontinuous and continuous analysis, and for a better understanding we shall treat both possibilities. A more general and phenomenological classification of automatic chemical analysis together with their finishing methods has been given elsewhere[1].

Discontinuous analysis must in the first place and at a recorded moment provide for automatic sampling directly from the process product or stream; in the second place, and by direct connection, the analytical measurement must be carried out in a fully automatic manner.

In the systems theory, the analysis can be considered as a process itself, producing information and for that purpose consisting of a chain of four steps (system elements or subsystems)[2]:

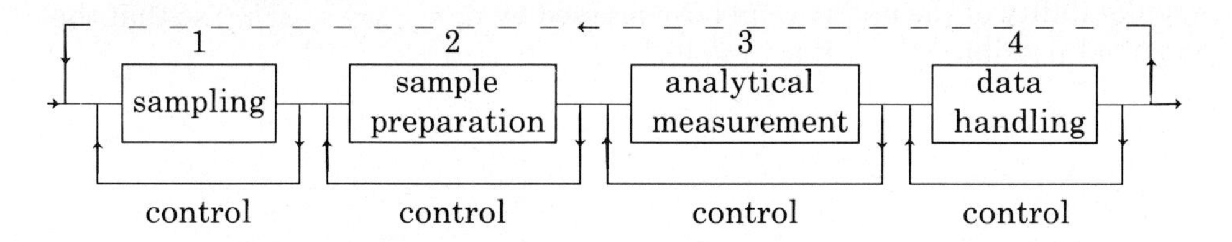

In view of the automation each subsystem needs its own feedback control, just as the entire system has an overall feedback control.

For those who do not need to be familiar with analytical measurement proper, subsystem 3 may be a "black box", but for the analytical chemist its functioning must be fully understood; to him the periodical introduction of a

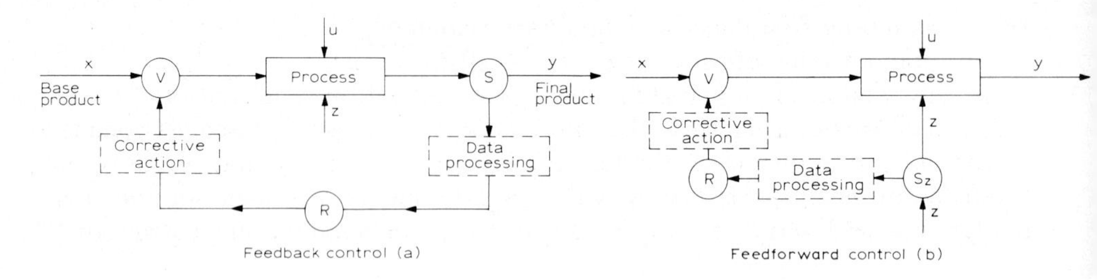

Fig. 5.3. Process regulation: (a) feedback control, (b) feedforward control.

calibration sample into the system offers a suitable means of keeping the entire system under control and of accounting for a possible drift in precision. Nevertheless, the resulting signal will be distributed to some extent by noise, which will affect the sensitivity of the measurement, i.e., the lowest detectable level of measurement, for which one in general uses 3 times the noise level. There are many possibilities, depending on the specific type of measurement, for improving the sensitivity.

As we intend to apply discontinuous automatic analysis as a sensor in process regulation, we must now first consider the manufacturing process as a "black box" also, and next discuss how the analytical sensor functions.

Fig. 5.3 a and b illustrates a process regulation scheme with either (a) feedback or (b) feedforward control, where S = sensor, V = regulation valve and R = regulator, e.g., one of the following types: P = proportional, PD = proportional and differential, PI = proportional and integrating and PID = proportional, integrating and differential; on–off regulator with time-proportioning and other more sophisticated (preferably computer) types; x = input (concentrations, amounts, identities) $(x_1, \ldots, x_i, \ldots, x_n)$ and y = output (reading) $(y_1, \ldots, y_i, \ldots, y_n)$, representing x–y relations; u and z are input variables, i.e. u are controllable and z non-controllable variables (disturbances), which lead to (stochastic) fluctuations in the output[3], as illustrated by Fig. 5.4.

Let σ_p be the standard deviation of the non-regulated process and σ_ε that of the regulated process; with positively acting regulation $\sigma_\varepsilon < \sigma_p$, whereas the regulatability of the process can be expressed by $r_p = \sqrt{\sigma_p^2 - \sigma_\varepsilon^2/\sigma_p^2}$, so that the maximal possible $r_p = 1$ (for $\sigma_\varepsilon = 0$).

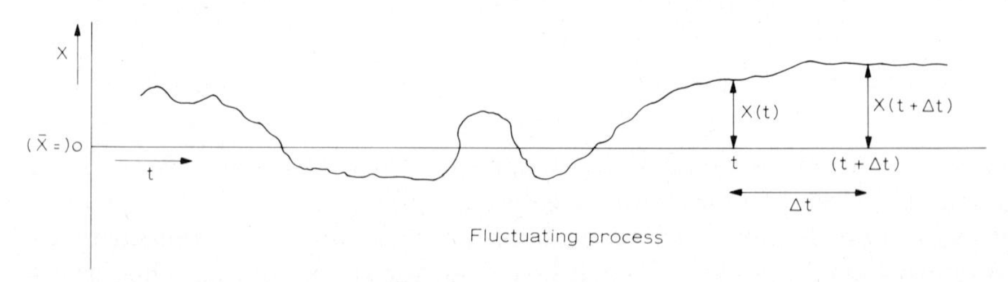

Fig. 5.4. Signal of a fluctuating process.

The relationship between x in Fig. 5.4 and x_i as a result of the analytical measurement at moment t is as follows: the variance $\sigma_x^2 = \Sigma_{i=1}^{n}(x_i - \bar{x})^2/(n - 1)$, so that if n is large as a consequence of a high sampling frequency, we may substitute n for $(n - 1)$; further, if we plot $\bar{x}$ as $\bar{x} = 0$, we obtain the curve in Fig. 5.4, where

$$\sigma_x^2 = \sum_{i=1}^{m} X^2/n = \bar{X}^2 \tag{5.1}$$

As expected for large n and for a non-fluctuating process, accordingly yielding a completely constant product, $\sigma_x^2 = \bar{X}^2$ must be zero; this remains also when the process fluctuates but shows no drift in the long term, as depicted in Fig. 5.4; however, within short lapses of time there are considerable asymmetric fluctuations, during which the final product may deviate unacceptably from the standard quality. Fortunately, such local asymmetric fluctuations can be traced by means of the so-called correlation function, because, as shown in Fig. 5.4, a correlation can occur between $X(t)$ and $X(t + \Delta t)$, if Δt is not large. Averaging the products $X(t) \cdot X(t + \Delta t)$ as a function of time, we obtain the covariance function

$$\Phi_{XX}(\Delta t) = \overline{X(t) \cdot X(t + \Delta t)} \tag{5.2}$$

For $\Delta t = 0$ the function is equal to the variance $\sigma_x^2 = \bar{X}^2$ (equation 5.1), but for $\Delta t \rightarrow \infty$ its value approaches zero because of the increasing probability of products of both positive and negative values, the summation of which becoming zero. Normalization of eqn. 5.2 by dividing both members by σ_x^2 yields the correlation function:

$$\frac{\Phi_{XX}(\Delta t)}{\sigma_x^2} = \frac{\overline{X(t) \cdot X(t + \Delta t)}}{\sigma_x^2} = e^{-\frac{\Delta t}{\tau_c}} \tag{5.3}$$

(see Fig. 5.5). The third member of the equation has been added as in many instances in practice we have to deal with an exponential curve, characterized by the correlation time constant τ_c. The equation has a value of unity for

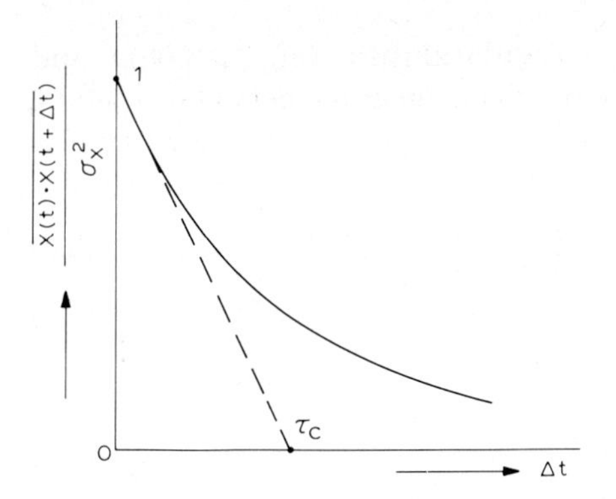

Fig. 5.5. Correlation function of a fluctuating process.

$\Phi_{XX}(\Delta t) = \sigma_x^2 = \bar{X}^2$, where there is a complete correlation between two x values; for $\Delta t = \infty$ there can no longer, be any correlation. The tangent at the curve in point ($x = 0, y = 1$) intersects the abscissa at $\Delta t = \tau_c$, which forms a useful measure of the frequency of the fluctuations.

The evaluation of the curve in Fig. 5.4 in order to obtain the correlation curve in Fig. 5.5 necessitates extensive calculations, especially for a strongly fluctuating process. As an example, we may choose a relatively short constant analysis interval, T_a, and plot the measurement results on a time basis for each space of 24 h; the curve for one date (as in Fig. 5.4) can then be evaluated for various Δt values, so that the τ_c concerned becomes known. For a fluctuating process, which is stable in the long term, τ_c can vary from day to day, but $\bar{x}$ remains constant; for an unstable process one finds a τ_{drift}, although much less than τ_c, with a real value also. Process dynamics studies are used to clarify (possible by process simulation also) how the process reacts with its input variables, and to indicate accordingly what kind of analytical measurements should be performed by the sensor; it will be simplest to have a key test, in which the process reacts most sensitively via a feedback. It is self-evident that for the above-described evaluation of process regulation, the calculation work can best be carried out using a computer, either on-line if necessary or with time sharing.

Let us now consider a fluctuating process with time constant $\tau_c = T_p$, which should be regulated by a sensor for an analytical key test with standard deviation $\sigma_x = \sigma_z$; the fluctuation value of the process, which can also be obtained from the fluctuation curve in Fig. 5.4, may be σ_p.

In 1965, Van der Grinten[4] introduced the interesting concept of measureableness a as qualification of an analytical method in relation to the process to be regulated, with its specific fluctuation velocity. In this connection he also defined

$$r = r_p m_{tot} \tag{5.4}$$

and

$$m_{tot} = m_d m_a m_n \tag{5.5}$$

where r = quality of regulation, r_p = regulatability (cf., p. 306), and m_{tot} = total measurableness, the following relationships between specific measurableness m_d, m_a and m_n and T_d and T_a, σ_z and T_p and σ_p, respectively, having formulated:

$$m_d = e^{-T_d/T_p} \tag{5.6}$$

$$m_a = e^{-\frac{1}{2}T_a/T_p} \tag{5.7}$$

and

$$m_n = 1 - \frac{\sigma_z}{\sigma_p}\sqrt{\frac{T_a}{T_p}} \tag{5.8}$$

where the subscript n represents the near-by expression of accuracy.

Considering $\frac{1}{2}T_a$ in eqn. 5.7, it must be realized that at a moment just before sampling the value of x is only known from a prediction over the analysis interval T_a directly after the value last measured; therefore, the average value of x during that interval must be obtained by a prediciton over $\frac{1}{2}T_a$. Further, it is evident that an analytical sensor can only satisfy the requirements of a positive process regulation, if its T_d, T_a and σ_z values are small in comparison with T_p and σ_p; the smaller they are, the more m_d, m_a and hence m_{tot} will increase from zero to unity as the extreme limit, analogous to what has been said already for a positive regulatability r_p; in the same way, $r = r_p m_{tot}$ (eqn. 5.4) as a definition of the quality of the regulation loop is the better, the more its value increases from zero to unity.

On the basis of the Van der Grinten approach, Dijkstra[5] compared several types of discontinuous analysis in connection with the regulation of a manufacturing process with $T_p = 75\,\mathrm{min}$ and $\sigma_p = 10\%$; he simplified the comparison by taking $T_a = T_d$, a common procedure when each new analysis starts at the end of the preceding one. Without going into detail, it can be said that a higher analysis frequency with lower accuracy prevails over the reverse.

A more complicated situation in process regulation occurs when among various chemical properties there is not one that can explicitly be indicated as a key test, so that more than one sensor (1, 2, ..., n) has to be used, in which case one refers to "multivariable systems in process control"[6].

$$r_1 = r_{p_1} m_{tot_1} \tag{5.4.1}$$

$$r_2 = r_{p_2} m_{tot2} \tag{5.4.2}$$

$$r_n = r_{p_n} m_{tot_n} \tag{5.4.n}$$

In the above, we concentrated on the analytical measurement proper; however, automation of the entire system necessarily implies the specific automation of all subsystems from sampling up to data handling. In the laboratory, the sample introduction is usually a manual operation and accordingly the analysis as a whole is better referred to as semi-automatic. In a plant however, we need a completely automated analysis, including sampling procedures in preceding subsystems and more severely controlled procedures in the remaining subsystems. In view of this, the rapid, efficient consultation of literature data on available variants of such procedures may be considerably advanced by the representation of analytical methods subdivided into a sequence of standard symbols for distinguishable procedures, as proposed by Malissa and coworkers[7]; these workers later added examples of analytical procedures in computer language to facilitate computer storage and consultation.

Continuous analysis offers another very useful possibility of completely automated chemical control, especially in manufacturing processes, but also in analytical processes such as separational flow techniques where the analytical measurement proper acts as a sensor, usually called the detector. As long as a physical or physico-chemical constant yields a sufficiently accurate and specific

References pp. 358–364

indication of the most important stream component, it represents a very simple and therefore attractive method of continuous chemical control*. However, when such a constant is not available or lacks selectivity, a suitable preceding chemical reaction must be used to convert the analyte of interest into a selectively detectable compound. Even if applied in a by-pass, in plant control such a procedure may have the disadvantage of high cost of chemicals consumed and/or the impossibility of recycling in the process; these inconveniences can be partly overcome in a process that is not strongly fluctuating by being satisfied with a discontinuous sampling hence a repetitive analysis of the continuous process stream under control. In analytical separational flow techniques, the analyte concentrations are so low that we can accept the waste of the conversion chemicals concerned; in chromatography, pre- or post-column derivatization to enhance detection is a common technique[8].

The transfer of an automated analysis from the laboratory to the plant will often require special precautions; for instance, while turbidities in a process stream can cause a loss of selective absorptivity in a spectrophotometric measurement, in potentiometric methods fouling of the electrodes, potential leakage in metal containers or tubing and loss of signal in remote control may occur (see later).

The character and the degree of automation in chemical control may have been covered in the above treatment of semi-automatic or completely automatic, and of discontinuous or continuous analysis, but something more should be said about the means by which automation proper has been performed in recent times. Whereas in the past automated analysis involved the use of merely, mechanical robots, to-day's automation is preferably based on computerization in a way which can best be explained with a few specific examples. Adjustment knobs have been increasingly replaced with push-buttons that activate an enclosed fully dedicated microcomputer or microprocessor in line with the measuring instrument; the term microcomputer is applicable if, apart from the microprocessor as the central processing unit (CPU), it contains additional, albeit limited, memory (e.g., 4K), control logics and input and output lines, by means of which it can act as satellite of a larger computer system (e.g., in laboratory computerization); if not enclosed, the microcomputer is called on-line.

When one uses a laboratory or general-purpose computer (e.g., a DEC PDP 11/10), and the measuring instrument as such has not been prepared for automation and computerization, then interfaces have to be planned in order to convert the instrument signals into signals acceptable to the computer and, conversely, computer signals acceptable to the instrument; here the computer is on-line again with the instrument, although not dedicated, which means batch processing for incidental connection and priority for continuous connection.

*For a continuously sampled process stream, it means that T_a in eqn. 5.7 is zero and so $m_a = 1$; eqns. 5.6 and 5.8 are still valid and reduce eqn. 5.5 to $m_{tot} = m_d m_n$.

The interfacing usually involves the conversion of either voltage or current from analog to digital form by means of an A/D converter (if necessary after preceding amplification of the instrument signal) and from digital to analog from by means of a D/A converter (in order to provide an analog signal back to the instrument); disturbances by noise with remote control can often be avoided by signal transport in binary form only, either parallel or in series.

The software is the most difficult element of the automated system to implement as the chemist concerned may not be sufficiently familiar with the programming technique or choice of computer language. This is not such an obstacle for routine laboratory practice where many automated analytical systems consisting of apparatus with a dedicated computer including software are available, such as in gas (GC) and liquid chromatography, NMR and mass spectrometry (MS), GC–MS and multi-channel blood-serum analysis. In plant and environmental analysis, however, tailor-made systems are often lacking, and in view of plant regulation or the ecological measures to be taken the data processing must yield a clear display of actual situations and trends; the entire system naturally becomes more complex, especially when there is more than one property under control. In these circumstances, it is a long process to develop not only the sequence of problem definition, systems analysis, planning of a flow diagram, equipment development and programming, but to demonstrate also that the system meets the requirements.

5.3. ROLE OF ELECTROANALYSIS IN AUTOMATED CHEMICAL CONTROL

Among the various advantages of electroanalysis in general, as mentioned in the general introduction (p. XV), we can stress again, considering automation, the direct* accessibility to electronic and hence automatic control even at a distance, simple automatic data treatment and simple insertation, if desired, into a process regulation loop.

The way in which automation of electroanalysis can be achieved depends very much on the specific requirements of the application. In order to illustrate this we have selected a number of typical examples. However, in doing so, we did not consider normal automation inherent to the nature of the analytical method, e.g., automatic scanning of the voltammetric curve in polarography and other voltammetric techniques, in addition to many additional refinements within these methods such as those treated already in Chapter 3; therefore, the selection of the examples in this chapter cannot be other than arbitrary**, where the borderline between the common and the uncommon in the future certainly will shift towards the former.

*An electric signal is directly available so that transducers are not needed.

**Literature data on automation of electroanalysis from 1972 to 1980 were kindly provided by Mr. A. A. Deetman of AKZO Zout Chemie, Hengelo, The Netherlands (see Preface, p. IX).

In order to arrive at a more or less systematic treatment of selected examples of automated electroanalysis, we have followed the sequence of systems involving the following:

(I) laboratory control, divided into (1) autonomic electroanalytical systems and (2) systems with electroanalytical detection; and

(II) plant and environment control, divided into (1) sampling and pre-cleaning systems and (2) systems with electroanalytical detection.

5.4. AUTOMATED ELECTROANALYSIS IN LABORATORY CONTROL

In the laboratory, electroanalysis is used for two main purposes, either for direct measurement of a physico-chemical property that is informative with respect to the identity and/or amount of the analyte, or for detecting the course of conversion of the analyte or indicating the separate appearance of analyte components, which is informative with respect to their identity and amount. In the former instance we are dealing with conductometry, voltammetry and coulometry and in the latter with various titrations and mostly separational flow techniques such as chromatography and flow injection analysis.

5.4.1. Automated autonomic electroanalytical systems in laboratory control

As typical examples of automated potentiometers we can mention the following microprocessor-controlled pH meters: Orion Models EA 920 and EA 940 pH/ion analyzers (Cambridge, U.S.A.); Metrohm Model 654 pH meter (Herisau, Switzerland); Knick Models 762, 763 and 764 pH meters (Berlin, F.R.G.); Philips (electrochemistry program)* PW 9422 pH meter (Eindhoven, The Netherlands, or Pye Unicam, Cambridge, U.K.); Radiometer PHM Autocal pH meter (Copenhagen, Denmark). All types are digital and display data inputs, measurement stages, possible failures and output results; as their working principles and operations are much the same, we have confined outselves to the following concise description of the PW 9422 (Figure 5.6)**:

Modes and ranges: pH (– 19.999 to 19.999), mV (– 1999.9 to 1999.9); manual compensation of temperature, – 10 to 200° C; automatic compensation of temperature, – 10 to 200° C, by means of a Pt 100 resistance thermometer; the resolutions are pH 0.001, mV 0.1 and temperature 0.1° C.

Display: main, five-character, seven-segment LED for results and program prompts eight bar lamps indicating pH, mV, ° C, ∇ shift key, CAL, slope, EA and

*In this series of instruments for analytical electrochemistry, Philips also supplies the microprocessor-controlled PW 9527 digital conductivity meter with 16 push-buttons and on the rear an analogue output for connection to a recorder and a 25-way connector providing a two-way RS 232 serial connection (see Philips leaflet 9498 362 9326).

**For the more specific purpose of ion measurement with ion-selective electrodes there is the Philips microprocessor-controlled PW 9415 ion-selective meter (see Philips leaflet 9498 365 92211).

Fig. 5.6. Philips PW 9422 microprocessor pH meter.

ISO, where CAL = calibration, EA = asymmetry potential and ISO = isopotential.

Input: international BNC sockets for pH, glass or combination electrodes, 4-mm sockets for RE electrode and Pt 100; input at 25° C, current < 0.1 pA, resistance $> 10^{13}\Omega$.

Keyboard: sixteen push-buttons for entry of data and instructions for manual or automatic calibration, based on the three National Bureau of Standards (NBS) buffers of pH 4, 7 and 9, whose values as a function of temperature have been permanently and separately stored to three decimal places and at intervals of 0.1° C. Hence there is no need to enter pH values, as the electrode automatically identifies the buffer in use and the apparatus immediately retrieves and displays the temperature-adjusted buffer value; the isopotential adjustment is defaulted to pH 7.000.

Output: a built-in RS 232 C digital output provides a connection, if desired, to a printer or to data logging and computer systems, so both single-shot and continuous printout of pH, mV and temperature are possible.

"Err" signal: errors are automatically shown on the LED display in the case of a defective pH electrode or faulty temperature sensor (Pt 100) and for incorrect buffer; two set points can be set over any part of the pH, mV or 0° C scale, which when exceeded starts an alarm and/or allows readjustment of dosing valves or pumps via an interface circuit.

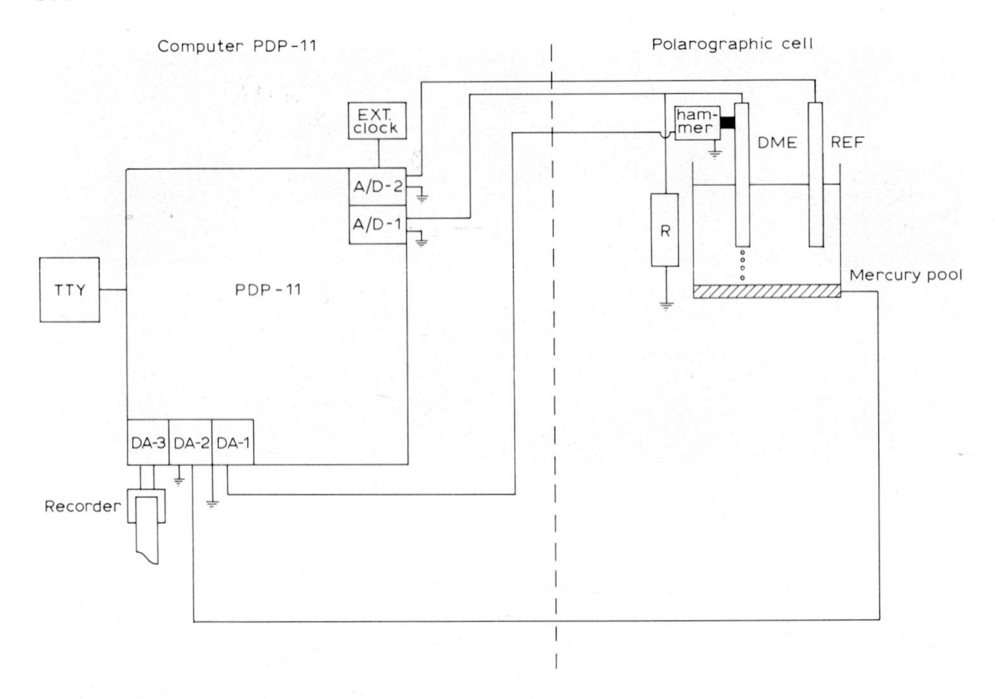

Fig. 5.7. Computer DC polarograph.

With their Models 763 and 764 microprocessor-controlled pH meter, Knick deliver and adapted Model 790 serial printer; this yields in a simple way by means of an AC-synchronous clock and two adjustment knobs a hard copy of measurements at intervals of 2, 5, 10, 30 or 60 s or minutes printed out in a 32.5-mm rule of 12 positions on a small paper roll.

The above-described automation of zero-current potentiometry is fairly simple and straightforward because it concerns a one-dimensional technique; however, automation of polarography/voltammetry becomes complex as a consequence of its two-dimensionality and the many additional refinements (cf., Table 3.4); treatment of the achievements of automation in these field suffers from the difficulty already mentioned, of the vague borderline between commercial and often sophisticated apparatus and what must next be considered as advanced automation. In this connection we have followed the approach of Donche[9], who considers automation as the situation where the method has been adapted to microprocessor or even computer control. For instance, among the earliest computerized systems in polarography/voltammetry* are those by Perone and co-workers[10] in 1968–1970, soon followed by the development of more specific applications such as in DC polarography[11,12], cyclic voltammetry[13–15], stripping voltammetry[16–20], normal- and differential-pulse polarography[21–24], AC polarography[25], Kalousek polarography[26] and staircase voltammetry[27,28]. In order to

*Dr H. Donche (see Preface, p. IX) kindly provided relevant literature and data.

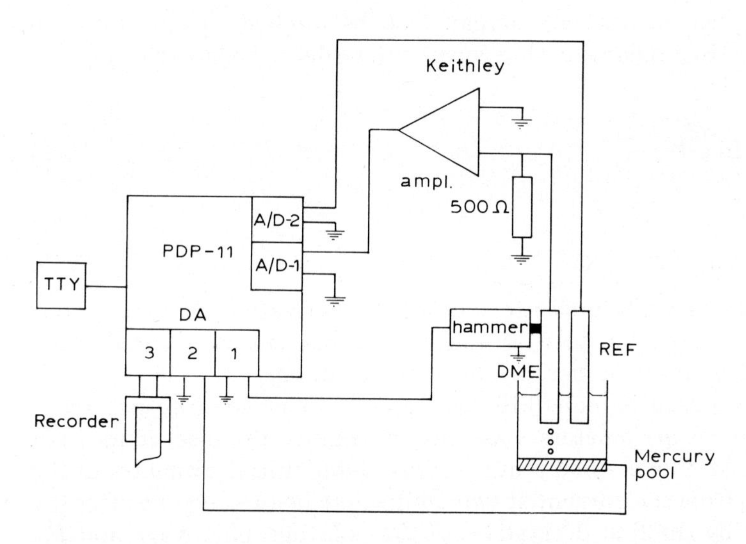

Fig. 5.8. Computer Kalousek polarograph.

provide some insight into the way in which on-line computerization of such specific techniques can be achieved, we describe the methods of Bos applied to DC[11] and Kalousek[26] polarography (Figs. 5.7 and 5.8, respectively) in a concise manner (for experimental details, see the original papers).

In both instances Bos employed a Digital Equipment Corporation PDP 11 computer on-line with a Radiometer polarographic stand with drop-life timer, a three-electrode system (DME, reference and Hg pool), two A/D and three D/A converters, together with a DEC writer and a recorder. For the "computer polarograph" he used a Radiometer PO_4 polarograph and an external clock, and the computer had to perform the following functions:

(1) maintain the potential of the DME at a set point vs. the reference electrode by changing the potential of the auxiliary electrode (Hg pool);

(2) change the set point of the DME linearly with time;

(3) measure and store the cell current with a relative accuracy of $\leqslant 0.1\%$ of its full-scale value;

(4) synchronize the measurement of the cell current with Hg drop life; and

(5) dislodge the Hg drop at fixed intervals.

Further, the operator must be able to choose the drop lifetime and the scan parameters, viz., the starting potential, direction (cathodic or anodic), rate and end potential, together with the sensitivity of the current measurement and the amplification in the ohmic cell resistance compensation circuit. Convenient additional facilities are (a) display of the polarogram on an oscilloscope, (b) delivery of hard copy of the polarograms on a chart recorder and (c) repeated recording of the polarographic curve for the same sample.

The software required was based on a flow diagram of the program for digital control and a flow chart of the program for processing the polarographic data.

In addition to its use in analysis, it can also be applied to elucidate the electrodic process. Most polarographic waves can be described by the equation (cf., eqn. 3.43)

$$E = E_{1/2} + \frac{RT}{\alpha nF} \ln \left(\frac{i_d - i}{i} \right) \tag{5.9}$$

rewritten as

$$i = i_d / \{\exp [(E - E_{1/2})/S] + 1\} \tag{5.10}$$

where $S = RT/\alpha nF$. As can be understood from their derivation, these equations can be applied only to the polarography of substances whose reduction products are soluble either in mercury or in the medium.

For curve fitting based on eqn. 5.10, the rigorous least-squares method of Wentworth[29] for three parameters was chosen, mainly because of its high computational speed. The method requires reasonable initial estimates of the parameters, viz., i_d from the current at two points, one before and one after the rising part (check by the first derivative) of the polarographic wave and $E_{1/2}$ from the zero crossing point in the second derivative of the wave. Good agreement was found for Cd(II) and K(I) with data from the conventional analog PO_4 instrument; the accuracy of the computerized system was $i_d \pm 2\%$, $E_{1/2} \pm 2$ mV, slope of logarithmic plot $\pm$ 2 mV.

With the Kalousek computer polarograph, a programmable real-time clock (with rates up to 1 MHz) and an operational amplifier were used, and the computer had to perform the following functions:

(1) control the experiment and measurement (cf., DC computer polarograph);

(2) reduce the experimental data to current vs. square-wave amplitude tables;

(3) average the current vs. square-wave amplitude curves over successive scans;

(4) display the polarograms on a strip-chart recorder or in digital form; and

(5) apply curve-fitting to the polarograms to obtain i_d, $E_{1/2}$ and the slope of the logarithmic plot.

In the real-time program initiated by the software clock, the number of pulses was restricted to the following set: 96, 48, 32, 24, 16, 12, 6, 4, 3 and 2 pulses per drop time; current and cell voltage measurements were made simultanuously, but only during a time slice of 40 ms every 120 ms because of the nature of the operating system. One should consult in the original paper[26] the various timing diagrams (cf., Fig. 3.49) and the flow chart for processing polarographic data; the latter is based on the general Koutecký equations (e.g., eqn. 3.42) modified by Ružić[30] with respect to the mean current vs. potential relationships for the various types of Kalousek measurements; also, eqn. 5.10 was used with $\alpha = 1$ and i measurement vs. the foot of the wave as the zero line. The accuracy for determinations of Cd(II), K(I) and Li(I) at concentrations down to 10^{-5} M was $\pm 5\%$ for pulse rates up to 25 Hz.

Bos[31] treated the subject of on-line computers in classical chemical analysis in general; near the end of this chapter (see Section 5.4.3) we shall return to this topic, and considering the above examples we confine ourselves here to the following remarks: considerable time can be saved in both techniques (DC[11] and Kalousek[26] polarography) by processing the experimental data by computer, and in the Kalousek technique, moreover, this is still possible with a standard polarograph even though not originally designed for this purpose.

Automation of coulometry, i.e., controlled potential coulometry, is uncommon; on pp. 217–218 we mentioned as the principle of a possible automation method the use of the exponential curve according to eqn. 3.108 in order to avoid the long-lasting procedure of total coulomb integration. In this respect Ehlers and Sease[32] described as early as 1959 the potentiostatic coulometry of nitro and halo compounds in methanol (99%) with graphic end-point prediction; so the next obvious step in shortening the analysis time even more and at the same time to achieve automatic coulometry was computerization. However, one can observe that the curve is irregular at the onset, where the current at the working electrode (WE) in the stirred analyte solution still has to be built up and to arrive at equilibrium; soon afterwards the current values are registered as a function of time and introduced into an on-line computer; next, for many combinations of at least three points at mutually short distances checks are made along a major part of the curve in order to obtain by means of data processing a smoothed exponential curve from which the total coulomb consumption can finally be predicted. Thus Jackson et al.[33] performed under remote control highly selective automatic controlled-potential coulometric determinations of plutonium in nuclear-fuel reprocessing materials.

Since 1978, the microprocessor has been increasingly used in electrochemical apparatus[34]. For instance, the above-mentioned microprocessor-controlled pH meters appeared on the market in about 1979 and the first application of a microprocessor in the differential pulse technique was even reported in 1977[35], in stripping voltammetry in 1978[36] and in combinations* of various techniques[37–41] from 1980 to 1982.

In the same period, more automated apparatus with microprocessor technology was developed as a replacement for classical apparatus in techniques such as square-wave voltammetry[42], chronoamperometry or LSV[43], pulse polarography[44], AC polarography[45], stripping voltammetry[46] and combinations of DC, AC and pulse polarography[47].

Commercial autonomic apparatus with microprocessor technology, but with more emphasis on routine analysis, also became available, viz., the Metrohm 646 VA processor and the PARC Model 384B polarographic analyser; the main features of these are worthy of brief consideration here.

The Metrohm 646 VA processor and 647 VA stand is based on a polarographic/voltammetric analyser with method memory and automatic curve evaluation, combined with a multi-mode stand (see Fig. 5.9). The following four determination techniques for polarography, voltammetry and stripping

*In this list the borderline between microprocessor and microcomputer control cannot be sharply drawn.

Fig. 5.9. Metrohm 646 VA processor and 647 VA stand.

voltammetry are incorporated: DP (differential pulse), DC_T (direct current test), NP (normal pulse) and SQ (square wave). The processor contains an alphanumeric keyboard, a video display unit and a recorder/printer, which allow* the following: data input; clear control/operation structure, i.e., continuous monitoring (by video) of the instrument's status (running values, signaling etc.); menu selection of stored determination processes and consultation of tables for all necessary analytical commands and allocations; complete hard-copy documentation of analysis (curves, results, sample data); complete parameter printout including date, hour and name of operator; and first evaluation (if desired) by display of unsmoothed measured values (as a curve and as a table). Using the method memory, several complete determination processes (including instructions and instrument/electrode settings) can be permanently stored as individual and customer-specified analytical instructions [when switching off, the memory contacts are not lost owing to a non-volatile memory, and are immediately available again when switching on; when operating the instrument for the first time one of several standard determination processes directly available (so-called firmware) can be selected and started at once]. Automatic curve evaluation enables: elimination of spikes, smoothing of the curve, recognition of peaks and steps, determination of peak

*According to the manufacturer's literature (P646 + 647 e/a, 82.05).

voltage and height (even with an unfavourable baseline and noise), and finally programmable conversion of determined results into desired units.

The stand contains a multi-mode electrode (MME) system for mercury electrodes, where DME/SMDE + HMDE are combined in one design (with operation control under inert gas overpressure and with an inexpensive glass capillary for one-way exchange, the whole being readily accessible and visually inspectable), while interchangeable electrode tips of various materials can be fitted to the threaded flange of the built-in stirrer, even by way of a rotating disk electrode (otherwise the built-in motor with a PTFE rod stirrer provides hydrodynamically optimized stirring); an analog electronic system is directly incorporated in the screening VA stand, which can occasionally detect extremely weak electrode signals and amplify them directly on location, thus yielding a far better signal-to-noise ratio and accordingly lower detection limits. The stand represents a compact measuring assembly with easy accessibility to electrodes, vessel (glass or PTFE), etc., under a double-hinged cover; there is a central inert gas connection and a plastic tray takes up any splashed liquids; the VA stand as a whole is enclosed in metal all round (Faraday cage) and is free from distortion and vibration. Several Dosimats (e.g. the microprocessor Dosimat 655; see under automatic titration) can be connected to the 647 VA stand for automated standard additions; moreover, one or two 647 VA stands can be connected to a 646 VA processor (alternative operation), for example, one stand with an MTFE and the other with a DME.

The PARC Model 384B polarographic analyser and Model 303A SMDE combination allows* completely automated voltammetry, including data processing, concentration calculations and output operations; although the instrument's set-up is autonomic, an RS 232C port facilitates two-way communication (transmission rates up to 9600 baud) with a host computer, and it can also be interfaced with a digital plotter (RE 0093) or printer and up to four model 303A SMDEs (the 384B-4 system); moreover, the CO 175 cable allows it to function with other electrochemical cells.

The Model 384B (see Fig. 5.10) offers nine voltammetric techniques: square-wave voltammetry, differential-pulse polarography (DPP), normal-pulse polarography (NPP), sampled DC polarography, square-wave stripping voltammetry, differential pulse stripping, DC stripping, linear sweep voltammetry (LSV) and cyclic staircase voltammetry.

The 384B console has a 40-character display and 74 touch keys on the front panel; a built-in floppy disk stores the operating software (nine complete analytical methods and nine voltammograms); the front-panel touch keys are functionally arranged and colour-coded as follows: grey, analyte identification (blank, standard or sample); brown, run control and keystroke programming (run and stop the analysis, hold, then continue, advance to next step in analysis, e.g., from purge to scan, play back and label voltammogram, define and

*According to the manufacturer's literature (EG & G Princeton Applied Research Corp., U.S.A.) (T 433; 38/11-M52-TP, 1983).

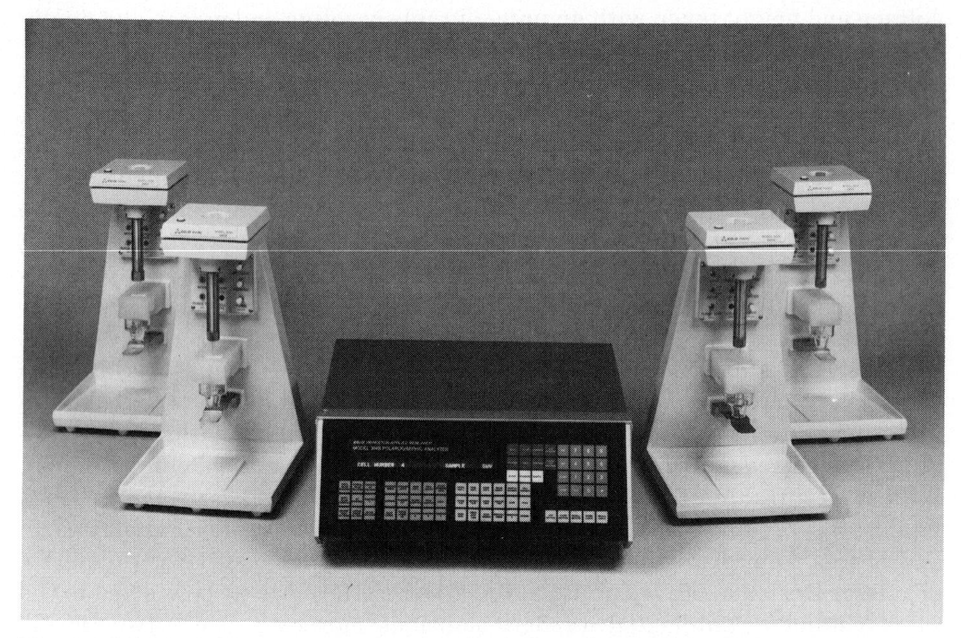

Fig. 5.10. PARC Model 384B polarographic analyzer and Model 303A static mercury drop electrode.

execute keystroke programs); blue, numerical keypad [in order to enter numerical information, yes and no keys request method or curve storage and recall (all overrides are yes/no functions)]; green, technique selection (one key each for the nine above-mentioned techniques); orange, data manipulation (choice of blank subtract, tangent fit, calibration graph or standard addition, least-squares or polynomial fit for calibration graph, curve derivative, etc.); yellow, experimental parameters (parameter keys are used with numerical keys to set initial and final potential, pulse height and scan rate, and purge, deposition and drop times); and beige, method and curve storage (keys for storage and recall of method and of curve; list method key generates complete method print out on the digital plotter).

The Model 303A static mercury drop electrode (SMDE) has received continuous design improvements (cf., Fig. 3.28 and associated explanation), e.g., the larger drop size afforded by the 303A design has yielded a further increase in sensitivity and the flip of a switch converts it into an extremely stable HMDE for stripping and/or square-wave voltammetry.*

The Model 303A together with the Model 303A/99 accessory offers a DME operational mode for traditional DC polarography or special techniques that require a thin-bore capillary approach. The Model 305 stirrer, an instant off/on

*The combination of stripping with square-wave voltammetry[47a] (see p. 154) is a very attractive procedure, especially for the PARC Model 303A HMDE.

stirrer that does not produce heat, utilizes an electronically generated rotating magnetic field to stir at a very reproducible rate. The optional Model 308 Multiplex Module, when connected directly to the rear panel of the 384B console, interfaces via a CO 199 cable as many as four Model 303A electrodes.

Some experimental parameters are as follows:

(a) potential ranges: $\pm$ 4.094 V, with resolution of 2 mV and maximum scan range of $\pm$ 2 V from initial potential;

(b) pulse height: 0.002 to 0.250 V, with resolution of 2 mV;

(c) purge, deposit, conditioning and equilibration time: 0 to 9999 s;

(d) frequency (applicable to square wave only): 1 to 120 Hz;

(e) drop/step time: 0.1 to 20 s for all modes except CV (cyclic voltammetry, 0.004 to 9.999 s, square wave not applicable);

(f) scan increment: 0, 2, 4, 6, 8, 10 mV;

(g) scan rate: calculated from scan increment and drop/step time, except in the square wave mode, where it is calculated from scan increment and frequency;

(h) replications: 1, 2 or 3; if more than 1 selected, results are averaged and there is automatic 30 s purge between replication.

Some hardware specifications are as follows:

(a) potentiostat compliance: $\pm$ 12 V at 2 mA;

(b) I/E converter: fully autoranging from 100 nA to 1000 μA (10 nA scale accomplished in software);

(c) AD converter: 13-bit, bi-directional and dual slope;

(d) memory system: static RAM, 22 kbytes; EPROM, 2 kbytes; disc storage, 89.6 kbytes.

PARC have also developed the Model 368 AC impedance systems, containing a research-quality, computer-controllable potentiostat/galvanostat (mainly for corrosion measurements), which can also be used for other types of electrochemical experimentation such as those envisaged by Donche[9]. For instance, the Model 368-1 system is based on the Model 173 potentiostat/galvanostat, which uses the Model 276 Interface to communicate with an Apple IIe computer; it incorporates a Model 5206 EC lock-in amplifier, also controlled via an IEEE 488 interface by the host computer. The even more accurate Model 368-2 system employs a Model 272/273 potentiostat/galvanostat and a 5301 EC programmable lock-in amplifier. Apparently the Model 276 Interface for computer-assisted electrochemistry represented one of the few commercially available approaches to Donche's research project[9].

Another development by PARC of purely analytical interest is the extension of the Model 384B to the Model 384B-7 containing a nebulizer with measuring cell and a sample charger (for a maximum of 76 samples[47b]). In the nebulizer, combined with an HMDE and the required RE (and/or AE), the sample is sprayed under nitrogen pressure into the 2.5-ml cell, so that it becomes quickly and completely free of oxygen, and then a stripping analysis with square-wave voltammetry can be carried out; between the successive measurements the cell is rinsed once or a few times, if desired, with sample solution. The system allows the determination of one or more elements in one sample per minute, e.g., 45 samples per hour with two elements such as Pb and Cd, or 35 samples per hour with four elements. Once a collection of samples has been introduced into the sample changer, the 384 B-7 works completely automatically without supervision and thus represents a cost-efficient technique in comparison with other laboratory methods such as AAS.

5.4.2. Systems with electroanalytical detection in laboratory control

As the main application areas of electroanalytical detection, which has become a subject of ever increasing importance, we shall now treat titrations and separational flow techniques.

5.4.2.1. *Automated titrations with electroanalytical detection*

Automated titrations can be divided into discontinuous and continuous, the former representing a discrete sample analysis, as a batch titration is the usual laboratory technique and the latter a flow technique, which is used less frequently in the laboratory, e.g., in kinetic studies, but is of greater importance in plant and environment control.

5.4.2.1.1 *Automated discontinuous titration*

The introduction of the piston burette gave a great impulse to the automation of titrations, especially after the motor-driven continuous delivery of titrant connected with recorder paper displacement was replaced by a step wise delivery by means of a stepping motor. The latter makes one quarter of a revolution per current pulse, thus permitting, in connection with a pulse generator, a digitized titrant delivery and recorder drive depending on the number of pulses admitted; one pulse generator, e.g., the Mettler Impulsomat 614, may serve several burettes.

In routine analysis, often a one-dimensional so-called end-point titration can be automatically carried out up to a pre-set pH or potential value and with a previously chosen overall titration velocity; in order to avoid overshoot, the inflection point should be sufficiently sharp and the titrant delivery must automatically diminish on the approach to that point in order to maintain equilibrium, and stop in time at the pre-set value. For instance, the Metrohm 526 end-point titrator changes both the dosing pulse length and its velocity by means of a pulse regulator in accordance with the course of the titration curve; in fact, the instrument follows the titration two-dimensionally, but finally reports only a one-dimensional result. The Radiometer ETS 822 end-point titration system offers similar possibilities. However, automated titrations mostly represent examples of a two-dimensional so-called eqilibrium titration, where the titration velocity is inversely proportional to the steepness of the potentiometric titration curve; hence the first derivative of the curve can usually also be recorded as a more accurate means of determining the inflection point. Examples of commercial apparatus are the Metrohm 536 Potentiograph, the Radiometer RTS 822 recording titration system and the Tacussel Titrimax.

For conductimetric incremental titrations, large rugged analogue conductivity meters have also become available, e.g., the Metrohm 518 conductometer (in connection with the Model 536 potentiograph its yields a rapid recording of the curve together with end-point indication) and the Philips PW 9505 analogue conductivity meter (in addition to a recorder output and an output for electrode re-platinization, there is a choice of manual or, by use of a Pt 100 resistance thermometer, automatic temperature compensation).

From the literature on more or less automated discontinuous titrations published since 1970, we can select some with the following types of detection: conductometric[48], potentiometric[49–53], amperometric (dead-stop)[54] and coulometric[55,56]. Ebel and Seuring[51] published an interesting review on automated potentiometric titrations, in which the aspects of end-point titration and titration systems with analogue registration and/or digital treatment were extensively discussed. Also for automated potentiometric titrations with ion-selective electrodes, we draw attention to the computer-controlled equivalence point determinations by Frazer et al.[50] on the basis of the Gran (antilog) plot (see pp. 92–94). Another example in which curve evaluation by means of an on-line computer has appeared to be essential is that by Bos and co-workers[57] already treated on p. 92; it concerns the situation where the potential jump between weak acids or bases is so faint that only by curve evaluation well before and beyond the jump can the equivalence point be determined.

Other frequent applications of automation are the Karl Fischer dead-stop titration of water (see pp. 204–205), e.g., in the Metrohm Automatic 633 KF titrator and the Radiometer ETS 850 KF titration system, and many coulometric titrations; the latter, representing constant-current coulometry, are of great importance owing to their accuracy and sensitivity, so that they are often applied as microcoulometric titrations[58]. The electrode processes, whether it is conversion of the analyte itself or generation of the titrant, represent an autonomic electroanalytical system, but for the detection any suitable sensor can be used; however, no transducer is required in electrochemical detection and especially then the method is directly adaptable to automation[56] and remote control. Cadwgan and Curran[55] published a coulometric titration of acids with digital timing circuits, digital readout of the electrolysis time with six-figure resolution and timing ranges of 1, 10, 100 and 1000 s, a logic-operated analogue switch for control and cell current and a comparator input circuit for end-point detection systems[59]; the latter is adaptable to any end-point detection providing a voltage signal. Commercial apparatus intended specifically for coulometry (both controlled-potential and constant-current) is the combination of the Metrohm E 524 Coulostat and E 525 integrator.

The above so-called automated titrations still require manual sample introduction directly into the measuring cell[60]; in order to avoid this in series analysis, a few manufacturers added automatic samplers, which in laboratory practice require only the previous introduction of samples into a series of cups or vessels placed on the sampler. Radiometer, for instance, call it "total automation" by means of their DTS 834 digital titration system, which combines automatic titration with automatic sampling (55 samples in sequence); the autopipetting operation includes sample vessel transport, pipetting of sample, pipetting of one or two extra reagents, titration and rinsing of the titration vessel. The computerized DTS 834 facilitates the automatic detection of one, two or more equivalence points, while sample number, titrant consumption and pH, pX or mV value of the equivalence points are printed out*. The heart of the system, i.e. without the sampler changer, is in fact the DTS 833 digital titration system, which is an extremely versatile apparatus in itself owing to the built-in microcomputer.

In 1976, Radiometer[61] presented for the first time a microprocessor-controlled titration system. Since then, the microprocessor has been used preferentially and as a fully integrated part (in line) in electroanalytical instruments as a replacement for the on-line microcomputer used before. Bos[62] gave a comprehensive description of the set-up and newer developments with microprocessors in relation to microcomputers and indicated what they can do in laboratory automation. Many manufacturers are now offering versatile microprocessor-controlled titrators such as the Mettler DL 40 and DL 40 RC MemoTitrators, the Metrohm E 636 Titroprocessor and the Radiometer MTS 800 multi-titration system. Since Mettler were the first to introduce microprocessor-controlled titrators with their Model DK 25, which could be extended to a fully automated series analysis via the ST 80/ST 801 sample transport and lift together with the CT 21/CT 211 identification system, we shall pay most attention to the new Mettler MemoTitrators, followed by additional remarks on the Metrohm and Radiometer apparatus.

Considering the related Mettler DL 40 and DL 40 RC MemoTitrators**, the DL 40 can be used for ten different volumetric and potentiometric methods: titration to a pre-selected absolute (EPA) or relative (EPR) end-point; equilibrium titration (EQU)–recording titration (REC)–incremental titration (INC); Karl Fischer water determination (KF)–controlled dispensing (DOSE); pH and pX measurements (pX/E); multi-level titrations and back-titrations with automatic calculation (CALC); and manual titration (MAN)–automatic calibration of electrodes (CAL).

The instrument has a splash-proof alphanumeric keyboard with 20 dual-function (yellow/green) touch keys and an additional green key for function choice (yellow or green); it allows an interchangeable memocard to be attached in front above the keyboard, up to three interchangeable burettes to be tightly installed on top of the titrator and with connected supply bottles behind , thus allowing also sequential titrations in one batch, and finally the titrations vessel with stirrer (0–3500 rpm), electrodes, titrant (and inert gas) inlet tubes

*Description as given by Radiometer (Copenhagen, Denmark) in leaflet 916-273, Interprint A/S, Copenhagen 1979/9B.

**Abbreviated description according to Mettler brochures 1.7367.72B and 1.7499.72A.

and a quickly closing gas-tight titration head, together to be attached at the right upper side of the titrator. Owing to the built-in microprocessor, every determination in the MemoTitrator is carried out completely automatically on the basis of a program that can be previously introduced on request via the keyboard or is recalled from storage in the high-capacity memory of the microcomputer; naturally, the (interactive) program will ask from the operator the required occasional input data such as sample weight and titrant normality, before it continues; manual operation remains possible, e.g., in a learning titration. The operating steps are shown on the 20-place alphanumeric display.

With no power supplied to the instrument, the stored contents remain available for more than 3 months; six standard methods and nineteen free and fully defined methods can be stored, in addition to two standard reagent entries and ten freely definable reagent entries. When the titrator is switched on, the most recently used method is automatically activated; all other stored routine methods can be recalled under their number of up to five digits. If connected to a Mettler balance, the sample weight is directly transferred; titration results are calculated in desired units ($g\,l^{-1}$, ppm, etc.); if connected to a Mettler GA 40 alphanumeric printer (twenty characters per line) the results are printed on a small paper band (in duplicate if desired); connection to a GA 14 or 15 flat-bed recorder is also possible. The MemoTitrator only accepts instructions that are meaningful for a given range of applications; all numerical entries are checked for plausibility, which ensures a high degree of reliability. Photometric endpoint indication by connecting a DK 18 or DK 19 Phototitrator also includes the large field of colour change titration. Another addition to the DL 40 titrator even permits remote control.

Although the DL 40 was capable of performing Karl Fischer water titrations and Mettler developed a separate microprocessor-controlled push-button operated DL 18 KF titrator, they also introduced as an all-purpose apparatus the improved DL 40 RC (see Fig. 5.11) with a dual titration head and with a modified software program to handle the new two-component titrants for Karl Fischer titration (see pp. 204–205). The instrument can also be expanded into an automatic series titrator by connecting the RT 40 sample transport for 16 samples and storage of 50 sample weights from a connected balance; this series routine can be interrupted at any time after completion of the titration in progress.

The Metrohm E 636 Titroprocessor (Fig. 5.12) offers in principle the same possibilities as the Mettler MemoTitrators, but with some additional features* concerning its operation, illustrated by the schematic diagram in Fig. 5.13, using the usual microprocessor terminology[62]: 1, CPU; 2, RAM; 3, ROM; 4, Bus; 5, I/O interface; 6, 16-place alphanumeric display; 7, numerical push-buttons; 8, function and control buttons; 9, card reader (for routine analysis); 10, alphanumeric thermoprinter for curves (normal and first derivative) and data on

*See colour picture of E 636 Titroprocessor in Metrohm brochure, and P. Gilgen and H. Kobler, Moderne Potentiometrische Titrier Praxis, Metrohm, Herisau, Switzerland.

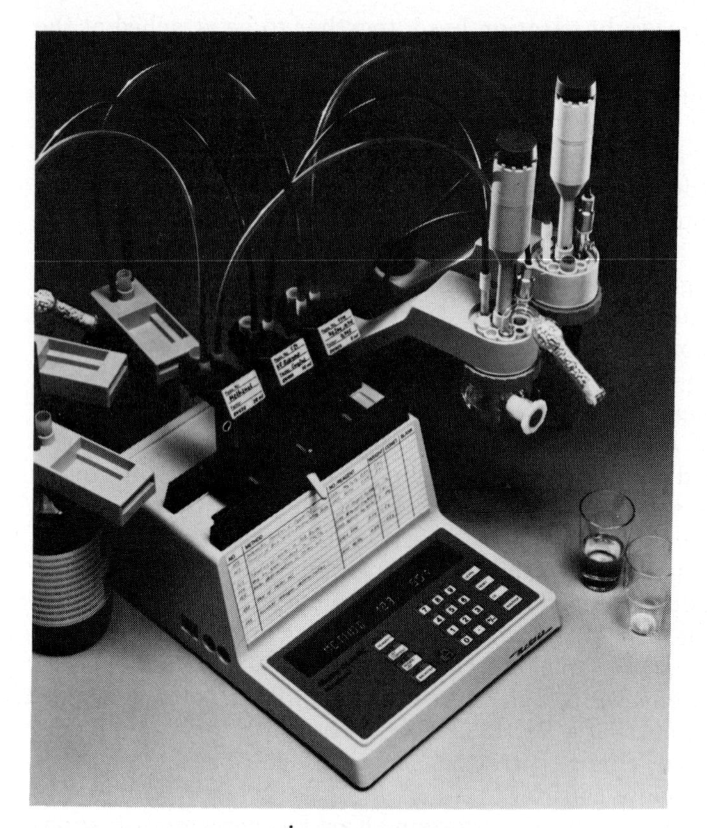

Fig. 5.11. Mettler DL 40 RC MemoTitrator.

folding paper (DIN A4); 11, Opamp (operational amplifier) for potential measurements; 12, IE; 13, RE; 14, Pt 100 Ω; 15, interface to burette; 16, microprocessor-controlled Dosimat E 635 burette; 17, E 552 exhange unit for titrant bottle with Teflon tubes, flat top for automatic change-over and EA 1118 burette top with microvalve; 18, low-voltage stirrer; 16, 17 and 18 together represent the specially designed E 638 titration stand. The attraction of the whole set-up may be summarized as follows: (1) the instrument in itself fully combines the conveniences of both analogue and digital titration, accomplished with independent numerical end-point establishment; and (2) the characteristic program for the instrument as a titrator has been permanently stored (ROM); the additional specific instructions about what should be carried out and by which parameters are introduced (RAM) either by button operation only or in combination with a card inserted into the optical reader. There are two types of cards: EA 1122/1 for routine determination and EA 1122/2 for calculation of titration results in the units desired; on these cards the required functional squares must be marked with a black pencil.

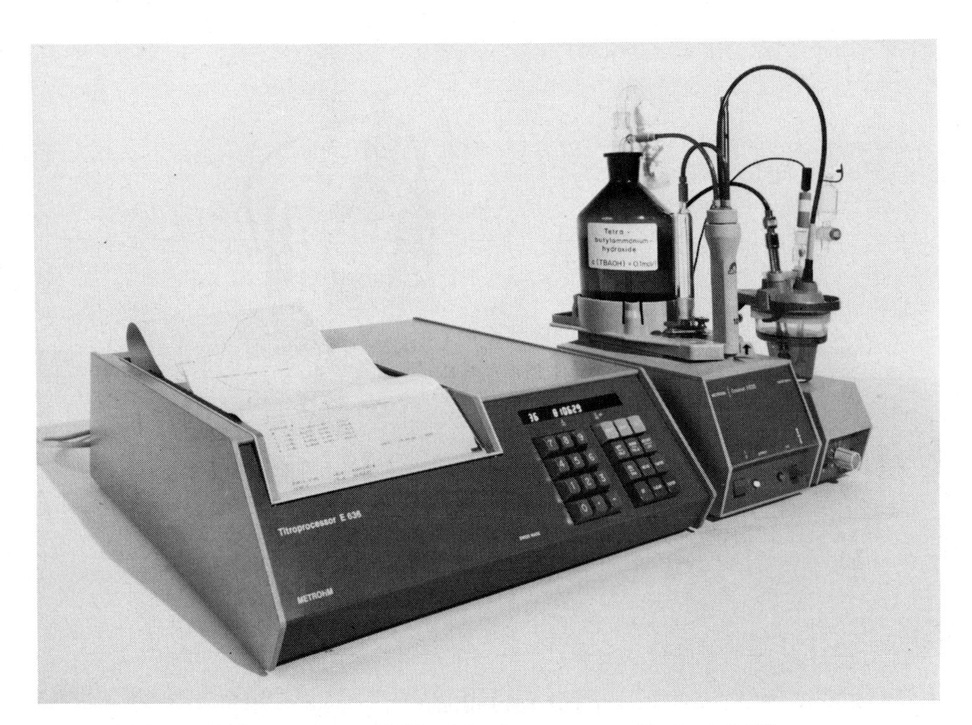

Fig. 5.12. Metrohm Titroprocessor E 636 with microprocessor Dosimat E 635.

Although the E 636 allows Karl Fischer water determinations, as any other titration, a separate microprocessor-controlled 658 KF processor has been developed, and there is also the microprocessor-controlled 652 KF coulometer (see pp. 221–222).

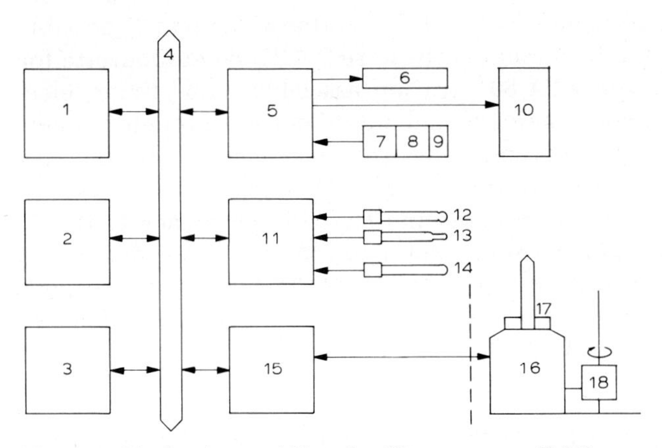

Fig. 5.13. Block scheme of Metrohm Titroprocessor E 636.

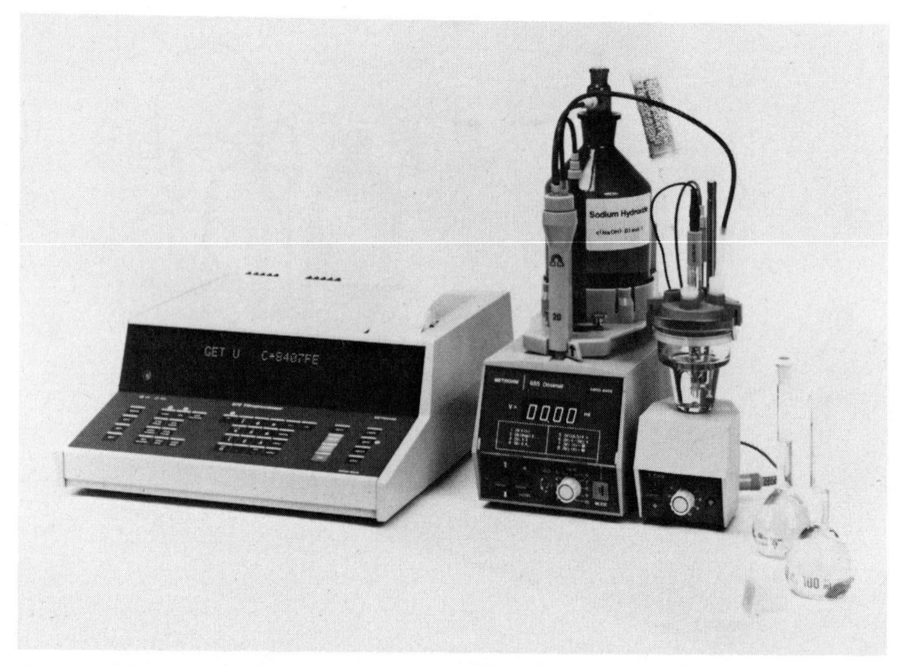

Fig. 5.14. Metrohm 672 Titroprocessor with 655 Dosimat.

Metrohm have also designed the 672 Titroprocessor with the 655 Dosimat (Fig. 5.14 and p. 319)* as a rapid routine titrator with three functions: SET as an end-point titration, GET as a titration to find one or more equivalence points and KFT for Karl Fischer or other titrations with polarized electrodes; it can be expanded into an automatic series titrator by connecting the 624 Sample Changer for ten samples.

The Radiometer MTS 800 multi-titration system is a modular system** consisting of the following instrumentation: TTT 35 titrator with up to 31 possible method programs, which can be permanently stored, ABU 80 autoburette for accurate titrant delivery, and TTA 80 titration assembly with stirrer, electrodes and titration vessel; both analogue and digital outputs provide connection to computers, printers or recorders. With this instrument again one has the same possibilities as with the Mettler and Metrohm apparatus, viz., three titration modes (end-point, inflection point and Karl Fischer); each of the 31 methods is executed at the push of a single key; operator prompting occurs via a 20-character alphanumeric display; a patented algorithm for dynamic reagent addition facilitates rapid and accurate results; and pre-dose control minimises the titration time for batch analysis.

---

*A further expanded and more recent type is the 682 Titroprocessor with the 665 Multi-Dosimat.
**Extract from Radiometer (Analytical Division), 916-389 Copenhagen, Kelly/Group 83/8.

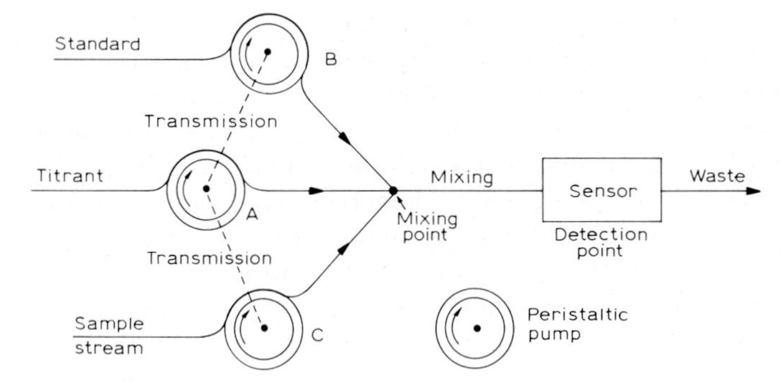

Fig. 5.15. Continuous injection titration.

By connecting the PRS 12 Alpha Printer, complete documentation of titration results, stored methods or other user input is available. The SAC 80 sample changer can be connected for fully automatic titrations or pH measurements on 20 samples, while the sample weights can be keyed into the TTT 85 titrator.

5.4.2.1.2. *Automated continuous titration*

This wording may be considered as duplication, because one can hardly think of continuous titration without automation; however, the intention is simply to stress its character as an alternative to automated discontinuous titrations. The principle of continuous titration can be illustrated best by Fig. 5.15[1]; it applies to a steady stream of sample (C). Now, let us assume at first that the analyte concentration is on specification, i.e., it agrees with the analyte concentration of the standard (B). If, when one mixes the titrant (A) with the sample stream (C), the mass flow (equiv./s) of titrant precisely matches the mass flow of analyte, then the resulting mixture is on set-point. However, when the analyte concentration fluctuates, the fluctuations are registered by the sensor; it is clear that the continuous measurement by mixing A and C is only occasionally interrupted by alternatively mixing A and B in order to check the titrant for its constancy.

In the above-described measurement, which we call the absolute method, all pumps have equal speeds (rpm) owing to interconnection to the same drive-shaft. In order to express, if required, a deviation registered for the analyte concentration, one must calibrate with a standard by varying its rpm (B) with respect to that of the titrant (A): a B/A rpm ratio greater than unity means a proportionally lower concentration and vice versa. In general, the absolute method serves to control a sample stream with nearly constant analyte concentration; as a sensor one uses not only electroanalytical but often also optical detectors. However, with considerably varying analyte concentrations the differential method is more attractive; its principle is that in the set-up in Fig. 5.15 and with the sensor adjusted to a fixed and most sensitive set-point, the rpm of the sample stream (C) is varied with respect to that of the titrant (A) by a feedback control (see Fig. 5.3a) from the sensor via a regulator towards the peristaltic pump (C); therefore, the analyte concentration equals the A/C rpm ratio times the standard concentration, which also means that this ratio must be registered. A problem may be that the mass flows in the regulation of such an analytical process are so small that the mutual pump variation becomes insufficiently accurate. However, in some situations, as in acid–base titrations, this difficulty can be overcome by feedback coulometric generation of either base or acid (oxidant or reductant, respectively) within the entering sample stream (C); in other words, only the deviations from a standard are micro-coulometrically determined, and all pumps have the same rpm by means of one common drive-shaft. In such a regulation loop, potentiometric detection is the most attractive method owing to its selectivity and direct voltage signal; the point that hydrodynamically we have to deal with flow-through detection will be considered in Section 5.4.2.2.

In 1982 Fehér et al.[63] reviewed flow-through titrimetric techniques published since 1970. Up to then one had been satisfied simply by end-point titrations, but subsequently the aim became to obtain a complete titration curve; this could be done by varying according to a certain program one of the following three parameters: concentration of the titrant, flow-rate of the titrant or flow-rate of the sample.

A linear gradient of the titrant concentration, representing a gradient titration technique described first by Eichler[64] and investigated more closely by Fleet and Ho[65], can yield a potentiometric titration curve; however, the evaluation of its starting point suffers from uncertainties, which can be avoided by the triangle potentiometric titration developed by Pungor and co-workers[66]. Its principle is that one changes the titrant concentration according to an isosceles triangle, so that the sample is titrated twice, i.e., forwards (1) and backwards (2); the analytical information is obtained from the time interval (Q) of overtitration between the end-points 2 and 1 vs. the total time duration (2τ) of the program, viz.,

$$Q = t_{E2} - t_{E1} = 2\tau - 2\frac{a}{bn} \cdot C_s v \quad \text{or} \quad C_s = \frac{b}{2a} \cdot \frac{n}{v}(2\tau - Q) \tag{5.11}$$

where

C_s = concentration of sample ($\text{mol}\,\text{l}^{-1}$);
b/a = stoichiometric factor of the titration reaction [between (a) sample and (b) titrant];
n = slope of titrant mass flow vs. time ($\text{mol}\,\text{s}^{-2}$);
v = sample flow-rate ($\text{l}\,\text{s}^{-1}$);
2τ = total duration of titrant addition program.

In this automatic system, the authors preferably used coulometric generation of titrant (cf., microcoulometric determination of deviations in the above end-point titration[1]), e.g., H^+, OH^-, Ag^+, Hg^{2+}, Br_2, I_2, $Fe(CN)_6^{3-}$ (cf., Table 1 in ref. 63). The detection method may be potentiometric (logarithmic signal), amperometric (linear signal), biamperometric, conductometric, oscillometric, etc. Moreover, the authors evaluated triangle programmed titration curves by

Fourier transformation[67] and also automated such potentiometric titrations with an on-line computer (a Hungarian EMG 666 desk-top computer)[68]. A linear variation of the titrant flow-rate can be considered as an alternative of the above varying concentration method, as both concern a linear variation of titrant mass flow. The linear variation of the sample flow rate simply means the opposite and has been proposed for feedback control in the differential method of end-point titration in Fig. 5.15. Ashworth et al.[69] applied an increase in sample flow-rate for discrete sample titration via injection into a continuous stream of water or another solvent; in the so-called controlled dynamic titrator the sample, being sucked within a short time (e.g., 90 s) from a sample changer, is injected into the solvent with a piston burette, by means of which the sample solution is transferred under a linearly increasing flow-rate to the mixing cell, where it meets a constant flow (w) of titrant (with titre c), passes through a delay tube in order to reach equilibrium and is next detected by the sensor. The concentration ratio of sample vs. titrant should be such that during the mixing programme the equivalence point is passed (e.g., at time t'). When at time t' the sample flow-rate is v', we may write $cw = xv' = xkt'$, so the sample concentration x can be calculated. By connecting the sensor with a printer the titration curve is automatically and completely registered, no matter whether the detection has been conductometric, potentiometric, amperometric, voltammetric (see polarovoltry, p. 206) or photometric.

Another interesting development, in which continuous flow was combined with discrete sample titration, is continuous flow titration by means of flow injection analysis (FIA) according to Růžička and co-workers[70]. Fig. 5.16 shows a schematic diagram of flow injection titration, where P is a peristaltic pump, S the sample injected into the carrier stream of diluent (flow-rate f_A), G a gradient chamber of volume V, R the coil into which the titrant is pumped (flow-rate f_B), D the detector and W waste.

The detection signal as a function of time yields the dispersion pattern of a peak, which in a long, thin gradient can be described by a Gaussian curve; however, in practical use with a gradient device of shorter holding time (tube length up 25 cm and inner diameter up to 1.5 mm) the peak becomes skewed, which can be described by the statistical function of a tank-in-series model. Here mathematical derivation has shown that the relationship becomes simplified in the special case when $f_A = f_B$; this approach, resulting in eqn. 5.12, can be used experimentally either in a two-channel instrument (f_A and f_B

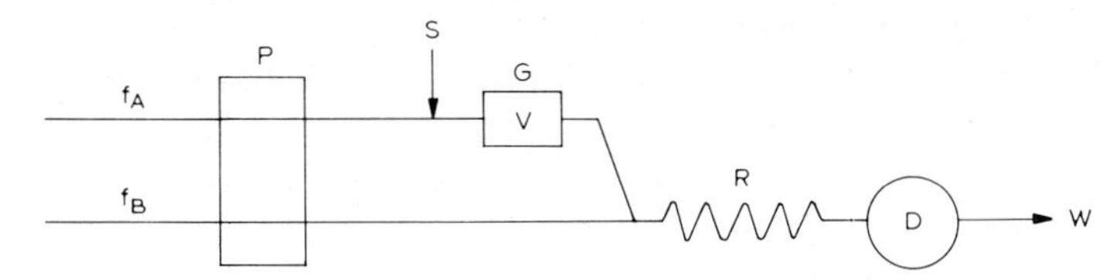

Fig. 5.16. Flow injection titration.

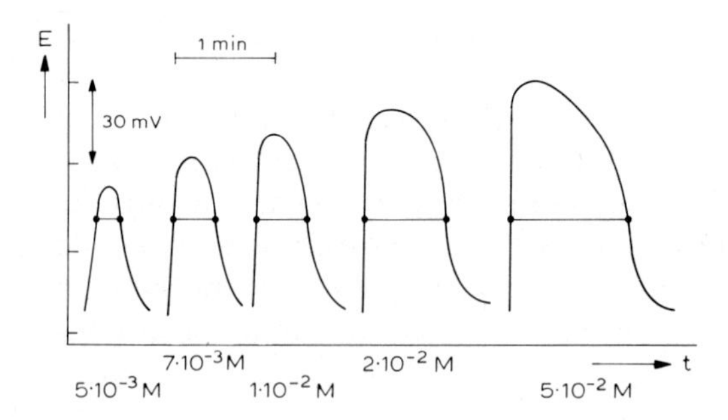

Fig. 5.17. Flow injection potentiometric titration of Ca^{2+}.

identical) or in a one-channel instrument where the titrant stream simultaneously serves as the diluent.

$$t_{eq} = (V/f) 2.3 \log (C_A^0/C_B) n \qquad (5.12)$$

Fig. 5.17 shows the curves for the potentiometric titration of Ca^{2+} in the range $5 \cdot 10^{-3}$–$5 \cdot 10^{-2}$ M with a titrant carrier stream of $5 \cdot 10^{-4}$ M EDTA using a calcium ion-selective electrode; each titration is initiated by an abrupt increase in the potential, followed by an S-shaped decrease in which the inflection point marks the end of titration. According to eqn. 5.12, where the titration product is AB_n, the mixing volume V, the original concentration of A in the sample C_A^0 and the titrant concentration C_B, t_{eq} can be calculated. In the experiments in Fig. 5.17 the sample volume was 200 μl and $f = 0.84$ ml min^{-1}; by choosing a fixed potential, t_{eq} was read off between points ●——● of each curve and plotted against the concentration of the injected sample, yielding a straight line in agreement with eqn. 5.12. The examples in Fig. 5.17 show that the one-channel system allows the automatic microtitration to be performed in less than 1 min. Spectrophotometric titrations can also be carried out, but the indicator colour change at the equivalence point does not cover such a wide range of ionic activities as does the electrode measurement.

Another rapid and simple flow-through technique is the so-called single-point titration, based on mixing, for instance, a titrand base solution to a mixture of different weak acids of such a composition that there is a linear relationship between pH over a large range and the amount of base admixed. The method was originally introduced by Leithe[71], who called it "Einpunkt-Titration"; it was worked out theoretically by Johansson and Backén[72] and developed to continuous flow-through titration on the basis of direct pH measurement in a potentiometric sensor by Åström[73].

Recently, Van der Schoot* and Bergveld[74] developed an ISFET-based microlitre titrator by combination with a microcoulometric pH actuator. As a result of integrated circuit technology they could integrate ten pH-sensitive ISFETs

*Dr B. H. van der Schoot (see Preface, p. IX) kindly made available to the author a full description of the experiments.

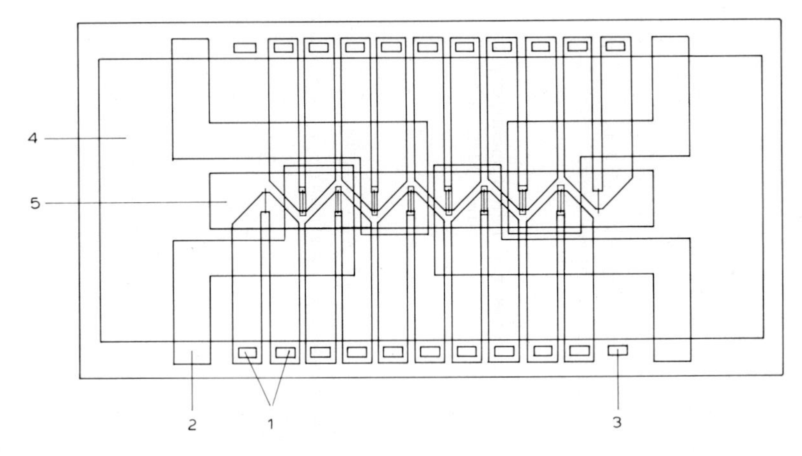

Fig. 5.18. ISFET-based microlitre titrator (× 5).

with four noble-metal electrodes (actuators), evaporated on to the same substrate (1 × 2 cm p-Si), into a chemical pH sensor–actuator system [Fig. 5.18: (1) source and drain connection of first ISFET; (2) one of the four gold electrodes surrounding two ISFET gates; (3) bulk connection; (4) polyimide rim on which cover is placed; and (5) sample channel (for ISFET construction and functioning see pp. 80–82)]. Such miniaturization, offering the advantages of small sample size and less reagent use and leading to rapid analysis, has also been pursued by others, e.g., Růžička and Hansen[75] for FIA (in a small transparent PVC block) and Terry et al.[76] at Standford University for gas chromatography (design on a silicon wafer). As an example, Van der Schoot titrated a 0.002 *M* solution of acetic acid in the presence of quinhydrone (cf., pp. 00–00) as a redox couple in order to avoid evolution of hydrogen or oxygen gas bubbles. The base titrant was coulometrically generated at the gold cathodes vs. a suitable counter electrode. The extremely short response time[77] of the ISFETs has been shown to be essential for the realization of high-speed titration; moreover, the extremely small size of the ISFET sensing surface allows a local detection of the pH actuation.

Three modes of titration operation were carried out:

(1) Static measurements (stationary solution). After a coulometric pulse of specific magnitude, the resulting pH step is measured. Repeating the experiment with different pulses allows the construction of the titration curve.

(2) Dynamic measurements (stationary solution). The pH change during one continuous pulse (up to 150 μC) is registered. Owing to the small volume involved, diffusion times have only a limited effect and the recording gives a fair approximation of the titration curve.

(3) Continuous measurements (flowing solution). All generation electrodes acting as cathodes produce OH^- ions and the solution is now titrated as it passes through the analyser. The ISFETs along the channel (1.5 mm wide and 12 mm long; total volume of sample 0.54–1.8 μl depending on the coverage of Si

TABLE 5.1

CHARACTERISTIC CONCOMITANT DEVELOPMENTS IN NON-SEPARATIONAL FLOW TECHNIQUES

Method principles	Measurements aids	Additional field effects
Continuous flow analysis (CFA) (Skeggs), since 1960 Segmented flow	Improved ISEs Tubular electrodes (Blaedel) Adapted ISFETs	Special sampling requirements in plant and environment control (Sections 5.5 and 5.6)
Flow injection analysis (FIA) (Růžička and Hansen), since 1975 In continuous flow, stopped flow or with merging zones (FIA scanning or intermittent pumping)	Adapted voltammetric electrodes Membranes* for: Partial dialysis Membrane amperometry (Clark) Differential techniques (Donnan) Computerization, including microprocessors	Special measuring requirements in plant control (to avoid voltage leakage, etc., Section 5.5)

* The methods concerned can still be considered as non-separational, because the membranes are used for phase separation but not for a mutual separation of the individual analyte conponents.

with the SiO_2 layer, spaced at 30–100 μm from the analyser chip) register the consecutive points of the titration curve. The titration depends on the flow-rate of the solution, while calculations showed that at all flow-rates only about 25% of the solution takes part in the reaction. The authors concluded that the small dimensions allow chemical analysis within a few seconds; they consider that this application of integrated circuit technology may form the basis for a new group of chemical transducers (for details see their publications).

5.4.2.2. *Flow techniques with electroanalytical detection*

Flow techniques have become of considerable importance, not only in routine titrations but also in other analytical methods; as automated analytical processes they all need to be under the control of a detector, often called a sensor and sometimes a biosensor. We can divide the techniques into the following:

(1) non-separational flow techniques, which in themselves are not separational, but where the sensor is selective for the analyte (e.g., with directly working ion-selective electrodes or the usual electrodes) or becomes so via an separational function (e.g., with indirectly working ion-selective electrodes such as enzyme or gas-sensing electrodes);

(2) separational flow techniques, where the sensor does not need selectivity so much, but above all should have a high quantitative sensitivity to the type of analyte components being separated.

5.4.2.2.1. *Non-separational flow techniques*

In fact continuous titration belongs to this class, but has already been treated above on the basis of the use of the sensor merely as an end-point indicator of the titration reaction. For the remaining non-separational flow techniques, such a multiplicity of concomitant developments has occured since 1960 that in a survey we must confine ourselves to a more or less personal view based substantially on the information obtained from some important reviews and more specific papers presented at a few recent conferences[78–82], or from leaflets offered by commercial instrument manufacturers. The developments are summarized in Table 5.1.

Continuous flow analysis (*CFA*). In this technique, introduced by Skeggs[82a] in 1957 and developed by Technicon (U.S.A.) into their well-known Auto-Analyzer, air segmentation (see Fig. 5.19) has been always an essential element in the proper functioning of the measurement within the detector; the sample is automatically sucked in a certain period of time out of a vessel on the sample changer and usually added alternately with wash solution to the reagent line. The combined stream is then segmented with air bubbles, passes the mixing coil, whose windings have a fairly horizontal axis, and finally after being debubbled is measured in the detector at a later moment at a constant maximum signal.

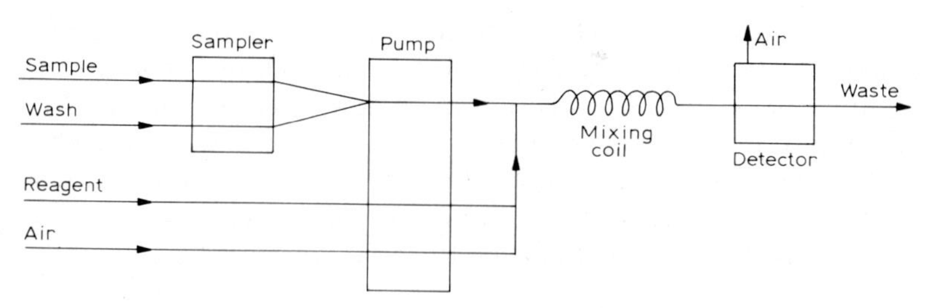

Fig. 5.19 Continuous flow analysis with air segmentation (Skeggs).

The air segmentation, together with with position of the coil windings, reduces the risk of laminar flow and thus achieves homogeneous mixing within each fluid segment; it also hampers intermixing of adjacent segments. The optimal procedure is the one where the sensor detects the signal during the complete sample passage in order not only to check the proper functioning of the flow system, i.e., whether the signal curve shows a flat portion, but also to ensure that precisely the maximum steady-state signal is registered for analytical evaluation.

Although absorptiometric detection via preceding colour reactions is still mainly used, CFA has shown increasing versatility not only in detection choice but also in sample pre-treatment, such as (partial) dialysis, distillation, evaporation to dryness, digestion, long-term incubation and coated-tube separation; a review was published by Snyder[83].

However, the greatest application of CFA is in Technicon AutoAnalyzers of the so-called SMA (sequential multiple analysis) type, especially for clinical analysis in hospitals. For instance, in the SMA 12/60 (which can be extended to the SMA 18/60 by interfacing with an SMA 6/60), a blood serum sample is divided over twelve CFA channels each with its own pre-treatment procedure, colour reaction and detection. Details of the mainly biochemical reactions with optical detection are not of interest in the context of this chapter, but some remarks on the character of the automation and information involved are worthwhile. Each sample receives three adhesive labels, to be affixed to (1) the sample tube immediately after the blood is drawn, (2) the requisition slip and (3) a special sample cup that is inserted into the sampler; the last, larger, label utilizes a specially designed bar code; thanks to the Technicon (Tarrytown, NY, U.S.A.) IDee system of identification, the bar code is read optically at the moment that the sample is taken, the coded information is stored and the sample number is transferred by a printing device to the Serum Chemistry Graph and to the Technilogger II teletype output device. During the analysis the detection signals are written by a one-pen recorder on the graph in a fixed sequence of the twelve channels; each channel has its own scale with the usual appropriate units, while a certain range of it has been hatched; in this way the patient's serum condition can be directly monitored by the physician. The most conspicuous SMA features are the indissoluble connection between the sample, patient code and Serum Chemistry Graph, the high frequency and speed of analysis (60 samples per hour, including six blanks with about a 10-min dead time, thus yielding 432 twelve-test profiles per 8 h), a monitoring system with visual display of the continuously measured signal for each channel and easily exchangeable analytical cartridges for each type of determination. As a consequence, the SMA 12/60 found rapid acceptance, especially in larger hospitals. For smaller clinical laboratories the automation of twelve determinations on each serum was often considered inefficient. To a certain extent this problem has been solved by the use of automatic discrete sample analysers such as the Parallel analytical system and the Perspective Analyzer, both from American Monitor Corp. (Indianapolis, IN, U.S.A.) and both using a bar code, the Beckman Astra Systems Multichannel Analyzers: Astra 4, 8 and Ideal with corresponding capacities (Beckman, Brea, CA, U.S.A., Bulletin 6112 Int., 1983), and the DuPont Automatic Clinical Analyzer (ACA) (DuPont, Instrument Products Division, Wilmington, DE, U.S.A.); the latter analyser uses a binary code; its unique working may follow from the simple operating instructions: (1) fill in identification card and attach to sample; (2) fill cup with sample, cover and place in input tray; (3) place desired test pack(s) in tray following sample; (4) push operate button; (5) obtain results of first test in about 7 min (the disposable plastic test pack functions as a flexible cuvette in which both chemical reaction and photometric analysis take place).

In these analysers, by means of a code affixed to the sample tube, a programmable choice can be made for only those determinations which have been actually requested. This possibility was realized by Technicon via the smaller second-generation AutoAnalyzers such as the SMA II and the third-generation instruments, the SMAC (SMA plus computer) and the more expanded SMAC II. These generations of analysers then acquired for determinations of Na, K, Cl, F, NH_3, etc., options for ion-selective electrode tests including pH and conductivity. Further, it has been shown that in the Technicon AutoAnalyzer system other electrochemical, e.g., polarographic[84–86], methods of detection can also be applied, but then inert gas segmentation must be used to prevent interference by oxygen; even a pulsed voltammetric sensor[87] has been used in CFA according to the Skeggs principle[82a].

Some investigators[88] tried to replace the air segmentation by solvent segmentation, which is valuable especially for reactions with residence times up to 23 min, e.g., in the dansylation of pharmaceutically important primary and secondary amines. Plugs of an organic solvent such as dichloroethane minimize band broadening in the CFA and act as both a reagent carrier and a product carrier, so that continuous flow dansylation provides a sensitive fluorimetric detection method for AutoAnalyzer systems and high-performance liquid chromatography (with post-column derivatization).

However, around 1983 Technicon introducted their "random access fluid" technology, with retention nevertheless of the air segmentation, the whole being realized in the Technicon RA-1000 system.

We confine ourselves here to a brief description of the capabilities and working principles of the instrument, referring the reader for details to the

Technicon literature*. In the RA-1000 a single probe aspirates and dispenses all samples, while another handles all reagents without the need for time-consuming and error-prone wash cycles. It provides random access, allowing multiple tests, organ panels, single tests and STATs at 240 sets per hour (and electrolytes at 720 test per hour) at any time, in any combination, with results available in as little as 2 min; push-button operation with microprocessor control, and virtually no start-up of shut-down time required; bench-top convenience with no water source or drain required; total flexibility, allowing changes of resident chemistries in less than 1 min, and choice of methodology for zero-order, first-order and end-point reactions with or without blanks (available at 30 to 37° C) and with pre-calibration for zero-order or single-point calibration for first-order and end-point reactions; microsampling and economical reagent consumption, i.e., 2–30 μl of sample (1–30 samples on a circular sample tray), 350–375 μl of a single reagent and 30–50 μl of a second reagent (1–12 reagents and 2 blanks on a circular reagent tray), all reagents conveniently packaged to minimize waste; computer compatibility with any laboratory host computer via standard serial output (RS 232C); results for each sample reported (on a paper band) with time, date and tray-cup number, results being flagged if outside the normal limits.

The working principle of the instrument is as follows. The Technicon Random Access Fluid "encapsulates" the sample and reagent separately and after being air-segmented the inert fluid carries them individually, whilst protecting the sample and reagent from contact with the external and internal surfaces of their respective probes, to the reaction tray cuvette where they meet for reaction; as the fluid drops to the bottom (with ridges cast in the base), the air escapes and finally the absorbance is measured horizontally via one of the six narrow band-pass filters (in an eight-position filter wheel) against an air reference path by means of a beam splitter. The disposable reagent tray is prepared off-line; the disposable reaction tray consists of 100 rectangular cuvettes, the reagent being dispensed first and then the sample, these being mixed by the motion of the tray itself while being heated by an air bath at 30 or 37° C.

Fig. 5.20 shows the instrument with the addition of the ion-selective electrode (ISE) module, which performs Na, K and CO_2 analyses (240 samples or 720 tests per hour) as requested and without interruption of the other analyses. The sample for electrolyte determination is also placed on the sample tray and a third probe aspirates diluted sample from the reaction tray for processing by the ISE module. Finally, the Technicon SRA-2000 system is a computer-controlled network of the subsystems SMAC II and RA-1000.

As a development of experience with the RA-1000 instrument, Technicon recently introduced their most advanced Chem 1 capsule chemistry system. It consists of a one-channel CFA with a capacity of 1800 tests per hour and provided with eight sequential optical detectors along its pathway, which

*Technicon RA-1000, Technicon SRA 2000 and Technicon Random Access Fluid, leaflet 6603-R/3-84 9.C.

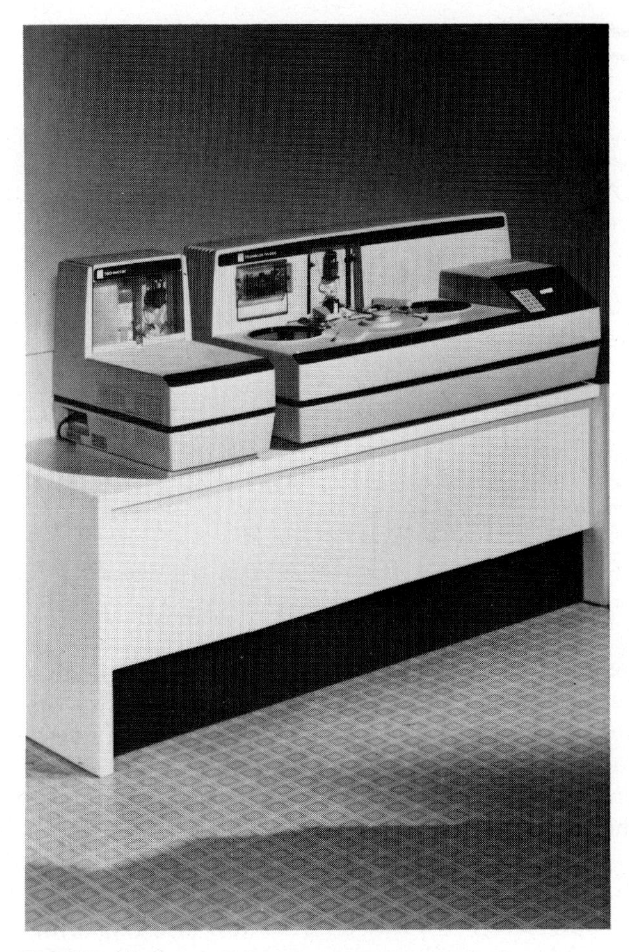

Fig. 5.20. Technicon RA-1000 system with ISE module.

means 225 samples per hour or 4 samples per minute ($T_a = 0.25\,\text{min}$) and a residence time (T_d) of 15 min; there is a random access to 32 optical tests and 3 additional ISE tests (for K, Na and CO_2). An instantaneous view inside and along the single optical channel of 1.5-mm bore Teflon tubing, internally coated with Technicon Random Access Fluid, would show from the single probe to the final peristaltic pump the following sequence:

$$\overbrace{\text{air, }\underbrace{\text{R}_2\text{, air, R}_1 + \text{S}}_{\text{test 3}}\text{, air,}}^{\substack{\text{pre-vanish}\\ \text{test capsule}}}\ \overbrace{\text{buf., air, }\underbrace{\text{R}_1 + \text{S} + \text{R}_2}_{\text{test 2}}\text{ air, buf.,}}^{\substack{\text{vanish zone}\\ \text{(increase in bore size)}}}\ \overbrace{\text{air, }\underbrace{\text{R}_1 + \text{S} + \text{R}_2}_{\substack{\text{test 1}\\ \text{(detectors 1–8)}}}}^{\substack{\text{post-vanish}\\ \text{test capsule}}}$$

$\xrightarrow{\text{flow}}$

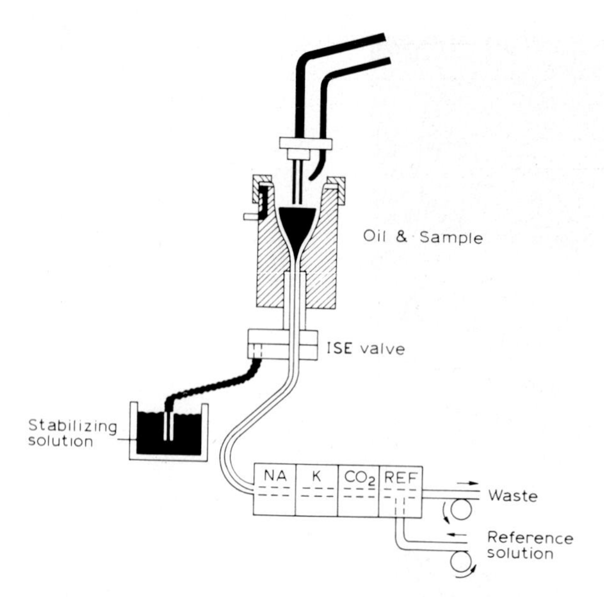

Fig. 5.21. ISE channel of Technicon Chem 1 capsule chemistry system.

where buf. is buffer, R_1 reagent 1, S sample and R_2 reagent 2; for more details the reader should consult the Technicon* literature. The system maintains the test records in the memory and can be integrated with a host computer; Chem 1 prints out complete laboratory reports in the format requested, either in real time or organized at the end of a run. In the context of electroanalytical detection, Fig. 5.21 shows the ISE channel.

Flow injection analysis (*FIA*). In this technique, introduced by Růžička and Hansen, a small amount of sample is injected into a liquid flow (see Fig. 5.16), which apart from being automated is normally continuous, but can include the use of stopped-flow, merging zones extraction techniques in addition to FIA scanning and methods based on intermittent pumping[89]. The principles of FIA and the versions just mentioned will now be briefly discussed on the basis of the excellent review of Růžička and Hansen[89] in order to understand the applicational possibilities of electrochemical detection in this technique.

Fig. 5.22, where R is reagent, P pump, S sample, D detector and W waste, shows the usual and simplest set-up of the method. The sample is injected into the reagent as the carrier by sluicing-in via the rotary valve and its bypass without causing any pulsation of the continuous stream. The laminar flow through the 0.5-mm-bore Teflon tube will yield on detection a peak whose height, width and shape depend on the sample volume, tube length and pumping rate. The square-wave concentration profile, originally present within the valve, at a constant pumping rate is converted into a peak more or less

*Technicon Chem 1 system, 1985; The AutoAnalyst, Technicon Special Cardiff Edition, May 1985.

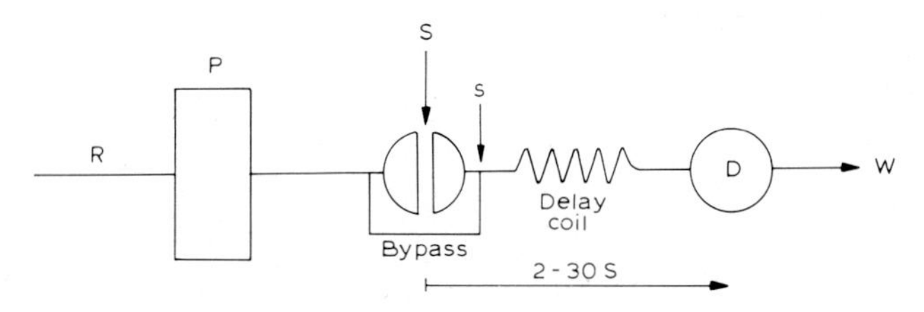

Fig. 5.22. FIA according to Růžička and Hansen.

asymmetric as a consequence of the carrier flow, i.e., the peak is compressed (steep) at the start of the scan and extended (sloping) at the end (cf., p. 331). High sampling frequency is, of course, preferred, but there is a maximum to the pumping rate in order to prevent either (with a long residence time) peak spreading, which means an expressive length and intermixing with the next oncoming sample zone, or (with a short residence time) an excessively small peak, which means incomplete chemical reaction. However, a high sampling frequency is possible for a system with limited dispersion, i.e., if only the original composition of the undiluted sample is to be assayed; thus pH measurements by injecting 30 μl of sample into 500 μl of a carrier stream of $2 \cdot 10^4$ M phosphate buffer in 0.14 M NaCl (pH 6.64) in 10 cm of a 0.5-mm-bore tubing could take place at a rate of 240 samples per hour with a reproducibility of 0.01 pH unit (see Fig. 5.23)[90] by means of a flow-through capillary glass electrode vs. a calomel reference electrode. In the same way, pH and pCa could be determined in a serum at a rate of 100 samples per hour when a calcium electrode in a specially designed flow-through cell is added[91].

For the required limited dispersion in the above pH and pCa determinations, the analytical read-out had to be made within a residence time of only 5 s; however, in the gradient flow injection titration illustrated in Fig. 5.16 and 5.17, and explained on pp. 331–332, the residence time increased from about 0.2 to 6 min (Fig. 5.17) or more.

From FIA theory[92] with regard to pumping rate and peak spreading (dispersion), as noted earlier, one should, with a slow reaction, decrease the flow-rate rather than increase the length of the reaction (delay) coil. An attractive means of limiting dispersion is stopped-flow FIA, whose principle can be described as follows: when the sample is injected, an electronic timer is activated by a microswitch positioned on the injection valve; the time from injection to stopping the pumping (delay time) and the length of the stop time can both be pre-set on the timer. Further, a variant of this technique, but with increased sampling frequency, is operation with intermittent pumping; here a pump I accomplishes the stopped flow, just described, and after the peak maximum has been recorded a pump II is activated to wash the sample out of the system from coil to waste at a higher pumping rate; after a pre-set time, pump I is restarted and pump II stopped, thus permitting a new sample injection. Continuous pumping may be wasteful, especially with an expensive reagent or an enzyme as a catalyst. The merging zone principle avoids this by injecting the sample

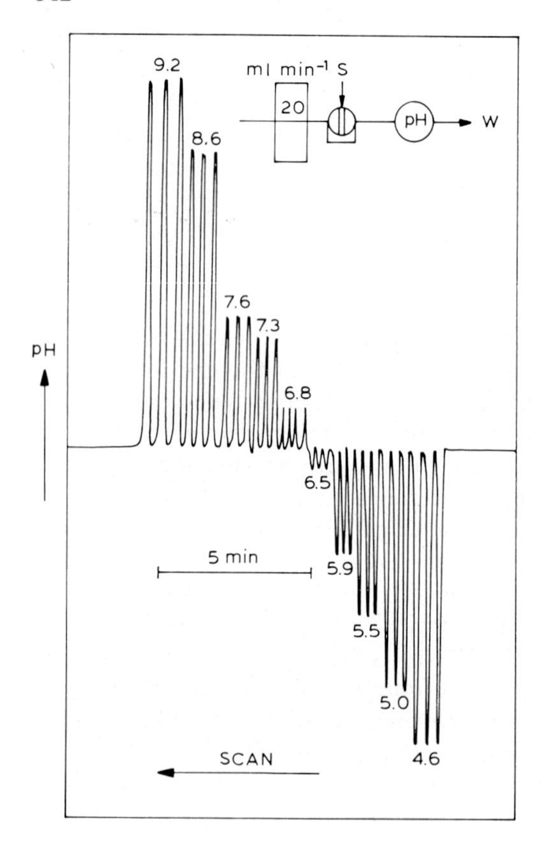

Fig. 5.23. Flow injection pH measurements.

and introducing the reagent solutions in such a way that the sample zone meets the selected section of the reagent stream in a controlled manner[89], while the remainder of the system is filled with wash solution or pure water. This can be achieved by the intermittent pumping just mentioned, or by multiple injection; here, by means of a multi-injection valve[93,94], sample and reagent zones are injected into two separate carrier streams with the same speed, which then meet. The carrier streams are either continuous or intermittent (stopped flow). Enzyme consumption under the latter conditions became considerably lower in glucose measurements[90].

Also of great interest is the so-called FIA scanning as a method for investigating, for instance, the influence of pH on the solvent extraction of metal dithizonates[95] by controlled-potential continuous alteration of the pH of the carrier stream. Many other investigations can thus be made, such as the catalytic activity of enzymes and the influence of pH on ion-selective electrodes.

More recent developments in FIA have been reported by Van der Linden. For instance, with regard to manipulation of dispersion[96] use has been made of packed bed reactors[97], but their disadvantage is a high pressure drop across the reactor, which requires high-pressure pumps and very tight connections. However, this difficulty can be overcome with the single bed string reactor (SBSR)[98], a tube filled with impervious glass beads having a diameter of 60–80% of the inner diameter of the tube; it shows a dispersion about ten times smaller than for open tubes of the same dimensions and yields a peak height almost independent of flow-rate (in the range of 0.2–1.5 ml min^{-1}). A reactor length up to 1.5 m accommodating 2–3 samples at a time (residence time up to 1 min) allows operation with a normal peristaltic pump; instantaneous mixing in the first part of the reactor can be accomplished by supply of sample and reagent(s) in parallel streams which then meet in a simple Y-piece.

Electrochemical detection in FIA has not been confined to potentiometry at ion-selective electrodes; amperometry, polarography and voltammetry in general, stripping analysis and coulometry have been used. For comprehensive literature data up to 1980 and a detailed discussion, the reader should consult the review paper by Růžička and Hansen[89] (especially Table 1 on "Instrumental analytical methods and techniques used with FIA" and Table 2 on "Flow injection references listed according to area of application") and that by Tóth et al.[99] on the application of electroanalytical detectors in CFA. It has appeared that in general optical detectors were and still are preferred in CFA and FIA, especially in clinical analysis; however, interest in electrochemical detection is growing. In this connection, a few investigations published since 1980 and reported mainly in symposia may be mentioned. For instance, Morf and Simon[100] reported on potentiometric flow-through detectors and their clinical applications, paying attention to the choice of reference electrolyte (liquid-junction potential), influence of streaming potentials, response time and electrode life. Mottola et al.[101] discussed potentiometric and amperometric detection in flow injection enzymatic determinations involving monitoring of protons and dissolved oxygen levels. Tóth et al.[102] developed an automated polarographic and photometric system for serial analysis, which amalgamates batch and flow operation so that both retain their advantages and overcome the drawbacks, and they tested the system on pharmaceutical preparations. Dieker and Van der Linden[103] determined Fe(II) and Fe(III) by FIA and amperometric detection with a glassy carbon electrode at levels down to 10^{-6} *M*. Kågevall et al.[104] determined water in organic solvents by FIA with Karl Fischer reagent, coulometrically generated and potentiometrically detected, at a rate of 120 samples per hour in the range of 0.01–5% (v/v) with a relative standard deviation of less than 0.5% (v/v).

It is understood that FIA particularly requires a small detector that is fairly sensitive and has a very short response time; according to Ramsing et al.[105], who were the first to demonstrate the applicability of IFSETs in flow analysis systems, these sensors fulfil the requirements just cited and show an improved signal-to-noise ratio[106] together with reduced dispersion in comparison with the usual ISEs. Bousse et al.[107] studied Al_2O_3 ISFETs in FIA (see adapted ISFETs on p. 352)[77] and Bergveld et al.[108] developed a microprocessor-controlled coulometric system for stable pH control on the basis of an ISFET compared with a glass electrode. A few examples of microprocessor control had already been reported[89] since 1978, e.g., for automated multiple FIA (AMFIA)[109]; such control is also commercially available for CFA in the Technicon RA-1000 (see p. 338); in the latter instrument and in the Technicon Chem 1 system, the sampling frequency has now become as high as in FIA.

5.4.2.2.2. *Separational flow techniques*

In these flow systems a certain kind of separation, be it pre-concentration or a more sophisticated separation such as chromatography, of individual analyte components preceeds the detection; in treating the subject we shall distinguish between the techniques for gaseous samples and those for liquid samples, while concentrating on electrochemical detection.

Separational techniques for gaseous samples. Illustrative examples of importance are the well known Keidel and Hersch cells. In the Keidel[110] cell, an electrolytic system for the determination of water in gas, the water-containing gas (10–1000 ppm) passes at a constant rate (about 100 ml min^{-1}) through a 0.12-mm bore Teflon tube containing two intertwined Pt electrodes, isolated from each other by a Teflon coil and coated with a thin film of anhydrous phosphoric acid; a DC voltage, permanently applied across the electrodes, electrolyses any water absorbed into hydrogen and oxygen, thus yielding a faradaic current directly proportional to the water content of the gas. Commercial trace moisture analysers, as panel-mounted or portable models, all with rhodium cells, are available from Beckman[111], by means of which moisture contents even down to 1 ppm can be determined in hydrogen chloride, hydrogen, oxygen, Freon, protective atmospheres of argon or nitrogen, natural gas, etc.; ammonia, amines, methanol and olefins that may polymerize interfere.

In the Hersch[112] cell, an internal electrolytic system for the determination of oxygen in a gas, the oxygen-containing gas (down to 1 ppm) passes at a constant rate (about 100 ml min^{-1}) over an Ag spiral cathode partly submerged in a 2 *N* KOH solution with a Pb or Cd plate anode at the bottom, both electrodes being short-circuited over an ammeter. At the cathode $\frac{1}{2}O_2$ is reduced to $2OH^-$, and at the anode Pb or Cd is oxidized in turn to plumbite ions or cadmium hydroxide; as a result, the actual OH^- concentration does not change much during use, so that the cell has a long life. Although the cathodic current efficiency is constant, as a consequence of a side-reaction it is lower than the reaction $\frac{1}{2}O_2 + H_2O + 2e \rightarrow 2OH^-$ indicates, so that periodical calibration is required by means of known amounts of oxygen generated in a separate electrolytic cell.

Similar coulometric cells have been developed for sulphur dioxide, chlorine and nitrogen oxides; e.g., SO_2 is determined by anodic oxidation on lead dioxide electrodes[113] in air passing through 1 *N* $HClO_4$ solution (1 ppm of SO_2 at a flow-rate of 250 ml min^{-1} of air yielded an anodic current of 33 μA). In other methods SO_2 is oxidized by coulometrically generated iodine[114], and bromine or iodine[115]; Cl_2 is determined (together with another oxidizing component if any) by passing the gas through an absorbing solution of bromide or iodide, while the Br_2 or I_2 produced is measured by coulometric cathodic reduction[116]. In a discontinuous on-line controlled-potential coulometer for chlorine gas[117], the Cl_2 disolves in a thin film of hydrochloric acid wetting a Pt gauze and the current required for reduction is integrated; CO in air is continuously measured (down to 0.1 ppm by volume) by passing the air over I_2O_5 at about 145° C and then absorbing the I_2 produced in a phosphate-buffered electrolyte, where it is amperometrically determined via coulometric reduction at a cathode of Pt on graphite vs. an anode of electrolyte paste around active carbon[118].

Numerous commercial oxygen analysers are available, based on the principle of the Hersch cell, but all being protected from the analyte medium, either gaseous or liquid, by means of a membrane (usually Teflon). This membrane detector is known as the Clark cell (see later under membranes as measurement aids, p. 352); for gas analysis we may mention the Beckman Models 715 (also for liquids), 741, 743 (for flue gas), 755 and 778[119].

The most important separation technique for gaseous samples is certainly gas chromatography (GC); however, although many detector types are available, electrochemical detection is limited. As an example we can mention the portable continuous chromatographic coulometric sulphur emission analyser[120] for stack gases, based on automated coulometric titration, where in a 10-min cycle a gas sample via a sample loop is flushed by the carrier gas (nitrogen or dried and filtered air) into the chromatographic column for separation into SO_2, H_2S and CH_3SH, which next are successively oxidized by Br_2 generated at Pt wire electrodes (vs. calomel reference electrode) in the coulometric detector. Some calibration is required as the overall bromination reactions of the analyte components are not completely clear.

Further, mention should be made of solid-cell coulometric detectors for halogens and halogen compounds[121] and for sulphur compounds[122], where for example in the latter instance a solid cell of $Pt|Ag|AgI|Ag_2S|Pt$ interacts with passing gaseous sulphur compounds.

Separational techniques for liquid samples. For direct application to CFA, rapid polarography[123] can be used with a potential scan rate of up to 200 mV s^{-1} and a cycle time of up to 2–3 s. As a rapid discontinuous detector with an increased selectivity without a significant loss of response speed, it has advantages over a constant-potential polarographic detector, and so it can be used for liquid chromatography. Pungor and co-workers[124] applied silicone rubber-based graphite electrodes in CFA for the determination of many organic and inorganic compounds in the potential range -0.5 to $+1.5$ V (vs. SCE), because at constant potential the current exhibited a linear dependence on the depolarizer concentration and on the square root of the flow-rate.

Stripping analysis with inherent pre-concentration seems attractive to CFA, but until recently such a procedure appeared rather exceptional. It has been used in the automatic determination of heavy metals in water by anodic stripping voltammetry (ASV[125]) in a continuous flow cell with a mercury-covered graphite electrode, having the advantage that one can distinguish between complexed and free metal ions. Discontinuous methods have involved ASV (with a vitreous carbon rotating electrode[126] for trace metals in water, and with an RDE[127] for Ag^+) and CSV (with an HMDE for water-soluble mercaptans[128]). For liquid samples, commercial oxygen analysers of the membrane type (see p. 000) are available from Philips (PW 9600 and 9610)[129] and Beckman (Models 715, 7001, 7002 and 7003)[130].

Electrochemical detection in liquid chromatography (LC) has been developed. A review on potentiometric flow-through detectors and their clinical applications was given by Morf and Simon[131], who discussed particularly the key influence of the liquid-junction potential at the reference electrode, the occurrence of a streaming potential owing to electrokinetic effects (with a small channel diameter, high flow velocity, low electric conductivity and large distance between the ISE and RE) and the influence of the response time and limited life-time of liquid and solid-state membrane electrodes. Kemula and Kutner[132] discussed amperometric flow-through detection comprehensively on the basis of a very extensive list of references. Especially in high-performance liquid chromatography (HPLC) intensive development of liquid chromatography electrochemistry (LCEC) has taken place, so that the authors, instead of giving an exhaustive treatment of the subject, preferred to limit themselves to aspects only of amperometric detection in LC elaborated in their laboratory, viz., in its application to electrochemical detectors (ref. 132, Table 1) and of hydrodynamic systems of amperometric flow-through detectors with a solid sensor electrode (ref. 132, Table 2) (cf., our discussion on voltammetry at hydrodynamic electrodes, pp. 186–206).

An interesting paper on the topical significance of electrochemical sensors in clinical chemistry was recently published by Czaban[133], dealing with potentiometric and amperometric detection, especially with potentiometric and amperometric gas sensors, in addition to CHEMFETs and ISFETs [see later under adapted ISFETs (p. 352) and membranes (p. 352). However, the appearance of many papers on LCEC since 1980 has shown that the design of a properly functioning electrochemical detector for LC remains a very difficult problem, not only from a theoretical but also from a constructional point of view. In this connection, Hoogvliet[134] recently made the following apposite remark: "Those investigators dealing with theoretical problems are seldom aware of problems which may arise in practical situations, whilst investigators working in the application field are rarely interested in theoretical questions and often rely on the trial-and-error type of experiments". Therefore, in the context of our limited treatment of the subject, we have chosen to refer the interested reader to an arbitrary choice of some publications that contain numerous useful references. They concern the following aspects: amperometric and differential pulse voltammetric detection in HPLC[135]; pulse polarography and electrochemical flow-through detection[136]; a ring-disk flow-through detector for LC[137] (see RRDE, pp. 186–191); thin-layer electrochemical detectors in LC[138]; optimization of amperometric detectors[139]; a differential amperometric detector[140] for HPLC; design and optimization of electrochemical detectors[141]

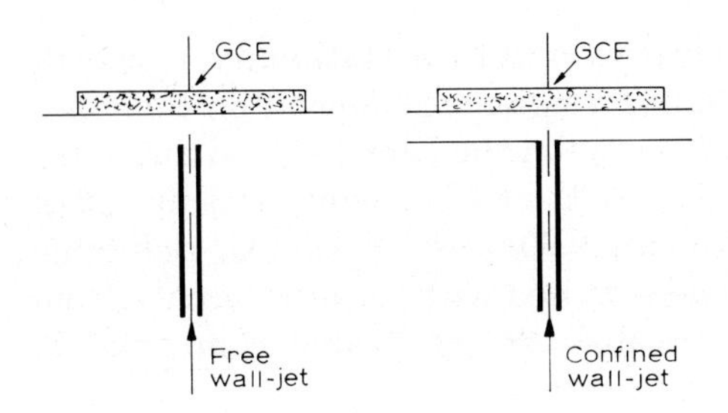

Fig. 5.24. Wall-jet GCE detectors for LC.

for HPLC (and FIA); a differential amperometric detector in anion-exchange chromatography[142]; noise and signal-to-noise ratio in electrochemical detectors[143]; noise and drift phenomena in amperometric and coulometric detectors[144] for HPLC (and FIA); performance of liquid-phase flow-through detectors[145]; polarographic continuous-flow detection (in HPLC)[146]; LC with a copper electrode amperometric detector[147]; and electrochemical detection in HPLC review)[148]. We shall devote more specific attention to confined wall-jet and dual-electrode and also computerized fast-scan VA, in LC.

Considering the wall-jet or impinging-jet type of detectors, there has been some confusion in the literature about the names of the different types[149]; Fig. 5.24 shows the difference between the free and the confined wall-jet detectors, where the flow jet is appreciably smaller than in the impinging-jet types.

Hanekamp[150] carried out experiments with the confined wall-jet detector from Metrohm (VA detector 641, EA 1096; see p. 192). Hoogvliet and coworkers[151] considered the hydrodynamics and mass-transfer characteristics of the confined wall-jet type and developed their BP-2 detector with a three-point spacer. In connection with the optimization of the PB-2, Hoogvliet and coworkers concluded that the main advantage of the free or confined wall-jet detector over a thin-layer detector is the enhanced mass transport in the impingement region, the noise level being approximately proportional to the electrode surface area and independent of the distance between the working (WE) and auxiliary electrodes (AE) in the range 25–500 μm; therefore, the WE radius can be equal to the inner radius of the inlet capillary (e.g., 150 μm), so that the WE can be placed opposite an AE-containing inlet capillary, at a distance of about 25 μm. In this configuration with an exceptionally small cell volume (about 2 nl), the detector becomes adapted to monitoring in micro-HPLC.

Another interesting method of amperometric detection for LC is dual-electrode electrochemical detection. Instead of a single WE, one can place two WEs in series, parallel to or opposite each other. The series configuration is mostly used, mainly in the collection mode, i.e., the electroactive substance entering the detector is converted at the upstream (generator) electrode into a product that either is or is not detected at the downstream (indicator) electrode, depending on the potential of the latter. Hoogvliet et al.[137,152] were easily able to rebuild their PB-2 confined wall-jet detector into a stationary ring-disk electrode (RDE); its collection and shielding efficiencies depend on the volumetric flow-rate, v, and on the distance, b, between the ring-disk and auxiliary electrode (cf., analogous efficiency definition of the RRDE, eqns. 3.93 and 3.94); relatively high collection efficiencies (about 0.60) were obtained; reducible compounds such as vitamin K_1 could be determined without interference from oxygen, if present, provided that at the disk an oxidizable compound is generated.

Sometimes one requires a fully two-dimensional voltammetric curve during an LC procedure in order to elucidate the electrochemical behaviour of the eluting compounds or to study peaks with overlapping phenomena; for this purpose one needs fast-scanning voltammetry. This can be perfomed in two ways either by repetitive scans (chromatovoltammetry), yielding a three-dimensional chromatovoltammogram, or by a single scan at the top of each chromatographic peak (peak scanning). Hoogvliet[153] discussed the choice between these methods as a question of preference, simply depending on the type of information specifically required. With today's computers and microprocessors, fast-scanning voltammetry has become much easier than at the time of Forina's experiments[123] (see above), in which rapid polarography was applied with a rapid DME (t_d = 0.1 s) at a scan rate of 200 mV s^{-1} in a 2–3-s cycle. Hoogvliet achieved scan rates of up to 1200 mV s^{-1} with the use of the HMDE or SMDE in the PARC Model 310 polarographic detector* and of a developed computer program (WASCAN, largely written in RT-11 Fortran) for on-line control of the instrumentation (including data sampling and processing). Comparison of the three techniques DSCV (DC staircase VA), CSV (cyclic-staircase VA) and SWV (square-wave VA) revealed that SWV has hardly any advantages over DCSV at scan rates above 150 mV s^{-1} and a time constant of 10 ms or more. The electrochemical behaviour of nitroprusside and platinum(II) complexes of alkali metals (Na and K) was elucidated by computerized fast-scan detection.

In Table 5.1 we stressed characteristic concomitant developments in non-separational flow techniques, which does not mean that these would not play that role in separation techniques; on the contrary, their influence may on the one hand be simplified by the previous separation of analyte compounds, but the detection requirements may on the other hand be increased as a consequence of the slight amounts and concentrations of components even passing through at high speed. Some specific remarks now follow on measurement aids, whilst additional field effects will be discussed separately later.

Improved ISEs. In 1980, Ammann et al.[154] reported on clinical and biological applications of liquid membrane electrodes based on an ion-selective component and a suitable plasticizer in a PVC matrix for the determination of Na^+, K^+, Ca^{2+}, Cl^- and H^+ in blood serum or whole blood and Na^+ and K^+ in urine. They gave extensive information on membrane compositions and selectivity

*EG & G Princeton Applied Research Corp., Princeton, NJ, U.S.A., T433; 38/11-M52-TP, 1983.

factors, $\log K_{ij}^{pot}$ (cf., K_{ij} in the comprehensive Nikolski eq. 2.86), of their electrodes, which allows measurements of the above-mentioned ions, in both the intra- and extracellular fluids already cited. Intracellular measurements of these ions, including Li^+, using the same ion-selective components in microelectrodes were possible. An H^+-selective liquid membrane electrode with ETH 301 synthetic ionophore (3-hydroxy-N-dodecylpicolinamide)[155] showed a linear regression for its pH response, yielding a slope of the glass electrode slope of 99.1%.

ISEs are becoming more widely applied, as they allow fast and reliable local measurements of ion activities; in this connection we may refer to a continuous ion-selective and electrochemical–enzymatic haemoanalysis of Na^+, K^+, Ca^{2+} and β-D-glucose directly in the streaming blood of a patient[156]. In this bedside-analyser the blood coming from the patient is passed continuously by means of a peristaltic pump, via a catheter system with extra-corporal intravasal heparinization (for anticoagulation), a spiral dialyser, and after being treated with protamine chloride (for antagonization of excess of heparin) is reinfused back into the patient. On the other (analyser) side of the sterilised dialyser membrane a buffer solution passes in counter-current and then flows through a channel along the sensors in series, viz., ISEs, the Ag/AgCl reference electrode and a first and a second pO_2 sensor, and finally re-enters the dialyser, thus closing the analyser cycle loop. Between the pO_2 sensors there is a delay coil into which a small amount of glucose oxidase (GOD) is pumped, which means that according to the reaction

$$\text{glucose} + O_2 + H_2O \overset{\text{GOD}}{\rightleftharpoons} \text{gluconic acid} + H_2O_2$$

the glucose content of the blood can be derived from the potential difference between the pO_2 sensors; the immobilization of GOD within the coil instead of addition is an attractive alternative[157]. The system allows continuous measurement during and after intravenous glucose infusion; it can be completed also by a delay coil for urease addition, so that from a subsequent NH_4^+-selective electrode potential the urea content of the blood can also be derived. Apparently good results with this haemoanalysis could be obtained with the use of ion-selective disc electrodes with active phases of carrier antibiotics (see p. 66) and synthetic neutral carriers.

Some other investigations on ISEs can also be mentioned, as follows.

(1) A direct potentiometric measurement of NO_2 in air appeared possible by means of a properly activated Fe–chalcogenide glass electrode[158] with the approximate composition $Se_{60}Ge_{28}Sb_{12}$ doped with 1.7–2% of Fe; it yields a positive potential shift when equilibrated with NO_2 in the presence of air, being insensitive to NO, SO_2, CO and CH_4.

(2) Ion-selective electrodes responsive to anionic detergents[159] consist of a tip of a 0.7-mm-thick Pt wire, fused at the end to make a ball ca. 1.5 mm in diameter and dipped several times in a coating mixture; after an initial conditioning by soaking it for 30 min in a $10^{-4}\,M$ solution of the anion to be measured, it is

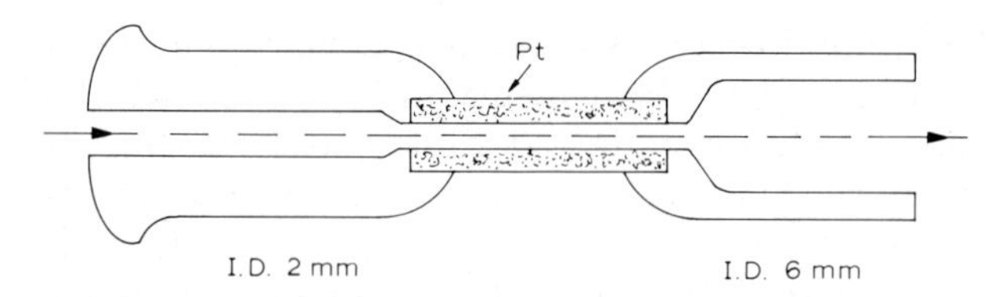

Fig. 5.25. Tubular Pt electrode according to Blaedel.

stored in air when not in use and reconditioned in this anion solution 5 min before use.

(3) A continuous potentiometric determination of sulphate in a differential flow system[160] consisted of a flow cell with two Pb^{2+}-selective electrodes in series. All solutions contained 75% of methanol and were adjusted to pH 4; a standard solution of Pb(II) passes the first sensor and, after being mixed with the sulphate sample stream, yielding a $PbSO_4$ precipitate in addition to excess of Pb(II), it passes the second sensor; from the potential difference between the sensors the sulphate content of the sample can then be derived.

Tubular electrodes. Blaedel et al.[161] were the first to introduce the TBE, a tubular electrode. It was constructed by melting a Pt cylinder in a glass capillary (see Fig. 5.25), the total length of Pt being 25.5 mm and its diameter 0.75 mm. The authors originally used it for an enzymatic determination of glucose by means of differential amperometry based on the following sequential reactions[162]:

$$\text{glucose} + O_2 + H_2O \xrightarrow{\text{oxidase}} \text{gluconic acid} + H_2O_2$$

$$H_2O_2 + 2K_4Fe(CN)_6 \xrightarrow{\text{peroxidase}} 2\,KOH + 2\,K_3Fe(CN)_6$$

all in a pH 7.5 buffer solution (of mono- and diphosphate).

Fig. 5.26 shows the measuring system, where 1 = peristaltic pumps, 2 = mixing T-tube, 3 = pulsator, 4 = induction delay tube, 5 = upstream Pt TBE, 6 = delay tube between TBEs and 7 = downstream Pt TBE. The system is connected to one saturated calomel electrode acting as a reference, and a constant potential is applied across each indicator electrode and the reference, so that as a consequence of the enzymatic reactions intermediately evolving in delay tube 6 a difference between the current intensities of both TBEs occurs, from which the glucose content of the sample can be derived via calibration with standards. Because the above amperometric determination is based on the reaction at initial velocities, linear proportionality can be obtained for concentration of up to 100 ppm of glucose with a relative standard deviation of 1%. We consider the method to be a classical and early example of the application of tubular electrodes. Since then, many publications on continuous analysis with TBEs and based also on different methods have

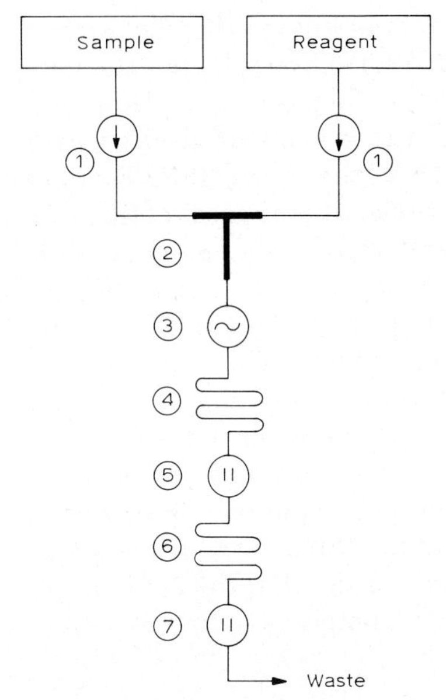

Fig. 5.26. Kinetic enzymatic amperometric determination of glucose.

appeared, not only by Blaedel and co-workers but also by others, a review of which has been given by Kékedy and Kormos[163]. As an illustration we shall briefly mention here a few of them:

(a) continuous amperometric determination (with a tubular gold electrode) of cyanide[164] in a stream of dilute base, in which HCN previously volatilized from plant tissues had been absorbed;

(b) anodic-stripping voltammetry at a tubular mercury-covered graphite electrode (TMCGE) in the flowing system used for the determination of Tl[165], Zn in seawater[166] (see also the remarks on TBE on p. 192); the subsequent operation only concerns the pre-concentration (plating) step, and it is followed by the usual anodic stripping;

(c) anodic-stripping voltammetry at a pair of tubular electrodes in series with collection at thin mercury films was applied by Schieffer and Blaedel[167] in a typical cell consisting of two glassy carbon tubular working electrodes separated by a thin Teflon spacer; trace metal cations were stripped from the upstream TBE while being collected at the downstream TBE at a constant potential, which allowed the measurement of sub-nanomole Pb concentrations; to a certain extent the procedure is comparable to the anodic-stripping voltammetry with collection (ASVWC) of silver at a ring-disc (glassy carbon–Pt) system by Johnson and Allen[168] (cf., p. 189);

(d) Blaedel and co-workers also reported on current–voltage curves, mainly for dilute $K_3Fe(CN)_6$ solutions, at a gold micromesh electrode[169] and a turbulent tubular electrode[170].

Adapted ISFETs. From early investigations it was known that field-effect transistors (FETs) can become pH sensors (H^+-ISFETs) by replacing the gate metal with silicon nitride or polymeric pH-selective membranes[171]. However, according to de Rooy[172], the silicon dioxide of the transistor itself, which normally insulates the gate, is also pH sensitive; therefore, ISFETs were investigated as pH sensors in rapid acid–base titrations, where speeds of titration at least five times greater than those with glass electrodes appeared possible with accuracies of $\pm 1\%$[173]. Further, on a new type of pH-sensitive FET an Al_2O_3 insulating layer was deposited, thus yielding an Al_2O_3 ISFET, the pH sensor mechanism of which is based on surface reactions[174]. The time response of pH for this type of ISFET was studied in a flow-injection analysis system using stepwise concentration changes[107], where for response times down to 15 ms there was a linear relationship between response time and inverse flow-rate, mainly caused by the non-ideality of the applied stepwise concentration change. However, there have been indications that the intrinsic response speed is of the order of 1 ms or less. For the moment one can say that the application of the Al_2O_3 ISFET in rapid acid–base titrations[175] compares favourably with that of a glass electrode.

Adapted voltammetric electrodes. Depending on various experimental conditions, voltammetric electrodes have to be adapted especially for use in flowing systems. We have considered this situation already under other headings above, such as tubular electrodes and their various modes, e.g., a tubular mercury-covered graphite electrode (TMCGE); mostly these are used amperometrically, but often also voltammetrically in stripping analysis. Other examples are the silicone rubber-based graphite voltammetric electrodes of Fehér et al.[176]; their anodic oxidation range is extensive, thus permitting the voltammetric determination of many organic compounds[177]; moreover, they have almost no memory effect, which makes them suitable for repeated and continuous measurements without electrode-surface renewal. A silicone rubber-based graphite electrode was used in a recirculating dissolution measuring cell (for drugs from pharmaceutical preparations)[178] and in a turbulent voltammetric cell (for amino acid analysis)[179], both cells having an Ag/AgCl reference electrode. A completely different example is that of a mercury pool indicator electrode in a micro-cell for continuous-flow pulsed voltammetric analysis[87].

Membranes. Apart from the role of membranes[180] in ISEs, there are at least three important applications of membranes as measurement aids in flow analysis, viz., as diffusion membranes in (1) (partial) dialysis and in (2a) membrane amperometry (MEAM) and (2b) membrane voltammetry (MEVA), and as Donnan membranes in (3) differential ionic chromatography.

Neither the usual membrane ISEs nor the gas-sensing electrodes, in which their internal indicator electrode functions as a zero-current potentiometric half-cell, are under consideration here.

(1) (Partial) dialysis in flow analysis. The sample solution flows along one side of the membrane, while the analyser solution passing (often in countercurrent) on the other side takes up the diffused components from the sample. A dynamic equilibrium is reached (under steady-state conditions) in the leaving analyser solution, which is then analysed and from the result of which the analyte content can be derived via calibration with standard solutions treated in exactly the same way. This is a common procedure, e.g., in Technicon AutoAnalyzers, and has also been applied in haemoanalysis by Ammann et al.[154] as described above.

(2a) MEAM in flow analysis. The amperometric method is the most usual procedure and was originally applied by Clark[181] for the clinical determination of oxygen in blood; the electrolytic system, the same as in the open Hersch cell, is, as explained on p. 344, protected from the gaseous or liquid analyte by a membrane and owing to its success has become well known as the Clark cell. Naturally, the optimal (semi)permeability of the membrane will be different for gaseous and liquid samples, which explains that, e.g., Beckman supply a choice of oxygen analysers according to the type of analyte sample (cf., p. 345). In this connection, the detection of oxygen in a Clark cell has been investigated at low temperatures and low pressures[182] by studying the voltammetric wave and by varying the types of thin membranes, polyethylene, Teflon and silicone rubber being the best.

An amperometric membrane probe has been developed for the determination of free residual chlorine[183] in saline cooling waters; hypochlorous acid over the range of 0–5 $\mu g\,ml^{-1}$ gave a linear response, but chloramines can interfere. An interesting case of MEAM is represented by the anodic oxidation of CO at a Teflon-bonded diffusion (sensing) electrode[184] in 3.4 M H_2SO_4 solution with counter and reference electrodes; all three electrodes were Teflon-bonded and catalysed by Pt, while the sensing electrode as the anode was maintained at 1.2 V (vs. SHE), a potential optimal for CO oxidation, but avoiding O_2 reduction and H_2O oxidation; during operation the CO–air mixture passed at a constant flow-rate over the back (gas side) of the anode, while current values reproducible to $\pm 1\%$ in the range 21–385 ppm of CO were obtained. A similar membrane amperometric CO determination was developed by Bergman[185], but based on a membrane anode metallized on the internal side by evaporating or sputtering on to a non-porous but gas-permeable polymer membrane. The electrochemical cell consisted of two such electrodes vs. an auxiliary electrode in a control circuit; hence the CO passing has to diffuse from the outside of the anodes in order to become oxidized at their metallized insides.

(2b) MEVA in flow analysis. The voltammetric method is seldom used but the example discussed above[182] shows its usefulness for the detection of possibly interfering sample components.

(3) Differential ionic chromatography. In a potentiometric method for recording ion-exchange elution curves, a dual-channel membrane cell is used as a differential detector[186] for following the eluate composition in comparison with the eluent. In the chromatography of alkali metal ions over a

cation-exchange resin column, the electrochemical chain of the cell has the following initially symmetric composition:

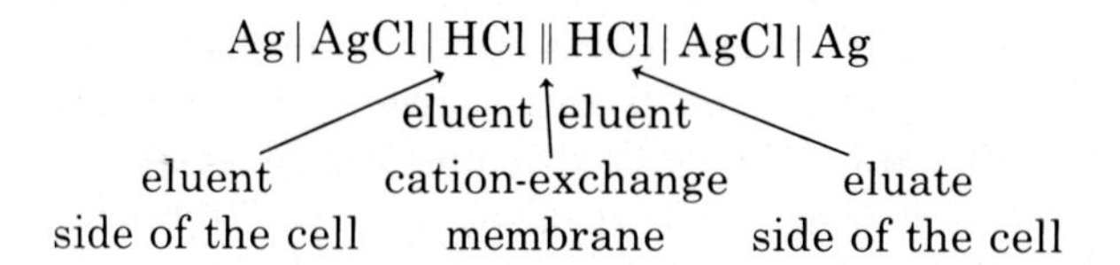

However, as soon as at the eluate-side H^+ ions are replaced with an equivalent amount of Na^+ or K^+ ions, which elute, the then asymmetric cell acquires a potential that reflects the Donnan equilibrium potential on the basis of the ion mobilities concerned. Hence the potential change as a function of time represents the ionic chromatogram and the peaks concerned yield the alkali metal ion contents via calibration.

5.4.3. Electroanalysis and computerization

The considerable importance of computerization in electroanalysis has been well illustrated by the many examples in the earlier part of this chapter; microprocessors in laboratory instruments and on line computers for automation have become common features. However, a few additional remarks on electroanalysis and its future computerization may still be useful.

In addition to the review by Bos[62] on microprocessor terminology and application (see p. 324), he also gave his opinion on the influence of on-line computers in classical chemical analysis[187]: "developments in digital electronics enable classical chemical analysis to regain some of the terrain lost to instrumental techniques; in chemical analysis, computerization can provide higher precision, higher speed and lower cost, while the value of interactive systems in routine work is emphasized". He predicted a considerable influence of increasing data treatment and enhanced application of mathematical procedures such as signal averaging, digital filtering, direct digital control and digital display of results; however, in the automation of the classical chemical analysis in particular, one must avoid the situation where the analytical technician is left only with sample preparations and book-keeping.

Wieck et al.[188] reported on what they called "a simple approach to microcomputer-controlled electrochemistry", which in fact represents a considerable amount of laboratory automation. In this direction, the "enhancement of the performance of analytical laboratories, a theoretical approach to analytical planning" by Janse and Kateman[189] represented a further step in laboratory management.

Whatever the future of electroanalysis really will be, it is worth considering what Faulkner recently said[190] at a meeting on "Electrochemistry Faces Reality", where he mentioned more explicitly the significant impact of LCEC, liquid chromatography electrochemistry (see p. 346) and cybernetic instrumentation.

5.5. AUTOMATED ELECTROANALYSIS IN PLANT CONTROL

Earlier in this chapter we considered the nature of chemical control (Section 5.1), the character and degree of automation (Section 5.2) with the choice between discontinuous and continuous analysis, the role of electroanalysis in automated chemical control (Section 5.3) and automated electroanalysis in laboratory control (Section 5.4).

The question that arises next is how automated laboratory methods can be applied in and/or should be adapted to plant control. In principle those methods can be devoted to this role[191–193] provided that some additional field effects (see Table 5.1) are taken into account, viz., special sampling and measuring requirements; we shall confine ourselves to these aspects and not consider extra safety measures against hazards such as contact explosions in plants.

5.5.1. Special sampling requirements in plant control

In both gaseous and liquid process streams, undissolved particles often occur and their mostly harmful disturbances must be prevented by filtration or other means. Sometimes such precautions may not be sufficiently effective and as a consequence interfering adherences to electrodes can occur, unless these phenomena are anticipated and inhibited. For instance, an on-line amperometric analysis system and method incorporating automatic flow compensation[194] is improved by ultrasonic and electrochemical cleaning of the electrodes and by compensating for the changes in flow-rate and temperature. For pH measurements in industry, Beckman[195] supply an ultrasonic electrode cleaner, in which an ultrasonic transducer is mounted directly on the flow or submersion assembly and is connected to an electronic exciter unit housed in a NEMA-4 waterproof enclosure; the exciter is equipped with an 18-min cycle timer (115 V, 50 Hz), which can be set to allow ultrasonic activation for any portion of that period or continuously. Polymetron[196] developed the Type 8346 pH/redox immersion probe with mechanical electrode cleaning, in which the electrodes, being gold for redox or antimony for pH measurements, are continuously cleaned by a rotating carborundum cleaning bar, while a special transformer ensures galvanic separation between the power supply and drive.

Another problem arises with refillable reference electrodes when process pressure fluctuations occur; for a minimal diffusion potential these electrodes must have a positive flow of electrolyte towards the analyte solution to prevent junction fouling or poisoning of the reference system. Yokogawa Electrofact[197] has solved this problem by means of their Bellomatic electrodes, the mounting kit (Fig. 5.27) of which maintains automatically the internal pressure higher than the process pressure by means of its bellows, thus preventing penetration of process liquid. Blaedel and Laessig[198] described a flow-through calomel electrode.

Special mention may still be made of the Ingold[199] range of industrial immersion and continuous-flow probes for pH measurements at high pressures, as

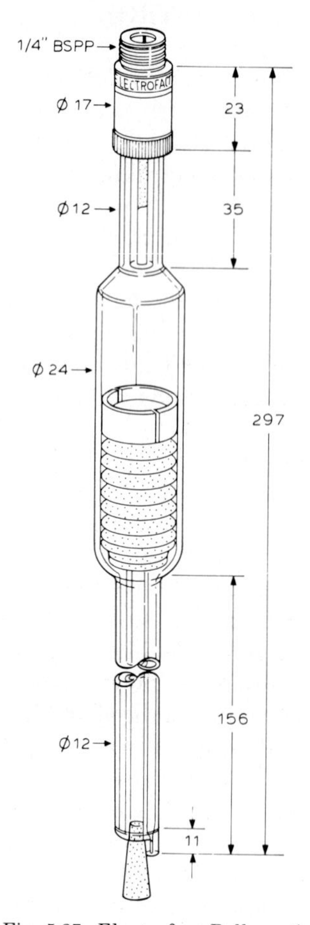

Fig. 5.27. Electrofact Bellomatic reference electrode. Sizes in mm.

well as of their retractable probes which enable pH, ion-selective, redox and Xerolyte reference electrodes to be serviced, exchanged and sterilized without interrupting the production process under control.

5.5.2. Special measurement requirements in plant control

Although the precautions mentioned in Section 5.5.1 ultimately concern measurement requirements also, special attention will be paid here to the voltage leakage between the indicator and reference electrodes, which may occur in industrial measurements and is often overlooked. This problem was solved by Beckman in 1970 with their differential input pH amplifier/transmitter, redesigned to use J-FET integrated circuit operational amplifiers[200], the whole being mounted on a single circuit board sealed into the top of the electrode station. This remote pH amplifier/transmitter (1) eliminates drift and

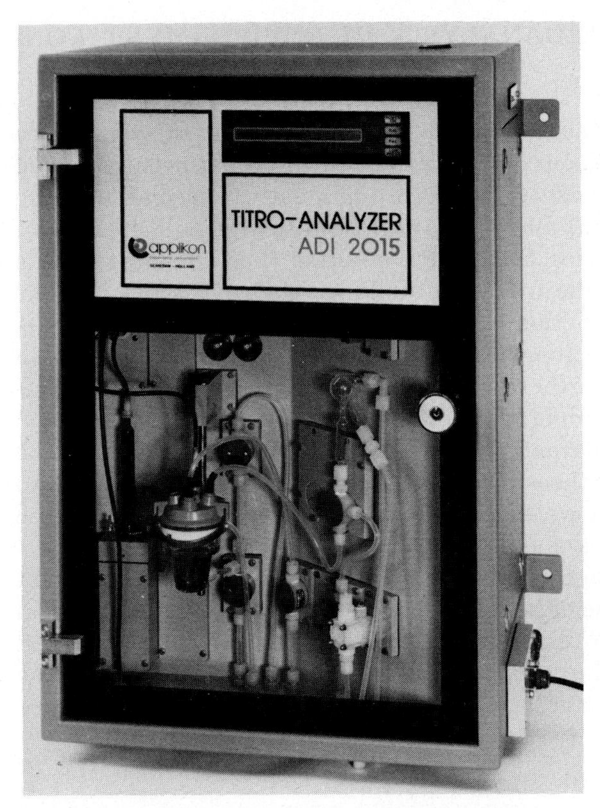

Fig. 5.28. Applikon Titro-Analyzer ADI 2015 (courtesy of Applikon Dependable Instruments).

noise, (2) reduces stringent installation and maintenance requirements, (3) allows the pH indicator to be separated at any distance from the electrodes and (4) eliminates the use of a costly co-axial cable between the measuring point and the pH indicator; the Model 940-A as an option permits mounting of the system integrally inside the case.

The above system of directly sensing a process stream without more is often not sufficiently accurate for process control; so, robot titration is preferred in that case by means of for instance the microcomputerized (64K) Titro-Analyzer ADI 2015 (see Fig. 5.28) or its more flexible type ADI 2020 (handling even four sample streams) recently developed by Applikon Dependable Instruments[201]. These analyzers take a sample directly from process line(s), size it, run the complete analysis and transmit the calculated result(s) to process operation (or control); they allow for a wide range of analyses (potentiometric, amperometric and colorimetric) by means of titrations to a fixed end-point or to a full curve with either single or multiple equivalent points; direct measurements with or without (standard) addition of auxiliary reagents can be presented in any units (pH, mV, temperature, etc.) required.

5.6. AUTOMATED ELECTROANALYSIS IN ENVIRONMENT CONTROL

In general, one may expect that laboratory methods of electroanalysis will be directly suitable for environement control. However, what has been indicated in Table 5.1 as additional field effects applies pre-eminently to special sampling requirements in environmental analysis. Apart from the difficult problem of obtaining representative samples, there is the question of pre-cleaning; in this connection, Epstein[202] made the following striking remark: "While an electrochemical system might function admirably in the laboratory under tap-water conditions, it still is a long way from having the system operate in sewage or natural water". Further, it will be clear that the method of pre-cleaning or pre-treating the sample and the choice of a suitable analysis depend considerably on the specific character of the sample and the component(s) to be determined. Epstein[202] treated the subject by dividing environment control into air pollution monitoring and water pollution monitoring, and within these fields he gave specific examples of analyte components and appropriate electroanalytic procedures. We close this chapter by mentioning a few more recent important papers, viz., the use of liquid chromatography (see electrochemical and conductivity detectors)[203], electrochemical gas monitors in occupational hygiene[204] and modulated polarographic and voltammetric techniques in the study of natural water chemistry[205].

REFERENCES

1 E. A. M. F. Dahmen, Analyse Chimique Continue et Automatique, 18e Série de Mises au Point de Chimie Analytique Organique, Pharmaceutique et Bromatologique, publiées sous la direction de J. A. Gautier et P. Malangeau, Masson, Paris, 1969, pp. 127–160.

2 D. L. Massart, L. Kaufman and A. Dijkstra, Evaluation and Optimization of Laboratory Methods and Analytical Procedures (Techniques and Instrumentation in Analytical Chemistry, Vol. 1), Elsevier, Amsterdam, 1978, 2nd reprint 1984.

3 Ref. 2, Part V; cf., A. Dijkstra, Analyticum 1969, Lunteren, pp. 237 and 243 and ref. 4.

4 P. M. E. M. van der Grinten, Control Effects of Instrument Accuracy and Measuring Speed, Part I, ISA J., 12, No. 12 (1965) 48; Part II, ISA J., 13, No. 1 (1966) 58; see also P. M. E. M. van der Grinten, Regeltechniek en Automatisering in de Procesindustrie, Prisma-Technica, Het Spectrum, De Meern, 1968; P. M. E. M. van der Grinten, Ingenieur (The Hague), 78, No. 27 (1966) 43.

5 A. Dijkstra, Chem. Weekbl., 64, No. 40 (1968) 19; for a discussion of minimizing analysis costs, see also F. A. Leemans, Anal. Chem., 43 (1971) 36A.

6 J. E. Rijnsdorp, in H. Schwarz (Editor), Proceedings of the 2nd IFAC (International Federation of Automatic Control) Symposium, Düsseldorf (F.R.G.), October 11–13, 1971, Vol. 4 (Survey Papers), North-Holland, Amsterdam; 3rd and 4th Symposia, Manchester (U.K.), 1974, and Fredericton (Canada), 1977, respectively.

7 H. Malissa and G. Jellinek, Z. Anal. Chem., 247 (1969) 1; H. Malissa and V. Simeonov, Z. Anal. Chem., 289 (1978) 257.

8 N. A. Parris, Instrumental Liquid Chromatography, a Practical Manual on High-Performance Liquid Chromatographic Methods (Journal of Chromatography Library, Vol. 27), Elsevier, Amsterdam, 2nd revised ed., 1984; J. Drozd, Chemical Derivatization in Gas Chromatography (Journal of Chromatography Library, Vol. 19), Elsevier, Amsterdam, 1981; J. F. Lawrence and

R. W. Frei, Chemical Derivatization in Liquid Chromatography (Journal of Chromatography Library, Vol. 7), 1976, 1st reprint 1983, Elsevier, Amsterdam, 1983.
9 H. Donche, Thesis, State University, Ghent, 1986.
10 S. P. Perone, J. E. Harrar, F. B. Stephens and R. E. Anderson, Anal. Chem., 40 (1968) 899; S. P. Perone, D. O. Jones and W. F. Gutknecht, Anal. Chem., 41 (1969) 1154; W. F. Gutknecht and S. P. Perone, Anal. Chem., 42 (1970) 906.
11 M. Bos, Anal. Chim. Acta, 81 (1976) 21.
12 J, Cipak, I. Ružić and Lj. Jeftić, J. Electroanal. Chem., 75 (1977) 9.
13 S. P. Perone, J. W. Frazor and A. Kray, Anal. Chem., 43 (1971) 1485.
14 S. C. Creason, R. J. Loyd and D. E. Smith, Anal. Chem., 44 (1972) 1159.
15 P. E. Whitson, H. W. Vanden Born and D. H. Evans, Anal. Chem., 45 (1973) 1298.
16 L. Krijger, D. Jagner and H. J. Skov, Anal. Chim. Acta, 78 (1975) 241; L Krijger and D. Jagner, Anal. Chim. Acta, 78 (1975) 251.
17 Q. V. Thomas, L. Kryger and S. P. Perone, Anal. Chem., 48 (1976) 761.
18 S. D. Brown and B. R. Kowalski, Anal. Chim. Acta, 107 (1979) 13.
19 P. F. Seelig and H. W. Blount, Anal. Chem., 51 (1979) 327.
20 J. T. Stock, J. Chem. Educ., 57 (1980) A125.
21 H. E. Keller and R. A. Osteryoung, Anal. Chem., 43 (1971) 342; J. A. Turner, V. Eisner and R. A. Osteryoung, Anal. Chim. Acta, 90 (1977) 25; J. A. Turner and R. A. Osteryoung, Anal. Chem., 50 (1978) 1496; T. R. Brumleve, J. J. O'Dea, R. A. Osteryoung and J. Osteryoung, Anal. Chem., 53 (1981) 702.
22 S. C. Rifkin and D. H. Evans, Anal. Chem., 48 (1976) 2174.
23 K. F. Drake, R. P. van Duyne and A. M. Bond, J. Electroanal. Chem., 89 (1978) 231; A. M. Bond and B. S. Grabarić, Anal. Chem., 51 (1979) 126.
24 P. F. Seelig and H. N. Blount, Anal. Chem., 51 (1979) 1129.
25 D. E. Smith, Application of On-Line Digital Computers in AC Polarography and Related Techniques, in J. S. Mattson, H. B. Mark and H. C. MacDonald (Editors), Computers in Chemistry and Instrumentation, Marcel Dekker, New York, 1972.
26 M. Bos, Anal. Chim. Acta, 103 (1978) 367.
27 J. J. Zipper and S. P. Perone, Anal. Chem., 45 (1973) 452; L. -H. L. Miaw, P. A. Boudreau, M. A. Pichler and S. P. Perone, Anal. Chem., 50 (1978) 1988.
28 J. A. Turner, J. H. Christie, M. Vukovic and R. A. Osteryoung, Anal. Chem., 49 (1977) 1904.
29 W. E. Wentworth, J. Chem. Educ., 42 (1965) 96.
30 I. Ružić, J. Electronal. Chem., 39 (1972) 111.
31 M. Bos, Anal. Chim. Acta, 122 (1980) 193.
32 V. B. Ehlers and J. W. Sease, Chapter 4, ref. 153.
33 D. D. Jackson, R. M. Hollen, F. R. Roensch and J. E. Rein, Analytical Chemistry of Nuclear Fuel Reprocessing, Proceedings of 21st Conference, Los Alamos, 1977, ORNL (Oak Ridge National Laboratory, Oak Ridge, 1978, pp. 151–158; C.A., 90 (1979) 13691j.
34 D. Gosden, M. Hayes, A. T. Kuhn and D. Whitehouse, J. Appl. Electrochem., 8 (1978) 437; U. v. Alphen, K. Graf and M. Hafendorfer, J. Appl. Electrochem., 8 (1978) 557.
35 A. M. Bond and B. S. Grabarić, Anal. Chim. Acta, 88 (1977) 227.
36 T. Anfält and M. Strandberg, Anal. Chim. Acta, 103 (1978) 379.
37 E. B. Buchanan and M. L. Buchanan, Talanta, 27 (1980) 947.
38 J. E. Anderson, R. N. Bagchi, A. M. Bond, H. B. Greenhill, T. L. E. Henderson and F. L. Walter, Amer. Lab., February (1981).
39 L. Kryger, Anal. Chim. Acta, 133 (1981) 591.
40 J. P. Price, S. L. Cooke and R. P. Baldwin, Anal. Chem., 54 (1982) 1011.
41 O. R. Brown, Electrochim. Acta, 27 (1982) 33.
42 E. B. Buchanan and W. J. Shelenski, Talanta, 27 (1980) 955.
43 H. J. Cheny, W. White and R. N. Adams, Anal. Chem., 52 (1980) 2445; W. S. Lindsay, B. L. Kizroot, J. B. Justice, J. D. Salamone and D. B. Neill, Chem. Biomed. Environ. Instrum., 10 (1980) 311.
44 P. Barett, L. J. Dairdowski and T. R. Copeland, Anal. Chim. Acta, 122 (1980) 67; J. E. Anderson and A. M. Bond, Anal. Chem., 53 (1981) 504.

45 A. M. Bond, J. Electroanal. Chem., 11 (1981) 381; J. E. Anderson and A. M. Bond, Anal. Chem., 53 (1981) 1394.
46 A. Granelli, D. Jagner and M. Josefson, Anal. Chem., 52 (1980) 2220.
47 J. E. Anderson and A. M. Bond, Anal. Chem., 55 (1983) 1934.
47a J. G. Osteryoung and R. A. Osteryoung, Anal. Chem., 57 (1985) 101A.
47b PARC Model 309 automatic analysis (Model 384B-7), EG & G Princeton Applied Research, Princeton, NJ, T-458, 1985; see also N. Yarnitzky, Anal. Chem., 57 (1985) 2011.
48 M. R. F. Asworth, W. Walisch, W. Becker and F. Stutz, Z. Anal. Chem., 273 (1975) 275.
49 J. Slanina, F. Bakker, A. J. P. Groen and W. A. Lingerak, Z. Anal. Chem., 289 (1978) 102.
50 J. W. Frazer, W. Selig and L. P. Rigdon, Anal. Chem., 49 (1977) 1250.
51 S. Ebel and A. Seuring, Angew. Chem., 89 (1977) 129.
52 R. W. Hendler, D. Songco, T. R. Clem, Anal. Chem., 49 (1977) 1908; R. W. Hendler, Anal. Chem., 49 (1977) 1914.
53 P. Legittimo, G. G. Peleggi and F. Pantani, Rass. Chim., 3/4, No. 2 (1976) 48.
54 A. de Robertis, A. Bellomo, C. D'Arrigo and D. De Marco, Ann. Chim. (Rome), No. 11–12 (1973) 63.
55 G. E. Cadwgan, Jr., and D. J. Curran, Mikrochim. Acta, II (1977) 461.
56 J. T. Stock, Trends Anal. Chem., 1, No. 3 (1981) 59.
57 M. Bos, Anal. Chim. Acta, 90 (1977) 61; W. B. Roolvink and M. Bos, Anal. Chim. Acta, 122 (1980) 81.
58 L. B. Jaycox, G. E. Cadwgan and D. J. Curran, Anal. Lett., 6 (1973) 1061.
59 L. B. Jaycox and D. J. Curran, Anal. Chem., 48 (1976) 1061.
60 L. M. Doane, J. T. Stock and J. D. Stuart, J. Chem. Educ., 56 (1979) 415.
61 H. Malmvig, Achema Lecture, Frankfurt, 1976.
62 M. Bos, Naturwissenschaften, 68 (1981) 14.
63 Zs. Fehér, G. Nagy, K. Tóth and E. Pungor, in E. Pungor, I. Buzás and G. E. Verres (Editors), Modern Trends in Analytical Chemistry (Proceedings of the Two Scientific Symposia, Mátrafüred, Hungary, 17–20 and 20–22 October 1982) (Analytical Chemistry Symposia Series, Vol. 18), Elsevier, Amsterdam, 1984, p. 285.
64 D. L. Eichler, Technicon Symposium 1960, Vol. 1, Mediad, New York, 1970, p. 51.
65 B. Fleet and A. Y. W. Ho, Anal. Chem., 46 (1974) 9.
66 G. Nagy, K. Tóth and E. Pungor, Anal. Chem., 47 (1975) 1460; G. Nagy, Zs. Fehér, K. Tóth and E. Pungor, Anal. Chim. Acta, 91 (1977) 87, 97 and 100 (1978) 181; G. Nagy, Z. Lendyel, Zs. Fehér, K. Tóth and E. Pungor, Anal. Chim. Acta, 101 (1978) 261; Zs. Fehér, G. Nagy, K. Tóth, E. Pungor and A. Tóth, Analyst (London), 104 (1979) 560; K. Tóth, G. Nagy, Zs. Fehér, G. Horvai and E. Pungor, Anal. Chim. Acta, 114 (1980) 45.
67 A. Bezegh, Zs. Fehér, K. Tóth and E. Pungor, ref. 63, p. 241.
68 M. Gratzl, Zs. Fehér, K. Tóth and E. Pungor, ref. 63, p. 297.
69 M. R. F. Ashworth, W. Walisch, W. Becker and F. Stutz, Z. Anal. Chem., 273 (1975) 275; S. M. Abicht, Thesis, Universität des Saarlandes, Saarbrücken, 1979; Anal. Chim. Acta, 114 (1980) 247.
70 J. Růžička, E. H. Hansen and H. Mosbaek, Anal. Chim. Acta, 92 (1977) 235; J. Růžička and E. H. Hansen, Flow Injection Analysis, Wiley, New York, 1981, p. 90; A. U. Ramsing, J. Růžička and E. Hansen, Anal. Chim. Acta, 129 (1981) 1.
71 W. Leithe, Chem. -Ing. -Tech., 36 (1964) 112.
72 G. Johansson and W. Backén, Anal. Chim. Acta, 69 (1974) 415.
73 O. Åström, Anal. Chim. Acta, 88 (1977) 17; 97 (1978) 259; 105 (1979) 67.
74 B. H. van der Schoot and P. Bergveld, Extended Abstract, "Conference Digest" Transducers '85, 3rd International Conference on Solid-State Sensors and Actuators, June 11–14, 1985, Philadelphia, PA, U.S.A., cf., Sensors Actuators, 8 (1985) 11; B. H. van der Schoot, Thesis, Twente University of Technology, Enschede, 1986.
75 J. Růžička and E. H. Hansen, Anal. Chim. Acta, 161 (1984) 1425.
76 S. C. Terry, J. H. Jerman and J. B. Angel, IEEE Trans. Electron Devices, 26 (1979) 1880.
77 M. Bos, P. Bergveld and A. M. W. van Veen-Blaauw, Anal. Chim. Acta, 109 (1979) 145; B. H. van der Schoot, P. Bergveld, M. Bos and L. J. Bousse, Sensors Actuators, 4 (1983) 267.

78 W. F. Smyth (Editor), Electroanalysis in Hygiene, Environmental, Clinical and Pharmaceutical Chemistry; Proceedings of a Conference Organized by the Electroanalytical Group of the Chemical Society, London, 17–20 April, 1979, (Analytical Chemistry Symposia Series, Vol. 2), Elsevier, Amsterdam, 1980.
79 Ref. 63, Part A, Electrochemical Detection in Flow Analysis (1st Symposium).
80 E. Pungor and I. Buzás (Editors), Ion-Selective Electrodes, 3, Proceedings of the 3rd Symposium, Mátrafüred, Hungary, 13–15 October 1980 (Analytical Chemistry Symposia Series, Vol. 8), Elsevier, Amsterdam, 1981.
81 T. Seyama, K. Fueki, J. Shiokawa and S. Suzuki (Editors), Chemical Sensors, proceedings of the International Meeting, Fukuoka, Japan, 19–22 September 1983 (Analytical Chemistry Symposia Series, Vol. 17), Elsevier, Amsterdam, and Kodamsha, Tokyo, 1983.
82 Flow Analysis, Proceedings of a Conference held in Amsterdam, 11–13 September 1979, Anal. Chim. Acta, 114 (1980).
82a L. T. Skeggs, Am. J. Clin. Pathol., 28 (1957) 311.
83 L. R. Snyder, Anal. Chim. Acta, 114 (1980) 3.
84 W. Lund and L. -N. Opheim, Anal. Chim. Acta, 79 (1975) 35; 82 (1976) 245.
85 L. F. Cullen, M. P. Brindle and G. J. Papariello, Adv. Autom. Anal., Technicon Int. Congr., 1972 (pub. 1973), 9,9; J. Pharm. Sci., 62 (1973) 1708.
86 A. Cinci and S. Silvestra, Farmaco, Ed. Prat., 27 (1972) 28.
87 P. W. Alexander and S. H. Qureshi, J. Electroanal. Chem., 71 (1976) 235.
88 C. E. Werkhoven-Goewie, U. A. Th. Brinkman and R. W. Frei, Anal. Chim. Acta, 114 (1980) 147.
89 J. Růžička and E. H. Hansen, Anal. Chim. Acta, 114 (1980) 19.
90 J. Růžička and E. H. Hansen, Anal. Chim. Acta, 106 (1979) 207.
91 E. H. Hansen, J. Růžička and A. K. Ghose, Anal. Chim. Acta, 100 (1978) 151.
92 J. Růžička and E. H. Hansen, Anal. Chim. Acta, 99 (1978) 37; J. M. Reijn, W. E. van der Linden and H. Poppe, Anal. Chim. Acta, 114 (1980) 105 and 126 (1981) 1; J. M. Reijn and H. Poppe, Anal. Chim. Acta, 145 (1983) 59.
93 H. Bergamin Filho, E. A. G. Zagatto, F. J. Krug and B. F. Reis, Anal. Chim. Acta, 101 (1978) 17.
94 J. Mindegaard, Anal. Chim. Acta, 104 (1979) 185.
95 O. Klinghoffer, J. Růžička and E. H. Hansen, Talanta, 27 (1980) 169.
96 W. E. van der Linden, Trends Anal. Chem., 1 (1982) 188.
97 J. H. M. van den Berg, R. S. Deelder and H. G. M. Egberink, Anal. Chim. Acta, 114 (1980) 91.
98 J. M. Reijn, W. E. van der Linden and H. Poppe, Anal. Chim. Acta, 123 (1981) 229.
99 K. Tóth, G. Nagy, Zs. Fehér, G. Horvai and E. Pungor, Anal. Chim. Acta, 114 (1980) 45.
100 W. E. Morf and W. Simon, ref. 79, pp. 33–48.
101 H. A. Mottola, Ch. -M. Wolff, A. Iob and R. Gnanasekaran, ref. 79, pp. 49–75.
102 K. Tóth, Zs. Fehér, G. Horvai, G. Nagy, Zs. Niegreisz and E. Pungor, ref. 79, pp. 167–187.
103 J. W. Dieker and W. E. van der Linden, Anal. Chim. Acta, 114 (1980) 267.
104 I. Kågevall, O. Åström and A. Cedergren, Anal. Chim. Acta, 114 (1980) 199.
105 A. U. Ramsing, J. Janata, J. Růžička and M. Levy, Anal. Chim. Acta, 118 (1980) 45.
106 A. Haemmerli, J. Janata and J. J. Brophy, J. Electrochem. Soc., 129 (1982) 2306.
107 L. J. Bousse, P. Bergveld and W. E. van der Linden, ref. 79, pp. 257–265.
108 P. Bergveld, B. H. van der Schoot and J. H. L. Onokiewiez, Anal. Chim. Acta, 15 (1983) 143.
109 K. K. Stewart, J. F. Brown and B. M. Golden, Anal. Chim. Acta, 114 (1980) 119.
110 F. A. Keidel (du Pont de Nemours), U.S. Pat. 2830845, 1958; H. H. Willard, L. L. Merritt J. A. Dean, Instrumental Methods of Analysis, Van Nostrand, New York, 5th ed., 1974, p. 784; for a discontinuous instrument see P. S. Gray, I. Gordon and N. Forsyth, J. Phys. Sci. Instrum., 6 (1973) 1145.
111 Beckman Instruments, Process Instruments Division, Fullerton, CA, Bulletin 4101-B.
112 P. Hersch, Instrum. Pract., 11 (1957) 817, 937; ref. 110, p. 784.
113 G. Bélanger, Anal. Chem., 46 (1974) 1576; F. Lindqvist, J. Air Pollut. Control Assoc., 28 (1978) 138.

114 A. D. Campbell, B. P. Hubbard and N. H. Tioh, Anal. Chim. Acta, 76 (1975) 483.
115 E. Schnell, Mikrochim. Acta, II (1977) 617.
116 A. Takahashi, Japan Kokai 74, 115 590; C.A., 83 (1975) 154789c.
117 W. H. Parth, ISA Trans., 12 (1973) 142.
118 F. Lindqvist, Galvanisch-coulometrische Methode voor de Continu Meting van Koolmonoxide in de (Buiten)lucht, IMG-TNO, Delft, publication No. 628, 1981.
119 Beckman Instruments, Process Instruments Division, Fullerton, CA, Bulletins 4091 E(715), 4171(741), 4122B(743) (1976), F 4182(755) (1977) and 4041-B(778).
120 R. J. Robertus and M. J. Schaer, Environ. Sci. Technol., 7 (1973) 849.
121 E. Cremer and E. Bechtold, Swiss Pat., 447 665, 1968.
122 E. Cremer and E. Bechtold, Ger. Pat., 1 598 141, 1966; C.A., 83 (1975) 52915k; Swiss Pat., 450 013, 1968.
123 M. Forina, Ann. Chim. (Rome), 63 (1973) 763.
124 E. Pungor and É. Szepesváry, Anal. Chim. Acta, 43 (1968) 289; E. Pungor, Zs. Hefér and G. Nagy, Anal. Chim. Acta, 51 (1970) 417; Zs. Fehér, G. Nagy, K. Tóth and E. Pungor, Analyst (London), 99 (1974) 699; M. Varadi and E. Pungor, Anal. Chim. Acta, 94 (1977) 351.
125 B. J. A. Haring, Automatic Determination of Heavy Metals in Water by Anodic Stripping Voltammetry, Chem. Biol. Afd. Rijksinst. Drinkwatervoorz., Den Haag, H_2O, 8 (1975) 146.
126 A. H. Miguel and C. M. Jankowski, Anal. Chem., 46 (1974) 1832.
127 D. C. Johnson and R. E. Allen, Talanta, 20 (1973) 305.
128 W. M. Moore and V. F. Gaylor, Anal. Chem., 49 (1977) 1386.
129 PW 9610 Dissolved Oxygen Sensor and PW 9600 Dissolved Oxygen Transmitter, 7600.32.9000.11, Philips, Eindhoven.
130 Beckman Instruments, Process Instruments Division, Fullerton, CA, Bulletins 4091 E(715), 4168(7001) (1976), 4169(7002) and 4170(7003).
131 W. E. Morf and W. Simon, ref. 79, pp. 33–48.
132 W. Kemula and W. Kutner, ref. 79, pp. 3–31.
133 J. D. Czaban, Anal. Chem., 57 (1985) 345A.
134 J. C. Hoogvliet, PhD Thesis, State University, Leiden, 1985, p. 14.
135 D. G. Swartzfager, Anal. Chem., 48 (1978) 2189.
136 W. P. van Bennekom, PhD Thesis, State University, Leiden, 1981.
137 J. C. Hoogvliet, F. Elferink, C. J. van der Poel and W. P. van Bennekom, Anal. Chim. Acta, 153 (1983) 149; K. Brunt, C. H. P. Bruins, D. A. Doornbos and B. Oosterhuis, Anal. Chim. Acta, 114 (1980) 257.
138 S. G. Weber, J. Electroanal. Chem., 145 (1983) 1.
139 J. M. Elbicki, D. M. Morgan and S. G. Weber, Anal. Chem., 56 (1984) 978.
140 K. Brunt and C. H. P. Bruins, J. Chromatogr., 16 (1978) 310.
141 K. Brunt, PhD Thesis, State University, Groningen, 1980.
142 K. Brunt and C. H. P. Bruins, J. Chromatogr., 172 (1979) 37.
143 D. M. Morgan and S. G. Weber, Anal. Chem., 56 (1984) 2560.
144 H. W. van Rooyen and H. Poppe, J. Liq. Chromatogr., 6 (1983) 2231.
145 H. Poppe, Anal. Chim. Acta, 145 (1983) 17.
146 H. B. Hanekamp, PhD Thesis, Free University, Amsterdam, 1981, Ch. 4.
147 W. Th. Kok, H. B. Hanekamp, P. Bos and R. W. Frei, Anal. Chim. Acta, 142 (1982) 31; W. Th. Kok, U. A. Th. Brinkman and R. W. Frei, J. Chromatogr., 256 (1983) 17; Anal. Chim. Acta, 162 (1984) 19; W. Th. Kok, G. Groenendijk, U. A. Th. Brinkman and R. W. Frei, J. Chromatogr., 315 (1984) 271.
148 K. Štulik and V. Pacáková, CRC Crit. Rev. Anal. Chem., 14 (1984) 297.
149 J. C. Hoogvliet, ref. 134, p. 19.
150 H. B. Hanekamp, ref. 146, Section 5.2.2, p. 114, H. B. Hanekamp and H. J. van Nieuwkerk, Anal. Chim. Acta, 121 (1980) 13.
151 A. J. Dalhuijsen, Th.H. van der Meer, G. J. Hoogendoorn, J. C. Hoogvliet and W. P. van Bennekom, J. Electroanal. Chem., 182 (1985) 295; ref. 134, Ch. 1 (especially Fig. 2) and Ch. 3.
152 J. C. Hoogvliet, ref. 134, Ch. 4.

153 J. C. Hoogvliet, ref. 134, Ch. 5.
154 D. Ammann, H. -B. Jenny, P. C. Meier and W. Simon, ref. 78, pp. 3–10.
155 D. Erne, D. Ammann and W. Simon, Chimia, 33 (1979) 88.
156 J. G. Schindler, R. Dennhardt and W. Simon, Chimia, 31 (1977) 404.
157 M. Nelboeck and D. Jaworek, Chimia, 29 (1975) 109.
158 G. G. Barna and R. J. Jasinski, Anal. Chem., 46 (1974) 1834.
159 T. Fujinaga, S. Okazaki and H. Freiser, Anal. Chem., 46 (1974) 1842.
160 M. Trojanowicz, Anal. Chim. Acta, 114 (1980) 293.
161 W. J. Blaedel, C. L. Olson and L. R. Sharma, Anal. Chem., 35 (1963) 2100.
162 W. J. Blaedel and C. L. Olson, Anal. Chem., 36 (1964) 343.
163 L. Kékedy and F. Kormos, Rev. Chim. (Bucharest), 30 (1977) 787.
164 D. B. Easty, W. J. Blaedel and L. Anderson, Anal. Chem., 43 (1971) 509.
165 W. R. Seitz, R. Jones, L. N. Klatt and W. D. Mason, Anal. Chem., 45 (1973) 840.
166 S. H. Liebermann and A. Zirino, Anal. Chem., 46 (1974) 20.
167 G. W. Schieffer and W. J. Blaedel, Anal. Chem., 49 (1977) 49.
168 D. C. Johnson and R. E. Allen, Talanta, 20 (1973) 305.
169 W. J. Blaedel and S. L. Boyer, Anal. Chem., 45 (1973) 258.
170 W. J. Blaedel and G. W. Schieffer, Anal. Chem., 46 (1974) 1564.
171 P. T. McBride, J. Janata, P. A. Comte, S. D. Moss and C. C. Johnson, Anal. Chim. Acta, 101 (1978) 239.
172 N. F. de Rooy, Thesis, Twente University of Technology, 1978.
173 M. Bos, P. Bergveld and A. M. W. van Veen-Blaauw, Anal. Chim. Acta, 109 (1979) 145.
174 L. J. Bousse, Thesis, Twente University of Technology, Enschede, 1982.
175 B. H. van der Schoot, P. Bergveld, M. Bos and L. J. Bousse, Sensors Actuators, 4 (1983) 267.
176 Zs. Fehér, G. Nagy, K. Tóth and E. Pungor, Analyst (London), 99 (1974) 699.
177 E. Pungor and E. Szepesváry, Anal. Chim. Acta, 43 (1968) 289.
178 E. Pungor, Zs. Fehér and G. Nagy, Anal. Chim. Acta, 51 (1970) 417.
179 M. Varadi and E. Pungor, Anal. Chim. Acta, 94 (1977) 351.
180 W. E., Morf, The Principles of Ion-Selective Electrodes and of Membrane Transport (Studies in Analytical Chemistry, 2), Elsevier, Amsterdam, 1981.
181 J. C. Clark, Jr., Trans. Amer. Soc. Artif. Internal Organs, 41 (1956) 2.
182 G. Halpert and R. T. Foley, J. Electroanal. Chem., 6 (1963) 426.
183 N. A. Dimmock and D. Midgley, Water Res., 13 (1979) 1101.
184 H. W. Bay, K. F. Blurton, J. M. Sedlak and A. M. Valentine, Anal. Chem., 46 (1974) 1837.
185 I. Bergman, Ann. Occup. Hyg., 18 (1975) 53.
186 H. G. Spencer and F. Lindstrom, Anal. Chim. Acta, 27 (1962) 573.
187 M. Bos, Anal. Chim. Acta, 122 (1980) 193.
188 H. J. Wieck, G. H. Herder, Jr., and A. M. Yacynych, Anal. Chim. Acta, 166 (1984) 315.
189 T. A. H. M. Janse and G. Kateman, Anal. Chim. Acta, 150 (1983) 219.
190 L. R. Faulkner, Anal. Chem., 57 (1985) 40A.
191 H. H. Willard, L. L. Merritt, Jr., and J. A. Dean, Instrumental Methods of Analysis, Van Nostrand, New York, 5th ed., 1974, Ch. 27.
192 H. L. Kies, Continuous Analysis based on Electrochemistry, International Quarterly Review Journal, Scientific Publications Division, Freund Publishing House, Tel-Aviv, 1973.
193 J. K. Foreman and P. B. Stockwell, Automatic Chemical Analysis, Ellis Horwood, Chichester, 1975, Ch. 2.
194 K. S. Fletcher, III (Foxboro Co.), U.S. Pat., 4 033 830, 1977; C.A., 87 (1977) 77935n.
195 Beckman Instruments, Process Instruments Division, Fullerton, CA, P76214-1075-15T, p. 11 (Bulletin 4111-D).
196 Physicochemical Control Systems for Industrial Applications, Polymetron, Hombrechtikon, Switzerland, TE863, 1981.
197 Bulletin GS 12B6J1-01E, Yokogawa Electrofact, Amersfoort, The Netherlands, 2nd ed., 1985, p. 3.
198 W. J. Blaedel and R. H. Laessig, Anal. Chem., 36 (1964) 1617.

199 Dr. W. Ingold AG, E-ELK-1-CH, E762/763-35-1-CH (pH); E-ISE-2-CH (ISE); 776-1-CH (pH and redox, both retractable); 764-50-2-CH (pH and redox, both sterilizable); short temperature response EQUI(THAL)-1-CH (pH); XER-2-CH and retractable 777-1-CH (Xerolyte reference).
200 Ref. 195, pp. 6 and 16.
201 Bulletin of On-line Analysis, Applikon B.V., Schiedam, The Netherlands, or Applikon Dependable Instruments, Austin, TX, U.S.A.
202 B. D. Epstein, in J. O'M. Bockris (Editor), Electrochemistry of Cleaner Environment, Plenum Press, New York, 1972, Ch. 6.
203 R. L. Grob and M. A. Kaiser, Environmental Problem Solving Using Gas and Liquid Chromatography (Journal of Chromatography Library, Vol. 21), Elsevier, Amsterdam, 1982, Ch. 6.
204 I. Bergman, ref. 78, pp. 167–174.
205 W. Davidson and M. Whitfield, J. Electroanal. Chem., 75 (1977) 763.

Appendix

Table of selected half-wave potentials for inorganic substances (courtesy of PARC)

Many handbooks like the CRC Handbook of Chemistry and Physics provide, on behalf of electrochemistry investigation, values of standard reduction potentials, listed either in alphabetical order and/or in potential order. These must be considered as potentials of completely reversible redox systems. In current analytical practice one is interested in half-wave potentials of voltammetric, mostly polarographic analysis in various specific media, also in the case of irreversible systems. Apart from data such as those recently provided by Rach and Seiler (Spurenanalyse mit Polarographischen und Voltammetrischen Methoden, Hüthig, Heidelberg, 1984), these half-wave potentials are given in the following table (Application Note N-1, EG & G Princeton Applied Research, Princeton, NJ, 1980).

KEY: All potentials are referenced to the saturated calomel electrode.
*. Potentials which have an asterisk are differential pulse peak potentials. All other potentials are half-wave potentials.
[]. Where a working electrode other than the dropping mercury electrode is used, the electrode material is indicated by brackets.
Ag(I). Ions in bold type are those that can also be determined using differential pulse anodic or cathodic stripping voltammetry.
(). Potential values in parentheses indicate that the electrode reaction is an oxidation.
Acetate (4.5). Numbers in parentheses next to a supporting electrolyte are the pH values for that electrolyte.
Supporting electrolytes are listed in approximate order of preference. These supporting electrolytes were selected from commonly available literature sources and are not meant to be all-inclusive. Abbreviated electrolytes are explained following the table.

Ion	Supporting electrolytes			
Ag(I)	0.1 *M* KNO_3: 0.10[Pt]	NH_3–NH_4Cl: − 0.24		
Al(III)	SVRS–Ac(4.5): − 0.46*	0.1 *M* TMAC: − 1.75		
As(III)	1 *M* HCl: − 0.42*, − 0.84*	1 *M* H_2SO_4: − 0.43, − 0.81	1 *M* NaOH: − 0.3	H_2SO_4–NaCl: − 0.20
As(V)	$HClO_4$–Pyrogallol: − 0.11			
Au(I)	0.1 *M* KOH: − 1.16			
Au(III)	1 *M* HCl: 0.37[GCE]			
Ba(II)	0.1 *M* LiCl: − 1.92			
Bi(III)	1 *M* HCl: − 0.09	Tartrate(4.4): − 0.14	NH_4 Cit(3): − 0.19	
Br^-	0.1 *M* KNO_3: (0.12)	KNO_3–MeOH: (0.15)[Ag]		
BrO_3^-	H_2SO_4–KNO_3: − 0.41	0.1 *M* KCl: − 1.78		
Cd(II)	NH_4 Cit(3): − 0.63*	Acetate(4.5): − 0.65	1 *M* HCl: − 0.64	NH_4 Tart(9): − 0.69*
Ce(III)	2 *M* K_2CO_3: (− 0.16)			
Ce(IV)	2 *M* K_2CO_3: − 0.16			
$\mathbf{Cl^-}$	0.1 *M* KNO_3: (0.25)	KNO_3–MeOH: (0.28)[Ag]		
ClO^-	0.5 *M* K_2SO_4(7): 0.08			
ClO_2^-	1 *M* NaOH: − 1.0			
CN^-	Borate(9.75): (− 0.27)*	0.1 *M* NaOH: (− 0.36)		
Co(II)	NH_3–NH_4Cl: − 1.30	Py–PyHCl: − 1.06	5 *M* $CaCl_2$: − 0.82	NH_4 Cit(9): − 1.39
Cr(III)	KSCN(3.2): − 7.5*			
Cr(VI)	NH_3–NH_4Cl: − 0.30*	1 *M* NaOH: − 0.84*	NH_4 Tart(9): − 0.24	
Cu(I)	NH_3–NH_4Cl: (− 0.22), − 0.50			
Cu(II)	NH_4 Cit(3): − 0.06*	Acetate(4.5): − 0.07	1 *M* HCl: − 0.22	NH_4 Tart(9): − 0.36*
Eu(III)	0.1 *M* NH_4Cl: − 0.67			
Fe(II)	Oxalate(4): (− 0.23)*	$Na_4P_2O_7$(9): (− 0.37)*		
Fe(III)	Oxalate(4): − 0.23*	NaOH–TEA: − 1.01*	NH_4 Tart(9): − 1.45*	$Na_4P_2O_7$(9): − 0.99
Ga(III)	1 *M* NaSCN(2): − 0.83			
Ge(II)	6 *M* HCl: − 0.45			
Hg(II)	1 *M* HCl: 0.44[Au]			
H_2O_2	PO_4–Cit(7): (0.18), − 1.0			
$\mathbf{I^-}$	0.1 *M* KNO_3: (− 0.03)	KNO_3–MeOH: (0.0)[Ag]		
In(III)	Acetate(4.5): − 0.71	1 *M* HCl: − 0.56		
IO_3^-	Phosphate(6.4): − 0.79	1 *M* KCl: − 1.16		
IO_4^-	K_2SO_4–H_2SO_4: − 0.12			

Ir(IV)	1 *M* HCl: 0.65[GCE]			
K(I)	0.1 *M* TBAOH: − 2.14			
Li(I)	0.1 *M* TBAOH: − 2.33			
Mn(II)	NH_3–NH_4Cl: − 1.66	NH_4Tart(9): − 1.55*		
Mo(VI)	0.3 *M* HCl: − 0.26, − 0.63	H_3Cit: 0.04, − 0.44		
Na(I)	0.1 *M* TBAOH: − 2.12			
Nb(V)	8 *M* HCl: − 0.46, − 0.70			
NI(II)	NH_3–NH_4Cl: − 1.10	NH_4Tart(9): − 0.98*	Py–PyHCl: − 0.75	
NH_2OH	1 *M* NaOH: (− 0.43)			
N_3^-	0.1 *M* KNO_3: (0.25)			
NO_2^-	DPA–SCN(1): − 0.54*	2 *M* Cit(2.5): − 1.06*	U(VI)–Ac–KCl(2): − 0.98*	
NO_3^-	U(VI)–Ac–KCl(2): − 0.98*			
O_2	0.1 *M* KNO_3: − 0.05, − 0.90			
Pb(II)	NH_4Cit(3): − 0.48*	Acetate(4.5): − 0.50	1 *M* HCl: − 0.44	NH_4Tart(9): − 0.52*
Pd(II)	NH_3–NH_4Cl: − 0.75			
Rb(I)	0.1 *M* TBAOH: − 2.03			
Rh(III)	NH_3–NH_4Cl: − 0.93			
Ru(IV)	1 *M* $HClO_4$: 0, 0.02, − 0.34			
S^{2-}	0.1 *M* NaOH: (− 0.78)*			
Sb(III)	6 *M* HCl: − 0.23*	1 *M* HCl: − 0.15		
Sb(V)	6 *M* HCl: − 0.23*			
Se(IV)	1 *M* HCl: − 0.10, − 0.40			
Sn(II)	1 *M* HCl: (− 0.1), − 0.47	NH_4Cit(3): (− 0.21), − 0.54	NH_3Tart(9): (− 0.53), − 0.77	1 *M* NaOH: (− 0.73), − 1.22
Sn(IV)	HCl–NH_4Cl: − 0.25, − 0.52	1 *M* HCl: − 0.1, − 0.47		
SO_3^{2-}	Acetate(5): (− 0.62)*			
$S_2O_3^{2-}$	Acetate(5): (− 0.21)*			
Te(IV)	NH_4Tart(9): − 0.71	NH_3–NH_4Cl: − 0.67		
Ti(III)	H_2Tart: (− 0.44)			
Ti(IV)	0.1 *M* HCl: (− 0.81)			
Tl(I)	Acetate(4.5): − 0.47	1 *M* HCl: − 0.48	Ac–EDTA(4.5): − 0.50	
U(VI)	0.1 *M* HCl: − 0.18, − 0.94			
V(V)	H_2SO_4–KSCN: − 0.52*			
W(VI)	10 *M* HCl: 0, − 0.60			
Zn(II)	NH_4Cit(3): − 1.05*	Acetate(4.5): − 1.1	NH_3–NH_4Cl: − 1.35	NH_4Tart(9): − 1.24*

Supporting Electrolytes: The list below includes only those supporting electrolytes which were abbreviated for the table.

1 Ac–EDTA(4.5): 0.1 *M* sodium acetate–0.1 *M* acetic acid–0.1 *M* disodium ethylenediamine tetraacetic acid, pH 4.5
2 Acetate(4.5): 0.1 *M* sodium acetate–0.1 *M* acetic acid, pH 4.5
3 Acetate(5): 0.1 *M* sodium acetate + acetic acid to pH 5
4 Borate(9.75): 0.1 *M* boric acid + NaOH to pH 9.75
5 2 *M* Cit(2.5): 2 *M* citric acid + NaOH to pH 2.5
6 DPA–SCN(1): 1.3×10^{-4} *M* diphenylamine (DPA)–0.01 *M* NaSCN–0.04 *M* $HClO_4$
7 H_3Cit: saturated citric acid
8 $HCl–NH_4Cl$: 1.0 *M* HCl–4 *M* NH_4Cl
9 $HClO_4$–Pyrogallol: 2 *M* $HClO_4$–0.5 *M* pyrogallol
10 $H_2SO_4–KNO_3$: 0.1 *M* H_2SO_4–0.2 *M* KNO_3
11 H_2SO_4–KSCN: 0.1 *M* H_2SO_4–0.1 *M* KSCN
12 H_2SO_4–NaCl: 2 *M* H_2SO_4–2 *M* NaCl
13 H_2Tart: saturated tartaric acid
14 KNO_3–MeOH: 0.1 *M* KNO_3 in 50% methanol
15 KSCN(3.2): 0.2 *M* NaSCN–0.2 *M* acetic acid, pH 3.2
16 $K_2SO_4–H_2SO_4$: 0.16 *M* K_2SO_4–1 *M* H_2SO_4
17 NaOH–TEA: 0.3 *M* triethanolamine–0.2 *M* NaOH
18 $Na_4P_2O_7$(9): 0.2 *M* sodium pyrophosphate + H_3PO_4 to pH 9
19 $NH_3–NH_4Cl$: 1 *M* NH_3–0.1 *M* NH_4Cl
20 NH_4Cit(3): 0.1 *M* citric acid + NH_4OH to pH 3
21 NH_4Cit(9): 0.1 *M* citric acid + NH_4OH to pH 9
22 NH_4Tart(9): 0.1 *M* tartaric acid + NH_4OH to pH 9
23 Oxalate(4): 0.1 *M* oxalic acid + NaOH to pH 4
24 Phosphate(6.4): 0.2 *M* sodium dihydrogen phosphate + NaOH to pH 6.4
25 PO_4–Cit(7): 0.1 *M* sodium dihydrogen phosphate–0.1 *M* sodium citrate adjusted to pH 7.0
26 Py–PyHCl: 0.1 *M* pyridine–0.1 *M* pyridine · HCl
27 SVRS–Ac(4.5): 0.1 *M* acetate buffer, pH 4.7–1.4×10^{-4}, Solochrome Violet RS–12% ethanol
28 Tartrate(4.4): 0.1 *M* Na_2 tartrate, pH 4.4
29 0.1 *M* TBAOH: 0.1 *M* tetrabutylammonium hydroxide
30 0.1 *M* TMAC: 0.1 *M* tetramethlammonium chloride
31 U(VI)–Ac–KCl(2): 20 ppm U(VI)–0.2 *M* KCl–0.1 *M* acetic acid

REFERENCES

EG & G Princeton Applied Research Application Briefs and Application Notes.
D. T. Sawyer and J. L. Roberts, Jr., Experimental Electrochemistry for Chemists, Wiley, New York, 1974.
L. Meites, Polarographic Techniques, Interscience, New York, 1955.
L. Meites, Handbook of Analytical Chemistry, McGraw-Hill, New York, 1963.

AUTHOR INDEX

The numbers in this index have the following significance: the first number indicates the text page on which the author/manufacturer is mentioned; the number in parentheses is the reference number; and the last number of each entry is the page of the reference list where the full reference is given.

SUBJECT INDEX